Second Edition

PROBLEM SOLVING IN GENERAL CHEMISTRY

RONALD A. DELORENZO
Middle Georgia College

Wm. C. Brown Publishers
Dubuque, Iowa•Melbourne, Australia•Oxford, England

To Mary, Joy, and Amy

Book Team

Editor *Craig S. Marty*
Developmental Editor *Elizabeth M. Sievers*
Publishing Services Coordinator *Julie Avery Kennedy*

Wm. C. Brown Publishers
A Division of Wm. C. Brown Communications, Inc.

Vice President and General Manager *Beverly Kolz*
National Sales Manager *Vincent R. Di Blasi*
Assistant Vice President, Editor-in-Chief *Edward G. Jaffe*
Marketing Manager *Christopher T. Johnson*
Advertising Manager *Amy Schmitz*
Managing Editor, Production *Colleen A. Yonda*
Manager of Visuals and Design *Faye M. Schilling*
Publishing Services Manager *Karen J. Slaght*
Permissions/Records Manager *Connie Allendorf*

Wm. C. Brown Communications, Inc.

Chairman Emeritus *Wm. C. Brown*
Chairman and Chief Executive Officer *Mark C. Falb*
President and Chief Operating Officer *G. Franklin Lewis*
Corporate Vice President, President of WCB Manufacturing *Roger Meyer*

Cover and interior design by Lachina Publishing Services

Copyediting, permissions, and production by Lachina Publishing Services

Library of Congress Catalog Card Number: 92–72543

ISBN 0–697–16411–X

Printed in the United States of America by Wm. C. Brown Communications, Inc., 2460 Kerper Boulevard, Dubuque, IA 52001

10 9 8 7 6 5 4 3 2 1

QUICK REFERENCE TABLE TO STANDARD TEXTS

All figures refer to chapter sections

Chapter in this Book	Kask Rawn	Koltz Purcell	Brown Lemay	Chang	Rock McQuarrie	Ebbing
1	app 1.3	app 1.5	app 1.4	app 1.8	app 1.5	app 1.3
2	1.2	app 1.4 1.6	1.3 1.5	1.9 1.7	1.4 1.7	1.4 1.6
3	1.2 5.1	1.4 6.2	1.3 5.1	1.7	app 1.4	1.4 6.1
4	3.1 3.2 3.3	2.4 2.6 3.5 3.6	3.5 3.6 3.8	2.3 2.4 2.4 2.6	5.1 5.2 5.3 5.4 5.5	4.1 4.2 4.3 4.5
5	10.1 10.2 10.3 10.4 10.5	12.1 12.2 12.3 12.7	10.1 10.2 10.3 10.4	5.1 5.2 5.3 5.4	7.1 7.2 7.3 7.4 7.6 7.12	5.1 5.2 5.3
6	3.4	4.7 5.2 12.4	3.1 3.3 3.5 3.6 3.8	3.1 3.6 4.1 4.2 5.5	3.1 3.2 5.6 5.7 5.9	4.6 4.7 4.8 5.4 6.5
7	3.5	14.1	4.1	4.4	6.1 6.2	4.9 12.4
8	12.4 12.6	14.3	13.5	12.8 12.9	15.1 15.3 15.4 15.5	12.6 12.7 12.8
9	13.1 13.2 13.4	16.1 16.2 16.3 16.4 16.5	15.1 15.2 15.3 15.6	14.1 14.2 14.3 14.5 14.6	17.1 17.2 17.3 17.5 17.7 17.8 17.9	15.1 15.2 15.4 15.5 15.6 15.7 15.8

Chapter in this Book	Kask Rawn	Koltz Purcell	Brown Lemay	Chang	Rock McQuarrie	Ebbing
10	15.1 15.2 15.3 15.4	17.1 17.3 17.4 17.5 18.2	16.2 16.3 16.4 16.5 16.6	16.1 16.2 16.6	18.1 18.3 18.5 18.6 18.7 20.2	16.1 16.2 16.3 16.4
11						
12	5.2 5.3 5.4 5.5	6.1 6.4 6.5 6.6 6.7 6.8	5.2 5.4 5.5 5.6	6.1 6.2 6.3 6.5 6.7 18.1 18.2	8.1 8.3 8.4 8.5 8.6 22.1 22.2 22.3 22.4 22.6	6.2 6.3 6.4 6.6 6.7 6.8 18.1
13	20.1 20.2 20.3 20.4	20.1 20.2 20.3 20.4 20.5 20.6 20.7	5.1	18.3 18.4	22.8 22.9 22.10	18.2 18.3 18.4 18.5 18.6 18.7
14	17.2 17.4 17.7 17.8 20.5	21.1 21.2 21.3 21.4 21.5 21.6	20.1 20.2 20.3 20.4 20.5 20.7	19.1 19.2 19.3 19.4 19.5 19.8	23.1 23.2 23.3 23.4 23.5 23.7 23.8	19.1 19.2 19.3 19.4 19.5 19.6 19.9
15	6.1 24.1 24.2 24.3 24.4 24.5 24.6	2.3 7.1 7.2 7.3 7.4 7.6 7.8 7.9	21.1 21.2 21.4 21.6 21.7 21.8	2.2 2.3 23.1 23.2 23.3 23.5 23.6	2.8 2.9 2.10 24.1 24.2 24.3 24.5 24.6 24.7 24.9 24.11 24.15	20.1 20.2 20.4 20.5 20.6 20.7

Contents

15 Nuclear Chemistry 343

Preface

The main goal of this text is to build lifelong skills and confidence in problem solving that will serve students in their future science-related courses and careers. In fact, the long-term usefulness of these skills is a primary tribute from many students who have used this book and who have then gone on to careers in areas such as medicine, pharmacy, science, and engineering. And with this book, students can build their problem-solving skills while relating chemistry to some unusual and interesting real-world applications, such as determining the temperature of hell, discussing the proper way to open beverage cans, and examining the possibilities of losing thousands of calories by simply drinking cold water.

This book is intended for use in preparatory or general college chemistry courses. No previous mathematical or chemical knowledge is assumed. All discussions begin at a low level and are developed in great detail.

Each chapter is composed of five parts: (1) an introduction, (2) discussions, (3) examples with detailed solutions, (4) self-tests, and (5) end-of-chapter problems. Many of the solutions contain nouns and verbs as part of their units, which make them even easier for students to follow. Answers are given for all problems so that students can check their work. Approximately 30% of the problems use SI units.

The first three chapters lay the basic foundation for the remainder of the text. Chapter 1 offers a stepwise development of mathematical skills associated with the handling of numerical calculations. Chapter 2 presents the dimensional analysis technique and focuses on the importance of units in guiding students to successful problem-solving methods. Chapter 3 is a slightly more advanced chapter on dimensional analysis that builds on the material of Chapters 1 and 2. One national review has noted that the first three chapters alone are worth the entire price of the book.

The remaining twelve chapters contain a multitude of example problems, each with detailed solutions, that illustrate most of the major topics discussed in general chemistry courses. These topics include the mole, stoichiometry, gas laws, concentration units, equilibrium, chemical thermodynamics, electrochemistry, and nuclear chemistry.

PEDAGOGICAL FEATURES

This text develops student communication, thinking, and problem-solving skills by encouraging students to participate by picking up a pencil, solving problems, and writing brief responses. Questions and problems that encourage a written response are italicized and separated from the text by horizontal lines. The correct numerical answer, solution details, or a written response is given in the paragraph or paragraphs following the italicized material.

Other pedagogical features include the following:

- Approximately 2500 problems and questions
- Dimensional analysis used in all problems
- A gradual change in problem difficulty within each chapter
- Detailed solutions to example problems, including the use of nouns and verbs in units
- The use of annotation and arrows in problem solutions, which mimic a teacher pointing out and clarifying ideas
- End-of-chapter problems that parallel the example problems
- Self-tests throughout each chapter to check comprehension
- Answers for all self-tests and end-of-chapter problems
- Approximately 150 sketches and photographs
- Chapter outlines at the beginning of each chapter
- Chapter introductions to provide an overview and to explain how the material ties in with previous and later chapters
- Useful hints on preparing for and taking examinations

I am particularly indebted to my wife, Mary, who originally suggested this project. Without her continuing efforts, patience, suggestions, and encouragement, this text would never have been written.

I would particularly like to thank Dr. Larry Krannich, University of Alabama in Birmingham. His extra effort, beyond that expected, and his blend of skills as a chemist and educator helped greatly in the development of this book.

I would also like to thank Dr. John Pasto, Department of Biology, Middle Georgia College, for his ideas and help with biologically related material.

It has been a pleasure to work with the personnel at W. C. Brown Publishers. In particular, I would like to thank Liz Sievers for the generous amounts of time, attention, and sound advice she has given to this probject. I am also grateful to Bob Olander, whose skills as a copyeditor never ceased to amaze me.

Ronald A. DeLorenzo

To the Student

Each chapter in this book is made up of five elements: (1) an introduction, (2) discussions, (3) example problems, (4) self-tests, and (5) end-of-chapter problems.

You will find many questions that are italicized and separated from the text by horizontal lines. To use this book effectively, cover each page containing an italicized question with a piece of paper in such a way that you see only the written material above and within the horizontal lines. *Write* your calculations and answers to these questions on a piece of paper before reading the correct answers, which follow all italicized questions. *Don't answer these questions mentally*. If you are asked for an explanation, *write* a brief paragraph before you read the correct answer given below the horizontal lines.

If your answer is correct, continue your reading. If your answer is incorrect, restudy the question, the correct answer, and the explanation given until you fully understand the material. Occasionally, you may also need to reread the discussion material preceding the question. Never proceed to a new question or to new material until you understand and can answer correctly any questions that you miss.

You will find several self-tests scattered throughout each chapter. The self-tests usually contain only about four problems. Answers to the self-tests are shown immediately following the last self-test problem. Cover these answers with a piece of paper until you have finished taking the self-test. Do not go beyond a self-test until you are sure that you know how to do all of the self-test problems correctly.

After you have studied a small group of three to five example problems, you may benefit even further by going back to the first example problem in that group, covering the solution with a piece of paper, and trying to solve it and the other examples in that group on a separate piece of paper.

A series of problems and questions appears at the end of each chapter. These end-of-chapter problems parallel the examples in the chapter. This means, for example, that the fifth problem at the end of any chapter is very similar to the fifth example in that chapter. If you should have trouble with Problem 5 at the end of a chapter, go back and look over Example 5 in the same chapter. The answers to the end-of-chapter problems are given at the end of the book so that you can check your work.

1 Basic Math Skills

1.1 INTRODUCTION

The following is an excerpt from a speech given by historian Linda Kerber at the University of Iowa:

"The college at which I studied bore some resemblance to this one. It required art or music, English, history and the social sciences, and laboratory sciences. It did not, to my great relief, require mathematics in addition to what I had studied in high school. I took chemistry courses—until I reached the point that math skills were needed—and dropped chemistry. I took experimental psychology and did well in it—until I reached the level that math was needed—and I dropped psychology. I took economics—and did very well in it—until math was needed—and then I dropped economics.

"This process of elimination made my choice of a major somewhat easier. It is no accident that I ended in history—that is, a field which seemed to promise I would never again have to contemplate a number. It is an irony I muse upon each week as I join our beginning graduate students in a course on statistics and computers for historians, struggling to learn techniques without which I now risk becoming hopelessly out of date."

It is possible to study chemistry and gain some appreciation and understanding of it without mathematics. However, the depth of your comprehension of chemistry (and almost every other subject) is directly related to your math skills. The topics reviewed in the following sections are the essential math skills that you should master before studying college chemistry.

1.2 EXPONENTIAL NOTATION

If every number fell between 1 and 10, we would not have to bother with exponential notation (also called scientific notation). Unfortunately, numbers such as 93 000 000 000 000 000 and 0.000 000 000 000 000 458 exist. We find them awkward to write and more awkward to work with. And so, instead of writing 93 000 000 000 000 000 (that's 93 followed by 15 zeros), we write 9.3×10^{16}. This compact form of the number (9.3×10^{16}) is called exponential, or scientific, notation. The 16 is called an exponent and it means that the decimal point between the 9 and the 3 (in 9.3) would have to be moved 16 places to the right to produce the original number (93 000 000 000 000 000). Don't you agree the 9.3×10^{16} is easier to write? Later you will see that it is also easier to work with.

There are other reasons why exponential notation is used. One is related to the idea of significant digits (see Section 1.5).

Here's a question that will give you a hint of things to come in the section on significant digits: What is the area of a rectangle whose sides, measured to the nearest inch, are found to be 2 inches and 20 inches?

If you wrote 40, you're wrong. (Did you actually pick up a pencil and write your answer as you should have?) Other incorrect answers typically given include 40 in., 40 inches^2, 40 in.^{2}, and 40. in.^{2}. The correct answer is 4×10^1 in.^{2}. We'll see why in the section on significant digits. But for now, it is important for you to realize that some numbers must be written using exponential notation. There is an important difference between 40 in.^{2} and 4×10^1 in.^{2}. However, we won't worry about this until we get to the section on significant digits.

You will appreciate another motive for learning exponential notation after you've read the section on skillful estimation techniques. The techniques, when mastered, will allow you to do most chemistry calculations in this book mentally. In fact, you'll be doing them faster than you could do them using a calculator. However, these techniques require a firm understanding of exponential notation.

Usually students who have trouble with exponential notation experience their trouble while performing calculations with exponential numbers (e.g., while adding or multiplying exponential numbers). We will go over several examples of such calculations. But first we will review how to write numbers in scientific notation and how to express numbers written in scientific notation as ordinary numbers.

A number expressed in exponential notation (such as 5×10^4) has three parts: (1) a nonexponential part, written first, which is usually a number between 1 and 10 (5 is the nonexponential part of 5×10^4); (2) an exponential part, which is 10 raised to a power (10^4 is the exponential part of 5×10^4); and (3) an exponent, which is the power of 10 (4 is the exponent in 5×10^4). The nonexponential part is always multiplied times the exponential part. In the number 5×10^4, the sign × represents multiplication.

nonexponential part ↓ exponential part ↓

$$5 \times 10^4 \leftarrow \text{exponent}$$

↑ multiplication

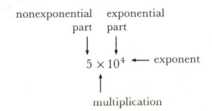

To express large numbers (e.g., 945) in exponential notation (9.45×10^2), move the decimal point in the large number (945.) to the left to produce the nonexponential part (a number between 1 and 10). While moving the decimal point to the left, count the number of places (digits you pass by) that the decimal point is moved. The number of places the decimal point is moved to the left is equal to the exponent (power of 10) in the exponential part.

945. ← The decimal point is moved two places to the left.
2 1

The number 945, expressed in exponential notion, is 9.45×10^2.

EXAMPLE 1 What are the exponential and nonexponential parts of 9.3×10^{16}? What is the value of the exponent in this number?

Solution The nonexponential part of 9.3×10^{16} is 9.3. The exponential part is 10^{16}. The exponent is 16.

EXAMPLE 2 Write the number 4700 in scientific notation.

Solution *Scientific notation* means the same thing as *exponential notation*.

4700. ← The decimal point is moved three places to the left to produce a number between 1 and 10 (4.700).
3 2 1

The nonexponential part of the number is 4.7, the exponent is 3, and the exponential part is 10^3. The number 4700, expressed in scientific notation, is 4.7×10^3.

EXAMPLE 3 Express 4 320 000 000 in scientific notation.

Solution

4 320 000 000. ← The decimal point is moved nine places to the left to produce a number between 1 and 10.
9 8 7 6 5 4 3 2 1

The nonexponential part is 4.32 and the exponent is 9. We say that 4.32×10^9 is the number 4 320 000 000 expressed in scientific notation.

We have been defining the nonexponential part of a number written in scientific notation as a number between 1 and 10. More precisely, it can be any number greater than or equal to 1 but less than 10. You should know that the number 10^0 is defined as 1. In fact, anything (except zero) raised to the zero power is defined as being equal to 1 (e.g., $4^0 = 1$). Therefore, we can express 1 in scientific notation as 1×10^0; 3 is 3×10^0.

EXAMPLE 4 Express one million in exponential notation.

Solution

$$1\,\underbrace{000}_{654}\,\underbrace{000}_{321}.$$

The number 1×10^6 is scientific notation for one million.

We said that scientific notation is also useful for expressing very small numbers. To express small numbers (e.g., 0.000 03) in exponential notation, move the decimal point to the right to produce a number between 1 and 10.

$$0.\underbrace{000\,03}_{123\,45} \text{ produces } 3.$$

This new number (3.) is the nonexponential part. We moved the decimal point five places to the right to produce 3. Because the movement was to the right (not to the left), the exponent is −5. (The exponent would have been +5 if the decimal point had moved to the left.) The exponential part is 10^{-5} and the exponential notation for 0.000 03 is 3×10^{-5}. The negative sign in front of the exponent (−5) tells you that the decimal point was moved to the right.

EXAMPLE 5 Express 0.3 in scientific notation.

Solution

$$0.\underbrace{3}_{1}$$

We moved the decimal point one place to the right to obtain a number between 1 and 10. Therefore, the nonexponential part is 3 and the exponent is −1. The number 0.3, expressed in scientific notation, is 3×10^{-1}.

EXAMPLE 6 Express 0.0052 in scientific notation.

Solution

$$0.\underbrace{005}_{123}2$$

To convert the given number (0.0052) to a number between 1 and 10, move the decimal point three places to the right. Therefore, the exponent is −3 and the nonexponential part is 5.2. We say that 5.2×10^{-3} is scientific notation for 0.0052.

SELF-TEST Make sure you can work the following problems correctly before going on to more advanced material:

1. Express 5.3 in scientific notation.
2. Express 0.1 in scientific notation.
3. Express 5 400 000 in scientific notation.
4. Express 0.000 000 21 in scientific notation.

ANSWERS **1.** 5.3×10^0 **2.** 1×10^{-1} **3.** 5.4×10^6 **4.** 2.1×10^{-7}

We have been showing how to express ordinary numbers (such as 62) in exponential notation (6.2×10^1). It is equally important for you to know how to take a number written in scientific notation and convert it into an ordinary number.

Numbers expressed in exponential notation with positive exponents (such as 5×10^3) are converted into ordinary numbers by moving the decimal point in the nonexponential part (5.) to the right a number of places equal to the power of 10 (3).

$$5.\underbrace{000}_{1\,2\,3} \text{ produces } 5000.$$

Therefore, $5 \times 10^3 = 5000$.

Numbers expressed in exponential notation with negative exponents (such as 4×10^{-2}) are converted to ordinary numbers by moving the decimal point in the nonexponential part (4.) to the left by a number of places equal to the power of 10 (2). The purpose of the negative sign in the exponent (−2) is to tell you the direction the decimal point must be moved (to the left).

$$\underbrace{04}_{2\,1}. \text{ produces } 0.04$$

Therefore, $4 \times 10^{-2} = 0.04$.

EXAMPLE 7 Express 3×10^4 as an ordinary number.

Solution The exponent is +4. It is positive, so you must move the decimal point to the right four places.

$$3.\underbrace{0000}_{1\,2\,3\,4} \text{ produces } 30\,000.$$

Thus, $3 \times 10^4 = 30\,000$.

EXAMPLE 8 Express 3×10^{-4} as an ordinary number.

Solution The exponent is −4. It is negative, so you must move the decimal point to the left four places.

$$\underbrace{0003}_{4\,3\,2\,1}. \text{ produces } 0.0003$$

Thus, $3 \times 10^{-4} = 0.0003$.

EXAMPLE 9 Express 54.3×10^2 in correct scientific notation.

Why isn't 54.3×10^2 considered to be in correct scientific notation?

Solution 54.3×10^2 is not considered to be written in correct scientific notation because the nonexponential part (54.3) is not a number between 1 and 10. The exponent (2) tells us that this number can be expressed as an ordinary number by moving the decimal point two places to the right.

$$54.\underbrace{30}_{1\,2} \text{ produces } 5430.$$

We can now express 5430. as 5.43×10^3 using our regular method of expressing large numbers in scientific notation.

EXAMPLE 10 Express 0.0043×10^4 in standard exponential notation.

Solution In its present form, 0.0043×10^4 is not in standard exponential notation because 0.0043 is not a number between 1 and 10. First, convert 0.0043×10^4 to an ordinary number by moving the decimal point four places to the right.

$$0.0043 \text{ produces } 43.$$

Thus, $0.0043 \times 10^4 = 43$. Now we can express 43. in exponential notation using our regular method: $43. = 4.3 \times 10^1$.

SELF-TEST Make sure you can work the following problems correctly before going on to more advanced material:

1. Express 3.4×10^{-4} as an ordinary number.
2. Express 8.5×10^3 as an ordinary number.
3. Express $0.000\,005\,2 \times 10^3$ in standard scientific notation.
4. Express $52\,000 \times 10^{-8}$ in standard scientific notation.

ANSWERS **1.** 0.000 34 **2.** 8500. **3.** 5.2×10^{-3} **4.** 5.2×10^{-4}

Let's look at the following numbers, all of which are equal:

$$3.4 \times 10^3 = 34 \times 10^2 = 340 \times 10^1 = 3400 \times 10^0 = 3400 = 34000 \times 10^{-1}$$
$$= 0.34 \times 10^4 = 0.034 \times 10^5 = 0.0034 \times 10^6 = 0.00034 \times 10^7$$

Notice how the exponent decreases by one every time we move the decimal point one place to the right. Also, the exponent increases by one every time the decimal point is moved one place to the left.

EXAMPLE 11 Express the number 27 eight different ways. Let some of your exponential notations contain positive exponents and let some contain negative exponents.

Solution

$$27 = 2.7 \times 10^1 = 0.27 \times 10^2 = 0.027 \times 10^3 = 0.0027 \times 10^4$$
$$= 27 \times 10^0 = 270 \times 10^{-1} = 2700 \times 10^{-2} = 27\,000 \times 10^{-3}$$

In the next section we will show how to add and subtract numbers written in exponential notation. Knowing how to change the value of the exponent by moving the decimal point to the left and to the right is an important part of the method used to add and subtract exponential numbers.

SELF-TEST Express each of the following numbers three different ways using exponential notation:

1. 50
2. 1
3. 0.02

ANSWERS **1.** 50×10^0, 5×10^1, 0.5×10^2 **2.** 1×10^0, 10×10^{-1}, 100×10^{-2}
3. 2×10^{-2}, 20×10^{-3}, 200×10^{-4}

1.2.1 ADDITION AND SUBTRACTION

Here's a trivial problem: Add 20 000 to 2000. What is the sum of these two numbers?

That's right, the sum of 20 000 and 2000 is 22 000.

Now for the hard part: Write 20 000 and 2000 in exponential notation and add them again. Try it and see what happens.

The numbers 20 000 and 2000, expressed in exponential notation, are 2×10^4 and 2×10^3. That part is easy. It is more difficult to add these two numbers in exponential form. Whenever you add or subtract numbers in exponential form, be sure that the exponential parts of the numbers are the same. The exponential parts of 2×10^4 and 2×10^3 are not the same ($10^4 \neq 10^3$). We can make the exponential parts the same by moving the decimal point in the nonexponential part.

Express 20 000 (2×10^4) in exponential notation with an exponent of 3.

You can express 2×10^4 in exponential notation with an exponent of 3 by moving the decimal point in 2 (2.) over one place to the right (20.). This produces 20×10^3. Now we can add 20×10^3 and 2×10^3 because they both have the same (equal) exponential parts. Their sum, 22×10^3, is obtained by adding the nonexponential parts ($20 + 2 = 22$) to get the nonexponential part of the answer. The exponential part of the answer is equal to the exponential part of either one of the numbers that are being added. (Remember: The numbers being added must always have equal exponential parts.)

•
•

EXAMPLE 12 Subtract 5×10^4 from 4×10^5.

Solution

$$\begin{array}{r} 4 \times 10^5 \\ -5 \times 10^4 \\ \hline \end{array} \xrightarrow{\text{make both exponents the same}} \begin{array}{r} 40 \times 10^4 \\ -\ 5 \times 10^4 \\ \hline 35 \times 10^4 \end{array}$$

We can express 35×10^4 in standard scientific notation by moving the decimal point in 35 (35.) one place to the left, giving 3.5×10^5, our final answer.

Here is another method for solving the same problem: Instead of changing 4×10^5 to 40×10^4, we could have changed 5×10^4 to 0.5×10^5.

$$\begin{array}{r} 4 \times 10^5 \\ -5 \times 10^4 \\ \hline \end{array} \xrightarrow{\text{make both exponents equal}} \begin{array}{r} 4.0 \times 10^5 \\ -0.5 \times 10^5 \\ \hline 3.5 \times 10^5 \end{array}$$

Both methods produce the same answer, 3.5×10^5.

•
•

EXAMPLE 13 Add 3×10^5 to 6×10^8.

Solution

$$\begin{array}{r} 6 \times 10^8 \\ +3 \times 10^5 \\ \hline \end{array} \xrightarrow{\text{make exponents equal}} \begin{array}{r} 6000 \times 10^5 \\ +\quad 3 \times 10^5 \\ \hline 6003 \times 10^5 \end{array}$$

$$6003 \times 10^5 = 6.003 \times 10^8$$

•
•

EXAMPLE 14 Subtract 5×10^2 from 8×10^6.

Solution

$$\begin{array}{r} 8 \times 10^6 \\ -5 \times 10^2 \\ \hline \end{array} \xrightarrow{\text{make both exponents equal}} \begin{array}{r} 80\,000 \times 10^2 \\ - \quad 5 \times 10^2 \\ \hline 79\,995 \times 10^2 \end{array}$$

$$79\,995 \times 10^2 = 7\,999\,500 = 7.9995 \times 10^6$$

SELF-TEST Perform the following calculations:

1. $3 \times 10^4 + 4 \times 10^3$
2. $6 \times 10^3 - 8 \times 10^2$
3. $4000 - 3.2 \times 10^2$
4. $6.1 \times 10^3 - 3.1 \times 10^{-1}$

ANSWERS **1.** 3.4×10^4 **2.** 5.2×10^3 **3.** 3.68×10^3 **4.** $6.099\,69 \times 10^3$

1.2.2 MULTIPLICATION AND DIVISION

Multiplying and dividing exponential numbers is easier than adding and subtracting exponential numbers. Why? Because you do not have to change the exponential parts of the numbers prior to multiplication or division the way you did with the addition and subtraction problems.

Whenever you multiply two or more numbers expressed in exponential notation, add the exponents to get the exponent of the final answer. Multiply the nonexponential parts to get the nonexponential part of the answer.

EXAMPLE 15 Express 20 and 200 in exponential notation and multiply them. (*Note:* $20 \times 200 = 4000$.)

Solution The values of 20 and 200, when expressed in exponential notation, are 2×10^1 and 2×10^2. The product of 2×10^1 and 2×10^2 is 4×10^3. The nonexponential part of the answer (4) is obtained by multiplying the nonexponential parts of the given number ($2 \times 2 = 4$). The exponential part of the answer (10^3) is obtained by adding the exponents of each given number ($1 + 2 = 3$).

EXAMPLE 16 Multiply 5×10^2 times 8×10^4.

Solution The correct answer is 4×10^7. We obtain the nonexponential part of the answer by multiplying the nonexponential parts of the given numbers ($5 \times 8 = 40$). We obtain the exponential part of the answer by adding the exponents of each number ($2 + 4 = 6$). The answer is 40×10^6 or, in standard scientific notation, 4×10^7.

In the next several problems you will be adding and subtracting both positive and negative numbers. Before going on, let's review how this is done.

Adding two numbers that have the same sign is relatively easy. Just add the numbers and place the correct algebraic sign in front of your answer.

EXAMPLE 17 Perform the following addition: $(+5) + (+2)$.

Solution The numbers $+5$ and $+2$ have the same algebraic sign (+). First add 5 and 2 ($5 + 2 = 7$) and then place the correct algebraic sign in front of your answer ($+7$). The final answer is $+7$.

The problems

$$5 + 2 = ? \quad \text{and} \quad (+5) + (+2) = ?$$

are identical. Although we always place a negative sign (−) in front of all negative numbers, it isn't always necessary to place a positive sign (+) in front of all positive numbers. Also, the parentheses around the numbers can be omitted if you don't find their omission confusing. For example,

$$(+5) + (+2) = (5) + (2) = 5 + 2 = 7$$

EXAMPLE 18 Add the following: $(-5) + (-2)$.

Solution The numbers (-5) and (-2) have the same sign (−). First add 5 and 2 $(5 + 2 = 7)$. Then, place the common sign in front of your answer to the first step (-7). The final answer is -7.

Adding two numbers that do not have the same sign is usually more difficult for students. To add two numbers with different signs (e.g., add $+3$ and -10), first determine which number is the larger number (10 is larger than 3) and then subtract the smaller number from it $(10 - 3 = 7)$. The sign of the answer is the same as the sign of the larger number. Therefore,

$$(+3) + (-10) = -7$$

EXAMPLE 19 Perform the following addition: $(-2) + (+5)$.

Solution The numbers (-2) and $(+5)$ have different signs. First determine which number is larger (5) and subtract the smaller number (2) from it $(5 - 2 = 3)$. Then, place the sign of the larger number (+) in front of the answer you got to the first part (3). The final answer is $+3$.

EXAMPLE 20 Perform the following addition: $(-8) + (+6)$.

Solution The signs of the two numbers being added are different. First, find the larger number (8) and subtract the smaller number (6) from it $(8 - 6 = 2)$. Then, place the sign of the larger number in front of your answer to the first step. The sign of the larger number is negative (−). The answer is -2.

As we've said, the positive signs and the parentheses are not always necessary. Let's look at an example dealing with this point.

EXAMPLE 21 Rewrite the addition problems in Examples 17, 18, 19, and 20, omitting unnecessary positive signs and parentheses.

Solution

17. $(+5) + (+2) = (5) + (2) = 5 + 2 = 7$
18. $(-5) + (-2) = -5 + (-2) = -5 + -2 = -7$
19. $(-2) + (+5) = (-2) + (5) = (-2) + 5 = -2 + 5 = 3$
20. $(-8) + (+6) = (-8) + (6) = (-8) + 6 = -8 + 6 = -2$

SELF-TEST Perform the following additions:

1. $(-5) + (-2)$
2. $(+5) + (+3)$
3. $(-5) + (+2)$
4. $(+5) + (-3)$

ANSWERS **1.** -7 **2.** $+8$ **3.** -3 **4.** $+2$

Subtraction problems are done by first changing the sign of the number to be subtracted and then following the rules for addition.

EXAMPLE 22 Perform the following subtraction: $8 - (+2)$.

Solution The number to be subtracted $(+2)$ has a positive sign. First, change its sign; that is, change $(+2)$ to (-2). Then, add the numbers: $(8) + (-2) = +6$. The answer is $+6$.

EXAMPLE 23 Perform the following subtraction: $8 - (-2)$.

Solution The number to be subtracted (-2) has a negative sign. First, change its sign; that is, change (-2) to $(+2)$. Then, add the numbers: $(8) + (+2) = +10$. The answer is $+10$.

EXAMPLE 24 Perform the following subtraction: $(-8) - (-2)$.

Solution The number to be subtracted (-2) has a negative sign. First, change its sign; that is, change (-2) to $(+2)$. Then, add the numbers: $(-8) + (2) = -6$. The answer is -6.

SELF-TEST Perform the following subtractions:

1. $(-5) - (-2)$
2. $(+5) - (+3)$
3. $(-5) - (+2)$
4. $(+5) - (-3)$

ANSWERS **1.** -3 **2.** $+2$ **3.** -7 **4.** $+8$

Now let's put to use all our knowledge of adding and subtracting numbers with different signs as we consider division problems involving numbers expressed in exponential notation. Consider the following problem:

$$\frac{8 \times 10^{12}}{4 \times 10^{3}} = 2 \times 10^{9}$$

To divide numbers expressed in exponential notation, first perform the indicated division of the nonexponential parts in the usual way ($8 \div 4 = 2$) to get the nonexponential part of the answer (2). Then subtract the exponents ($12 - 3 = 9$) to obtain the exponential part of the answer (10^9).

Let's apply these division rules to another problem, the reciprocal of the division we just considered.

$$\frac{4 \times 10^{3}}{8 \times 10^{12}} = 0.5 \times 10^{-9}$$

The nonexponential part of the answer (0.5) is obtained by performing the indicated division of the nonexponential parts of the given numbers ($4 \div 8 = 0.5$). The exponential part is obtained by subtracting the exponents $(3 - 12 = -9)$. The answer 0.5×10^{-9} is not in standard scientific notation because its nonexponential part (0.5) is not a number between 1 and 10. Move the decimal point one place to the right, changing 0.5 to 5. This also changes the exponent from -9 to -10. Our final answer is 5×10^{-10}.

EXAMPLE 25 Divide 60 000 by 300. Then express these two numbers in scientific notation and repeat the division.

Solution Dividing 60 000 by 300 gives 200. Expressing 60 000 and 300 using exponential notation gives us 6×10^4 and 3×10^2. Dividing 6×10^4 by 3×10^2 gives 2×10^2.

$$\frac{6 \times 10^4}{3 \times 10^2} = 2 \times 10^2$$

To obtain the nonexponential part of the answer (2), perform the indicated division of the nonexponential parts of the given numbers $(6 \div 3 = 2)$. Then subtract exponents $(4 - 2 = 2)$ to obtain the exponential part of the answer (10^2). The final answer is 2×10^2.

EXAMPLE 26 Divide 6×10^3 by 2×10^{-2}.

Solution

$$\frac{6 \times 10^3}{2 \times 10^{-2}} = 3 \times 10^5$$

The nonexponential part of the answer (3) is obtained by dividing 2 into 6. The exponential part of the answer (10^5) is obtained by subtracting -2 from 3 as follows: $(3) - (-2) = 5$.

EXAMPLE 27 Divide 3×10^{-9} into 6×10^{-2}.

Solution The answer is 2×10^7. The nonexponential part of the answer (2) is obtained by dividing 3 into 6. The exponential part of the answer (10^7) is obtained by subtracting -9 from -2 as follows: $(-2) - (-9) = +7$.

SELF-TEST Perform the following divisions:

1. $\dfrac{8 \times 10^6}{2 \times 10^4}$

2. $\dfrac{2 \times 10^4}{8 \times 10^6}$

3. $\dfrac{8 \times 10^{-6}}{2 \times 10^{-4}}$

4. $\dfrac{2 \times 10^{-4}}{8 \times 10^{-6}}$

ANSWERS **1.** 4×10^2 **2.** 2.5×10^{-3} **3.** 4×10^{-2} **4.** 2.5×10^1

Now you're an expert on adding, subtracting, multiplying, and dividing exponential numbers. If you don't feel like one, test yourself further with the problems at the end of this chapter. You can also make up your own problems and check your answers with the help of a pocket calculator.

1.2.3 ROOTS

Taking the root of 10 raised to an even power can be easy. Simply express the root as a fraction (e.g., the square root would be expressed as the fraction $\frac{1}{2}$, the cube or third root would be $\frac{1}{3}$, and the fourth root would be $\frac{1}{4}$) and multiply that fraction times the exponent of 10.

EXAMPLE 28 Find the square root of 10^4.

Solution The square root of 10^4 is 10^2 because $\frac{1}{2}$ of 4 is 2.

EXAMPLE 29 Find the cube root of 10^{12}.

Solution The cube root of 10^{12} is 10^4 because $\frac{1}{3}$ of 12 is 4.

EXAMPLE 30 Find the fourth root of 10^{24}.

Solution The fourth root of 10^{24} is 10^6 because $\frac{1}{4}$ of 24 is 6.

To find the square root of a number expressed in exponential notation, multiply the exponent of 10 by $\frac{1}{2}$ and take the square root of the nonexponential part of the number in the usual way.

EXAMPLE 31 Find the square root of 64×10^6.

Solution The square root of 64×10^6 is 8×10^3. The nonexponential part of the answer (8) is obtained by taking the square root of 64. The exponent is found by taking $\frac{1}{2}$ of 6 ($\frac{1}{2}$ of 6 is 3).

EXAMPLE 32 Find the cube root of 8×10^9.

Solution The cube root of 8×10^9 is 2×10^3. The nonexponential part of the answer (2) is found by taking the cube root of 8 ($2 \times 2 \times 2 = 8$, $2^3 = 8$). The exponential part of the answer (10^3) is found by taking $\frac{1}{3}$ of 9.

SELF-TEST

1. Find the square root of 10^{26}.
2. Find the square root of 4×10^{26}.
3. Find the cube root of 10^{90}.
4. Find the cube root of 27×10^{90}.

ANSWERS **1.** 10^{13} **2.** 2×10^{13} **3.** 10^{30} **4.** 3×10^{30}

Taking the square root of 10 raised to an odd power (e.g., 3, 5, 7, and 9) is more difficult. The square root of 10^5 is not found by taking $\frac{1}{2}$ of 5. Before halving an odd exponent, the exponent must be converted to an even number (e.g., 2, 4, 6, and 8). This is done by moving the decimal point of the nonexponential part of the number either to the left or to the right. The number 10^5 can be written 1×10^5 and the exponent can be turned into an even number by moving the decimal point of the 1 (1.) to the right, giving 10×10^4. Now the square root of 10 is about 3.16 and the square root of 10^4 is 10^2.

EXAMPLE 33 Find the square root of 6.4×10^3.

Solution The square root of 6.4×10^3 is 8×10^1. Because the exponent in the original number (3) is not an even number, we need to convert it into an even number (such as 2 or 4) by moving the decimal point in 6.4. If we move the decimal point in 6.4 one place to the right, then 6.4×10^3 becomes 64×10^2. We can halve the exponent (2) now ($2 \div 2 = 1$). The square root of 64 is 8. This gives us our answer, 8×10^1.

EXAMPLE 34 Find the square root of 6.4×10^3 by moving the decimal point to the left.

Solution When we move the decimal point one place to the left, the original number (6.4×10^3) becomes 0.64×10^4. The square root of 0.64 is 0.8. One half of the exponent (4) is equal to 2. This approach gives us an answer of 0.8×10^2 or 8×10^1. We get the same answer as we got in Example 33.

Negative exponents don't present a problem when finding the root of a number expressed in exponential notation. The rules are the same.

EXAMPLE 35 Find the fourth root of 16×10^{-40}.

Solution The fourth root of 16×10^{-40} is 2×10^{-10}. The nonexponential part of the answer is obtained by taking the fourth root of 16. The fourth root of 16 is equal to 2. The exponential part of the answer is found by taking $\frac{1}{4}$ of the exponent. One-fourth of -40 equals -10.

EXAMPLE 36 Find the cube root of 6×10^{-20}.

Solution The cube root of 6×10^{-20} is 3.9×10^{-7}. Because the exponent (-20) cannot be divided by 3 to give a whole number (e.g., 1, 2, 3, 4, 5, . . .), we must convert the exponent to a number that will give us a whole number when divided by 3. We do this by moving the decimal point of the nonexponential part (6.) one place to the right, giving 60×10^{-21}. Now, the cube root of 60 is about 3.9 and $\frac{1}{3}$ of -21 is equal to -7. Thus, our answer is 3.9×10^{-7}.

SELF-TEST

1. Find the square root of 16×10^{14}.
2. Find the square root of 6×10^5.
3. Find the cube root of 27×10^{-18}.
4. Find the cube root of 9×10^{16}.

ANSWERS **1.** 4×10^7 **2.** 7.7×10^2 **3.** 3×10^{-6} **4.** 4.5×10^5

1.2.4 POWERS

If you understand everything so far, then this section will be easy. To raise a number expressed in exponential notation (e.g., 2×10^2) to a power [e.g., $(2 \times 10^2)^3$], raise the nonexponential part of the given number to the power ($2^3 = 8$) and multiply the exponent times the power ($2 \times 3 = 6$). Therefore, $(2 \times 10^2)^3 = 8 \times 10^6$.

EXAMPLE 37 Raise 4×10^{-8} to the fourth power.

Solution Raising 4×10^{-8} to the fourth power gives 256×10^{-32}. The nonexponential part (256) is obtained by raising 4 to the fourth power. The exponential part (10^{-32}) is obtained by multiplying 4 times the exponent in the original number ($4 \times -8 = -32$). Note that 256×10^{-32} equals 2.56×10^{-30}.

EXAMPLE 38 Find the value of $(2 \times 10^7)^5$.

Solution The value of $(2 \times 10^7)^5$ is 32×10^{35}. The nonexponential part of the answer (32) is found by raising 2 to the fifth power ($2^5 = 32$). The exponential part of the answer is found by multiplying the exponent (7) times 5. A more accurate way of writing the final answer would be 3.2×10^{36}.

SELF-TEST Find the values of the following expressions:

1. $(2 \times 10^{-4})^3$
2. $(6 \times 10^2)^2$
3. $(8 \times 10^{-5})^4$

ANSWERS **1.** 8×10^{-12} **2.** 3.6×10^5 **3.** 4.096×10^{-17}

1.3 LOGARITHMS*

The solutions to many chemistry problems (e.g., *p*H and radiocarbon dating) require knowledge of logarithms and antilogarithms. We will define, discuss, and practice using logarithms in this section. We will also look at an easy way to derive a logarithm table, which could come in handy some day. The derivation serves as a good review of logarithms.

In our discussion of scientific notation (exponential notation), we showed that numbers such as 10 000 000 can be written in an abbreviated form (1×10^7). We can abbreviate 10 000 000 to an even greater degree by writing just the exponent (7). The exponent 7 is called the logarithm (or just log) of 10 000 000. We say, "The log of 10 000 000 equals 7." As you can see, a logarithm is just an exponent (power).

EXAMPLE 39 Find the logarithm of 1000.

Solution The log of 1000 is 3. The number 1000 can be written as 1×10^3 or simply 10^3. When 1000 is expressed in exponential notation, the power (exponent) of 10 is 3. The power is the logarithm. A logarithm is an exponent. The logarithm of 10^3 is 3.

EXAMPLE 40 Find the log of 0.000 01.

Solution The log of 0.000 01 is −5. The number 0.000 01 can be written in scientific notation as 1×10^{-5} or 10^{-5}. The exponent (power) of 10 is −5. The exponent is the logarithm.

---

*Adapted from Henry Freiser, "The Two-Place Logarithm Table: An Aid to Understanding and Use of Logarithms," *Journal of Chemical Education* 49 (May 1972):325.

EXAMPLE 41 What is the log of 1?

Solution The log of 1 is 0. The number 1 can be expressed in scientific notation as 1×10^0 or just 10^0. The exponent of 10 is 0 and the exponent is called the logarithm.

Unfortunately, not all numbers can be expressed as a simple power of 10 (i.e., 10 raised to a power that is a whole number). For example, the number 5 cannot be expressed as 10 raised to an integer (whole number) power. To find the logarithm of numbers that cannot be expressed as integer powers of 10, we must use a logarithm table or a calculator with logarithm capabilities.

1.3.1 SHORTCUT LOG TABLE

Like most students, you probably know how to use a calculator to find the log of a number. (If not, it's pretty easy. For example, to find the log of 10, enter the number 10, and then press the LOG key. Your calculator displays the result, which is 1.) Despite the availability and ease of use of calculators, many students have found the following material on the shortcut log table to be useful while taking exams for which they either forgot their calculator or during which their calculator batteries went dead. Also, the following discussion provides a greater comprehension of logs that goes beyond the push-button level, which requires little if any understanding. We will use the shortcut log table shown in Figure 1.1 for logarithm problems.

FIGURE 1.1 Shortcut log table

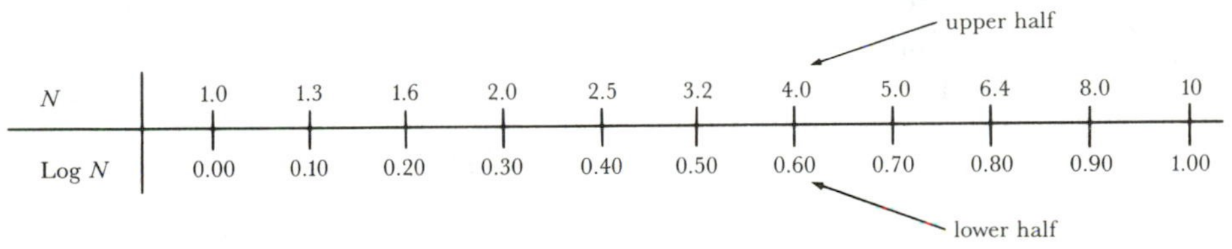

This shortcut log table is a two-place log table that is usually accurate to within 0.01 log units. It is sufficient for the logarithm problems encountered in general chemistry.

EXAMPLE 42 Find the log of 10 using the shortcut table.

Solution To find the log of 10, look at the upper part of the shortcut log table, labeled *N*, until you find the given number (10). Directly under the 10 is the value 1.00 (shown in the lower half of the table, labeled Log *N*). The value 1.00 is the logarithm of 10. The log of 10 equals 1.00. This is sometimes written log 10 = 1.00.

EXAMPLE 43 Use the table to find the logarithm of 4.0.

Solution The logarithm of 4.0 is 0.60. To obtain the value 0.60, find the given number (4.0) in the upper half of the log table, labeled *N*. Directly under the 4.0 is the value 0.60 (in the lower half of the table, labeled Log *N*). The value 0.60 is the logarithm of 4.0.

EXAMPLE 44 Find the log of 9.0.

Solution The log of 9.0 is 0.95. Finding the logarithm of 9.0 is not as straightforward as the earlier problems because the value 9.0 is not written in the upper half of the table. But 9.0 is midway between 8.0 and 10.0, as shown in the accompanying table:

N	8.0	9.0	10
Log N	0.90	0.95	1.00

The value of the log of 9.0 falls midway between 0.90 and 1.00. Thus, log 9.0 = 0.95.

EXAMPLE 45 *Find the log of 2.25.*

Solution The log of 2.25 is 0.35. Locate the position in the upper half of the table where 2.25 would be written. That position is midway between 2.00 and 2.50. Directly below that position is a point midway between 0.30 and 0.40 on the lower half of the table. This point is 0.35. The log of 2.25 is 0.35.

EXAMPLE 46 Find the log of 1.7.

Solution The log of 1.7 equals 0.23. This problem requires even more effort than the preceding problems and is representative of the more difficult logarithm value determinations you will have to perform with the shortcut log table. The majority of logarithm problems will be much easier than this example problem.

On the upper half of the log table, the number 1.7 lies about one-fourth of the way between 1.6 and 2.0, as shown in the accompanying table. The logarithm of 1.7 is a number one-fourth of the way between 0.20 and 0.30. Thus, log 1.7 = 0.225 = 0.23.

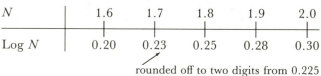

N	1.6	1.7	1.8	1.9	2.0
Log N	0.20	0.23	0.25	0.28	0.30

(0.23: rounded off to two digits from 0.225)

SELF-TEST Find the logarithms of the following numbers:

1. 10^2
2. 1000
3. 3.2
4. 1.8

ANSWERS **1.** 2 **2.** 3 **3.** 0.50 **4.** 0.25

A difficulty exists for users of both the shortcut log table and conventional logarithm tables when values of the number N do not lie between 1 and 10. For example, 50 doesn't appear on the upper half of the log table. To find the logs of numbers that are not between 1 and 10, we make use of the ideas we discussed in the section on exponential notation.

EXAMPLE 47 Find the log of 50.

Solution To find the log of 50, first express 50 in standard scientific notation (5×10^1). The next idea is a new one; it may be a while before this idea makes any sense. All three of the following ways of writing the log of 50 are correct:

$$\log 50 = \log(5.0 \times 10^1) = [\log 5.0 + \log 10^1]$$

Remember that we add exponents when multiplying two numbers with the same base. For example, $10^2 \times 10^3 = 10^5$. Similarly, we add logarithms when finding the log of a product. For example, $\log(5 \times 10^1) = \log 5 + \log 10^1$. Logs are exponents.

We find the log of 50 in this way. First, find the log of 5.0 and the log of 10^1. Then, add the two logs. The log of 5.0 is found in the log table (0.70). The log of 10^1 is 1. (Remember: The log of 10 raised to any integer power is equal to that power.)

In summary, here are the steps required to find log 50:

$$\begin{aligned} \log 50 &= \log(5.0 \times 10^1) \\ &= \log 5.0 + \log 10^1 \\ &= 0.70 + 1 \\ &= 1.70 \end{aligned}$$

The log of 50 is equal to 1.70.

EXAMPLE 48 Find the log of 50 000.

Solution The logarithm of 50 000 is 4.70. Follow the procedure given in the last example to obtain the answer.

$$\begin{aligned} \log 50\,000 &= \log(5.0 \times 10^4) \\ &= \log 5.0 + \log 10^4 \\ &= 0.70 \times 4 \\ &= 4.70 \end{aligned}$$

EXAMPLE 49 Find the logarithm of 0.000 45.

Solution The logarithm of 0.000 45 is −3.35.

$$\begin{aligned} \log 0.000\,45 &= \log(4.5 \times 10^{-4}) \\ &= \log 4.5 + \log 10^{-4} \\ &= 0.65 + (-4) \\ &= 0.65 - 4 \\ &= -3.35 \end{aligned}$$

EXAMPLE 50 Find the log of 950.

Solution Log 950 = 2.98. Here are the steps for obtaining the answer:

$$\begin{aligned} \log 950 &= \log(9.50 \times 10^2) \\ &= \log 9.50 + \log 10^2 \\ &= 0.98 + 2 \\ &= 2.98 \end{aligned}$$

The log of 9.50 can be found by expanding the shortcut logarithm table as shown. Notice that the value 0.975 is rounded off to 0.98 because this is a two-place log table.

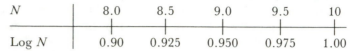

N	8.0	8.5	9.0	9.5	10
Log N	0.90	0.925	0.950	0.975	1.00

SELF-TEST Find the logarithms of each of the following numbers:

1. 32
2. 0.032
3. 90 000
4. 8×10^0

ANSWERS **1.** 1.50 **2.** −1.50 **3.** 4.95 **4.** 0.90

1.3.2 ANTILOGS

The log of 10^3 is 3. The antilog of 3 is 10^3. As you can see, finding antilogs is the reverse of the process of finding logarithms.

EXAMPLE 51 What is the antilog of 4?

Solution The antilog of 4 is 10^4 or 10 000. The log of 10^4 is 4.

In general, if log $X = N$ and you know the value of N, to find the antilog of N (the value of X), raise 10 to the Nth power. Thus, $X = 10^N$. Let's use this formula approach in the following examples.

EXAMPLE 52 What is the antilog of 7?

Solution In this problem, $N = 7$. To find X (the antilog of N), set $X = 10^N$. Thus, $X = 10^7$. The antilog of 7 is 10^7.

EXAMPLE 53 Find antilog 12.

Solution The antilog of 12 is 10^{12}. In this problem, $N = 12$. To find X (the antilog of N), set $X = 10^N$. Thus, $X = 10^{12}$.

EXAMPLE 54 Find the antilog of −5.

Solution The antilog of −5 is 10^{-5}. In this problem, $N = -5$ and $X = 10^N = 10^{-5}$. X is the antilog of N. Thus, $X = 10^{-5}$.

EXAMPLE 55 Find antilog 0.

Solution The antilog of 0 is 1. You may remember that the log of 1 is equal to 0, therefore the antilog of 0 must be equal to 1. Using the formula approach, we know that $N = 0$. Thus, the antilog of

N is equal to 10^0 which is equal to 1. We write: $X = 10^N = 10^0 = 1$. Any number raised to the 0 power equals 1 (except 0^0, which is undefined).

To find antilogs of numbers between 0 and 1, reverse the process for finding logarithms using the shortcut log table. For example, from the log table shown, we can see that the log of 2.0 is 0.30. The antilog of 0.30 must be 2.0. We write: antilog 0.30 = 2.0.

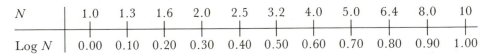

N	1.0	1.3	1.6	2.0	2.5	3.2	4.0	5.0	6.4	8.0	10
Log N	0.00	0.10	0.20	0.30	0.40	0.50	0.60	0.70	0.80	0.90	1.00

EXAMPLE 56 Find the value of antilog 0.50.

Solution The antilog of 0.50 is 3.2. To find the antilog of 0.50, look at the lower half of the log table (labeled Log N) until you find the given number (0.50). Directly above 0.50 is the value 3.2 (in the upper half of the table, labeled N). The value 3.2 is the antilog of 0.50. The log of 3.2 equals 0.50.

EXAMPLE 57 Find the antilog of 0.60.

Solution The antilog of 0.60 is 4.0. To obtain the value 4.0, find the given number (0.60) in the lower half of the table. Directly above the 0.60 is the value 4.0. The value 4.0 is the antilog of 0.60. We write: antilog 0.60 = 4.0.

EXAMPLE 58 Find the antilog of 0.45.

Solution The antilog of 0.45 is 2.85. This problem is a little more difficult than the preceding problems because 0.45 isn't written on the log table. But the value 0.45 is midway between 0.40 and 0.50 in the lower half of the table, so its antilog must lie midway between 2.5 and 3.2 in the upper half of the table.

To find antilogs of numbers that are not between 0 and 1, we use the following exponent property we made use of before: Exponents are added to find the product of two exponential numbers with the same base. For example, $10^6 \times 10^8 = 10^{14}$. Using our antilog formula approach to evaluate antilog 6.1, we get $N = 6.1$ and $X = 10^N = 10^{6.1} = 10^6 \times 10^{0.1}$. Always try to split (factor) noninteger exponents such as 6.1 into two numbers, one of which is an integer (whole number). The value of 10 raised to a power is equal to the antilog of the power, so $10^{0.1}$ must be equal to antilog 0.1. From the table, we know antilog 0.1 = 1.3. Therefore, $10^6 \times 10^{0.1} = 10^6 \times 1.3 = 1.3 \times 10^6$.

EXAMPLE 59 Find the antilog of 5.30.

Solution The antilog of 5.30 is 2.0×10^5. Here are the steps:

$$\text{antilog } 5.30 = 10^{5.30} = 10^5 \times 10^{0.30}$$

$$10^{0.30} = \text{antilog } 0.30 = 2.0 \text{ (from the log table)}$$

$$10^5 \times 10^{0.30} = 10^5 \times 2.0 = 2.0 \times 10^5$$

EXAMPLE 60 Find the value of antilog 4.40.

Solution The antilog of 4.40 is equal to 2.5×10^4.

$$\text{antilog } 4.40 = 10^{4.40} = 10^4 \times 10^{0.40}$$

$$10^{0.40} = \text{antilog } 0.40 = 2.5 \text{ (from the log table)}$$

$$10^4 \times 10^{0.04} = 10^4 \times 2.5 = 2.5 \times 10^4$$

EXAMPLE 61 Find the antilog of 3.65.

Solution The antilog of 3.65 is equal to 4.5×10^3.

$$\text{antilog } 3.65 = 10^{3.65} = 10^3 \times 10^{0.65}$$

$$10^{0.65} = \text{antilog } 0.65 = 4.5$$

$$10^3 \times 10^{0.65} = 10^3 \times 4.5 = 4.5 \times 10^3$$

EXAMPLE 62 Find the antilog of −7.20.

Solution The antilog of −7.20 is equal to 6.4×10^{-8}.

Finding the antilogs of negative numbers is usually more difficult for students than finding the antilogs of positive numbers. This difficulty exists for users of both the shortcut log table and the conventional log tables. Let's look carefully at the steps for solving this antilog.

We begin the same way we did before: antilog $-7.20 = 10^{-7.20}$. Our goal is to split (factor) $10^{-7.20}$ into two numbers, one of which is 10 raised to a negative whole number (integer) power and the other, 10 raised to a positive decimal number between 0 and 1. Thus $10^{-7.20}$ is equal to $10^{-8} \times 10^{0.80}$ because $(-8) + 0.8 = 0.8 - 8 = -7.2$.

To factor 10 raised to a negative exponent such as $10^{-7.20}$, let one of the factors be 10 raised to a negative whole number that is one less than the whole number part (−7) of the given exponent (−7.20). This gives us a factor of 10^{-8}. The second factor will always be equal to 10 raised to a power equal to 1 minus the decimal part (0.20) of the given exponent ($1 - 0.20 = 0.80$).

−7.20 is the given exponent.
−7 is the whole number part of the given exponent.
0.20 is the decimal part of the given exponent.
−8 is 1 less than −7.
0.80 equals 1 − 0.20.

$$10^{-7.20} = 10^{-8} \times 10^{0.80}$$

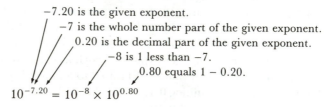

In summary, the following steps are taken to find the antilog of −7.20:

$$\text{antilog } -7.20 = 10^{-7.20} = 10^{-8} \times 10^{0.80}$$

$$10^{0.80} = \text{antilog } 0.80 = 6.4 \text{ (from the log table)}$$

$$10^{-8} \times 10^{0.80} = 10^{-8} \times 6.4 = 6.4 \times 10^{-8}$$

EXAMPLE 63 Find the antilog of −3.50.

Solution The antilog of -3.50 equals 3.2×10^{-4}.

$$\text{antilog } -3.50 = 10^{-3.50} = 10^{-4} \times 10^{0.50}$$

$$10^{0.50} = \text{antilog } 0.50 = 3.2 \text{ (from the log table)}$$

$$10^{-4} \times 10^{0.50} = 3.2 \times 10^{-4}$$

EXAMPLE 64 Find the value of antilog -4.95.

Solution The value of antilog -4.95 is 1.15×10^{-5}.

$$\text{antilog } -4.95 = 10^{-4.95} = 10^{-5} \times 10^{0.05}$$

$$10^{0.05} = \text{antilog } 0.05 = 1.15$$ ← This value (1.15) is read from the shortcut log table. The actual antilog of 0.05 is 1.12.

$$10^{-5} \times 10^{0.05} = 1.15 \times 10^{-5}$$

SELF-TEST Find the antilogs of each of the following numbers:

1. 6
2. -4
3. 0.5
4. 0.95
5. 3.30
6. -5.20

ANSWERS **1.** 10^6 **2.** 10^{-4} **3.** 3.2 **4.** 9.0 **5.** 2000 **6.** 6.4×10^{-6}

1.3.3 DERIVING A LOG TABLE

The following derivation requires that you first memorize the fact that the log of 2.0 is 0.30. Remembering this can be as easy as 1-2-3 if you make the following association:

1 2 3
log 2.0 = 0.30

Note: Log 2.0000 = 0.301 03, so there is a slight error in believing $\log 2.\bar{0} = 0.3\bar{0}$.

We begin our derivation by drawing a line marked with 11 positions, as shown earlier. Recall that $\log 1.0 = \log 10^0 = 0$ and $\log 10 = \log 10^1 = 1$, which supply values for the end points of our table. Enter these values for log 1.0, log 2.0, and log 10, as shown in the accompanying table:

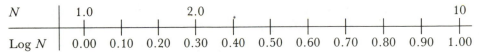

N	1.0			2.0							10
Log N	0.00	0.10	0.20	0.30	0.40	0.50	0.60	0.70	0.80	0.90	1.00

Look back at a completed shortcut log table (Figure 1.1) and notice that every time you move three positions to the right, the value of N doubles. You will soon be able to explain this observation.

As the next step in deriving the shortcut log table, consider the logic used to find the value of log 4.0. The log of 4.0 is the same as the log of (2.0×2.0); that is, $\log 4.0 = \log (2.0 \times 2.0)$.

$$\log 4.0 = \log (2.0 \times 2.0) \qquad \text{(factoring 4 into } 2 \times 2)$$

$$= \log 2.0 + \log 2.0 \qquad \text{(the log of a product equals the sum of the logs)}$$

$$= 0.30 + 0.30 \qquad (\log 2.0 = 0.30)$$

$$= 0.60$$

We have shown the log of 4.0 to be equal to 0.60. Place this information on your log table, as shown:

N	1.0			2.0			4.0				10
Log N	0.00	0.10	0.20	0.30	0.40	0.50	0.60	0.70	0.80	0.90	1.00

Now find the log of 8.0. (Hint: 8 = 2 × 4.)

The log of 8.0 is 0.90.

$$\log 8.0 = \log (2.0 \times 4.0)$$

$$= \log 2.0 + \log 4.0$$

$$= 0.30 + 0.60$$

$$= 0.90$$

Can you explain why the "move three-double N" rule works?

Doubling N (multiplying N by 2) is equivalent to adding 0.3 ($\log 2.0 = 0.30$) to log N. Adding 0.30 to a log N value is equivalent to moving three spaces to the right on the table.

Now let's explain why $N = 1.6$ (instead of 16) corresponds to the log $N = 0.20$ position.

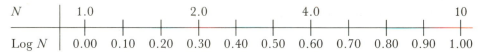

$$\log 1.6 = \log\left(\frac{2.0 \times 8.0}{10}\right)$$

← $2 \times 8 = 16$, which does not lie between 1 and 10. $\left(\frac{2 \times 8}{10}\right) = 1.6$, which does lie between 1 and 10.

$$= \log 2.0 + \log 8.0 - \log 10$$

← The log of a fraction is equal to the difference between the log of the parts of the fraction. Just as exponents are subtracted in division (e.g., $\frac{10^8}{10^2} = 10^{(8-2)} = 10^6$), logs (which are exponents) are subtracted. Thus, $\log \frac{10}{2} = \log 10 - \log 2$.

$$= 0.30 + 0.90 - 1.00$$

$$= 0.20$$

You should now be able to derive the rest of the log table. Here are a few words of warning, however. We mentioned earlier that the log of 2.0 is not 0.30; the log of 2.0000 equals 0.301 03. Only a small error results if you set log 2.0 equal to 0.30. A slightly larger error (although still small) results if you set log 4.0 to 0.60 (the log of 4.0000 is actually equal to 0.602 06). As we continue in this manner to increase log N values by 0.3 as we double N values, the error becomes increasingly significant. This is why we obtain some of our log table values by beginning at the $N = 10$ position. For example,

$$\log 5.0 = \log \frac{10}{2.0} = \log 10 - \log 2.0$$

$$= 1.00 - 0.3$$

$$= 0.70$$

Because the log of 5.00 equals 0.699, this approach keeps our error smaller than the alternative approach of finding the log of every value of N through a series of multiplications by 2.

1.4 CALCULATORS AND ESTIMATION TECHNIQUES

Now that you know everything you've ever wanted to know about exponential notation, we're going to sharpen your skills even further by showing you how to complement calculations done on an electronic pocket calculator with estimation techniques that utilize your facility with scientific notation.

1.4.1 THE ELECTRONIC POCKET CALCULATOR

This section is not an instruction manual for your electronic pocket calculator. Most college students already have had several years of experience in using pocket calculators. Unfortunately, many of these students have become calculator addicts. For them, using a calculator has become a habit. They use their calculators when doing simple arithmetic, such as $500 \div 2 = 250$. They mistrust their own ability to do any arithmetic with only a pencil and paper.

Pocket calculators do satisfy a real need. But they are also frequently misused. What we are proposing in this section is an approach that depends upon both your ability to use a pocket calculator and your ability to do math problems with just a pencil and a piece of paper. More on this in a minute.

First let's look at some serious errors that students make with calculators. In an earlier section we considered the multiplication of 2 in. × 20 in. and said that the answer is not 40 in.^{2} but 4×10^1 in.^{2}. We haven't given the reason for this yet, but we will shortly. For now just accept the fact that the number of digits shown in any answer is a significant part of the answer. Calculators don't recognize this fact! If a calculator can display ten digits, it will give an answer of $\frac{1}{3} = .333\,333\,333\,3$ even though the correct answer may be 0.3. In such cases, an estimated value is more accurate than an answer obtained using the most expensive calculator!

Another problem occurs when you use a calculator to check calculations you originally obtained with the calculator. If you've made an error in the original calculation (for example, incorrect sequence of button pushing on the calculator), you will most likely make the same error when you recheck your calculations. This human tendency is most effectively overcome by solving problems using two different methods.

1.4.2 COMPLEMENTARY METHOD

Solving problems by more than one method is a common practice in science. To be absolutely certain of an experimental result, you would use two completely different techniques. For example, to determine the concentration of iron in water, you would perform a volumetric analysis following by a gravimetric or spectrophotometric analysis.

Here is a simpler illustration: If you wanted to determine the mass of a water sample, you might weigh the water sample on a balance. Weighing the water sample a second time on the balance would not uncover any errors in your weighing technique or malfunctions in the balance. So you would gain little information by weighing the water sample a second time on the same balance. Another way to find the mass of the water sample (to check your first answer) is to measure its volume. One milliliter of water weighs 1 gram (1 g), so knowing the volume of water would allow you to determine the mass of the water sample. If both of these methods gave a value of 55 g for the mass of your water sample, you could be fairly certain that the mass of the water sample is 55 g.

We are going to develop a technique called skillful estimation. Once you have perfected this technique, you can use it to obtain answers more quickly than you can with a calculator. When practiced, this technique will help your mental alertness in much the same way that regular reading or jogging helps you to overcome the misuse of your television set or automobile.

Remember that the primary motive of skillful estimation is not to increase your intelligence or to replace your calculator. Rather, the motive is to introduce you to an alternative method that, although good in itself, can also serve as an effective check against calculator errors. Likewise, the calculator helps check against errors made with your estimations. The skillful estimation technique may even be useful in case your pocket calculator batteries fail during an examination.

We will discuss two methods for estimating answers. The first method is slower than the second, but it is easier for beginners. After you've had some experience with the first method, go on to the second method, which is merely a more efficient application of the first.

1.4.3 SKILLFUL ESTIMATION FOR BEGINNERS

We will demonstrate the basic techniques of skillful estimation with the following example:

EXAMPLE 65 Consider the problem of multiplying 417 × 589. Estimate the product using the technique of skillful estimation. Compare your estimated answer with the answer obtained with a calculator.

Solution Performing the multiplication 417 × 589 on a calculator yields an answer of 245 613. The answer obtained with a calculator may or may not be correct as we will see in the section on significant figures (see Section 1.5).

The first step in estimating the product of 417 × 589 is to express each number using exponential notation. The given numbers, expressed in exponential notation, are 4.17×10^2 and 5.89×10^2.

The second step in estimating the product of 417 × 589 is to discard all digits to the right of the decimal point, rounding off where necessary (see Section 1.6). This reduces our numbers to 4×10^2 and 6×10^2. Notice that 5.89 rounds off to 6.

The final step is to multiply the two exponential numbers according to the rules developed in the section on exponential notation (Section 1.2). This gives us an estimated answer of 24×10^4, 2.4×10^5, or 240 000. As we will see in the section on significant figures, this estimated answer may be a more accurate answer than the 245 613 figure obtained with a calculator.

EXAMPLE 66 Find the product of 908 432 × 0.000 721 9 using the estimation technique described in Example 65. Then, check your estimate with a calculator.

Solution The estimated answer is 63×10^1, 6.3×10^2, or 630. This answer is obtained by first converting each of the given numbers to exponential notation, giving $9.084\,32 \times 10^5$ and 7.219×10^{-4}. Next, round off each of these two exponential numbers to one digit, giving 9×10^5 and 7×10^{-4}. These two numbers are then multiplied to yield the estimated answer 630. The calculator gives the answer 655.797 060 7.

EXAMPLE 67 Divide 222 345 into 78 992 using the estimation technique. Then, check your estimate with your pocket calculator.

Solution The approximate answer is 0.4. Start by expressing both given numbers in exponential form, giving 2×10^5 and 8×10^4. Then, carry out the division: $(8 \times 10^4) \div (2 \times 10^5) = 4 \times 10^{-1} = 0.4$. The calculator gives the answer 0.355 267 714.

SELF-TEST Estimate the answers to the following problems:

1. 63×407

2. 0.0005×511

3. $\dfrac{893}{29}$

4. $\dfrac{8135.68}{0.020\,431}$

ANSWERS **1.** 2.4×10^4 **2.** 2.5×10^{-1} **3.** 3×10^1 **4.** 4×10^5

Now let's try a more difficult problem involving both multiplication and division.

EXAMPLE 68 Estimate the answer to the following calculation:

$$\frac{7941.32 \times 0.5214}{21.8}$$

Solution Each of the given numbers is replaced with a one-digit number written in exponential notation. The problem now looks like this:

$$\frac{(8 \times 10^3)(5 \times 10^{-1})}{(2 \times 10^1)}$$

Next, multiply and divide the nonexponential parts of each number ($8 \times 5 \div 2$), which gives 20. Finally, calculate the exponential part of the answer by combining the exponents in the problem according to the rules you have learned: $3 + (-1) - 1 = 1$. The estimated answer is 20×10^1, or 200.

EXAMPLE 69 Estimate the answer to the following calculation:

$$\frac{189 \times 406.2}{3186.4}$$

Solution Express the given problem with one-digit numbers in exponential notation and carry out the indicated mathematics:

$$\frac{(2 \times 10^2)(4 \times 10^2)}{(3 \times 10^3)} = \frac{8}{3} \times 10^1 = 3 \times 10^1 = 30$$

Notice that $8 \div 3 = 2.7$, but we estimated the division to get 3 ($9 \div 3 = 3$, so $8 \div 3$ must also equal about 3).

If you thought that these last two examples were simple, you'll be happy to know that the vast majority of the problems that you will encounter in general chemistry do not involve more than one or two multiplication and division operations. So these last problems are typical of the mathematical calculations that you will be performing.

We estimated an answer of 200 for Example 68. A calculator gives an answer of 189.935 974 6. Very few things are known (or need to be known) this accurately, even in science. In everyday life, knowing that a city is about 200 miles away is usually good enough. Knowing that the city is 189.935 974 6 miles away isn't practical. And, as we said before, this ten-digit answer may also be less correct than the estimated answer of 200.

Use your pocket calculator and practice doing a few more multiplications and divisions with this technique. When you think you have mastered the technique, check yourself further with the Self-Test and then graduate to the faster estimation technique described in the next section.

SELF-TEST Estimate the answers for each of the following problems:

1. $\dfrac{127 \times 691}{0.21}$

2. $\dfrac{0.000\,404\,816}{5.007}$

3. $\dfrac{8\,143\,276 \times 0.000\,104}{0.002\,146\,8}$

ANSWERS **1.** 4×10^5 **2.** 8×10^{-5} **3.** 4×10^5

1.4.4 SKILLFUL ESTIMATION FOR ADVANCED PLAYERS

The advanced way to estimate answers is actually a shortcut for the first method. It takes a few paragraphs to explain the technique, but it is really quite efficient and simple to do once you get the idea. It is all done mentally; nothing has to be written down except the final answer.

There are two rules for this advanced method. The first rule requires you to look only at the first digit of each number in your problem and carry out the required multiplication or division. As an example, consider the multiplication 893×31.54. To apply the first rule, look at the 8 and the 3. They have to be multiplied together. However, the 8 should be rounded off to a 9 (because it is followed by a digit that is 5 or greater, as discussed in Section 1.6). So we have $9 \times 3 = 27$. Write down the 27. You now have the first half of your answer.

The second half of your answer will be 10 raised to some power (exponent). The second rule shows how to find the exponent of 10. To find the exponent of 10, mentally move the decimal point of each number so that there is only one digit to the left of the decimal point. For the number 893, move the decimal point two places to the left, counting "one, two" as you do so:

$$\underset{2\,1}{8\underbrace{93}}.$$

For the second number, 31.54, move the decimal point one place to the left, counting "three":

$$3\underset{3}{1}.54$$

For this problem (893×31.54), the total number of places the decimal point is moved to the left in both numbers equals the power of 10. Thus, 893×31.54 is 27×10^3 (or 2.7×10^4).

Take a look at both methods, side by side, to see why the second method works:

$$\underset{2\,1}{8\underbrace{93}}. \times 3\underset{3}{1}.54 \qquad (9 \times 10^2) \times (3 \times 10^1)$$

↑ second method ↑ first method

Can you explain, by studying the preceding comparison, why the second method works?

The second method is just a shortcut for writing down and adding the exponential parts that are required by the first method. Both methods use essentially the same technique to determine the nonexponential part of the answer.

Let's try another problem. Multiply 0.000 17 × 104, using the second method. Check your estimation with a calculator.

Did you approximate 2×10^{-2}, without writing anything down? Your mental process should be: "2 times 1 is 2." At this point you can write down the 2. Then, count "−1, −2, −3, −4, −3, −2" as you mentally move the decimal point over four places to the right in the first number and then two places to the left in the second number, as shown:

$$0.000\,17 \times 104. \qquad (2 \times 10^{-4}) \times (1 \times 10^{2})$$

−1 −2 −3 −4 (under 0.000 1) −2 −3 (under 104)

second method — first method

Now let's try a more difficult problem involving both multiplication and division and both positive and negative exponents. Ready?

Try estimating the answer to the following problem using the second method. Then check your estimation with a calculator.

$$\frac{7941.32 \times 0.5214}{21.8}$$

The correct estimation is 20×10^1, 2×10^2, or 200. Here's the mental process that you should have followed: "$8 \times 5 = 40$, $40 \div 2 = 20$." Write down the 20. This is the nonexponential part of your estimated answer. To obtain the exponent of 10 in the exponential part of the answer, the mental process should go something like this: "1, 2, 3, 2, 1," as shown:

$$\frac{7941.32 \times 0.5214}{21.8}$$

3 2 1 (under 941) 2 (under 5) — 1 (under 1 of 21.8)

or

$$\frac{7941.32 \times 0.5214}{21.8} \qquad \frac{(8 \times 10^3)(5 \times 10^{-1})}{(2 \times 10^1)} = 20 \times 10^{3-1-1} = 20 \times 10^1$$

3 2 1 (under 941) −1 (under 5) — 1 (under 1 of 21.8)

Thus the power of 10 is 1, and the answer is 20×10^1.

Some people make up their own method by combining the ideas of the two methods discussed here. It is not important which method you use to estimate answers. It *is* important that you can mentally estimate an answer so that if your calculator should give you an answer of 853.543 234 6 when you're expecting the answer to be about 9, you will know that an error has been made. You should never blindly accept an impressive ten-digit answer displayed on a calculator. Practice this advanced technique some more by making up your own problems, estimating the answer, and checking your estimated answer with your calculator.

•
•

SELF-TEST Use the more efficient estimation technique to estimate the answers for each of the following calculations:

1. $\dfrac{127 \times 691}{0.21}$

2. $\dfrac{0.000\,404 \times 916}{5.007}$

3. $\dfrac{8\,043\,276 \times 0.000\,102}{0.002\,146\,7}$

ANSWERS **1.** 4×10^5 **2.** 7×10^{-2} **3.** 4×10^5

1.5 SIGNIFICANT DIGITS (ARE SIGNIFICANT)

By now you should have realized that there is something significant about how many digits you show in your answers. Recall, for example, that a rectangle that measures 2 in. × 20 in. (each measurement is known to the nearest inch) has an area of 4×10^1 in.^{2} and not 40 in.^{2}. The number of significant digits in a number is the number of digits that are known with some degree of accuracy.

We say that the sun is 93 000 000 miles away from the earth. Is it exactly 93 000 000.000 miles away? Are all of the digits in this number significant (accurately known)? Actually, the number 93 000 000 miles only has two significant digits and it would be more correct to write it as 9.3×10^7. This is one of the advantages of using scientific (exponential) notation. When a number is expressed in scientific notation, there is usually little doubt as to how many of the digits are significant.

Before we get into the rules of significant digits, try the following multiplication: 7×3.

Based on the rectangle problem, you may have come up with the incorrect answer of 2×10^1. The correct answer is 21. Let's see why $7 \times 3 = 21$ but 2 in. × 20 in. = 4×10^1 in.^{2}.

1.5.1 COUNTING, MEASURED, AND DEFINED NUMBERS

Numbers without units (e.g., inches) are called *counting numbers* (also called *whole numbers* or *integers*). Numbers that do have units associated with them are called *measured numbers*. Measured numbers are obtained with a measuring device, such as a balance (or scale), a meter stick (or yardstick), or a graduated cylinder (or measuring cup). We'll discuss defined numbers later.

A counting number is known exactly. If you are holding 5 pencils, you have exactly 5 pencils (represented by the notation $5.\bar{0}$, which means that there are an infinite number of 0's to the right of the 5.0). But a measured number is not known exactly. There is a limit to how many digits you can obtain when measuring anything with a measuring device. If something is measured within 0.1 in. (e.g., 27.4 in.), we know that the true length of the object may lie somewhere between 27.3 in. and 27.5 in.; that is, the object measures 27.4 ± 0.1 in. Its length is not $27.4\bar{0}$ in. The measured number 27.4 in. contains three significant digits. The counting number 5 (pencils) can be thought of as containing an infinite number of digits.

EXAMPLE 70 While jogging to your chemistry class, you pass 3 people, each drinking 1 qt of soda. Is the number 3 a measured number or an integer? Why?

Solution The number 3 is an integer (counting number). There are exactly 3 people, not 3.04 or 2.98 people. You can count the number of people to determine exactly how many people are present.

EXAMPLE 71 What about the volume of the soda they are consuming? Is the 1 (quart of soda) an integer or a measured number? Why?

Solution The number 1 is a measured number because it refers to a volume (measured quantity determined with a measuring device) of soda. One quart is a volume that must be measured. What you see may actually be 1.1 qt or 0.98 qt of soda. It will probably not be $1.\bar{0}$ qt.

•
•

Some numbers appear to be measured numbers because they have units of mass, length, or volume that are usually determined with measuring devices. The numbers 16 oz per pound, 12 in. per foot, and 2 pt per quart are such numbers. We will refer to these as *defined numbers* because somebody defined them to be exactly true. One foot contains 12 in. by definition and the definition 12 in. contains an infinite number of significant digits. There are exactly $12.\bar{0}$ in. in 1 ft. Also, in this definition, 1 ft contains an infinite number of significant digits ($1.\bar{0}$). The definition says that exactly 1 ft contains exactly 12 in.

When you perform multiplications and divisions that involve measured numbers, the numbers of digits the answer may contain is limited by the number of digits in the measured numbers being multiplied and divided. There are specific rules that deal with multiplications and divisions of measured numbers, but before we get involved with them, let's see how to determine the number of significant digits in a measured number.

•
•

EXAMPLE 72 Consider the following measured numbers: 23 in., 4.48 qt, 234.5353 g, 889 mi, and 4 lb. How many significant digits are in each of these measured numbers?

Solution

Measured Number	Number of Significant Digits
23 in.	2
4.48 qt	3
234.5353 g	7
889 mi	3
4 lb	1

It is relatively easy to determine the number of significant digits in these measured numbers. All you have to do is count the number of digits in each of the given numbers.

•
•

1.5.2 LEADING, MIDDLE, AND TRAILING ZEROS

Many students have trouble trying to determine the number of significant digits in measured numbers that contain zeros. All numbers can have as many as three different kinds of zeros: (1) leading zeros, (2) middle zeros, and (3) trailing zeros. The following two numbers have all three types of zeros:

000 430 086.0 and 0.004 300 860

The preceding two numbers differ only by the position of the decimal point. Both contain the same number of zeros and digits. The zeros before the 43 in both numbers are called leading zeros. The zeros between 43 and 86 in both numbers are called middle zeros. The zeros following 86 in both numbers are trailing zeros. Both numbers contain three leading zeros, two middle zeros, and one trailing zero:

three leading zeros	two middle zeros	one trailing zero		three leading zeros	two middle zeros	one trailing zero
000	430 086	.0	and	0.004	300	860

EXAMPLE 73 Consider the number 0.000 034 056 000. Identify the three types of zeros in this number. How many of each type are there?

Solution All five zeros to the left of 34 are leading zeros, the single zero between the 34 and the 56 is a middle zero, and the three zeros following the 56 are trailing zeros.

Leading zeros are not counted as significant digits. All middle zeros are significant digits. Trailing zeros are usually significant digits, but trailing zeros can be ambiguous.

The trailing zeros in 93 000 000 mi (the sun's distance from the earth) are obviously not significant. The trailing zeros in 5 000 000 000 people (the population of the earth) are obviously not significant. The trailing zeros in 400 ft (the distance between two objects on a road) may or may not be significant. This ambiguity can be cleared up if you (and others) will make it a habit of always using decimal points and/or scientific notation when expressing measured numbers. For example, it is better to say that the sun is 9.3×10^7 mi away (because the nonsignificant trailing zeros have vanished).

If two objects are exactly 400 ft apart, it is better to express the distance as 4.00×10^2 ft (three significant digits). If two objects are approximately 400 ft apart, it is better to express the distance as 4×10^2 ft (using only one significant digit).

A decimal point can also remove the ambiguity. For example, if the objects are 400.0 ft apart, all three trailing zeros are significant and this number (400.0) contains four significant digits.

EXAMPLE 74 How many significant digits does the measured number 20. in. contain?

Solution The measured number 20. in. contains two significant digits. The decimal point removes the ambiguity of the trailing zero.

EXAMPLE 75 If 0.000 434 034 500 is a measured number, how many significant digits does it contain?

Solution The number 0.000 434 034 500 contains nine significant digits. The first four (leading) zeros are not counted as significant digits.

EXAMPLE 76 Consider the following two numbers:

000 000 145 800 003 452 340 000 000.0

and

0.000 001 458 000 034 523 400 000 000

How many significant digits does each of these numbers contain?

Solution Both numbers contain 22 significant digits. All of the digits following the 6 leading zeros are significant.

SELF-TEST Determine the number of significant digits in each of the following numbers:

1. 1 penny
2. 1 in.
3. 0.004 mi
4. 4.00 ft
5. 30.005 qt

ANSWERS **1.** infinite **2.** one **3.** one **4.** three **5.** five

Now that you know how to find the number of significant digits in any number, let's discuss the rules that govern multiplying and dividing measured numbers. Let's look again at Example 68 and assume that all the numbers are measured numbers:

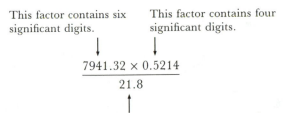

This factor contains six significant digits. (↓ 7941.32)

This factor contains four significant digits. (↓ 0.5214)

$$\frac{7941.32 \times 0.5214}{21.8}$$

(↑ 21.8) This factor contains three significant digits. This is the factor that contains the least number of significant digits. This factor limits the number of digits that can be present in the answer.

RULE: When you multiply or divide measured numbers, the answer must contain the same number of significant digits as are contained in the factor with the least number of significant digits. In the preceding example, the answer can contain only three significant digits because 21.8 is the factor with the fewest significant digits (three significant digits). A calculator value of 189.935 974 6 is meaningless. The answer that must be reported is 190. (see Section 1.6, "Rounding Off"). The value 190. contains three significant digits, the same number of significant digits as are in the factor with the least number of significant digits (21.8).

•
•

SELF-TEST Perform the following operations and express your answers with the correct number of significant digits. Assume all numbers shown are measured numbers.

1. 60.00×3.00
2. 12×100.0
3. 20.4×0.003
4. $100.0 \div 0.0010$
5. $\frac{68.0 \times 29.4}{2}$

ANSWERS **1.** 180. (three significant digits) **2.** 1.2×10^3 (two significant digits) **3.** 6×10^{-2} (one significant digit) **4.** 1.0×10^5 (two significant digits) **5.** 1×10^3 (one significant digit)

The following demonstration* may help you to understand the reasons behind the rule that limits the number of significant digits in an answer. Five students were asked to measure the length, width, and thickness of a stack of papers to the nearest 0.01 in. Then, the students were asked to calculate the volume of the stack of papers. Table 1.1 shows the measured lengths, widths, thicknesses, and calculated volumes as determined by the students.

TABLE 1.1

Length (inches)	Width (inches)	Thickness (inches)	Volume (cubic inches)
11.01	8.50	0.55	51.471 75
11.02	8.50	0.56	52.455 20
11.01	8.50	0.56	52.407 60
11.02	8.51	0.55	51.579 11
11.00	8.49	0.54	50.430 60

Which of the three measured values (length, width, or thickness) contains the least number of significant digits?

The thickness values contain only two significant digits. The thickness values contain the least number of significant digits.

Although each student calculated the volume of the papers to eight digits (with a pocket calculator), each student should have reported a volume with two digits. The volume of the stack of papers is 51 $in.^3$. This volume (51 $in.^3$) implies that the true volume lies between 50 $in.^3$ and 52 $in.^3$. Notice that all of the calculated volumes are in this range. When compared in this way, all of the calculated volumes now appear to be in better agreement. The seven significant digits shown in the table are meaningless.

This is the point we tried to make in our previous discussion of pocket calculators (see Section 1.4.1). For this demonstration, an estimated value of 54 $in.^3$ $[(1 \times 10^1) \times (9 \times 10^0) \times (6 \times 10^{-1}) = 54]$ is more meaningful and correct than a value such as 52.455 20 $in.^3$ because 54 $in.^3$ implies an accuracy in the measurements of 0.01 in. (which is the case) while 52.455 20 implies that each measurement is accurate to within 0.000 001 in. (a far cry from the truth).

1.5.3 ACCURACY VS. SIGNIFICANT DIGITS

As we have seen, whenever we multiply or divide measured numbers, we must be concerned about the number of significant digits in the final answer. The number of significant digits in the final answer cannot exceed the number of significant digits in the least significant number used in the multiplication or division.

When adding or subtracting measured numbers, we do not consider the number of significant digits. Instead, we consider only the accuracy of the numbers being added or subtracted. What do we mean by accuracy and how does accuracy differ from significant digits?

Consider the rectangle that measures 2 in. × 20 in. Both the 2 and the 20 have the same accuracy; that is, each of these two measurements is known accurately to the nearest inch. But these two measured numbers have a different number of significant digits: 20 has two significant digits and 2 has one significant digit. If you want to multiply these two numbers together, your answer can only contain one significant digit. But if you are going to add these two numbers together, the sum cannot have a greater accuracy than the accuracy of the least accurate of the two

*Adapted from Larry C. Stack, "Calculators and Significant Figures," *Journal of Chemical Education* 54 (Mar. 1977):177.

numbers. Since both numbers have the same accuracy (both are known accurately to the nearest inch), the sum of 20 in. and 2 in. equals 22 in. The sum 22 in. is also known accurately to within the nearest inch. When adding and subtracting numbers, examine the accuracy of each number being added or subtracted to determine how many digits will appear in the answer.

> **RULE:** The result of addition or subtraction cannot be more accurate than the accuracy of the least accurate number being added or subtracted.

•
•

EXAMPLE 77 Find the sum of 8.0 in. and 34.77 in.

Solution 8.0 in. + 34.77 in. = 42.8 in. The measured number 8.0 in. is known with an accuracy of 0.1 in. The answer (sum) cannot be known more accurately than to the nearest 0.1 in. If you are not sure why this must be so, consider the following demonstration:

•
•

Fill an empty 1-qt milk carton with water. How many liters of water are in the milk carton? How accurate is your answer?

Approximately 1 L of water is in the milk carton. The accuracy of this value is not very high. We can be fairly certain that the true volume of water lies somewhere between 0 L and 2 L. Our volume is only known to the nearest liter.

Before continuing with our demonstration, express the volume of water contained in the 1-qt milk carton in milliliters. (Note: 1 L = 1000 mL.)

There is 1×10^3 mL (*not* 1000 mL) of water in the milk carton. We would be in error to say that there is 1000 mL of water in the milk carton because the figure 1000 implies that we know the volume to be somewhere between 999 mL and 1001 mL, that is, 1000 mL ± 1 mL. But we don't know the volume of water that accurately; we only know that we have about 1 L of water.

Now, to the milk carton filled with water, add 1 mL of water. What is the sum of 1 L and 0.001 L? Explain your answer.

The sum of 1 L and 0.001 L is 1 L. One reason for this is that the sum cannot be more accurately known than the accuracy of the least accurate of the two numbers being added together. The value 1 L is a measured number known to the nearest units place (known to the nearest liter), while the value 0.001 L is a measured number known accurately to the nearest 0.001 L. The sum of these two numbers cannot be known more accurately than to the nearest units place (to the nearest liter). The sum of 1 L and 0.001 L is *not* 1.001 L.

Here's another way to analyze this demonstration. Since the actual water volume in the milk carton may have been 0.9 L, surely it would make no sense to say that 0.9 L of water plus 0.001 L of water equals 1.001 L of water?

•
•

EXAMPLE 78 Which of the following three numbers contains the greatest number of significant digits: 1023 g, 11.4 g, or 0.345 g?

Solution The first number, 1023 g, contains the largest number of significant digits. It contains four significant digits. The remaining numbers each contain three significant digits.

•
•

EXAMPLE 79 Which of the measured numbers in Example 78 is the least accurately known?

Solution Again, the answer is the first number, 1023 g. It contains more significant digits than the two other measured numbers, but it is known accurately only to the nearest gram (nearest units place). The last number, 0.345 g, has fewer significant digits than the first number but it is the most accurately known of the three numbers. The last number is accurately known to the nearest 0.001 g. The second measured number is known accurately to the nearest 0.1 g.

EXAMPLE 80 Thirty-one years ago the newspapers said that the pyramids were 5000 years old. How old are they today?

Solution The pyramids are still 5000 years old. We must assume that the measurement 5000 years is accurate only to the thousands place. If the two numbers were not measured numbers, their sum would be 5031. Since they are measured numbers, the value 5031 must be rounded off (see Section 1.6) to the nearest thousand years, which is the accuracy of the less accurately known of the two measured numbers.

SELF-TEST Perform the following operations and express your answers with the correct accuracy. Assume that all numbers shown are measured numbers.

1. 27 mi + 1 mi
2. 1 mL + 0.001 mL
3. 28 qt − 1.3 qt
4. 1.000 in. + 0.433 in.

ANSWERS **1.** 28 mi **2.** 1 mL **3.** 27 qt **4.** 1.433 in.

1.6 ROUNDING OFF

You should now know how to determine the number of significant digits that should be present in the answers to your calculations. Next, you have to learn how to eliminate any nonsignificant digits that may be present in your answers.

When you eliminate nonsignificant digits from an answer, the process is called rounding off your answer to the correct number of significant digits. Fortunately, there are only two very easy rules to follow when you need to round off an answer to the correct number of significant digits.

> **RULE 1:** If the digit to the right of the last digit you want to keep is less than 5, simply remove all the unwanted digits to the right of the last digit you want to keep.

EXAMPLE 81 Round off 0.427 to one significant digit.

Solution

0.427

This first significant digit is the only digit we want to keep.

This second digit is the digit to the right of the significant digit we want to keep. This second digit is less than 5.

Because the digit to the right of the last digit we want to keep is less than 5 (2 is less than 5), we can remove all the unwanted digits (27) to the right of the last digit we want to keep. This gives us the number 0.4, which has just one significant digit. When we eliminate the unwanted digits (27), we say that we have rounded off 0.427 to one significant digit (0.427 is rounded off to 0.4, which has one significant digit).

EXAMPLE 82 Round off 0.581 395 to three significant digits.

Solution

0.581 395

This third significant digit (counting from the left) is the last significant digit we want to keep.

This fourth digit is the digit to the right of the last significant digit we want to keep. This fourth digit is less than 5.

Since the digit to the right of the last digit we want to keep is less than 5 (3 is less than 5), we can remove all the unwanted digits (395) to the right of the last digit we want to keep. This gives us the number 0.581, which has three significant digits. When we eliminate the unwanted digits (395) from 0.581 395, we say that we have rounded off 0.581 395 to three significant digits (0.581 395 is rounded off to 0.581).

EXAMPLE 83 A rectangle has a length of 4.53 in. and a width of 2.14 in. Find the area of the rectangle and round off the area to the correct number of significant digits.

Solution The area of a rectangle is obtained by multiplying its length times its width. With the aid of a calculator, we find

$$4.53 \text{ in.} \times 2.14 \text{ in.} = 9.694\,200\,000 \text{ in.}^2$$

Our rules on significant digits tell us the answer can only have three significant digits (see Section 1.5.2). The calculator gave us an answer with ten digits.

$9.694\,200\,000 \text{ in.}^2$

This third significant digit (counting from the left) is the last significant digit we can keep.

This fourth digit is the digit to the right of the last significant digit we can keep. This fourth digit is less than 5.

The number $9.694\,200\,000 \text{ in.}^2$ must be rounded off to 9.69 in.^2.

> **RULE 2:** If the digit to the right of the last digit you want to keep is equal to or greater than 5, add 1 to the last digit you want to keep and remove all the unwanted digits to the right of the last digit you want to keep.

EXAMPLE 84 Round off 0.427 839 634 to five significant digits.

Solution

0.427 839 634

This fifth significant digit (counting from the left) is the last significant digit we want to keep.

This sixth digit is the digit to the right of the last significant digit we want to keep. This sixth digit is greater than 5.

Since the digit to the right of the last significant digit we want to keep is greater than 5 (9 is greater than 5), we add 1 to the last digit we want to keep ($1 + 3 = 4$) and discard all the digits to the right of the last digit kept. We say that we have rounded off 0.427 839 634 to 0.427 84.

EXAMPLE 85 Round off 846 271.491 537 to 11 significant digits. Then, round it off to numbers with 10, 9, 8, 7, 6, 5, 4, 3, and 2 significant digits. Finally, round it off to a single significant digit.

Solution Table 1.2 shows the original number rounded off to different numbers of significant digits. The table also indicates which rule (Rule 1 or Rule 2) was followed when rounding off the *original* number.

TABLE 1.2

Number of Significant Digits	Number	Rule
11	846 271.491 54	2
10	846 271.491 5	1
9	846 271.492	2
8	846 271.49	1
7	846 271.5	2
6	846 271.	1
5	8.4627×10^5	1
4	8.463×10^5	2
3	8.46×10^5	1
2	8.5×10^5	2
1	8×10^5	1

Always use the entire (original) number when you round off. That is, round off 846 271.491 537 to 846 271.; do not round off 846 271.5 to 846 272.

Round off the entire number 846 271.491 537 to 8×10^5; do not round off 8.5×10^5 to 9×10^5.

SELF-TEST Round off each of the following numbers to three significant digits:

1. 0.000 405 491
2. 45.371 907
3. 458 153 684.358 761
4. 93 982 376

ANSWERS **1.** 0.000 405 **2.** 45.4 **3.** 4.58×10^8 **4.** 9.40×10^7

PROBLEM SET 1

The problems in Problem Set 1 parallel the examples in Chapter 1. For example, if you should have trouble working Problem 5, go back to Example 5 in this chapter to get help. The correct answers to these problems are given at the end of this book.

1. What are the exponential and nonexponential parts of 7.4×10^2? What is the exponent in this number?
2. Write the number 5200 in scientific notation.
3. Express 8 340 000 in scientific notation.
4. Express one billion in exponential notation.

5. Express 0.7 in scientific notation.
6. Express 0.0018 in scientific notation.
7. Express 5×10^3 as an ordinary number.
8. Express 4×10^{-5} as an ordinary number.
9. Express 32.7×10^4 in standard scientific notation.
10. Express 0.0076×10^4 in standard scientific notation.
11. Express the number 54 in eight different ways, letting some of your nonstandardized exponential notations contain positive exponents and some, negative exponents.
12. Subtract 4×10^3 from 3×10^4.
13. Add 6×10^4 to 3×10^7.
14. Subtract 6×10^2 from 7×10^5.
15. Express 30 and 300 in exponential notation and multiply the exponential forms of these two numbers.
16. Multiply 3×10^2 times 4×10^3.
17. Perform the following addition: $(+3) + (+4)$.
18. Add the following: $(-8) + (-3)$.
19. Perform the following addition: $(-3) + (+7)$.
20. Perform the following addition: $(-9) + (+7)$.
21. Rewrite the addition problems in Problems 17, 18, 19, and 20, omitting unnecessary positive signs and parentheses.
22. Perform the following subtraction: $7 - (-3)$.
23. Perform the following subtraction: $6 - (-4)$.
24. Perform the following subtraction: $(-5) - (-3)$.
25. Divide 60 000 by 200 by first expressing these numbers in scientific notation and then performing the division using the numbers in scientific notation.
26. Divide 8×10^4 by 2×10^{-3}.
27. Divide 4×10^{-8} into 8×10^{-4}.
28. Find the square root of 10^6.
29. Find the cube root of 10^9.
30. Find the fourth root of 10^{32}.
31. Find the square root of 25×10^8.
32. Find the cube root of 27×10^9.
33. Find the square root of 1.6×10^3.
34. Find the square root of 0.16×10^6.
35. Find the fourth root of 16×10^{-80}.
36. Find the cube root of 7×10^{-14}.
37. Raise 3×10^{-4} to the fourth power.
38. Find the value of $(3 \times 10^8)^4$.
39. Find the logarithm of 10 000.
40. Find the log of 0.000 001.
41. What is the log of 1?
42. Find the log of 5, using the shortcut log table.
43. Find the log of 1.3, using the shortcut log table.
44. Find the log of 4.5.
45. Find the log of 2.25.
46. Find the log of 1.9.
47. Find the log of 40.
48. Find the log of 40 000.
49. Find the log of 0.000 045.
50. Find the log of 9500.
51. What is the antilog of 5?
52. What is the antilog of 8?
53. Find antilog 14.
54. Find the antilog of −6.
55. Find antilog 0.
56. Find the value of antilog 0.70.
57. Find the antilog of 0.80.
58. Find the antilog of 0.95.
59. Find the antilog of 5.40.
60. Find the value of antilog 3.30.
61. Find the antilog of 7.95.
62. Find the antilog of −8.2.
63. Find the antilog of −4.50.
64. Find the value of antilog −8.95.
65. Consider the problem of multiplying 423×610. Estimate the product, using the technique of skillful estimation.
66. Find the product of $863\,294 \times 0.000\,672\,4$, using the technique of skillful estimation.
67. Divide 218 458 into 79 125, using the estimation technique.
68. Estimate the answer to the following calculation:

$$\frac{7894.25 \times 0.4983}{22.1}$$

69. Estimate the answer to the following calculation:

$$\frac{204 \times 398.4}{2984.7}$$

70. There are 10 students in your chemistry class. Is the number 10 a measured number or an integer?
71. Each student in your class is 6 ft tall. Is the number 6 a measured number or an integer?
72. How many significant digits are in the measured number 5.32 in.?
73. Consider the number 0.001 043 200. How many leading, middle, and trailing zeros are there in this number?
74. How many significant digits does the measured number 400. in. contain?
75. If the number 0.001 043 200 is a measured number, how many significant digits does it contain?
76. How many significant digits are in 000 345 400 001 345.870 000?
77. Find the sum of 4.0 in. and 230.41 in.
78. Which of the following three numbers contains the greatest number of significant digits: 1530 g, 11.1 g, or 0.228 g?
79. Which of the three measured numbers in Problem 78 is the least accurately known?
80. Find the sum of the three measured numbers in Problem 78.
81. Round off 0.349 to one significant digit.
82. Round off 0.286 487 to three significant digits.
83. A rectangle has a length of 8.41 in. and a width of 4.72 in. Find the area of the rectangle and round off the area to the correct number of significant digits.
84. Round off 0.294 827 945 to five significant digits.
85. Round off 2 482 858.547 to nine significant digits. Then, round it off to eight, seven, six, five, four, three, and two significant digits. Finally, round it off to a single significant digit.

2 Dimensional Analysis

2.1 INTRODUCTION: COMMUNICATION, KNOWLEDGE, AND INTELLIGENCE

Approximately three million years ago, the brain size of our ancestors doubled (see Figure 2.1). Another increase in brain size occurred about one million years ago. But during these two periods of increased intelligence no corresponding advances were made in human technology or art. How was the newly acquired intelligence being used, if not for advanced technology or increased artistic skills? Why did humans become more intelligent?

A related mystery presented itself a few years ago when primates (monkeys, gorillas, and chimpanzees) were taught to "speak" using human sign language. We suddenly became acutely

FIGURE 2.1 Human development. (a) Cave drawing (art) and stone tool (technology) that have been found with skeletal remains before brain enlargement (c. 3 million years B.C.). (b) Similar cave drawing (art) and stone tool (technology) have been found with skeletal remains after brain enlargement (c. 1 million years B.C.).

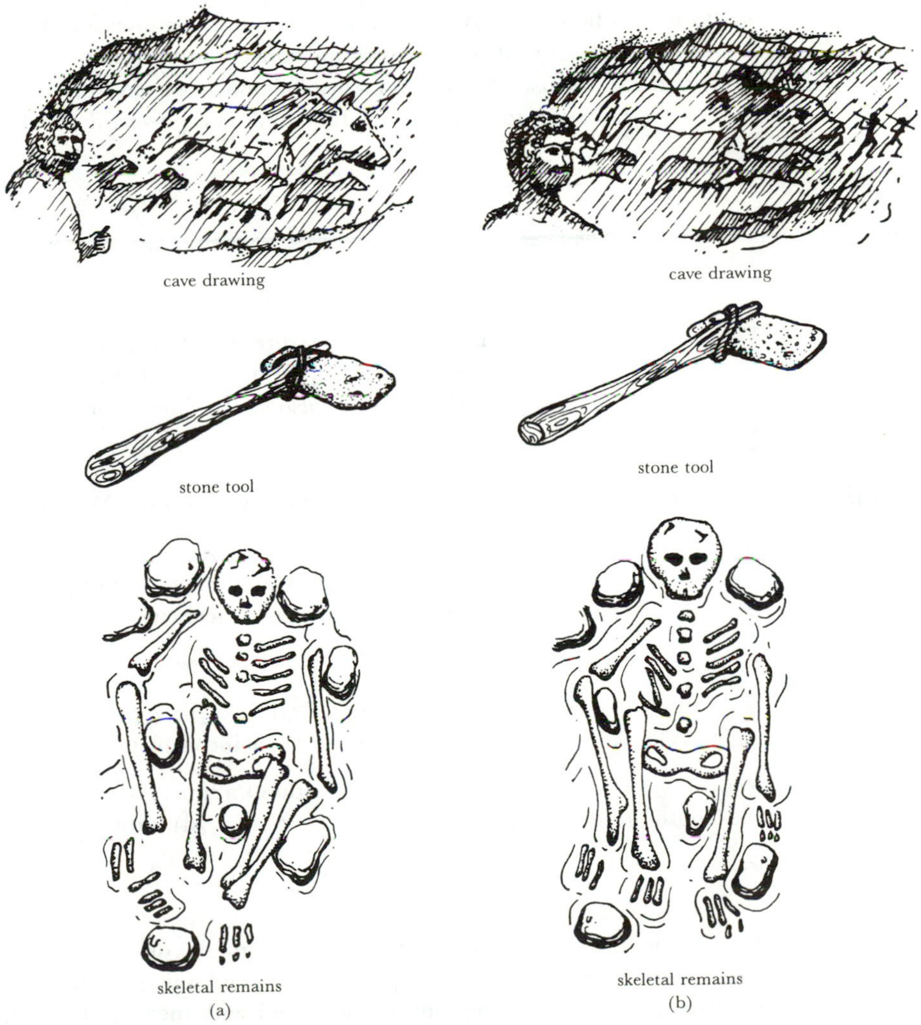

aware of the intelligence level of our fellow primates, but we were hard-pressed to explain what purpose the intelligence served. Although primates can and do use primitive tools, their intelligence exceeds that needed for the simple tasks they perform.

Wasted intelligence is uncharacteristic of an economical nature, so we can be relatively certain that primate intelligence does serve some purpose. Why did intelligence develop?

It is believed that intelligence developed for social reasons. Groups cooperate to achieve common goals more effectively than individuals can. Groups have a survival advantage over individuals, but functioning in a group requires intelligence. Members working within a group must learn how to use restraint, persuasion, tact, submission, perception, and, for humans, both verbal and written skills for effective communication. Intelligence is required for effective communication.

In 1978 the American Chemical Society announced that the ability to effectively communicate chemical knowledge and the chemical knowledge itself are of about equal importance. The society urges all college chemistry faculty to grade student papers on not only technical content but also basic communication skills including spelling, punctuation, sentence structure, and style. (Perhaps you can better understand the importance of your written answers to the questions found in this book. Avoid answering those questions mentally.)

It is the goal of this chapter to introduce you to dimensional analysis and help you gain proficiency in the use of units. Dimensional analysis is an important skill because it is an effective communication tool and it also functions as a powerful problem-solving technique.

2.2 UNITS

What are units? Well, whenever you measure something (for example, your height), the result of your measurement is usually made up of two parts: (1) a number (for example, 72) and (2) a unit (for example, inches). Units are also called dimensions.

•
•

EXAMPLE 1 Your father asks you to measure the length of your house. You find the house to be 100 ft long. Identify the two parts of your measurement.

Solution The result of your measurement is 100 ft. The number part of your measurement is 100 and the unit or dimension part of your measurement is feet (ft).

•
•

EXAMPLE 2 Your mother asks you to measure the length of a piece of cloth. You do so and tell her that it is 36. Why is your answer considered poor or incomplete?

Solution Your answer consists of just a number without a unit. Although it may be obvious to you that the length of the cloth is 36 in. (and not 36 ft or 36 yd), numbers without units can be confusing.

•
•

Unfortunately, many students go through school without using units on a regular basis. Math courses usually don't use units because math courses deal primarily with numbers, not measurements (which are made up of numbers and units). However, when you apply your math knowledge in everyday life, you usually work with measured quantities.

In elementary science courses, many students (and teachers) are more interested in getting correct answers (usually by memorizing a formula) than they are in understanding what they are doing. It's so much easier to write $2 \times 3 = 6$ than it is to write the same calculation with units accompanying each number. But this laziness has a price that you begin paying when you take more advanced science courses where units are a very important part of the calculations.

2.3 BASIC TECHNIQUE OF DIMENSIONAL ANALYSIS

Once you begin writing measurements using both numbers and dimensions (units), applying the technique of dimensional analysis follows almost naturally. Before we define what we mean by dimensional analysis, you should be aware that units, such as pounds, feet, and quarts, can be treated as if they are numbers in operations, such as addition, subtraction, multiplication, and division.

EXAMPLE 3 Multiply 2 times 2.

Solution

$$2 \times 2 = 2^2 = 4$$

EXAMPLE 4 Multiply seconds times seconds.

Solution

$$\text{seconds} \times \text{seconds} = \text{seconds}^2$$

EXAMPLE 5 Work the following calculation: $(2 \times 3)/2$.

Solution

$$\frac{2 \times 3}{2} = \frac{\cancel{2} \times 3}{\cancel{2}} = \frac{(\cancel{2})(3)}{(\cancel{2})} = 3$$

EXAMPLE 6 Divide (feet × seconds) by feet.

Solution

$$\frac{(\text{feet} \times \text{seconds})}{\text{feet}} = \frac{(\cancel{\text{feet}})(\text{seconds})}{(\cancel{\text{feet}})} = \text{seconds}$$

EXAMPLE 7 Divide $(2^2 \times 3)$ by 2.

Solution

$$\frac{2^2 \times 3}{2} = \frac{(\cancel{2})(2)(3)}{(\cancel{2})} = (2)(3) = 6$$

EXAMPLE 8 Divide (feet^2 × pounds) by feet.

Solution

$$\frac{\text{feet}^2 \times \text{pounds}}{\text{feet}} = \frac{(\cancel{\text{feet}})(\text{feet})(\text{pounds})}{(\cancel{\text{feet}})} = (\text{foot})(\text{pounds}) = \text{foot pounds}$$

Notice that we can change units from plural to singular (feet to foot) and vice versa (foot to feet) without changing the validity of an expression. Also notice that the symbols indicating multiplication (for example, the × and the parentheses) can be omitted between units: (foot)(pounds) = foot × pounds = foot pounds.

EXAMPLE 9 Multiply (feet/second) times (seconds/foot).

Solution

$$\frac{\text{feet}}{\text{second}} \times \frac{\text{seconds}}{\text{foot}} = \frac{(\cancel{\text{feet}})(\cancel{\text{seconds}})}{(\cancel{\text{second}})(\cancel{\text{foot}})} = 1$$

A measurement such as 2 in. is considered a multiplication calculation (2 × inches). We can say: 2 inches = 2 × inches = (2) (inches) = inches × 2. The order of the numbers in multiplication does not affect the final answer. For example, 2 × 3 × 4 = 2 × 4 × 3 = 24. The same is true for units; that is, units may be separated from their associated number. Therefore, (2 inches) × (3 inches) = 2 × inches × 3 × inches = 2 × 3 × inches × inches = 6 inches^2.

EXAMPLE 10 The area of a square is equal to the product of any two sides. Find the area of a square that measures 2 in. on each side.

Solution

$$\begin{aligned}(2\text{ in.}) \times (2\text{ in.}) &= 2 \times \text{inches} \times 2 \times \text{inches} \\ &= (2 \times 2) \times (\text{inches} \times \text{inches}) \\ &= 2^2 \times \text{in.}^2 \\ &= 4 \times \text{in.}^2 \\ &= 4\text{ in.}^2\end{aligned}$$

The area is 4 square inches (4 in.^2). It is not necessary (or even desirable) to go through all of the steps shown. The steps are shown to emphasize that units can be treated like numbers.

EXAMPLE 11 The volume of a cube is equal to the product of the length of any three of its sides. Find the volume of a cube that measures 3.0 in. on each side.

Solution

$$\begin{aligned}(3.0\text{ in.}) \times (3.0\text{ in.}) \times (3.0\text{ in.}) &= 3.0 \times 3.0 \times 3.0 \times \text{inches} \times \text{inches} \times \text{inches} \\ &= 3.0^3 \times \text{inches}^3 \\ &= 27\text{ in.}^3 \text{ (read } \textit{27 cubic inches}\text{)}\end{aligned}$$

EXAMPLE 12 Add 4 lb to 2 lb.

Solution

$$4\text{ lb} + 2\text{ lb} = (4 \times \text{pounds}) + (2 \times \text{pounds})$$

We can factor out the unit *pounds* (lb), giving

$$\begin{aligned}(\text{pounds}) \times (4 + 2) &= (\text{pounds}) \times (6) \\ &= 6\text{ lb}\end{aligned}$$

Normally you would not go through all of these steps to add 4 lb to 2 lb. The steps are shown to emphasize that units can be treated like numbers.

EXAMPLE 13 Subtract 2 in. from 5 in.

Solution

$$(5\ \text{in.}) - (2\ \text{in.}) = (5 \times \text{inches}) - (2 \times \text{inches})$$

Factor out inches, giving

$$(\text{inches}) \times (5 - 2) = \text{inches} \times 3$$
$$= 3\ \text{in.}$$

SELF-TEST

1. Add 12 sec to 8 sec, treating the units as numbers.
2. Multiply 3 ft times 2 ft, treating the units as numbers.
3. Divide 8 sec into 32 ft-sec.
4. Perform the following calculation:

$$\frac{(2\ \text{in.})(3\ \text{ft})(4\ \text{lb})}{(6\ \text{in.})}$$

ANSWERS **1.** 20 sec **2.** 6 ft^2 **3.** 4 ft **4.** 4 ft lb

DIMENSIONAL ANALYSIS DEFINED: We define dimensional analysis as the technique of treating units as numbers for the purpose of solving problems. When using the dimensional analysis technique, we usually manipulate units mathematically to get rid of unwanted units and to introduce units that are wanted.

EXAMPLE 14 You know that you live 3.0 mi away from a shopping center and wish to calculate the distance in feet. There are 5280 ft in each mile. Use dimensional analysis.

Solution You are given the measurement 3.0 mi. You wish to express this distance (3.0 mi) in feet. That is, you want to get rid of the unwanted unit *mile* and introduce the wanted unit *feet*. The following calculation (manipulation) will accomplish this:

The wanted unit is introduced.

The answer can only contain two significant digits.

The unwanted unit cancels out.

Our answer contains only the wanted unit *feet*.

$$(3.0\ \cancel{\text{mi}})\left(\frac{5280\ \text{ft}}{\cancel{\text{mi}}}\right) = 1.6 \times 10^4\ \text{ft}$$

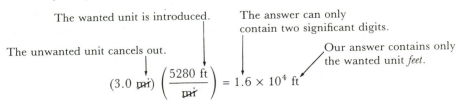

EXAMPLE 15 How many minutes will it take you to do a 2.0-hr job? Use dimensional analysis.

Solution You are given the measurement 2.0 hr. You wish to express this time (measurement) in minutes. This means that you want to change the unit of your measurement from hours to minutes. Dimensional analysis is a technique that manipulates units mathematically to get rid of unwanted units (hours) and to introduce units that are wanted (minutes). The following multiplication (manipulation) will accomplish this:

$$(2.0\ \cancel{\text{hr}})\left(\frac{60\ \text{min}}{\cancel{\text{hr}}}\right) = 1.2 \times 10^2\ \text{min}$$

per

EXAMPLE 16 If it takes 2.0 hr to walk to your friend's house, how many days does this trip require?

Solution

$$(2.0\ \cancel{\text{hr}})\left(\frac{1\ \text{day}}{24\ \cancel{\text{hr}}}\right) = 0.083\ \text{day}$$

The trip to your friend's house will take 0.083 day. Notice that 1 (day) and 24 (hours) each have an infinite number of significant digits, as they are defined numbers. The number of significant digits in our answer (0.083) is limited by the two significant digits in our given measurement (2.0 hr).

One of the advantages of using dimensional analysis to solve problems is that you always know if your approach to the problem is correct. If your approach is correct, your answer will have the correct (expected, desired, wanted) units. If you do not set up the problem correctly, your answer will have unwanted, unrealistic, or weird units.

•
•

EXAMPLE 17 Assume, for a moment, that you are going to work Example 16 over again but you are not sure whether to multiply the given measurement (2.0 hr) by 24 hr/day (24 hours per day) or to divide 2.0 hr by 24 (that is, multiply 2.0 hr by 1 day/24 hr). Try both approaches and decide which approach is the correct approach by observing the units in the answer that each approach produces.

Solution ***First Approach:***

$$(2.0\ \text{hr})\left(\frac{24\ \text{hr}}{\text{day}}\right) = \frac{48\ \text{hr}^2}{\text{day}} \longleftarrow$$ The unit hr^2 per day is weird.

Second Approach:

$$(2.0\ \text{hr})\left(\frac{1\ \text{day}}{24\ \text{hr}}\right) = 0.083\ \text{day} \longleftarrow$$ We wanted days as the unit in our answer.

The first calculation gives us an answer that has the unit *square hours per day*. We were looking for the number of days (or fraction of a day) that is represented by 2.0 hr. The answer obtained by the second approach, 0.083 day, appears to be correct because it contains the desired unit (days). Therefore, you could conclude, to convert hours to days you must multiply hours by 1 day/24 hr.

•
•

EXAMPLE 18 There are 3600 sec in 1 hr. You wish to find out how many hours are required to run for 5000 sec. Using dimensional analysis, decide whether you have to multiply 5000 sec by 3600 (3600 sec/hr) or to divide 5000 sec by 3600 (i.e., multiply 5000 sec by 1 hr/3600 sec).

Solution The two possible approaches are shown.

First Approach:

$$(5000\ \text{sec})\left(\frac{3600\ \text{sec}}{\text{hr}}\right) = \frac{1.800 \times 10^7\ \text{sec}^2}{\text{hr}}$$

Second Approach:

$$(5000\ \text{sec})\left(\frac{1\ \text{hr}}{3600\ \text{sec}}\right) = 1.389\ \text{hr}$$

Because we are looking for the number of hours in 5000 sec, the second approach must be correct because it provides an answer with the unit *hours*.

SELF-TEST In each of the following problems, try to obtain a correct answer by both multiplying and dividing by the given conversion factor. Decide which of the two approaches is correct by examining the units produced by each approach.

1. You weigh 200 lb. The conversion factor to convert pounds to tons is 0.0005 ton/lb (1 lb = 0.0005 ton). How many tons do you weigh?
2. You are 6 ft tall. The conversion factor to convert feet to miles is 1.894×10^{-4} mi/ft (1 ft = 1.894×10^{-4} mi). How many miles high are you?
3. Your car is 20 ft long. The conversion factor to convert feet to inches is 12 in./ft. What is the length of your car in inches?

ANSWERS **1.** 0.100 ton **2.** 1×10^{-3} mi **3.** 2.4×10^{2} in.

2.3.1 COOKBOOK DIMENSIONAL ANALYSIS*

Many students can understand the basic ideas of dimensional analysis and can follow their instructors when their instructors use dimensional analysis to solve problems. But these same students stumble with dimensional analysis when left on their own. A frequent comment is, "It looks easy when my professor does it, but I don't know where to start when I try it on my own."

2.3.2 FOUR-STEP RECIPE*

Here are four rules (steps) that are applicable to most of the problems you will encounter in chemistry as well as in other areas. The purpose of these four rules is to help you get started.

1. Write down the given number with its units (dimensions).
2. Write a ratio with the given unit in the denominator (at the bottom) and the unit sought in the numerator (on top).
3. Insert numbers into the ratio so that the numerator and denominator are equal.
4. Multiply Steps 1 and 3 together.

EXAMPLE 19 Five weeks ago you mailed a letter to a friend and are still waiting for a reply. How many days have you waited? Use the four-step recipe and show all four steps.

Solution **Step 1** You are given the measurement 5.0 weeks and wish to convert it to days. Write the given number (5.0) and its unit (weeks). That's all there is to Step 1. The first step has produced

given number → 5.0; given unit → weeks

5.0 weeks

Step 2 The given unit from Step 1 is weeks. The unit sought is days. Form a ratio with the given unit (weeks) in the denominator (at the bottom) and the unit sought (days) in the numerator (on top):

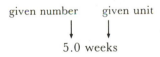

$$\frac{\text{days}}{\text{weeks}}$$

days ← unit sought in numerator

weeks ← unit given in denominator

*Adapted from Ronald DeLorenzo, "Cookbook Dimensional Analysis," *Journal of Chemical Education* 53 (Oct. 1976): 633.

Step 3 Insert numbers into the ratio (from Step 2) to make the numerator and the denominator equal. One week is equal to 7 days, so place a 1 in the denominator and a 7 in the numerator, giving

$$\frac{7 \text{ days}}{1 \text{ week}} \quad (7 \text{ days} = 1 \text{ week})$$

(per)

Step 4 Multiply the results of Steps 1 and 3 together; that is, multiply 5.0 weeks times 7 days/week:

$$(5.0 \cancel{\text{weeks}})\left(\frac{7 \text{ days}}{1 \cancel{\text{week}}}\right) = 35 \text{ days}$$

This ratio is called a conversion factor.

The unwanted unit, weeks, cancels out. The unit we are after, days, remains. The ratio produced by Steps 2 and 3 can be thought of as a conversion factor. This conversion factor converts the given unit (weeks) to the unit we want (days).

•
•

EXAMPLE 20 An object weighs 3.0 lb. What is the weight of this object in ounces?

Solution **Step 1** 3.0 lb (3.0 ← given number; lb ← given unit)

Step 2 $\frac{\text{oz}}{\text{lb}}$ (oz ← unit sought on top; lb ← unit given at the bottom)

Step 3 $\frac{16 \text{ oz}}{1 \text{ lb}}$ (per) (16 oz = 1 lb)

Step 4 $(3.0 \cancel{\text{lb}})\left(\frac{16 \text{ oz}}{1 \cancel{\text{lb}}}\right) = 48 \text{ oz}$ (Step 1 times Step 3)

•
•

EXAMPLE 21 You weigh 200 lb. (*Remember:* 1 ton = 2000 lb.) How many tons do you weigh?

Solution **Step 1** 200 lb

Step 2 $\frac{\text{tons}}{\text{lb}}$

Step 3 $\frac{1 \text{ ton}}{2000 \text{ lb}}$

Step 4 $(200 \cancel{\text{lb}})\left(\frac{1 \text{ ton}}{2000 \cancel{\text{lb}}}\right) = 0.100 \text{ ton}$

•
•

EXAMPLE 22 You are 6 ft tall. (1 ft = 1.894×10^{-4} mi.) How many miles high are you?

Solution **Step 1** 6 ft

Step 2 $\frac{\text{mi}}{\text{ft}}$

Step 3 $\frac{1.894 \times 10^{-4}\ \text{mi}}{1\ \text{ft}}$

Step 4 $(6\ \cancel{\text{ft}})\left(\frac{1.894 \times 10^{-4}\ \text{mi}}{1\ \cancel{\text{ft}}}\right) = 1 \times 10^{-3}\ \text{mi}$

SELF-TEST Use the cookbook dimensional analysis four-step recipe to solve the following problems:

1. A tree is measured and found to be 20 ft high. One mile is equal to 5280 ft. How many miles high is this tree?
2. A baby was born 250 days ago. What is the age of this baby in hours?
3. A package weigh 352 oz. How many pounds does it weigh?

ANSWERS **1.** 3.8×10^{-3} mi **2.** 6.00×10^{3} hr **3.** 22.0 lb

2.3.3 NOAH'S ARK, PART I

Some people believe that the remains of Noah's ark are buried in a glacier near the top of 17,000-ft Mt. Ararat in eastern Turkey, about twenty miles from the Armenian border. The object is 450 ft long, 75 ft wide, and 45 ft high. This is impressive if you consider that a football field is only 300 ft long. We're talking about a boat that is one and a half football fields in length sitting on top of a mountain three miles high! This 50-ton boat is made of unidentifiable hand-tooled wood and waterproofed with wood tar. There are no trees within a five-hundred-mile radius of this structure. What makes all of this even more interesting is that the Bible says that Noah left the ark on the Mountains of Ararat after the flood (Gen. 8:4). The ark's measurements are given as 300 cubits by 50 cubits (Gen. 6:15). Assume that 1.0 cubit is about 1.5 ft.

Do the measurements of this object (450 ft × 75 ft) agree with the measurements given in the Bible?

Yes, the measurements in feet of this object do agree with the measurements given in Genesis.

To show that the measurements agree, convert the given measurements (450 ft × 75 ft) to cubits. Another approach is to convert the measurements given in the Bible (300 cubits × 50 cubits) to feet. If you haven't done this, try it now. Try both approaches.

First Approach:

Step 1 450 ft long and 75 ft wide

Step 2 $\frac{\text{cubit}}{\text{ft}}$

Step 3 $\dfrac{1.0 \text{ cubit}}{1.5 \text{ ft}}$

Step 4 $$(450 \text{ ft long})\left(\frac{1.0 \text{ cubit}}{1.5 \text{ ft}}\right) = 3.0 \times 10^2 \text{ cubits long}$$

$$(75 \text{ ft wide})\left(\frac{1.0 \text{ cubit}}{1.5 \text{ ft}}\right) = 5.0 \times 10^1 \text{ cubits wide}$$

Second Approach:

Step 1 300 cubits long and 50 cubits wide

Step 2 $\dfrac{\text{ft}}{\text{cubits}}$

Step 3 $\dfrac{1.5 \text{ ft}}{1.0 \text{ cubit}}$

Step 4 $$(300 \text{ cubits long})\left(\frac{1.5 \text{ ft}}{1.0 \text{ cubit}}\right) = 4.5 \times 10^2 \text{ ft long}$$

$$(50 \text{ cubits wide})\left(\frac{1.5 \text{ ft}}{1.0 \text{ cubit}}\right) = 7.5 \times 10^1 \text{ ft wide}$$

2.3.4 COMMUNICATION TECHNIQUE

We mentioned that using units (dimensions) in problem solving is an effective communication technique. Let's pursue this idea a bit further. Assume that another student shows you his calculations for a homework problem. This is what he shows you:

$$2 \times 3 = 6$$

This calculation communicates less information than either of the following calculations:

$$(2 \text{ yd deep}) \times \left(\frac{3 \text{ ft}}{\text{yd}}\right) = 6 \text{ ft deep}$$

Note the use of adjectives and verbs to communicate more clearly.

$$(2 \text{ candy mints bought}) \times \left(\frac{3 \text{ cents paid}}{\text{candy mint bought}}\right) = 6 \text{ cents paid}$$

The numbers in all three calculations are the same. Adding units effectively communicates what the calculations are all about.

Whenever you solve a problem, the use of units will allow you to complete your own calculations more easily and, perhaps just as important, it will help your instructor, your fellow students, or your supervisor better understand what you are doing. Making verbs and adjectives units in problem solving has another important advantage. When a solution includes verbs and adjectives, you can convert it to a written explanation of how you solved the problem. For example, you could write that you bought 2 candy mints and, since you paid 3 cents per candy mint purchased, you paid 6 cents.

2.4 CONVERSION FACTORS: SINGLE-STEP PROBLEMS

Most of the problems that you will encounter in chemistry can be solved by converting the units of a given measurement into different units. Many of these problems can be solved in a single step (the fourth step of the cookbook dimensional analysis recipe). We will refer to such problems as single-step problems. The conversion factor in such problems is the third step of the recipe. Steps 2 and 3 produce the conversion factor, which is then multiplied (one single multiplication, one single step) by the given measurement to produce the desired answer.

Practicing dimensional analysis with single-step problems will be a relatively easy way for you to gain experience using dimensional analysis as a problem-solving technique. You will probably be able to do many of these problems without using dimensional analysis, but eventually your practice will pay off when you encounter more difficult problems. By practicing dimensional analysis on simpler problems, you can concentrate more of your efforts on the dimensional analysis method itself than on the complexities of the problems. Later, the opposite will be true.

2.4.1 DAILY LIFE APPLICATIONS

Previous examples were straightforward conversion problems, such as the conversion of feet to inches, minutes to seconds, pounds to ounces, and days to hours. The following examples are also considered conversion problems because they can be solved by converting the units of the given information into different units using a conversion factor (Step 3 of the recipe).

•

•

EXAMPLE 23 You just found a recipe for a spaghetti sauce that serves 8 people. You would like to make just enough to serve only yourself and a friend. The original recipe calls for 4 lb of hamburger. How many pounds of hamburger do you have to buy to make enough of the recipe for just 2 people?

Solution In this problem, you can think of 2 people as the given number and unit. You wish to convert from the unit *people* to the unit *pounds of hamburger*. The four steps of the cookbook dimensional analysis approach are shown:

Step 1 2 people eat

Step 2 $\dfrac{\text{lb hamburger needed}}{\text{people eat}}$

Step 3 $\dfrac{\text{4 lb hamburger needed}}{\text{8 people eat}}$

Translation: 4 lb of hamburger are needed for every 8 people who eat.

(This is the conversion factor that converts the given unit, people, to pounds of hamburger.)

Step 4 $(2\ \cancel{\text{people eat}})\left(\dfrac{\text{4 lb hamburger needed}}{8\ \cancel{\text{people eat}}}\right) = \text{1 lb hamburger needed}$

given — conversion factor — answer

This is the single step (multiplication) that given us our answer.

Notice that the phrase *hamburger needed* is part of the unit (pounds of hamburger needed vs. pounds). Without the phrase *hamburger needed*, it might not be clear that you mean pounds of hamburger and not pounds of onions. As a general rule, whenever you use units of quantity (for example, gallons, quarts, and pounds), identify the substance to which the quantity unit refers (for example, gasoline, milk, and onions). This is an important way of using dimensional analysis effectively as a communication tool. Adding a verb (e.g., *needed*) helps even more. Some of the problems in this book contain verbs when their presence is helpful.

•
•

EXAMPLE 24 You just bought a new high-mileage car that can get 80 mi per gallon of gasoline. Your gas tank only has 5 gal of gasoline in it. You wish to make a trip of 500 mi (5.00×10^2 mi). How far can you go with the available gasoline?

Solution This problem gives you more information than you need to get the requested answer. Such a situation is common in both chemistry problems and everyday problems. You don't always need all the information that is available. You have to learn how to pick out the important (necessary) information and convert it to information you desire to have. This particular problem, in terms of units given and units sought, boils down to this: you are given 5 gal of gasoline and you wish to find out how many miles you can travel. The units given are gallons of gasoline and the units sought are miles of travel.

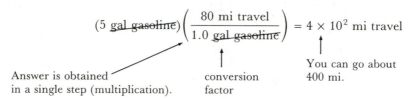

$$(5\ \cancel{\text{gal gasoline}})\left(\frac{80\ \text{mi travel}}{1.0\ \cancel{\text{gal gasoline}}}\right) = 4 \times 10^2\ \text{mi travel}$$

Answer is obtained in a single step (multiplication).

conversion factor

You can go about 400 mi.

•
•

EXAMPLE 25 Use the information given in Example 24 to find out how much gasoline you need to make a 500-mi trip.

Solution In terms of units given and units sought, this problem gives you 500 mi of travel and seeks the number of gallons of gasoline required.

$$(500\ \text{mi travel})\left(\frac{1.0\ \text{gal gasoline}}{80\ \text{mi travel}}\right) = 6.3\ \text{gal gasoline}$$

•
•

EXAMPLE 26 You would like to build and sell houses as a part-time job. This particular model house that appeals to you requires exactly 6000 lb of lumber. You have 25 000 lb of lumber on hand and would like to build 23 houses. How much lumber do you need to build 23 houses?

Solution This is another problem giving you more information than you need to obtain an answer. You want to find out how many pounds of lumber to buy. Pounds of lumber is the unit sought. You need this lumber to build 23 houses. The given measurement (number plus unit) is 23 houses. You are to convert the unit *houses* into the unit *pounds of lumber*.

$$(\cancel{\text{build}}\ 23\ \cancel{\text{houses}})\left(\frac{6000\ \text{lb lumber}}{1\ \cancel{\text{house built}}}\right) = 1.380 \times 10^5\ \text{lb lumber}$$

conversion factor

•
•

EXAMPLE 27 Use the information given in the previous problem to find how many houses you can build with the lumber you have in stock.

Solution You are given 25 000 lb of lumber. You wish to determine the number of houses that can be built.

$$(25\,000\ \cancel{\text{lb lumber}})\left(\frac{1\ \text{house can be built}}{6000\ \cancel{\text{lb lumber}}}\right) = 4\ \text{houses can be built}$$

(the fraction bar is labelled "per")

SELF-TEST

1. A recipe that serves 5 people requires 3.0 lb of onions. You wish to make enough of the recipe to feed 7 people. How many pounds of onions do you need?
2. A house requires 3000 nails. You have 45 348 nails on hand. How many houses can you build?
3. Your car can only go 20 mph. You wish to travel to a city that is 236 mi away. How many hours will it take you to reach your destination?
4. You realize that you are very tired and can only drive for 3.5 hr before you will have to stop and rest. How many miles can you travel in this time? Use the information given in Problem 3.

ANSWERS **1.** 4.2 lb **2.** 15 houses **3.** 12 hr **4.** 70 mi

2.4.2 FORMULA APPROACH: PROS AND CONS

If you need 10 gal of gasoline that sells for \$2.00 per gallon, it would cost you \$20.00. You can probably do a calculation like this mentally with very little conscious effort. We will use this easy problem to compare problem solving by memorized formulas and by dimensional analysis. We will discuss the advantages and disadvantages of each approach.

Here is a formula that can be used to calculate the cost of a gasoline purchase:

$$\left(\begin{array}{l}\text{number of gallons}\\ \text{purchased}\end{array}\right) \times \left(\begin{array}{l}\text{cost in dollars}\\ \text{per gallon}\end{array}\right) = \text{cost of gasoline in dollars}$$

With this formula you can always calculate the cost of a gasoline purchase by simply multiplying the cost of the gasoline in dollars per gallon times the number of gallons purchased. It is important that you can do this quickly and conveniently. Speed and convenience are the main advantages of using memorized formulas to solve problems. For example, if you buy 5.0 gal of gasoline, then

$$(5.0\ \cancel{\text{gal gasoline}}) \times \left(\frac{\$2.00}{\cancel{\text{gal gasoline}}}\right) = \$10$$

There is one major disadvantage with formulas: Unless you are very careful with the units, formulas don't always work. For example, if you needed 3 qt of gasoline that costs 25¢ a pint, you might obtain an incorrect answer with the preceding formula by multiplying 3 (quarts of gasoline) times 25 (cents per pint of gasoline) and obtaining an answer of 75 (units?). You probably would not make such a foolish mistake with this particular problem because of its simplicity, but you will not always understand chemistry problems with the same clarity as you understand this problem.

EXAMPLE 28 You need exactly 3 qt of gasoline in order to start your vacation with a full tank. Your gas station sells gasoline by the pint for 25¢ per pint. How much money will it cost you to top off (completely fill) your gas tank?

Solution This problem is a little more difficult than the preceding problems because it is actually made up of three separate problems. We'll spend a lot of time on problems like this one in Section 2.5, "Conversion Factors: Multi-Step Problems." At that time we will discuss an efficient method of solving problems like this one. For now, let's approach this as three separate problems. The first problem requires us to find out how many pints of gasoline you need. In this first problem, you need to convert the unit *quarts* into the unit *pints*.

$$(3\ \cancel{\text{qt}}\ \text{gasoline})\left(\frac{2\ \text{pt}}{\cancel{\text{qt}}}\right) = 6\ \text{pt gasoline}$$

The second problem is to calculate the cost of the gasoline in cents. You must convert the unit *pints of gasoline* to the unit *cents*.

$$(6\ \cancel{\text{pt gasoline}})\left(\frac{25\ \text{cents}}{\cancel{\text{pt gasoline}}}\right) = 150\ \text{cents}$$

The third problem is to convert the cost in cents to the cost in dollars. You must convert the unit *cents* to the unit *dollars*.

$$(150\ \text{cents})\left(\frac{1\ \text{dollar}}{100\ \text{cents}}\right) = 1.50\ \text{dollars} = \$1.50$$

2.4.3 I.Q. AND HINDSIGHT

Once you master the techniques of dimensional analysis, you will be equipped with a powerful tool that will be very valuable to you even after you finish this chemistry course. The method works with most problems, not just chemistry problems. Also, as we've shown, the method is a valuable communication device. You can communicate what you are doing (to both yourself and to others) more effectively when your calculations show both numbers and units.

Another reward that comes with the mastery of dimensional analysis is that you will find yourself solving problems in areas in which you are relatively ignorant. Your one method of attack (converting given units into desired units) can sometimes be enough to solve difficult and unfamiliar problems. Taken together, your ability to solve unfamiliar problems and your ability to solve problems without memorized formulas increase your effective intelligence.

Add hindsight to your mastery of dimensional analysis and you will find that you can "derive," very quickly, the formulas that Newton, Kepler, and Einstein took decades to derive. And you can do it without having to understand the complicated physics and mathematics and without having to do the laboratory experiments required to formally derive these equations. For example, in Einstein's equation $E = mc^2$, the units *energy* and *mass times velocity squared* must be equal. No other combination of these three terms could work. Such insights and abilities come with the experience gained from faithful use of dimensional analysis in all of your calculations.

2.5 CONVERSION FACTORS: MULTI-STEP PROBLEMS

Most problems can be thought of in terms of converting a given measurement (a given number and its units) into another measurement (a new number with different units). Fortunately, this is frequently done in a single step. Unfortunately, however, there are times when several steps (several conversion factors) are required to change the given measurement into a measurement with the desired units. This is shown in Example 28, the problem involving gasoline that sells for 25¢ per pint.

We will look at several multi-step conversion factor examples and discuss ways for you to approach these problems. We will start with a simple example that is actually a one-step problem and then gradually expand it to a problem requiring several steps.

EXAMPLE 29 Your mother told you that you were able to walk when you were about 2 years old. How many days old were you?

Solution This is a single-step conversion factor problem that is solved as follows:

one significant digit

$$(2 \text{ yr old})\left(\frac{365 \text{ days}}{\text{yr}}\right) = 7 \times 10^2 \text{ days old}$$

This single step (one multiplication) gives us our answer.

One conversion factor is used.

EXAMPLE 30 You began to walk when you were about 2 years old. How many hours old were you when you took your first step?

Solution Not many people know the number of hours in 1 year, so they would not be able to solve this problem in a single step as is shown:

only one significant digit

$$(2 \, \cancel{\text{yr}} \text{ old})\left(\frac{8760 \text{ hr}}{\cancel{\text{yr}}}\right) = 2 \times 10^4 \text{ hr old}$$

For most people, this problem is a two-step problem. The first step determines the number of days in 2 years.

Step 1 Multiplication by the first conversion factor:

$$(2 \, \cancel{\text{yr}} \text{ old})\left(\frac{365 \text{ days}}{\cancel{\text{yr}}}\right) = 7 \times 10^2 \text{ days old}$$

The second step determines the number of hours in 7×10^2 days.

Step 2 Multiplication by the second conversion factor:

$$(7 \times 10^2 \, \cancel{\text{days}} \text{ old})\left(\frac{24 \text{ hr}}{\cancel{\text{day}}}\right) = 2 \times 10^4 \text{ hr old}$$

There is a more efficient method for solving this problem. If you stop and think about it, you'll realize that you waste time calculating the intermediate answer in Step 1 because that intermediate answer (7×10^2 days old) is not going to be reported in the final answer. Rather than carry out the multiplications in two separate steps, we can combine Steps 1 and 2 into a single chain multiplication:

$$(2\ \cancel{\text{yr}}\ \text{old})\left(\frac{365\ \cancel{\text{days}}}{\cancel{\text{yr}}}\right)\left(\frac{24\ \text{hr}}{\cancel{\text{day}}}\right) = 2 \times 10^4\ \text{hr old}$$

One chain multiplication involving two conversion factors

The first conversion factor converts years to days.

The second conversion factor converts days to hours.

EXAMPLE 31 For some odd reason, you wonder how many minutes old you were when you began walking (around the age of 2 years).

Solution This is a three-step problem. It is possible to solve this problem in three steps.

Step 1: Convert your age in years to your age in days.
Step 2: Convert your age in days to your age in hours.
Step 3: Convert your age in hours to your age in minutes.

These steps are as follows:

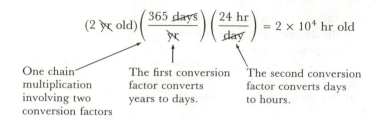

Step 1 $$(2\ \cancel{\text{yr}}\ \text{old})\left(\frac{365\ \text{days}}{\cancel{\text{yr}}}\right) = 7 \times 10^2\ \text{days old}$$

Step 2 $$(7 \times 10^2\ \text{days old})\left(\frac{24\ \text{hr}}{\text{day}}\right) = 2 \times 10^4\ \text{hr old}$$

Step 3 $$(2 \times 10^4\ \text{hr old})\left(\frac{60\ \text{min}}{\text{hr}}\right) = 1 \times 10^6\ \text{min old}$$

A better, more efficient approach is to combine all of these calculations into a single chain calculation with three conversion factors, as shown:

$$(2\ \cancel{\text{yr}}\ \text{old})\left(\frac{365\ \cancel{\text{days}}}{\cancel{\text{yr}}}\right)\left(\frac{24\ \cancel{\text{hr}}}{\cancel{\text{day}}}\right)\left(\frac{60\ \text{min}}{\cancel{\text{hr}}}\right) = 1 \times 10^6\ \text{min old}$$

The first conversion factor converts years to days.

The second conversion factor converts days to hours.

The third conversion factor converts hours to minutes.

Notice that each number, your age in years (2 years old) and your age in minutes (1 million minutes old), contains one significant digit.

EXAMPLE 32 Assume gold costs about \$500 per ounce. In the entire world there are only about 80 000 tons of gold that have been mined. You want to know how much money all this gold is worth.

Solution This is another multi-step problem. You would have to convert (1) tons of gold to pounds of gold, (2) pounds of gold to ounces of gold, and (3) ounces of gold to dollars. Here are the three steps in detail:

Step 1 $$(80\,000\ \cancel{\text{tons}}\ \text{gold})\left(\frac{2000\ \text{lb}}{\cancel{\text{ton}}}\right) = 2 \times 10^8\ \text{lb gold}$$

Step 2 $(2 \times 10^8 \text{ lb gold})\left(\frac{16 \text{ oz}}{\text{lb}}\right) = 3 \times 10^9 \text{ oz gold}$

Step 3 $(3 \times 10^9 \text{ oz gold})\left(\frac{\$500}{\text{oz gold}}\right) = 2 \times 10^{12} \text{ dollars (2 trillion dollars)}$

It would be more efficient (in terms of time and energy) not to calculate each of the intermediate answers (2×10^8 lb and 3×10^9 oz). Solve the problem as one large chain multiplication as follows:

$$(80\,000 \text{ tons gold})\left(\frac{2000 \text{ lb}}{\text{ton}}\right)\left(\frac{16 \text{ oz}}{\text{lb}}\right)\left(\frac{\$500}{\text{oz gold}}\right) = 1 \times 10^{12} \text{ dollars}$$

The first conversion factor converts tons to pounds.
The second conversion factor converts pounds to ounces.
The third conversion factor converts ounces of gold to dollars.

Notice that a different (and more correct) answer is obtained when the problem is solved as one chain multiplication because we do not have to round off our answer three times in the chain multiplication.

•
•

EXAMPLE 33 Your back yard measures 60.0 ft long and 30.0 ft wide. How many square miles of land do you have in your back yard?

Solution You could solve this problem in a series of steps by (1) calculating the number of square feet in your back yard, (2) calculating the number of square feet in a square mile, and (3) converting your back yard dimensions from square feet to square miles.

Step 1 $(60.0 \text{ ft} \times 30.0 \text{ ft}) = 1.80 \times 10^3 \text{ ft}^2$

Step 2 $(1 \text{ mi}^2)\left(\frac{5280 \text{ ft}}{\text{mi}}\right)\left(\frac{5280 \text{ ft}}{\text{mi}}\right) = (1 \text{ mi}^2)\left(\frac{(5280)^2 \text{ ft}^2}{\text{mi}^2}\right) = 2.78784 \times 10^7 \text{ ft}^2$

Step 3 $(1.80 \times 10^3 \text{ ft}^2)\left(\frac{1 \text{ mi}^2}{2.78784 \times 10^7 \text{ ft}^2}\right) = 6.46 \times 10^{-5} \text{ mi}^2$

The area of your back yard is 6.46×10^{-5} mi^2. We can get this same answer more efficiently by combining all of the preceding calculations (steps) into one chain calculation as follows:

$$(60.0 \text{ ft} \times 30.0 \text{ ft})\left(\frac{1 \text{ mi}}{5280 \text{ ft}}\right)^2 = 6.46 \times 10^{-5} \text{ mi}^2$$

Note: $\left(\frac{1 \text{ mi}}{5280 \text{ ft}}\right)^2 = \left(\frac{1 \text{ mi}}{5280 \text{ ft}}\right)\left(\frac{1 \text{ mi}}{5280 \text{ ft}}\right) = \frac{(1)^2 \text{ mi}^2}{(5280)^2 \text{ ft}^2} = \frac{1 \text{ mi}^2}{2.78784 \times 10^7 \text{ ft}^2}$

•
•

EXAMPLE 34 Your living room measures 3.00×10^2 in. by 1.50×10^2 in. How many square miles of living room area do you have?

Solution You can calculate the answer to this problem by completing the following series of steps: (1) calculate the number of square inches of living room area, (2) convert the area from square inches to square feet, and (3) convert the area from square feet to square miles.

Step 1 $(300.\ \text{in.} \times 150.\ \text{in.}) = 4.50 \times 10^4\ \text{in.}^2$

Step 2 $$(4.50 \times 10^4\ \text{in.}^2)\left(\frac{1\ \text{ft}}{12\ \text{in.}}\right)\left(\frac{1\ \text{ft}}{12\ \text{in.}}\right) = (4.50 \times 10^4\ \text{in.}^2)\left(\frac{1\ \text{ft}}{12\ \text{in.}}\right)^2$$
$$= 3.13 \times 10^2\ \text{ft}^2$$

Step 3 $$(3.13 \times 10^2\ \text{ft}^2)\left(\frac{1\ \text{mi}}{5280\ \text{ft}}\right)\left(\frac{1\ \text{mi}}{5280\ \text{ft}}\right) = (3.13 \times 10^2\ \text{ft}^2)\left(\frac{1\ \text{mi}}{5280\ \text{ft}}\right)^2$$
$$= 1.12 \times 10^{-5}\ \text{mi}^2$$

The same problem could be solved more efficiently in one chain calculation, as shown:

$$(300.\ \text{in.})(150.\ \text{in.})\left(\frac{1\ \text{ft}}{12\ \text{in.}}\right)^2\left(\frac{1\ \text{mi}}{5280\ \text{ft}}\right)^2 = 1.12 \times 10^{-5}\ \text{mi}^2$$

:

2.5.1 CAR CATALYSTS AND FOOTBALL FIELD UNITS

It is important for you to realize that dimensional analysis is not limited to scientific units of mass, length, and volume. In solving future problems, we will use units such as girls, gold coins, and tons of brontosaurus. In the following example problem we will use the unit *football fields*.

Many American cars have catalytic converters that help convert poisonous materials in the cars' exhaust to harmless chemicals. Only 0.100 oz of a palladium and platinum mixture is used in many car catalysts. The surface area of the catalyst is equivalent to the area of 59.0 football fields. This raises the question, "How thick is the catalyst layer if only 0.100 oz of catalyst is spread out over an area equal to that of 59.0 football fields?"

You might guess that the thickness of the catalyst layer is measured in atoms, that is, the layer might be one or two atoms thick. To calculate the catalyst thickness, we will assume that the catalyst is primarily platinum and calculate the number of platinum atoms per square angstrom of surface area. An angstrom is about 4×10^{-9} in. The angstrom unit is frequently used in chemistry because the diameter of most atoms, including platinum, is about one angstrom (1 Å). Therefore, if we find that there is one platinum atom per square angstrom, the platinum layer is about one atom thick.

We will do this calculation in three steps: (1) we will calculate the number of atoms of platinum in 0.100 oz of platinum, (2) we will calculate the number of square angstroms in 59.0 football fields, and (3) we will divide the results of Step 1 by the results of Step 2 to obtain the number of platinum atoms per square angstrom. This final calculation is the approximate thickness of the platinum layer. Before we begin our calculations, you should know that the symbol for platinum is Pt and that there are 6.02×10^{23} Pt atoms in 6.876 oz of platinum.

Now for the first part: How many platinum atoms are present in 0.100 oz of Pt?

There are 8.76×10^{21} Pt atoms in 0.100 oz of Pt.

$$(0.100\ \text{oz Pt})\left(\frac{6.02 \times 10^{23}\ \text{Pt atoms}}{6.876\ \text{oz Pt}}\right) = 8.76 \times 10^{21}\ \text{Pt atoms}$$

A football field measures 100 yards by 50 yd. Calculate the area of 59.0 football fields in square angstroms. The abbreviation for the unit angstrom *is* Å.

If you got an answer of 2×10^{25} Å^2, you're doing very well. Here's one way to show that there are 2×10^{25} Å^2 in 59.0 football fields.

$$(59.0\ \cancel{\text{football fields}})\left(\frac{100\ \text{yd} \times 50\ \text{yd}}{\cancel{\text{football field}}}\right) = 2.95 \times 10^5\ \text{yd}^2$$

$$(2.95 \times 10^5\ \cancel{\text{yd}^2})\left(\frac{3\ \cancel{\text{ft}}}{\cancel{\text{yd}}}\right)^2\left(\frac{12\ \cancel{\text{in.}}}{\cancel{\text{ft}}}\right)^2\left(\frac{1\ \text{Å}}{4 \times 10^{-9}\ \cancel{\text{in.}}}\right)^2 = 2 \times 10^{25}\ \text{Å}^2$$

How many platinum atoms are there per square angstrom?

There is less than 1 Pt atom per square angstrom.

$$\frac{8.76 \times 10^{21}\ \text{Pt atoms}}{2 \times 10^{25}\ \text{Å}^2} = \frac{4 \times 10^{-4}\ \text{Pt atoms}}{\text{Å}^2}$$

Dividing the numerator and the denominator by 4×10^{-4} gives

$$\frac{1\ \text{Pt atom}}{3 \times 10^3\ \text{Å}^2} \doteq \frac{1\ \text{Pt atom}}{50\ \text{Å} \times 50\ \text{Å}}$$

Our result means that there is 1 Pt atom for every 3×10^3 square angstroms. An area of 3×10^3 Å^2 is that area found in a square that measures approximately 50 Å per side. We can approximate this situation by visualizing every Pt atom in the center of a square that is 50 Å on each side as shown in Figure 2.2. (Remember that the diameter of each Pt atom is about 1 Å.) All of the Pt atoms are about 50 Å away from their closest neighboring Pt atoms. The layer of Pt atoms is therefore 1 atom thick and the Pt atoms are not touching.

FIGURE 2.2

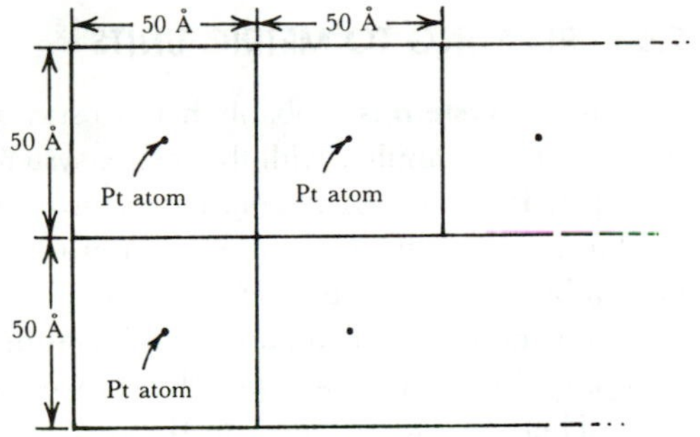

Can you think of a possible reason why each Pt atom is surrounded by so much vacant area?

A large vacant area surrounds each Pt atom to accommodate the relatively bulky gasoline molecules on which the Pt acts. Platinum would be wasted if the Pt atoms were placed too close together because only one Pt atom can work on one molecule in gasoline at a time, and the molecules in gasoline are larger than the Pt atoms (molecules are made up of atoms).

•
•

SELF-TEST

1. How many seconds are in 1.00 yr?
2. What fraction of a year is equivalent to 22 min?
3. The earth weighs 6.586×10^{21} tons (including the weight of the atmosphere). Express this weight in ounces.

ANSWERS **1.** 3.15×10^{7} sec **2.** 4.2×10^{-5} yr **3.** 2.108×10^{26} oz

2.6 METRIC SYSTEM

Several systems of units of measure are in use today throughout the world. Most scientists use either the metric system (with units such as grams, liters, and meters) or the International System of Units, which is commonly referred to as SI units (with units such as kilograms, cubic decimeters, and meters). The United States also uses the British, or English, system (with units such as pounds, quarts, and feet).

Most of the world seems to use metric units (more on this later). In the United States, as this book goes to press, President Bush is pushing all federal agencies to require suppliers to quote their bids in metric units by October 1992. Many are against this requirement because of the high expenses of retooling to meet metric standards. It is interesting to note that George Washington, some 200 years ago, first asked Congress to convert the United States to a metric country. Reasons given for converting America to the metric system include (1) most of the world uses this system and (2) it is an easier system to work with than the English system.

While the United States debates the pros and cons of the metric system, many scientists are now leaning toward the adoption of SI units. Although many non-American chemistry textbooks use SI units, in the United States most chemistry textbooks use both metric and SI units. We will work examples in both SI and metric units throughout this book.

2.6.1 RELATING TO METRIC UNITS

The metric system is probably here to stay and you should become as familiar with it as possible. Becoming familiar with the metric system does not mean memorizing conversion factors (for example, 454 g = 1 lb). Memorizing conversion factors does not give you a "feel" for what a gram is; that is, it is difficult to relate to a gram by just knowing that a gram is $\frac{1}{454}$ pound. You will have a better feel for the metric system if you can relate to it without going through mathematical conversions. You can do this by relating metric units, such as the gram, to objects in your everyday life, such as a paper clip. (Two paper clips weigh about 1 g.)

There are many units in the metric system, but usually only three are used frequently in introductory chemistry courses: (1) the meter (length), (2) the gram (mass or weight), and (3) the liter (volume). A meter is a little longer than 1 yard (1 m is about 1.1 yd). A liter is a little more than a quart (1 L is about 1.06 qt). A gram has about the same mass as a BiC ballpoint pen cap or two paper clips (454 g is about 1 lb, 28 g is about 1 oz). A nickel (five-cent piece) has a mass of about 5 g.

In addition to these three basic units of measurement, the metric system has several prefixes, such as *milli-*, *kilo-*, and *centi-*, which are shown in Table 2.1. Before you look at Table 2.1 and become discouraged, only three of the prefixes (*milli-*, *kilo-*, and *centi-*) are frequently used in introductory chemistry courses.

Let's relate more metric units with everyday objects.

TABLE 2.1 Prefixes used with metric units

Prefix	Abbreviation	Numerical Value
deci-	d	10^{-1}
centi-	c	10^{-2}
milli-	m	10^{-3}
micro-	μ	10^{-6}
nano-	n	10^{-9}
pico-	p	10^{-12}
femto-	f	10^{-15}
atto-	a	10^{-18}
deka-	da	10^{1}
hecto-	h	10^{2}
kilo-	k	10^{3}
mega-	M	10^{6}
giga-	G	10^{9}
tera-	T	10^{12}

- A milliliter is about 20 drops of water from an eye dropper. Five milliliters is about a teaspoon. (1 mL $\doteq \frac{1}{5}$ tsp.)
- A kilogram is a little more than 2 lb (2.2 lb $\doteq$ 1 kg).
- A kilometer is a little farther than $\frac{1}{2}$ mi (1 km $\doteq$ 0.62 mi).
- A millimeter is the diameter of the wire used for paper clips.

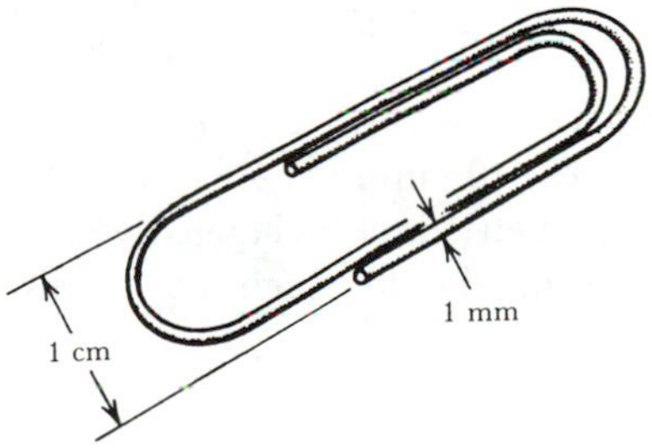

- A centimeter is about the width of a paper clip (1 in. $\doteq$ 2.54 cm; 1 cm $\doteq$ 0.4 in.). A paper clip turns out to be quite a handy object because (1) two of them weigh about 1 g, (2) its width is about 1 cm, and (3) the diameter of its wire is about 1 mm.

2.6.2 CONVERSION PROBLEMS

This section contains several problems that illustrate conversions from (1) one metric unit to another metric unit, (2) a metric unit to an English unit, and (3) an English unit to a metric unit.

•
•

EXAMPLE 35 The moon is about 239 000 mi away from the earth. Express this distance in kilometers.

Solution

$$(2.39 \times 10^5\ \cancel{\text{mi}})\left(\frac{1\ \text{km}}{0.62\ \cancel{\text{mi}}}\right) = 3.9 \times 10^5\ \text{km}$$

original (given) measurement with unit miles

conversion factor based on 1 km ≐ 0.62 mi

•
•

EXAMPLE 36 Two objects are separated by about 32 000 m. Express this distance in miles.

Solution

$$(3.2 \times 10^4\ \cancel{\text{m}})\left(\frac{1\ \cancel{\text{km}}}{1000\ \cancel{\text{m}}}\right)\left(\frac{0.62\ \text{mi}}{\cancel{\text{km}}}\right) = 2.0 \times 10^1\ \text{mi}$$

given measurement

The first conversion factor converts meters to kilometers.

The second conversion factor converts kilometers to miles.

•
•

EXAMPLE 37 The sun is 1.5×10^{13} cm from the earth. Express this distance in miles.

Solution

$$(1.5 \times 10^{13}\ \text{cm})\left(\frac{1\ \cancel{\text{m}}}{100\ \cancel{\text{cm}}}\right)\left(\frac{1\ \cancel{\text{km}}}{1000\ \cancel{\text{m}}}\right)\left(\frac{0.62\ \text{mi}}{1\ \cancel{\text{km}}}\right) = 9.3 \times 10^7\ \text{mi}$$

given measurement

Three conversion factors are being used here.

•
•

EXAMPLE 38 A fast runner can run 4.3×10^5 mm in 1 min. How many kilometers can he run in 1 min?

Solution

$$\left(\frac{4.3 \times 10^5\ \cancel{\text{mm}}}{\text{min}}\right)\left(\frac{1\ \cancel{\text{m}}}{1000\ \cancel{\text{mm}}}\right)\left(\frac{1\ \text{km}}{1000\ \cancel{\text{m}}}\right) = 0.43\ \text{km/min}$$

Notice that in converting from millimeters to kilometers only the power of 10 (exponent) is changed (4.3×10^5 mm vs. 4.3×10^{-1} km). This is true for all metric conversion problems in which one metric unit is changed to another metric unit. This conversion advantage (easier arithmetic) is missing for English to English conversion problems (e.g., 1.0×10^1 ft vs. 1.20×10^2 in.).

•
•

EXAMPLE 39 A Concorde jet can fly more than 2000 ft/sec. Express this speed in kilometers per second (km/sec).

Solution

$$\left(\frac{2 \times 10^3\ \cancel{\text{ft}}}{\text{sec}}\right)\left(\frac{1\ \cancel{\text{mi}}}{5280\ \cancel{\text{ft}}}\right)\left(\frac{1\ \text{km}}{0.62\ \cancel{\text{mi}}}\right) = 0.6\ \text{km/sec}$$

•
•

EXAMPLE 40 The earth has a mass of 5.98×10^{24} kg. Express the earth's mass in grams.

Solution

$$(5.98 \times 10^{24}\ \cancel{\text{kg}}\ \text{earth})\left(\frac{1000\ \text{g}}{\cancel{\text{kg}}}\right) = 5.98 \times 10^{27}\ \text{g earth}$$

Notice again that just the exponent (power of 10) part of the number changes (5.98×10^{24} kg vs. 5.98×10^{27} g).

•
•

EXAMPLE 41 If a professor has a mass of 1.00×10^2 kg, what is her mass in milligrams?

Solution

$$(1.00 \times 10^2\ \cancel{\text{kg}}\ \text{professor})\left(\frac{1000\ \cancel{\text{g}}}{\cancel{\text{kg}}}\right)\left(\frac{1000\ \text{mg}}{\cancel{\text{g}}}\right) = 1.00 \times 10^8\ \text{mg professor}$$

Again, compare 1.00×10^2 kg with 1.00×10^8 mg. Only the exponent part of the number changes. The nonexponential part remains constant.

EXAMPLE 42 Express the mass of a 2000-mg paper clip in megagrams.

Solution

$$(2 \times 10^3\ \cancel{\text{mg}}\ \text{paper clip})\left(\frac{1\ \cancel{\text{g}}}{1000\ \cancel{\text{mg}}}\right)\left(\frac{1\ \text{Mg}}{1 \times 10^6\ \cancel{\text{g}}}\right) = 2 \times 10^{-6}\ \text{Mg paper clip}$$

EXAMPLE 43 A black hole, about the size of a speck of dust, is thought to have a mass of 6×10^{21} Mg. Express this mass in pounds.

Solution

$$(6 \times 10^{21}\ \cancel{\text{Mg}}\ \text{black hole})\left(\frac{1 \times 10^6\ \cancel{\text{g}}}{\cancel{\text{Mg}}}\right)\left(\frac{1\ \text{lb}}{454\ \cancel{\text{g}}}\right) = 1 \times 10^{25}\ \text{lb black hole}$$

EXAMPLE 44 An electron has a mass of 9.1×10^{-31} kg. Express the mass of an electron in picograms.

Solution

$$(9.1 \times 10^{-31}\ \cancel{\text{kg}}\ \text{electron})\left(\frac{1000\ \cancel{\text{g}}}{\cancel{\text{kg}}}\right)\left(\frac{1 \times 10^{12}\ \text{pg}}{\cancel{\text{g}}}\right) = 9.1 \times 10^{-16}\ \text{pg electron}$$

EXAMPLE 45 A positron (positively charged antiparticle) has a mass of 3.25×10^{-29} oz. Express the mass of a positron in nanograms.

Solution

$$(3.25 \times 10^{-29}\ \cancel{\text{oz}}\ \text{positron})\left(\frac{28\ \cancel{\text{g}}}{\cancel{\text{oz}}}\right)\left(\frac{1\ \text{ng}}{1 \times 10^{-9}\ \cancel{\text{g}}}\right) = 9.1 \times 10^{-19}\ \text{ng positron}$$

EXAMPLE 46 A standard basketball has a volume of 7.47 L. Find the volume of a basketball in milliliters.

Solution

$$(7.47\ \cancel{\text{L}}\ \text{basketball})\left(\frac{1000\ \text{mL}}{1\ \cancel{\text{L}}}\right) = 7.47 \times 10^3\ \text{mL basketball}$$

EXAMPLE 47 Two grams of hydrogen occupies about 2.24×10^4 mL under certain conditions. Express this volume of hydrogen gas in liters.

Solution

$$(2.24 \times 10^4\ \cancel{\text{mL}}\ \text{hydrogen})\left(\frac{1\ \text{L}}{1000\ \cancel{\text{mL}}}\right) = 22.4\ \text{L hydrogen}$$

EXAMPLE 48 A quantity of oxygen was found to occupy 20.0 qt. Express this volume in microliters.

Solution

$$(20.0\ \cancel{\text{qt}}\ \text{oxygen})\left(\frac{1\ \cancel{\text{L}}}{1.06\ \cancel{\text{qt}}}\right)\left(\frac{1\ \mu\text{L}}{1 \times 10^{-6}\ \cancel{\text{L}}}\right) = 1.89 \times 10^7\ \mu\text{L oxygen}$$

SELF-TEST

1. Express 20.4 g in kilograms.
2. Express 82 mL in liters.
3. Express 55 oz in milligrams.
4. Express 14 nm in kilometers.
5. Express 5 mi in picometers.

ANSWERS **1.** 2.04×10^{-2} kg **2.** 8.2×10^{-2} L **3.** 1.6×10^{6} mg **4.** 1.4×10^{-11} km **5.** 8×10^{15} pm

2.6.3 METRIC VS. ENGLISH: PROS AND CONS

After studying Examples 35–48, can you describe an advantage of working with metric units (as opposed to English units)?

Converting from one metric unit to another always involves moving the decimal point or changing the power of 10. For example, converting grams to kilograms, we would have 1000. g = 1.000 kg or 10^3 g = 10^0 kg. In the English system, conversion from one unit to another always involves more than just a change in the power of 10. In the English system, it is necessary to memorize conversion factors that are not simple multiples or powers of 10. For example, there are 5280 feet per mile, 12 inches per foot, and 3 feet per yard. Mathematically it is more difficult to multiply and divide by 5280 than it is to move a decimal point or to change a power of 10. However, there are both pros and cons to be considered.

Students are usually given the impression that, except for the United States, the world is 100% metric. This is not true. There are many scientific and nonscientific measurements made today that are not metric and will probably never be metric. These include

1. the 60-second minute
2. the 60-minute hour
3. the 24-hour day
4. the 7-day week
5. the 31-day month
6. the 365-day year

None of these units is based on powers of 10. In addition, the majority of world trade is also based upon nonmetric measures, for example:

7. World oil production is measured in barrels (there are 44 gal per U.S. oil barrel).
8. Lumber is sold by the board foot.
9. All farm products marketed throughout the world are measured in bushels (there are 8 gal in a bushel), in pecks (2 gal per peck), and by the hundredweight (2000 lb equals 20 hundredweight [20 cwt] equals 1 short ton).
10. Silver, gold, platinum, copper, iron, and zinc are sold by the troy ounce (12 troy ounces equals 1 troy pound).
11. Eggs are sold by the dozen, not in groups of 10.

Many people believe that in everyday use the Fahrenheit temperature scale is superior to the Celsius (centigrade) temperature scale because the Fahrenheit scale, between 0°F and 100°F, spans almost the entire range of hot and cold in a temperate climate. The same range on a centigrade scale spans a range that doesn't quite go low enough. Most winter temperatures in a temperate climate fall below 0°C, the freezing point of water, which necessitates the frequent use of negative temperatures. The temperature of a temperate climate rarely goes over 40°C, leaving 60% of the Celsius temperature scale unused in everyday life.

Many measures of length are derived from the human body, making them somewhat universal. For example, almost every civilization (including China, Egypt, and Grcccc) has measured in terms of feet.

It would be very expensive to convert all our machines, tools, and hardware (e.g., $\frac{3}{4}''$ drills, $\frac{7}{8}''$ wrenches, and $\frac{1}{2}''$ bolts) to the metric system. Many of the tools you currently use would become obsolete. (Do you think that you would need to purchase a metric hammer to drive your metric nails?)

On the other hand, the English system can be quite confusing with its five different kinds of miles and . . . What? You thought that there was just one kind of mile? Besides the statute mile (5280 ft), we have the international nautical mile (6076.1 ft), the British nautical mile (6080 ft), the U.S. survey nautical mile (6076.103 ft), and the U.S. survey statute mile (5279.98 ft).

While we're on the subject, let's dispel another myth. You probably thought that an ounce of gold was just as heavy as an ounce of lead. Foolish you. An ounce of gold is heavier because gold is measured in troy ounces (1 troy ounce = 31.157 g) and lead is measured in avoirdupois ounces (1 ounce avdp = 28.348 g).

PROBLEM SET 2

The problems in Problem Set 2 parallel the examples in Chapter 2. For example, if you should have trouble working Problem 5, go back to Example 5 in this chapter to get help. The correct answers to these problems are given at the end of this book.

1. You told a lie and your nose grew 4 ft. Identify the two parts of this measurement.
2. Your friend asks you your age. You tell him that you are 20. Why is your answer considered poor or incomplete?
3. Multiply 3 times 3 by adding together the exponent part of each number.
4. Multiply inches times inches.
5. Work the following calculation by canceling out common numbers in the numerator and the denominator:

$$\frac{4 \times 5}{4}$$

6. Divide (feet × pounds) by pounds.
7. Divide ($3^2 \times 4$) by 3 by canceling out common numbers in the numerator and the denominator.
8. Divide (feet × seconds^2) by seconds.
9. Multiply (inches/foot) times (feet/yard).
10. Find the area of a square that measures 4.0 ft on each side.
11. Find the volume of a cube that measures 1 in. on each side.
12. Add 3 oz to 2 oz.
13. Subtract 4 ft from 6 ft.
14. You live 4.0 mi away from your friend. There are 5280 ft in a mile. Use dimensional analysis to calculate the distance in feet between your house and your friend's house.
15. How many minutes will it take you to take a 3.0-hr walk? Use dimensional analysis.
16. If you sleep 8.0 hr every night, how many days do you sleep each night?
17. Rework the preceding problem by assuming that you are not sure whether to multiply the given measurement (8.0 hr) by 24 hr/day or divide 8.0 hr by 24 (i.e., multiply 8.0 hr by 1 day per 24 hr). Try both approaches and decide which approach is the correct approach by observing the units in the answer that each approach produces.
18. There are 3600 sec in 1 hr. You wish to find out how many hours are required to run for 6000 sec. Using dimensional analysis, decide whether you have to multiply 6000 sec by 3600 sec per hour (3600 sec/hr) or divide 6000 sec by 3600 (i.e., multiply 6000 sec by 1 hr/3600 sec).
19. Use the four-step recipe and show all four steps to determine how many days there are in a 2.0-week vacation.
20. Use the four-step recipe and show all four steps to determine the weight of a 2.0-lb object in ounces.
21. You weigh 150 lb and there are 2000 lb in 1 ton. Show all four steps of the four-step recipe to determine your weight in tons.
22. You are 5 ft tall. One foot is 1.894×10^{-4} mi. How many miles high are you? Show all four steps of the four-step recipe to determine your height in miles.
23. You just found a recipe for raw snail sauce that serves 5 people. You would like to just make enough to serve only yourself and 2 friends. The original recipe calls for 2.0 lb of snail sauce enhancer. How many pounds of snail sauce

enhancer do you have to use to make enough of the recipe for just 3 people? Use the four-step cookbook dimensional analysis approach and show all four steps.

24. Your motorcycle can get 200 mi per gallon of gasoline. The gas tank on the motorcycle can hold 2.5 gal of gasoline. You wish to make a trip of 400 mi. How far can you go with a single tankful of gasoline?
25. Use the information given in the preceding problem to find out how much gasoline you need to make a 430-mi trip.
26. You would like to make and sell paper dolls as a part-time job. The doll you know how to make requires 3.4 oz of paper. You have 550 oz of paper and would like to make 175 dolls. How many ounces of paper do you need to make 175 dolls?
27. Use the information given in the previous problem to find out how many dolls you can make with the paper you have on hand.
28. You need 2.00 qt of gasoline in order to start your vacation with a full tank. Your gas station sells gasoline by the pint for 30¢ per pint. How much money will it cost you to top off your gas tank?
29. Your pet turtle is 3 yr old. How many days have you had it if you raised it from birth?
30. How many hours has the 3-yr-old turtle been in your care?
31. How many minutes has the 3-yr-old turtle been in your care? Do this calculation in a single step (chain calculation) by starting with the turtle's age in years and ignoring your answers from the last two problems.
32. Gold sells for about $500 per ounce. Your rich old uncle has left you 2 tons of gold in his will. How much money is this worth? Do this problem in a single step (one chain calculation).
33. Your bedroom measures 12.0 ft by 15.0 ft. How many square miles of bedroom do you have? Do this with one chain calculation.
34. Your bedroom closet measures 50 inches by 300 in. How many square miles of bedroom closet area do you have? Solve this problem with one chain calculation.
35. The sun is about 93 000 000 mi away from the earth. Express this distance in kilometers.
36. Two objects are separated by about 60 000 m. Express this distance in miles.
37. The moon is 3.9×10^{10} cm from the earth. Express this distance in miles.
38. You can run 3.1×10^{4} mm in 1 min. How many kilometers can you run in 1 min?
39. Superman can fly 3000 ft/sec. How fast can he fly in kilometers per second?
40. Your Uncle Harvey has a mass of 3×10^{2} kg. Express his mass in grams.
41. Your Aunt Tillie has a mass of 1.4×10^{2} kg. What is her mass in milligrams?
42. Express the mass of a fourteen-hundred-milligram cockroach in megagrams.
43. A black hole, about the size of a speck of dust, is thought to have a mass of 1×10^{25} lb. Express this mass in megagrams.
44. An electron has a mass of 9.1×10^{-16} pg. Express the mass of an electron in kilograms.
45. A positron (positively charged antiparticle) has a mass of 9.1×10^{-19} ng. Express the mass of a positron in ounces.
46. A standard basketball has a volume of 7.47×10^{3} mL. Find the volume of a basketball in liters.
47. Two grams of hydrogen occupy about 22.4 L under certain conditions. Express this volume of hydrogen gas in milliliters.
48. A quantity of oxygen was found to occupy 1.89×10^{7} μL. Express this volume in quarts.

3 Preparatory Concepts and Definitions

3.1 INTRODUCTION

You may be wondering why the topic *percentage* is included in this chapter on preparatory concepts and definitions. Why wasn't it included in the first chapter on basic math skills? Perhaps you think of percentage calculations as number manipulations when, in fact, finding a percentage is a technique frequently applied to measurements. It turns out that you must have an appreciation for units and dimensional analysis before you can begin to work intelligently with percentage problems.

Think of this chapter as a slightly more advanced treatment of dimensional analysis using the basic concepts and definitions of percentage, density, specific gravity, calorie, temperature conversion, and heat of fusion to further illustrate and sharpen your dimensional analysis skills.

3.2 PERCENTAGE

Consider a solution made up of just salt and water. The solution is 10% salt and has a total weight of 40 oz. Find the weight of the salt present in this solution. See if you can use units with each number in your calculations.

You were probably able to determine, mentally, the weight of the salt in the given solution. In your mind you said, "Ten percent of 40 ounces is 4 ounces. Therefore there are 4 ounces of salt in 40 ounces of this solution that is 10 percent by weight salt." Many people who can go through this mental process cannot work out the same problem on paper, showing units with every number and having unwanted units cancel.

•
•

EXAMPLE 1 A 40-g solution made up of just salt and water is 10% by weight salt. Use dimensional analysis to find the weight of the salt in this solution.

Solution When we say that a solution is 10% by weight salt, we mean that there are 10 parts of salt for every 100 parts of solution. This means that we have 10 g of salt for every 100 g of solution. Thus

$$(40 \text{ g } \cancel{\text{soln}})\left(\frac{10 \text{ g salt}\overset{\text{per}}{/}}{100 \text{ g } \cancel{\text{soln}}}\right) = 4.0 \text{ g salt}$$

•
•

Notice that we do not multiply 10% times 40 g of solution. Such a multiplication would give

$$(10\%)(40 \text{ g soln}) = 400\% \text{ g soln}$$

We also avoided following the rule, "Convert a percentage to a fraction by dropping the percent sign and moving the decimal point to the left by two places." This rule would give us

$$(0.10)(40 \text{ g soln}) = 4.0 \text{ g soln}$$

which is an answer with an incorrect unit. The correct unit in the answer is grams of salt, not grams of solution.

•
•

EXAMPLE 2 Your body is about 90% water. If you weigh 85 kg, how many kilograms of water do you contain? Use units in your calculations.

Solution

$$(85\text{ kg }\cancel{\text{body weight}})\left(\frac{90\text{ kg water}}{100\text{ kg }\cancel{\text{body weight}}}\right) = 77\text{ kg water}$$

EXAMPLE 3 You find a pile of 4000 coins and are told that 30% of the coins are gold. How may gold coins are in the pile? Associate each of the numbers used in your calculations with a unit and make sure that unwanted units cancel.

Solution It is not clear how many significant digits there are in the number 4000. You will frequently run across such ambiguity in daily life. We will work this problem assuming that all four digits are significant.

$$(4000\ \cancel{\text{coins}})\left(\frac{30\text{ gold coins}}{100\ \cancel{\text{coins}}}\right) = 1.2 \times 10^3\text{ gold coins}$$

There are about 1200 gold coins in the pile of coins. The two significant digits in the number 30% limits our answer to two digits.

The word *percent* is an abbreviation of the earlier term *per centum*. *Per* means "division by" and *centum* means "one hundred." To represent the fraction $\frac{2}{5}$ as a percentage, multiply the numerator and the denominator by 100, as shown:

$$\left(\frac{2}{5}\right)\left(\frac{100}{100}\right) = \left(\frac{2}{5}\right)\left(\frac{100}{\text{centum}}\right) = \left(\frac{200}{5}\right)\text{ per centum} = 40\text{ per centum}$$
$$= 40\text{ per cent.}$$
$$= 40\text{ percent}$$
$$= 40\%$$

per (arrow pointing to the fraction bar of $\frac{100}{\text{centum}}$)

$\frac{100}{100} = 1$. Anything can be multiplied by 1 without changing its value.

centum = 100

EXAMPLE 4 Express the fraction $\frac{3}{6}$ as a percentage using the per centum definition.

Solution

$$\left(\frac{3}{6}\right)\left(\frac{100}{100}\right) = \left(\frac{3}{6}\right)\left(\frac{100}{\text{centum}}\right) = \left(\frac{300}{6}\right)\text{ per centum}$$
$$= 50\text{ per centum} = 50\%$$

per

$\frac{100}{100} = 1$

centum = 100

EXAMPLE 5 Using the techniques shown in Example 4, express $\frac{7}{10}$ as a percentage.

Solution

$$\left(\frac{7}{10}\right)\left(\frac{100}{100}\right) = \left(\frac{7}{10}\right)\left(\frac{100}{\text{centum}}\right) = \left(\frac{700}{10}\right)\text{ per centum}$$
$$= 70\text{ per cent.} = 70\%$$

EXAMPLE 6 A class contains five students, two of whom are girls. Calculate the percentage of girls in the class using the per centum idea.

Solution

$$\left(\frac{2\text{ girls}}{5\text{ students}}\right)\left(\frac{100\text{ students}}{\text{centum students}}\right) = \frac{40\text{ girls}}{\text{centum students}}$$

(per; per; per; centum = 100)

$= 40$ girls per centum students
$= 40$ girls per cent. students
$= 40$ girls per 100 students
$= 40\%$ girls

SELF-TEST Use units and dimensional analysis for all of your calculations in the following problems:

1. A 100-g sample of chicken soup was 2.5% by weight chicken. How many grams of chicken are in this sample of soup?
2. You weight 170 lb and are 20% fat. How many pounds of fat are you carrying around with you?
3. Express the fraction $\frac{1}{25}$ as a percentage.
4. Only 2 dogs in a group of 50 dogs have a white spot. What percentage of the dogs have white spots?

ANSWERS **1.** 2.5 g chicken **2.** 34 lb fat **3.** 4% **4.** 4%

3.3 DENSITY AND SPECIFIC GRAVITY

Units such as grams and liters do not have to be used in isolation. We can combine units to form new units such as grams per liter (g/L). We refer to the unit *grams per liter* as *density*. Grams and liters by themselves do not say anything unique about substances. On the other hand, the density of each element is unique and can be identified by this measurement. No two elements have the same density.

It is common for students to get density and specific gravity confused. We will discuss density, present example density problems, define specific gravity, and then discuss the relationship between specific gravity and density.

The density of an object is defined as its weight* per unit volume:

$$\text{density} = \frac{\text{weight of object}}{\text{volume of object}}$$

$$= \frac{\text{grams of object}}{\text{cubic centimeters of object}}$$

The most frequently encountered density units in chemistry are grams per cubic centimeter (g/cm^3) for solids, grams per milliliter (g/mL) for liquids, and grams per liter (g/L) for gases. *Note*: 1 cubic centimeter (1 cc or 1 cm^3) = 1 milliliter (1 mL).

*There is a difference between mass (the amount of matter) and weight (the force that matter exerts due to gravity). We will follow common practice, however, and use the terms interchangeably. Density is more correctly defined as mass per unit volume.

EXAMPLE 7 A piece of pure aluminum weighs 40.3 g and occupies a volume of 14.93 cm^3. What is the density of aluminum? The symbol for aluminum is Al.

Solution The density of aluminum is 2.70 g/cm^3.

$$\text{density of aluminum} = \text{mass of Al per unit volume of Al}$$

$$= \frac{40.3 \text{ g Al}}{14.93 \text{ cm}^3 \text{ Al}} = \frac{2.70 \text{ g}}{\text{cm}^3} = 2.70 \text{ g/cm}^3$$

EXAMPLE 8 Imagine that you find a second piece of pure aluminum that weighs 200 g and occupies a volume of 74.074 cm^3. Find the density of this second piece of aluminum.

Solution The second piece of aluminum also has a density of 2.70 g/cm^3, the same value as the density of the first piece of aluminum.

$$\text{density of Al} = \frac{\text{mass of Al}}{\text{volume of Al}} = \frac{200 \text{ g Al}}{74.074 \text{ cm}^3 \text{ Al}} = 2.70 \text{ g/cm}^3$$

Is it just a coincidence that the densities of the two aluminum samples are the same? Explain your answer.

The density of a pure substance is constant. The density of 1 ton of aluminum is the same as the density of 1 g of aluminum. Both samples have a density of 2.70 g/cm^3.

Although most chemists use the unit *grams per cubic centimeter*, density can be expressed in many other units. Understanding how to express density in other units will prove helpful when we get to the topic of specific gravity.

EXAMPLE 9 A piece of aluminum weighs 200 lb and has a volume of 1.19 cubic feet (ft^3). Find the density of aluminum in pounds per cubic feet (lb/ft^3).

Solution The density of aluminum is 168 lb/ft^3.

$$\text{density of Al} = \frac{\text{mass of Al}}{\text{volume of Al}} = \frac{200 \text{ lb Al}}{1.19 \text{ ft}^3 \text{ Al}} = \frac{168 \text{ lb}}{\text{ft}^3} = 168 \text{ lb/ft}^3$$

EXAMPLE 10 A sample of gold weighs 10.00 lb and occupies 0.008306 ft^3. Find the density of gold. The symbol for gold is Au.

Solution The density of gold is 1204 lb/ft^3. (This means that one cubic foot of gold weighs over half a ton.)

$$\text{density of gold} = \frac{\text{mass of Au}}{\text{volume of Au}} = \frac{10.00 \text{ lb Au}}{0.008306 \text{ ft}^3 \text{ Au}} = \frac{1204 \text{ lb}}{\text{ft}^3} = 1204 \text{ lb/ft}^3$$

EXAMPLE 11 A gold sample is found to weigh 55.31 lb. What volume does the sample occupy?

Solution

$$(55.31 \text{ lb Au}) \left(\frac{1 \text{ ft}^3 \text{ Au}}{1204 \text{ lb Au}} \right) = 0.04594 \text{ ft}^3 \text{ Au}$$

EXAMPLE 12 Find the weight of an aluminum cube that measures 10.00 cm on each side.

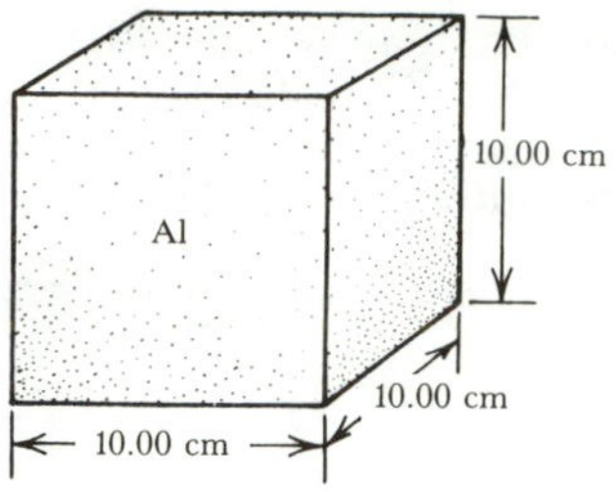

Solution The aluminum cube weighs 2.70×10^4 g.

$$\text{volume of Al cube} = (10.00\ \text{cm})(10.00\ \text{cm})(10.00\ \text{cm}) = 1000\ \text{cm}^3$$

$$(1000\ \text{cm}^3\ \text{Al})\left(\frac{2.70\ \text{g Al}}{\text{cm}^3\ \text{Al}}\right) = 2.70 \times 10^3\ \text{g Al}$$

EXAMPLE 13 The density of silver is 10.5 g/cm^3. You have been buying silver cubes from a man whose reputation for honesty is in doubt. Specifically, you would like to know if the cubes are hollow or solid without destroying a cube by sawing it in half.

How can you determine if your silver cubes, which measure 15.00 cm on each side and weigh 30.00 kg each, are solid silver or have hollow centers? (Ag is the symbol for silver.)

Solution Determine the density of one of the suspected silver cubes and compare it with the known density of pure silver.

$$\text{volume of cube} = (15.00\ \text{cm})^3 = 3375\ \text{cm}^3$$

$$\text{weight of cube} = (30.00\ \text{kg})\left(\frac{1000\ \text{g}}{\text{kg}}\right) = 3.000 \times 10^4\ \text{g}$$

$$\text{density of cube} = \frac{3.000 \times 10^4\ \text{g Ag}}{3375\ \text{cm}^3\ \text{Ag}} = \frac{8.888\ \text{g}}{\text{cm}^3} = 8.888\ \text{g/cm}^3$$

Your cubes are not 100% solid silver. The density of pure silver is 10.5 g/cm^3. Either your cubes have hollow centers or they are solid but not 100% pure silver (i.e., they could be an alloy of silver, which is silver mixed with a less dense and less expensive metal).

EXAMPLE 14 One gram of water has a volume of 1 cm^3. What is the density of water? (Remember: 1 cm^3 water = 1 mL water.)

Solution The density of water is 1 g/mL. For liquid, density is usually expressed in the unit *grams per milliliter*, not *grams per cubic centimeter*. Memorize the density of water and keep in mind that 1 g of water = 1 cm^3 of water = 1 mL of water.

All substances are compared to the density of water to obtain their specific gravities; that is, the specific gravity of a substance equals the mass of the substance divided by the mass of an equal volume of water.*

---

*To be precise, we should specify that the temperature of water must be 4°C when its volume is compared with another substance. However, the volume of 1 g of water is 1.00 cm^3 for the temperature range 0–30°C to within three significant digits.

EXAMPLE 15 The density of aluminum is 2.70 g/cm^3 and the density of water is 1.00 g/mL. Find the specific gravity of aluminum.

Solution

$$\text{specific gravity of aluminum} = \frac{\text{mass of Al}}{\text{mass of equal volume of water}}$$

$$= \frac{2.70\ \text{g}}{1.00\ \text{g}} = 2.70$$

•
•

Notice that the specific gravity of aluminum (2.70) does not have units. Why doesn't specific gravity have units?

Specific gravity is unitless because it is calculated by dividing two numbers with identical units, as shown in Example 15. The specific gravity can also be defined as the ratio of the density of an object to the density of water, i.e., the relative density of an object as compared to that of water. This is shown in the following equation:

$$\text{specific gravity of aluminum} = \frac{\text{density of aluminum}}{\text{density of water}} = \frac{2.70\ \text{g/cm}^3}{1.00\ \text{g/cm}^3} = 2.70$$

unitless number (2.70)

We can see that the density of aluminum is 2.70 times the density of water. Aluminum is 2.70 times as dense as water.

•
•

EXAMPLE 16 The density of gold is 19.3 g/cm^3. Find the specific gravity of gold.

Solution The specific gravity of gold is equal to the density of gold divided by the density of water.

$$\text{specific gravity of gold} = \frac{\text{density of gold}}{\text{density of water}} = \frac{19.3\ \text{g/cm}^3}{1.00\ \text{g/cm}^3} = 19.3$$

We can see that gold is 19.3 times as dense as water.

•
•

EXAMPLE 17 What is the density in grams per cubic centimeter of an object whose specific gravity is 15?

Solution The object is 15 times the density of water.

$$(15)\left(\frac{1\ \text{g}}{1\ \text{cm}^3}\right) = \frac{15\ \text{g}}{1\ \text{cm}^3} = 15\ \text{g/cm}^3$$

(water density: $\frac{1\ \text{g}}{1\ \text{cm}^3}$)

The density of the object is 15 g/cm^3.

•
•

Do not get the incorrect impression that specific gravity is always numerically equal to density (dropping the unit *grams per cubic centimeter*). Specific gravity is numerically equal to density only when the density is expressed in grams per cubic centimeter. This is another advantage of the metric system. To illustrate this point, consider what happens when density is measured in pounds per cubic foot.

•
•

EXAMPLE 18 Calculate the density of water in pounds per cubic foot. Remember that 1 g of water has a volume of 1 cm^3, that 454 g = 1 lb, and 2.54 cm = 1 in.

Solution The density of water is 62.4 lb/ft^3.

$$\left(\frac{1\text{ g H}_2\text{O}}{1\text{ cm}^3\text{ H}_2\text{O}}\right)\left(\frac{1\text{ lb}}{454\text{ g}}\right)\left(\frac{2.54\text{ cm}}{\text{in.}}\right)^3\left(\frac{12\text{ in.}}{\text{ft}}\right)^3 = \frac{62.4\text{ lb H}_2\text{O}}{\text{ft}^3\text{ H}_2\text{O}}$$

$$= 62.4\text{ lb/ft}^3$$

EXAMPLE 19 The density of aluminum is 168 lb/ft^3. Find the specific gravity of aluminum by comparing its density to the density of water using the density unit *pounds per cubic foot.*

Solution

$$\text{specific gravity of aluminum} = \frac{\text{density of Al}}{\text{density of H}_2\text{O}} = \frac{168\text{ lb/ft}^3}{62.4\text{ lb/ft}^3} = 2.70$$

The specific gravity of aluminum is 2.70 (unitless). This is the same unitless value that we calculated for aluminum when we compared aluminum's density of 2.70 g/cm^3 to water's density of 1 g/cm^3.

$$\text{sp gr Al} = \frac{\text{density of Al}}{\text{density of H}_2\text{O}} = \frac{2.70\text{ g/cm}^3}{1\text{ g/cm}^3} = 2.70$$

The unitless number 2.70 remains constant regardless of the units used to express density. Can you explain why?

Aluminum is 2.70 times as dense as water. This statement is not dependent upon units. If a volume of water weighed 100 tons, a similar volume of aluminum would weigh 2.70 times 100 tons, or 270 tons.

The important point made by Example 19 is that when density is expressed in units other than grams per cubic centimeter, the specific gravity and the density of the object are numerically different.

SELF-TEST

1. A 25-g object has a volume equal to 2.5 cm^3. What is the density of the object?
2. What is the specific gravity of a 30-g object whose volume equals 10 cm^3?
3. Calculate the density of water in pounds per cubic foot.
4. What is the specific gravity of an object whose density equals 124.8 lb/ft^3?

ANSWERS **1.** 1.0×10^1 g/cm^3 **2.** 3.0×10^1 **3.** 62.4 lb/ft^3 **4.** 2.00

3.4 THE JOULE AND THE CALORIE

The joule and the calorie are fundamental concepts in chemistry that are frequently used when discussing how much heat is given off during a chemical reaction. A heat effect accompanies all chemical reactions, and these heat effects are measured in units of both joules (abbreviated J) and calories (abbreviated cal).

Why do we represent quantities of heat with two units (joules and calories)? Before 1975, almost all American chemistry textbooks used only the calorie to express amounts of heat. Even today the calorie is in widespread use although the joule has been endorsed by the International Union of Pure and Applied Chemists (IUPAC). Whenever a change is suggested (e.g., the United States should go metric after 200 years of using English units), there is usually much debate as to whether the suggested change is worth the effort. During this period of debate and change, you will be exposed to both the joule and the calorie.

You should understand what is meant by temperature before you can fully appreciate the definition and concept of the joule or the calorie. Temperature is a measure of how hot an object is. More precisely, temperature is a measure of the kinetic energy of the atoms or molecules of an object. The higher the temperature, the faster molecules translate (move from one point to another), vibrate, and rotate. Temperature also determines the direction of heat flow. Heat always flows spontaneously from an object at a higher temperature to an object at a lower temperature. The reverse, however, is never true. Heat cannot flow spontaneously from a colder object to a warmer object.

The amount of heat flowing from a hotter object to a colder object is measured in units of joules and calories. The joule and the calorie are quantities (amounts) of heat. One calorie is the amount of heat required to raise the temperature of one gram of water one degree Celsius.* Since 1948 the National Bureau of Standards (NBS) has defined the calorie as exactly $4.184\bar{0}$ J. Therefore, 4.184 J is the amount of heat required to raise the temperature of 1 g water 1°C. As an approximation, just think of 1 J as the amount of heat required to raise the temperature of about $\frac{1}{4}$ g of water by 1°C.

•
•

EXAMPLE 20 You would like to raise the temperature of a 10.0-g water sample by 1.00°C. How many calories must you add to the water? How many joules?

Solution Ten calories of heat must be added to 10.0 g of water to raise the temperature of the water 1.00°C. One calorie is the amount of heat that will raise the temperature of 1 g of water 1°C. We have 10.0 g of water, so 10.0 cal is required to raise the temperature of the water by 1.00°, as shown.

$$(1.00^\circ\text{C temp. increase in 10.0 g } H_2O)\left(\frac{\text{1 cal of heat absorbed}}{1^\circ\text{C temp. increase in 1 g } H_2O}\right) = 10.0 \text{ cal absorbed}$$

Translation: 1 cal of heat is absorbed per 1°C of temperature increase in 1 g H_2O.

Ten grams of water must absorb 10.0 cal of heat to raise the temperature of the water by 1.00°C.

You would need to add 41.8 J of heat to 10.0 g of water to raise the temperature of the water by 1.00°C.

$$(1.00^\circ\text{C temp. increase in 10.0 g } H_2O)\left(\frac{\text{4.184 J of heat absorbed}}{1^\circ\text{C temp. increase in 1 g } H_2O}\right) = 41.8 \text{ J absorbed}$$

Translation: 4.184 J are absorbed by water for every 1 g of water and for every 1°C of temperature increase.

Ten grams of water must absorb 41.8 J of heat to raise the temperature of the water by 1.00°C.

You also could have used your answer in calories (10 cal needed to raise 10 g of water by 1°C) to determine the number of joules needed to raise the temperature of 10 g of water by 1°C, as shown.

---

*To be more precise, the calorie is defined as the amount of heat required to raise the temperature of 1 g of water from 14.5°C to 15.5°C.

$$(10.0 \text{ cal of heat absorbed})\left(\frac{4.184 \text{ J}}{\text{cal}}\right) = 41.8 \text{ J of heat absorbed}$$

It should seem reasonable that if 1 cal will increase the temperature of 1 g of water by 1°C, then 1 g of water that loses 1 cal will cool down by 1°C.

EXAMPLE 21 A beaker of 15 mL of water cools down from 22.0°C to 20.0°C. How many calories of heat does the water lose? How many joules does the water lose?

Solution *Remember*: 15 mL water = 15 g water.

$$(2.0°\text{C temp. decrease in 15 g } H_2O)\left(\frac{1 \text{ cal of heat lost}}{1°\text{C temp. decrease in 1 g } H_2O}\right) = 30 \text{ cal lost}$$

$$(2.0°\text{C temp. decrease in 15 g } H_2O)\left(\frac{4.184 \text{ J of heat lost}}{1°\text{C temp. decrease in 1 g } H_2O}\right) = 1.3 \times 10^2 \text{ J lost}$$

(about 130 J)

EXAMPLE 22 A tank containing 50 kg of water at 90°C cools down to 20°C. How many calories of heat does the tank of water lose? Express your answer in joules.

Solution

$$(50 \text{ kg } H_2O)\left(\frac{1000 \text{ g}}{\text{kg}}\right) = 5.0 \times 10^4 \text{ g } H_2O.$$

$$(70°\text{C temp. decrease in } 5.0 \times 10^4 \text{ g } H_2O)\left(\frac{1 \text{ cal of heat lost}}{1°\text{C temp. decrease in 1 g } H_2O}\right) = 3.5 \times 10^6 \text{ cal lost}$$

$$(3.5 \times 10^6 \text{ cal lost})\left(\frac{4.184 \text{ J}}{\text{cal}}\right) = 1.5 \times 10^7 \text{ J lost}$$

Large numbers of calories and joules (such as 3.5×10^6 cal and 1.5×10^7 J) are usually expressed in kilocalories (1000 cal = 1 kcal) and kilojoules (1000 J = 1 kJ).

$$(3.5 \times 10^6 \text{ cal})\left(\frac{1 \text{ kcal}}{1000 \text{ cal}}\right) = 3.5 \times 10^3 \text{ kcal}$$

$$(1.5 \times 10^7 \text{ J})\left(\frac{1 \text{ kJ}}{1000 \text{ J}}\right) = 1.5 \times 10^4 \text{ kJ}$$

SELF-TEST

1. How much heat is required to raise the temperature of 22 g of water by 1.0°C? Express your answer in calories and in joules.
2. How much heat is required to raise the temperature of 55.0 g of water by 10.0°C? Express your answer in calories and in joules.
3. How much heat is lost when 460 kg of water cools from 75°C to 25°C? Express your answer in calories and in joules.

ANSWERS **1.** 22 cal; 92 J **2.** 550 cal; 2.30×10^3 J **3.** 2.3×10^7 cal; 9.6×10^7 J

3.4.1 HEAT VS. TEMPERATURE

•
•

EXAMPLE 23 Consider two water samples. The first sample weighs 1 g and is at a temperature of 90°C (hot enough to burn you). The second water sample weighs 1000 g and is at a temperature of 11°C (it feels cool to the touch).

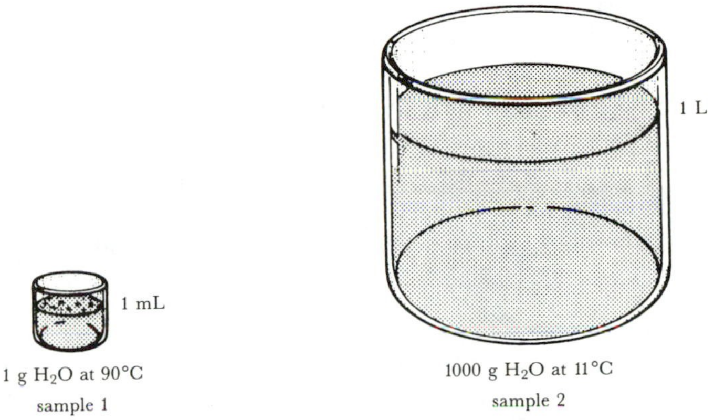

Although the second water sample is cooler than the first, the second water sample contains more heat (calories) than the first water sample. Let's try to explain why this is so.

Cool both water samples to the same temperature, 10°C, and calculate the amount of heat lost by each water sample.

Solution The 1-g water sample, when cooled from 90°C to 10°C, gives off 80 cal of heat. The 1000-g water sample, when cooled by only 1°C, gives off 1000 cal of heat.

1-g Sample:

$$(80^\circ \text{ temp. decrease in 1 g } H_2O)\left(\frac{\text{1 cal heat lost}}{1^\circ\text{C temp. decrease in 1 g } H_2O}\right) = \text{80 cal lost}$$

1000-g Sample:

$$(1^\circ\text{C temp. decrease in 1000 g } H_2O)\left(\frac{\text{1 cal heat lost}}{1^\circ\text{C temp. decrease in 1 g } H_2O}\right) = \text{1000 cal lost}$$

This last example should make you realize that a cooler object can contain more heat than a warmer object. Larger cold objects can contain more heat than smaller hot objects.

•
•

Let's look at another example in which the hotter substance contains less heat than a colder substance.

Sparks given off by a cigarette lighter are very hot, about 1000°C. (Any metal heated to about 1000°C will glow white hot.) But when sparks from a cigarette lighter touch your hand, you do not usually detect any warmth. Why?

Sparks from a cigarette lighter are very small particles of burning metal at a very high temperature. The particles are so small that only a small amount of heat (calories) is required to raise their temperature to 1000°C. Your body detects a warmer object by detecting the flow of heat (calories) from the warmer object to itself. Usually very hot objects give off many calories because the object had to absorb many calories to become hot. But more massive objects require a greater number of calories to warm up than less massive amounts of the same object. (It takes more heat to raise the temperature of 1 ton of gold by 1°C than it does to raise the temperature of 1 oz of gold by 1°C.)

The Pacific Ocean at 2°C contains more heat than the sparks produced by a cigarette lighter, which are at 1000°C. It would take the heat from 21 million hydrogen bombs to warm the Pacific Ocean by just 1°C! Conversely, if the Pacific Ocean were to cool down by 1°C, it would release an amount of heat equivalent to that produced by 21 million hydrogen bombs.

Don't lose sight of the fact that calories and temperature are related. Their relationship is expressed in the definition of the calorie, that is, 1 cal raises the temperature of 1 g of water by 1°C. Understanding this relationship will remove much of the mystery when you study more advanced topics, such as thermodynamics. One of the areas of concern in thermodynamics is determining the amount of heat (the number of calories) given off by a chemical reaction. Usually such chemical reactions are carried out underwater and enclosed in a metal container. Heat given off by the reaction warms up the surrounding water.

EXAMPLE 24

A chemical reaction takes place underwater and raises the temperature of the surrounding 500.0 g of water by 3.000°C. How much heat is released by this reaction?

Solution The chemical reaction releases 1500 cal.

$$(3.000°\text{C temp. increase in } 500.0 \text{ g } H_2O)\left(\frac{1 \text{ cal of heat absorbed}}{1°\text{C temp. increase in } 1 \text{ g } H_2O}\right)$$

$$= 1500 \text{ cal absorbed}$$

3.4.2 DRINK WATER TO LOSE WEIGHT: A MYSTERY

If you are trying to lose weight, should you drink your water at room temperature, with ice, or boiling hot?

You burn more of your excess calories (body fat) by drinking cold liquids. To see why, let's consider what happens when you drink a liter (a little more than a quart) of water that is ice-cold (0°C, 32°F). Your body has to raise the temperature of the ice-cold water to body temperature (37°C, 98.6°F).

EXAMPLE 25 How many calories does your body have to release to exactly 1 L of ice-cold water (0°C) to raise the water temperature to 37°C?

Solution Thirty-seven thousand calories of heat is required to raise the temperature of 1 L of water at 0°C to 37°C. One liter of water weighs 1000 g. To raise the temperature of 1000 g of water by only 1°C, 1000 cal of heat is required. To raise 1000 g of water to 37°C requires 37 000 cal of heat.

$$(37°\text{C temp. increase in } 1000 \text{ g } H_2O)\left(\frac{1 \text{ cal of heat absorbed}}{1°\text{C temp. increase in } 1 \text{ g } H_2O}\right)$$

$$= 3.7 \times 10^4 \text{ cal absorbed}$$

↑ about 37 000 cal

Now for the mystery:

EXAMPLE 26 According to medical specialists, the average American's diet produces about 2500 Calories per day. (Nutritionists use a capital *C* to spell the word *calorie*. Later in this example we'll discuss the difference between a *calorie* and a *Calorie*.) This means that the food we consume burns (oxidizes) in our bodies to produce on average 2500 Cal of energy. Our bodies use these 2500 Calories to power the various biochemical reactions that keep us alive. These reactions include breaking down nutrients to produce cellular energy, conversion of nutrients into compounds that our bodies need, and maintaining a constant body temperature, to name a few.

Medical authorities recommend that the average American drink six glasses of water each day for overall good health. Six glasses of water equals 1.5 quarts, or about 1.5 liters. The heat from our bodies warms the water we drink until the water reaches body temperature. How much heat would our bodies have to provide to warm 1.5 L of ice-cold (0°C) water to body temperature (37°C)?

Solution Our bodies would have to provide 55 000 cal of heat to warm 1.5 L of ice water to body temperature.

$$(37°\text{C temp. increase in 1500 g } H_2O)\left(\frac{1 \text{ cal heat absorbed}}{1°\text{C temp. increase in 1 g } H_2O}\right)$$
$$= 5.5 \times 10^4 \text{ cal absorbed}$$

Does this answer make sense? Does it make sense that each day we consume enough food to produce 2500 Cal and could drink six glasses of water that would consume 55 000 cal? What an amazingly easy way to lose weight! Unfortunately, it doesn't work.

Why? Where are we making an error?

In the area of nutrition, a *Calorie* (with a capital *C*) is what other scientists refer to as a *kilocalorie* (that is, 1000 calories). The average American diet is actually 2 500 000 calories per day (2500 kcal/day). This nutritional jargon of referring to a kilocalorie as a Calorie is very confusing and can lead to the type of misunderstanding that we have just encountered.

One and a half liters of ice-cold water does absorb 55 000 calories from our bodies, but that's only 55 kcal versus the 2500 kcal supplied by our daily diets. However, drinking cold beverages can be a worthwhile activity. Fifty-five kcal is significant when we realize that we have to walk more than two miles to burn 55 kcal of food or body fat.

3.4.3 HOT-BLOODED DINOSAURS

EXAMPLE 27 Dinosaurs roamed the earth for 140 million years (vs. 4 million years for the human race). About 70 million years before humans appeared, dinosaurs became extinct. Many theories have been proposed to explain why dinosaurs became extinct, and this is still an area of interest to scientists today. Some of the theories are based on the assumption that dinosaurs were cold-blooded animals (animals whose body temperature depends upon their surroundings). Some scientists question the popular assumption that all dinosaurs were cold-blooded animals, and this is still being debated today. There is an interesting argument based upon the calorie concept, and we can use a multi-step conversion factor dimensional analysis calculation to shed light on this matter.

The maximum heat of absorption from the sun by a brontosaurus is 4113 kcal per hour. (This is based on the total surface area of a brontosaurus.) Assume that we have a 40-ton brontosaurus that is 90% water, and we want to calculate how long it takes for this brontosaurus to raise its body temperature by 2°C.

Take this approach in determining the answer: First find how many grams of water are in a 40-ton brontosaurus, and then calculate the amount of time it would take to raise that amount of water by 2°C. You can break down this calculation into a series of steps, but, if you can, try a one-step chain multiplication approach using several conversion factors, as we illustrated earlier.

How long will it take for a 40-ton brontosaurus to raise its body temperature by 2°C?

Solution It would take 16 hours to raise the body temperature of a 40-ton brontosaurus by 2°C.

$$(\text{40-ton brontosaurus})\left(\frac{90 \text{ tons } H_2O}{100 \text{ tons brontosaurus}}\right)\left(\frac{2000 \text{ lb}}{\text{ton}}\right)\left(\frac{454 \text{ g}}{\text{lb}}\right)$$
$$\left(\frac{1 \text{ cal absorbed}}{1°\text{C temp. increase in 1 g } H_2O}\right)\left(2°\text{C temp. increase}\right)\left(\frac{1 \text{ hour}}{4\,113\,000 \text{ cal}}\right) = 16 \text{ hours}$$

A 40-ton brontosaurus whose body temperature decreases by 2°C has to lie in the sun for 16 hours before it can resume its normal activities because, for an animal to function normally, its body temperature must remain fairly constant—about 100°F (38°C). This is true for both warm-blooded animals and cold-blooded animals. On a cold day, for example, a bee cannot fly until it has raised its internal body temperature to about 100°F by warming itself in the sun. (Some insects also generate additional heat through a series of muscle contractions.)

Once a bee has elevated its body temperature to about 100°F, it can fly. But on cold days, the passage of cold air over the bee's body in flight lowers its body temperature and the bee becomes too sluggish to fly. It must once again raise its internal body temperature before it can continue to fly again.

Can you develop an argument, based upon this information, that would indicate that the brontosaurus was a warm-blooded animal?

Because a brontosaurus would be sluggish for long periods of time if many hours were required to raise its body temperature by just 1 or 2 degrees, it would be easy prey for its enemies.

Using the information given in Example 27, can you devise another argument that would indicate that the brontosaurus was a cold-blooded animal?

If it takes a long period of time for a brontosaurus to raise its body temperature by 2°C, it must also take a long period of time for its body to cool down. The enormous size of a brontosaurus would help it to maintain a nearly constant body temperature.

3.5 TEMPERATURE CONVERSIONS

The temperature at which a chemical reaction occurs is a very important consideration. Temperature affects both the rate (how fast) and the extent (how much of the reaction occurs) of a chemical reaction.

The phase (solid, liquid, or gas) of chemicals also depends upon the temperature. You should be familiar with the different ways temperature is measured and how to convert from one temperature scale to another. The purpose of this section is to help you develop that familiarity.

You have been exposed to enough science by now to know that there are two temperature scales in daily use, Fahrenheit and Celsius (also called centigrade). Metric countries usually use the Celsius, or centigrade, scale whereas the Fahrenheit scale finds its greatest use in the United States. When working with scientific formulas, chemists usually use the Kelvin as the absolute temperature scale (see Figure 3.1).

FIGURE 3.1 Comparing temperature scales

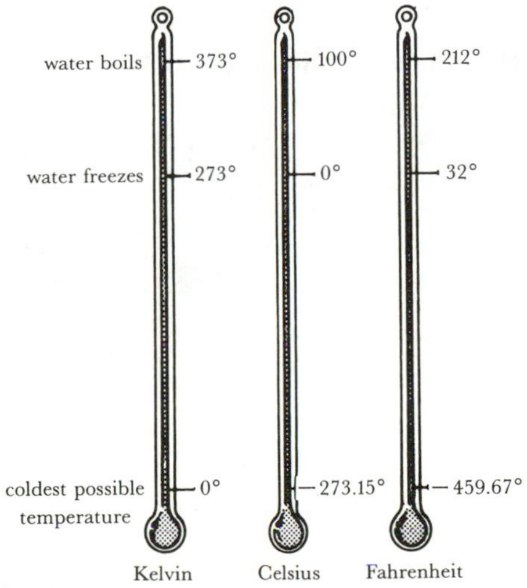

Unlike most of the problems solved in this book, temperature conversions do not always lend themselves easily to a dimensional analysis approach. Instead, the following formula is usually used to convert degrees Fahrenheit to degrees Celsius or vice versa:

$$\text{deg F} - 32 = \frac{9}{5}\,(°\text{C})$$

Note: deg F = °F

When the temperature increases 9°F, it increases 5°C. More on this later.

In keeping with the spirit of dimensional analysis, let's modify the preceding equation and use it in the following form:

$$\text{deg F} - 32°\text{F} = \frac{9°\text{F}}{5°\text{C}}\,(°\text{C}) \qquad \text{(equation A)}$$

To convert a temperature from degrees Fahrenheit to degrees Celsius, equation A is rearranged as follows:

$$\text{deg C} = \frac{5°\text{C}}{9°\text{F}}\,(\text{deg F} - 32°\text{F}) \qquad \text{(equation B)}$$

To convert from the Celsius to the Kelvin scales, add 273.15 to the given Celsius temperature, as follows:

$$\text{K} = \text{deg C}\left(\frac{1\ \text{K}}{1°\text{C}}\right) + 273.15 \qquad \text{(equation C)}$$

When the temperature increases 1°C, it also increases 1 K on the Kelvin scale.

To convert degrees Fahrenheit to degrees Kelvin, first convert degrees Fahrenheit to degrees Celsius, as shown earlier. Then, to the temperature expressed in degrees Celsius, add 273.15, as illustrated in the following example:

EXAMPLE 28 The melting point of aluminum is 1945°F. Find its melting point in degrees Celsius and in degrees Kelvin.

Solution The melting point of aluminum is 1063°C and 1336 K.

$$\text{deg C} = \left(\frac{5°\text{C}}{9°\text{F}}\right)(\text{deg F} - 32°\text{F})$$

$$= \left(\frac{5°\text{C}}{9°\text{F}}\right)(1945°\text{F} - 32°\text{F})$$

$$= \left(\frac{5°\text{C}}{9°\text{F}}\right)(1913°\text{F}) = 1063°\text{C}$$

$$\text{deg K} = (1063°\text{C})\left(\frac{1\ \text{K}}{1°\text{C}}\right) + 273.15\ \text{K} = 1336\ \text{K}$$

EXAMPLE 29 Liquid nitrogen, with its low boiling point of −319.9°F, was used to extinguish Kuwaiti oil wells that were ignited by retreating Iraqi soldiers in the Gulf War. Find the boiling point of liquid nitrogen in degrees Celsius and in degrees Kelvin.

Solution The boiling point of liquid nitrogen is −195.5°C and 77.7 K.

$$\text{deg C} = \left(\frac{5°\text{C}}{9°\text{F}}\right)(\text{deg F} - 32°\text{F})$$

$$= \left(\frac{5°\text{C}}{9°\text{F}}\right)(-319.9°\text{F} - 32°\text{F})$$

$$= \left(\frac{5°C}{9°F}\right)(-351.9°F)$$

$$= -195.5°C$$

$$\deg K = (-195.5°C)\left(\frac{1\ K}{1°C}\right) + 273.15\ K = 77.7\ K$$

Equation A can be arranged differently to find degrees Fahrenheit when the temperature is known in degrees Celsius.

$$\deg F = \left(\frac{9°F}{5°C}\right)(\deg C) + 32°F \qquad \text{(equation D)}$$

EXAMPLE 30 The boiling point of aluminum is 2940°C. Find the boiling point of aluminum in degrees Fahrenheit.

Solution The boiling point of aluminum is 5378°F.

$$\deg F = \left(\frac{9°F}{5°C}\right)(\deg C) + 32°F = \left(\frac{9°F}{5°C}\right)(2970°C) + 32°F$$

$$= \frac{(9°F)(2970°C)}{(5°C)} + 32°F$$

$$= 5346°F + 32°F$$

$$= 5378°F$$

EXAMPLE 31 The melting point of frozen oxygen is −218.8°C. Find the melting point of oxygen in degrees Fahrenheit.

Solution The melting point of oxygen is −361.8°F.

$$\deg F = \left(\frac{9°F}{5°C}\right)(\deg C) + 32°F = \frac{(9°F)(-218°C)}{(5°C)} + 32°F$$

$$= -393.8°F + 32°F$$

$$= -361.8°F$$

Note: The numbers 9, 5, and 32 are defined numbers with an infinite number of significant digits.

Although we resorted to using formulas in the preceding temperature conversion problems, notice that we did not abandon the use of units.

SELF-TEST

1. Convert 1063°C to degrees Kelvin and to degrees Fahrenheit.
2. Convert 90 K to degrees Celsius and to degrees Fahrenheit.
3. Convert −361.8°F to degrees Celsius and to degrees Kelvin.

ANSWERS **1.** 1945°F; 1336 K **2.** −183°C; −297°F **3.** −218.8°C; 54.4 K

Dimensional analysis can be used to convert a temperature increase on one scale to a temperature increase on another scale. Let's see how this is done.

Because an increase from 0°C (freezing point of water, see Figure 3.1) to 100°C (boiling point of water) on the Celsius scale corresponds to an increase from 32°F (freezing point of water) to 212°F (boiling point of water) on the Fahrenheit scale, a temperature increase of 100°C corresponds to a temperature increase of 180°F (212 − 32 = 180). It is from this observation that the fractions $\frac{5}{9}$ and $\frac{9}{5}$ arise in the conversion equations.

$$\frac{100°\text{C temp. increase}}{180°\text{F temp. increase}} = \frac{5°\text{C temp. increase}}{9°\text{F temp. increase}}$$

Because an increase of 100°C on the Celsius scale is identical to a temperature increase of 100 K on the Kelvin scale, we insert the factor 1 K/1°C into equation C.

Let's practice using dimensional analysis to convert a temperature increase on one temperature scale to a temperature increase on another temperature scale.

EXAMPLE 32 When the human body's internal temperature rises more than 5.4°F above normal, metabolism efficiency decreases. Over prolonged periods of time, such a fever can be detrimental. What are the corresponding increases in degrees Celsius and degrees Kelvin?

Solution Metabolism efficiency decreases when the internal body temperature increases 3.0°C or 3.0 K.

per

$$(5.4°\text{F temp. increase})\left(\frac{5°\text{C temp. increase}}{9°\text{F temp. increase}}\right) = 3.0°\text{C temp. increase}$$

$$(3.0°\text{C temp. increase})\left(\frac{1\text{ K temp. increase}}{1°\text{C temp. increase}}\right) = 3.0\text{ K temp. increase}$$

EXAMPLE 33 The difference between the boiling and melting points of gold is about 1900°C. Find this temperature difference in degrees Fahrenheit and degrees Kelvin.

Solution The difference between the melting and boiling points of gold is about 3400°F and 1900 K.

$$(1900°\text{C temp. difference})\left(\frac{9°\text{F temp. difference}}{5°\text{C temp. difference}}\right) = 3400°\text{F temp. difference}$$

2 significant digits

5 and 9 have an infinite number of significant digits

2 significant digits

$$(1900°\text{C temp. difference})\left(\frac{1\text{ K temp. difference}}{1°\text{C temp. difference}}\right) = 1900\text{ K temp. difference}$$

SELF-TEST

1. When the temperature decreases 3.0°C, what are the corresponding temperature decreases on the Kelvin and Fahrenheit scales?
2. When the temperature rises by 3400°F, what are the corresponding temperature increases on the Kelvin and Celsius scales?
3. When the temperature increases by 9°F, what is the increase on the Celsius scale?

ANSWERS **1.** 3.0 K; 5.4°F **2.** 1889 K; 1889°C **3.** 5°C (from 9°F/5°C)

Now let's tie in the ideas of temperature and calories with two final examples. The solution to the first problem also requires your ability to convert units by dimensional analysis using multi-step conversion factors and your knowledge of temperature conversions.

3.5.1 HEATING WATER THE HARD WAY

•
•

EXAMPLE 34 *Assume you're going to bathe. You turn on the hot-water faucet, walk away to disrobe, and return to find a tub filled with cold water (60.0°F). The hot-water heater is out of order! Fortunately, you remember reading that 1 g of candy contains 3.730 kcal and you just bought a 20-lb box of candy. To take a comfortable bath, you feel that the tub water should be at least 95.0°F. The tub contains 5.00 ft³ of water. How many pounds of candy must you consume to bring the bath water from 60.0°F to 95.0°F?*

Solution You must eat 1.63 lb of candy to raise the 5.00 ft^3 of bath water by 35.0°F.

$$(5.00\ ft^3\ H_2O)\left(\frac{12\ in.}{ft}\right)^3\left(\frac{2.54\ cm}{in.}\right)^3\left(\frac{1\ g\ H_2O}{1\ cm^3 H_2O}\right)$$

$$(95.0°F - 60.0°F)\left(\frac{5°C}{9°F}\right)\left(\frac{1\ cal\ absorbed}{1\ g\ H_2O\ 1°C}\right)\left(\frac{1\ kcal}{1000\ cal}\right) = 2.75 \times 10^3\ kcal\ absorbed$$

This calculation tells us that 5.00 ft^3 of water must absorb about 2750 kcal of heat to increase its temperature from 60.0°F to 95.0°F.

$$(2.75 \times 10^3\ kcal)\left(\frac{1\ g\ candy}{3.730\ kcal}\right)\left(\frac{1.00\ lb}{454\ g}\right) = 1.63\ lb\ candy$$

This calculation tells us that 1.63 lb of candy contains about 2750 kcal of heat, the amount of heat required to raise the temperature of 5.00 ft^3 of water from 60.0°F to 95.0°F. Unfortunately for the dental profession, few people heat their bath water this way!

•
•

3.5.2 TEMPERATURE OF HELL*

A study was conducted to determine the temperature of hell. The reasoning process used in the study is interesting because it involves both the knowledge and the logic with which you should be equipped.

The Bible (Rev. 21:8) tells us that hell is a lake of fire and brimstone.

*"Heaven Is Hotter Than Hell," *Applied Optics* 11:8 (Aug. 1972):A14.

What is brimstone?

Brimstone is sulfur.

In what physical phase (solid, liquid, or gas) does the sulfur exist in hell?

Sulfur must be molten (liquid phase) since the Bible says it is a lake.

Knowing that hell is liquid sulfur, how can you determine the temperature of hell?

You could start by looking up the melting and boiling points of sulfur. If sulfur is present as a liquid, its temperature must be somewhere between sulfur's melting and boiling points. The boiling point of sulfur is 444.6°C; its melting point is 119.0°C.

•
•

EXAMPLE 35 Calculate the temperature range for hell in both degrees Fahrenheit and degrees Kelvin. Hell melts (or freezes over) at 119.0°C and boils at 444.6°C.

Solution Hell can have a temperature range of from 246.2°F to 832.3°F or from 392.2 K to 717.8 K.

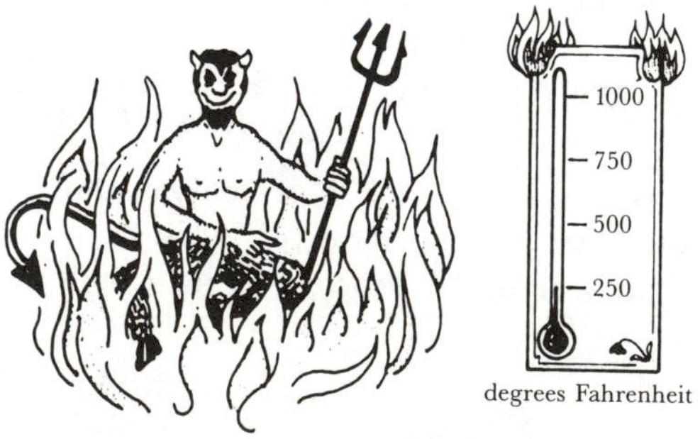

Minimum Temperature of Hell:

$$\text{deg F} = \left(\frac{9°\text{F}}{5°\text{C}}\right)(\text{deg C}) + 32°\text{F} = \frac{(9°\text{F})(119.0°\text{C})}{(5°\text{C})} + 32°\text{F} = 246.2°\text{F}$$

$$\text{K} = \left(\frac{1\ \text{K}}{1°\text{C}}\right)(\text{deg C}) + 273.15\ \text{K} = 119.0\ \text{K} + 273.15\ \text{K} = 392.2\ \text{K}$$

Maximum Temperature of Hell:

$$\text{deg F} = \left(\frac{9°F}{5°C}\right)(\text{deg C}) + 32°F = \frac{(9°F)(444.6°C)}{(5°C)} + 32°F = 832.3°F$$

$$K = \left(\frac{1\ K}{1°C}\right)(\text{deg C}) + 273.15\ K = 444.6\ K + 273.15\ K = 717.8\ K$$

The same study also determined the temperature in heaven. The Bible (Is. 30:26) tells us that in heaven the light of the moon is as the light of the sun. Also, the light of the sun is seven times the light of seven days on earth.

How much more light is received in heaven than is received on earth?

Heaven receives 50 times more light than the earth. Heaven gets 49 times the amount of light from the sun relative to the earth and an additional amount of light from the moon that equals the amount of light we on earth receive from the sun. So, all in all, heaven receives 50 times more light than we do on earth.

Assuming that the temperature of heaven remains constant, heaven must also lose by radiation 50 times as much heat as does the earth. The Stefan-Boltzmann fourth-power radiation law (discussed in advanced chemistry courses) predicts that heaven must be 977°F if it were to radiate this much heat.

EXAMPLE 36 Calculate the temperature of heaven in degrees Celsius and in degrees Kelvin if its temperature is 977°F.

Solution Heaven is 525°C and 798 K.

$$\text{deg C} = \left(\frac{5°C}{9°F}\right)(\text{deg F} - 32°F) = \left(\frac{5°C}{9°F}\right)(977°F - 32°F)$$

$$= \frac{(5°C)(945°F)}{(9°F)} = 525°C$$

$$\text{deg K} = (525°C)\left(\frac{1\ K}{1°C}\right) + 273.15\ K = 798\ K$$

Knowing that hell could be about 750°F cooler than heaven may be a comforting thought for some of us.

3.6 HEAT OF FUSION

Not all calories absorbed by matter increase the temperature of the matter that is absorbing the calories. Sometimes the absorbed heat breaks bonds or changes the phase of the matter (e.g., from solid to liquid).

One calorie will raise the temperature of 1 g of water by 1°C. But 1 cal will not raise the temperature of 1 g of ice at all. It would take 80 cal of heat to melt 1 g of ice whose temperature is 0°C. While the ice melts, its temperature remains at 0°C.

The 80 cal required to melt 1 g of ice at 0°C is called the heat of fusion of water. The heat of fusion is the quantity of heat that must be added to melt 1 g of material at the melting point.

EXAMPLE 37 How many calories would it take to melt a 5.0-g piece of ice whose temperature is 0°C? What is the temperature of the water after that number of calories has been added?

Solution

$$(5.0 \text{ g ice})\left(\frac{80 \text{ cal}}{\text{g ice}}\right) = 4.0 \times 10^2 \text{ cal}$$

The temperature of the resulting 5.0 g of water is 0°C.

EXAMPLE 38 How many kilojoules would it take to melt a 30.0-g piece of ice whose temperature is 0°C?

Solution

$$(30.0 \text{ g ice})\left(\frac{80 \text{ cal}}{\text{g ice}}\right)\left(\frac{4.184 \text{ J}}{\text{cal}}\right)\left(\frac{1 \text{ kJ}}{1000 \text{ J}}\right) = 10 \text{ kJ}$$

EXAMPLE 39 Gold has a heat of fusion of 3.03 cal/g. The melting point of gold is 1063°C. How many calories of heat does it take to melt 5 g of solid gold if the temperature of the gold is 1063°C?

Solution It takes 3.03 cal to melt 1 g of gold at its melting point. Therefore, about 20 cal are required to melt 5 g of gold at 1063°C.

$$(5 \text{ g Au})\left(\frac{3.03 \text{ cal}}{\text{g Au}}\right) = 2 \times 10^1 \text{ cal}$$

EXAMPLE 40 Aluminum has a heat of fusion of 3.55 cal/g. The melting point of aluminum is 660°C. How many calories are required to melt a 5-kg piece of aluminum whose temperature is 660°C?

Solution About 20 000 cal are required to melt a 5-kg piece of aluminum at its melting point.

$$(5 \text{ kg Al})\left(\frac{1000 \text{ g}}{\text{kg}}\right)\left(\frac{3.55 \text{ cal}}{\text{g Al}}\right) = 2 \times 10^4 \text{ cal}$$

SELF-TEST

1. How many calories of heat are required to melt a 20.0-g piece of ice? The temperature of the ice is 0°C.
2. How many kilocalories of heat are required to melt a 3000-g piece of ice whose temperature is 0°C?
3. Iron has a heat of fusion equal to 3.67 cal/g of iron. How many calories are required to melt a 10.0-g piece of iron at its melting point?

ANSWERS **1.** 1.6×10^3 cal **2.** 2×10^2 kcal **3.** 36.7 cal

3.6.1 DANGERS OF SNOW CONSUMPTION FOR EMERGENCY WATER

If you ever find yourself stranded in a snowstorm, you may be tempted to eat snow as an emergency water source. Although it is a reasonable assumption that snow would be an adequate substitute for water, such a practice could burn too many valuable calories, which you need to keep warm and to survive. How can that be?

One calorie (or 4.184 joules) will raise the temperature of 1 g of water by 1°C. However, 1 cal will not raise the temperature of 1 g of snow (or ice) at all. If you ate 1 g of snow, your body would have to burn 80 cal (435 J) just to melt the snow, assuming it is 0°C. While your body supplies the 80 cal to melt the snow, the temperature of the melted snow remains unchanged at 0°C. Your body still must provide additional heat to warm the melted snow from 0°C to normal body temperature, which is 37°C.

EXAMPLE 41 *If you were to consume 1000 g of snow (the equivalent of about 1 qt of water when melted), how many calories would your body have to burn just to melt the snow?*

Solution Your body would have to burn 80 000 cal just to melt the 1000 g of snow at 0°C, producing 1 L of ice water at 0°C.

$$(1 \times 10^3 \text{ g snow})\left(\frac{80 \text{ cal absorbed}}{\text{g snow}}\right) = 8 \times 10^4 \text{ cal}$$

EXAMPLE 42 *What is the total number of calories your body would have to supply to melt 1000 g of snow and then to raise the temperature of the resulting 1 L of water to 37°C?*

Solution The total number of calories required to convert 1000 g of snow at 0°C to water at 37°C is 1.17×10^5 cal. We've already shown that 8.0×10^4 cal is required to melt the snow. We have also shown (Example 25) that 3.7×10^4 cal is required to raise the temperature of 1 L of water from 0° to 37°C (see Section 3.4.2, "Drink Water to Lose Weight: A Mystery"). The sum of 8.0×10^4 cal and 3.7×10^4 is 1.17×10^5 cal.

Fortunately, there are several simple ways to get your water from snow and conserve survival calories so you don't freeze to death. Some people carry, as part of their car's winter emergency kit, a candle and a metal container, such as an empty coffee can, in which they can melt and warm snow. Another approach is to place snow in a container on a dark background and place this in the sun. The heat radiated from the dark background will melt the snow and warm the resulting water.

PROBLEM SET 3

The problems in Problem Set 3 parallel the examples in Chapter 3. For example, if you should have trouble working Problem 5, go back to Example 5 in this chapter to get help. The correct answers to these problems are given at the end of this book.

1. A 50-g solution made up of just salt and water is 5.0% by weight salt. Use dimensional analysis to find the weight of the salt in this solution.
2. Animals are about 90% by weight water. If a baby pig weighs 20 kg, how many kilograms of water does the pig contain? Use units in your calculations.
3. You find about 300 buttons and are told that 20% of the buttons are brass. How many brass buttons are in the pile? Associate each of the numbers used in your calculations with a unit and make sure that unwanted units cancel.
4. Express the fraction $\frac{2}{5}$ as a percentage using the per centum definition.
5. Express $\frac{3}{10}$ as a percentage using the per centum definition.
6. A barn contains eight chickens, two of which are brown. Calculate the percentage of brown chickens in the barn, using the per centum concept.
7. A piece of aluminum weighs 80.6 g and has a volume of 29.86 cm^3. What is the density of aluminum?
8. A second piece of pure aluminum weighs 400 g and has a volume of 148.148 cm^3. Find the density of this second piece of aluminum.
9. A piece of aluminum weighs 100 lb and has a volume of

0.595 ft^3. Find the density of aluminum in pounds per cubic foot.

10. A sample of gold weighs 2.500 lb and occupies 2.0765×10^{-3} ft^3. Find the density of gold.
11. A gold sample is found to weigh 48.45 lb. What volume does the sample occupy?
12. Find the weight of an aluminum cube that measures 3.000 cm on each side.
13. The density of silver is 10.5 g/cm^3. You have been buying silver cubes from a man whose reputation for honesty is in doubt. Specifically, you would like to know if the cubes are hollow or solid without destroying a cube by sawing it in half. Your cubes measure 20.00 cm on each side and weigh 84.00 kg each. Are the cubes solid silver or do they have hollow centers?
14. Ten grams of water has a volume of 10 cm^3. What is the density of water?
15. The density of a hypothetical metal is 8.45 g/cm^3, and the density of water is 1.00 g/mL. Find the specific gravity of the hypothetical metal.
16. The density of an unknown sample of wood is 0.75 g/cm^3. Find the specific gravity of this unknown wood sample.
17. What is the density in grams per cubic centimeter of an object whose specific gravity is 20?
18. Calculate the density of water in pounds per cubic foot. Remember that 1 g of water has a volume of 1 cm^3, that 454 g = 1 lb, and 2.54 cm = 1 in.
19. The density of a hypothetical solid is 200 lb/ft^3. Find the specific gravity of this hypothetical solid by comparing its density to the density of water (62.4 lb/ft^3 water) using the unit *pounds per cubic foot*.
20. You would like to raise the temperature of 10.0 g of water by 2.00°C. How many calories must you add to the water? How many joules?
21. A beaker of 20 mL of water cools down from 23.0°C to 19.0°C. How many calories of heat does the water lose? How many joules does the water lose?
22. A tank containing 60 kg of water at 90°C cools down to 20°C. How many calories of heat does the tank of water lose? Express your answer in joules.
23. Consider two water samples. The first sample weighs 2 g and is at a temperature of 80°C. The second water sample weighs 500 g and is at a temperature of 10°C. Both water samples are cooled to 5°C. Which beaker of water (the cooler beaker or the warmer beaker) loses more heat? How much heat is lost by both when cooled to 5°C?
24. A chemical reaction takes place underwater and raises the temperature of the surrounding 500.0 g of water by 4.000°C. How much heat is released by this reaction? Express your answer in calories and in joules.
25. How many calories does your body have to release to 2.0 L of ice-cold water (0°C) to raise the water temperature to 37°C?
26. How many calories are required to warm 0.5 L of water by 37°C?
27. The maximum heat of absorption from the sun by the hypothetical cold-blooded animal called the hyposaurus is 3598.9 kcal/hour. Assume the hyposaurus weighs 35 tons and is 85% by weight water. How long will it take this animal to raise its body temperature by 1.0°C?
28. The melting point of frozen oxygen is −361.8°F. Find its melting point in degrees Celsius and in degrees Kelvin.
29. The boiling point of aluminum is 5378°F. Find the boiling point of aluminum in degrees Celsius and in degrees Kelvin.
30. The melting point of aluminum is 1063°C. Find the melting point of aluminum in degrees Fahrenheit.
31. The boiling point of liquid oxygen is −183.0°C. Find the boiling point of liquid oxygen in degrees Fahrenheit.
32. When the outside temperature rises 10°F, what is the corresponding increase in degrees Celsius and in degrees Kelvin?
33. The difference between the boiling and melting points of a hypothetical metal is about 2500°C. Find this temperature difference in degrees Fahrenheit and degrees Celsius.
34. Assume you're going to bathe. You turn on the hot-water faucet, walk away to disrobe, and return to find a tub filled with cold water (50.0°F). The hot-water heater is out of order. Fortunately, you remember reading that 1 g of candy contains 3.730 kcal, and you just bought a 20-lb box of candy. To take a comfortable bath, you feel that the tub water should be at least 90°F. The tub contains 4.00 ft^3 of water. How many pounds of candy must you consume to raise the temperature of the bath water from 50.0°F to 90.0°F?
35. Hell has a minimum temperature of 392.2 K. What is this temperature in degrees Fahrenheit?
36. Heaven has a temperature of 798 K. What is its temperature in degrees Fahrenheit?
37. How many calories does it take to melt a 6.0-g piece of ice whose temperature is 0°C? What is the temperature of the melted water after that number of calories has been added?
38. How many kilojoules does it take to melt a 40.0-g piece of ice whose temperature is 0°C?
39. Gold has a heat of fusion of 3.03 cal/g. The melting point of gold is 1063°C. How many calories of heat would it take to melt 8 g of solid gold if the temperature of the gold were 1063°C?
40. Aluminum has a heat of fusion equal to 3.55 cal/g. The melting point of aluminum is 660°C. How many calories are required to melt a 9-kg piece of aluminum whose temperature is 660°C?
41. If you were to consume 500 g of snow (about two glasses when melted), how many calories would your body have to burn just to melt the snow?
42. What is the total number of calories your body would have to supply to melt 500 g of snow and then to raise the temperature of the resulting 500 mL of water to 37°C?

4 The Mole

4.1 INTRODUCTION

In spite of the relative simplicity of the mole concept, many students find themselves midway through their chemistry courses without a clear idea of what a mole is. The mole is an important concept that is frequently used in areas such as stoichiometry, concentration units, colligative properties, equilibrium, and molecular formulas.

4.2 THE MOLE AS A NUMBER

A mole is a number just as a dozen is a number. A dozen is 12 and a mole is 6.0221×10^{23}. For most chemistry problems, not much error results if you just remember that a mole is about 6×10^{23}, that is, 600 000 000 000 000 000 000 000.

Just as a dozen oranges means a collection of 12 oranges, a mole of hydrogen molecules means a collection of 6×10^{23} molecules of hydrogen. A mole of oxygen molecules (O_2) is 6×10^{23} oxygen molecules. A mole of oxygen atoms (O) is 6×10^{23} oxygen atoms. A mole of sodium ions (Na^+) is 6×10^{23} sodium ions.

A mole is usually used with particles (such as atoms and molecules) that are so small that it is only convenient to work with large numbers of them (about 6×10^{23}). However, there is nothing wrong with talking about a mole of pencils (6×10^{23} pencils) or a mole of chairs (6×10^{23} chairs).

EXAMPLE 1 A box contains 2 moles (2 mol) of sulfur dioxide (SO_3). How many SO_3 molecules are in the box?

Solution The box contains 1×10^{24} SO_3 molecules.

$$(2 \text{ mol } SO_3)\left(\frac{6 \times 10^{23} \text{ } SO_3 \text{ molecules}}{\text{mol } SO_3}\right) = 1 \times 10^{24} \text{ } SO_3 \text{ molecules}$$

EXAMPLE 2 How many oxygen molecules (O_2) are in a box that contains 0.5 mol of oxygen?

Solution The box contains 3×10^{23} molecules of oxygen.

$$(0.5 \text{ mol } O_2)\left(\frac{6 \times 10^{23} \text{ } O_2 \text{ molecules}}{1 \text{ mol } O_2}\right) = 3 \times 10^{23} \text{ } O_2 \text{ molecules}$$

EXAMPLE 3 If a container is filled with 24×10^{24} molecules of hydrogen (H_2), how many moles of hydrogen are in the container?

Solution The container has 40 mol of hydrogen.

$$(24 \times 10^{24} \text{ } H_2 \text{ molecules})\left(\frac{1 \text{ mol } H_2}{6.0 \times 10^{23} \text{ } H_2 \text{ molecules}}\right) = 4.0 \times 10^{1} \text{ mol } H_2$$

4.2.1 AVOGADRO'S NUMBER

Avogadro's number is 6.0221×10^{23}. A mole of particles is said to contain this number (Avogadro's number) of particles. Amedeo Avogadro was an Italian physics professor who never knew the value of the number that was eventually named after him. The value now referred to as Avogadro's number (6.0221×10^{23}) was determined after his death. It was Avogadro, however, who suggested that such a number exists. (We'll see how and why later.)

4.2.2 COMPREHENDING AVOGADRO'S NUMBER*

Scientists and science students use Avogadro's number (6×10^{23}) so frequently that they usually do not appreciate its magnitude.

•
•

EXAMPLE 4 Calculate how long it would take for a computer to count to 6×10^{23}. Assume that the computer can count 1 billion numbers every second.

Solution

$$(6 \times 10^{23})\left(\frac{1\text{ sec}}{1 \times 10^{9}}\right)\left(\frac{1\text{ min}}{60\text{ sec}}\right)\left(\frac{1\text{ hr}}{60\text{ min}}\right)\left(\frac{1\text{ day}}{24\text{ hr}}\right)\left(\frac{1\text{ yr}}{365\text{ days}}\right) = 2 \times 10^{7}\text{ yr}$$

It would take the computer about 20 million years to count to 1×10^{23} at the rate of 1 billion numbers per second.

•
•

EXAMPLE 5 Imagine each of the approximately 6 billion people on earth equally dividing 6×10^{23} pennies among themselves. Each one of these 6 billion people could spend 1 million dollars an hour, 24 hours a day, throughout their lifetime. Calculate the number of years it would take for 6 billion people to spend 6×10^{23} pennies if each person individually spends their pennies at the rate of a million dollars an hour.

Solution The task of spending all of that money could not be accomplished because it would take 171 years, more than double the average person's life expectancy.

$$\frac{(6 \times 10^{23}\text{ pennies})}{(6\text{ billion people})}\left(\frac{1\text{ dollar}}{100\text{ pennies}}\right)\left(\frac{1\text{ hr}}{10^{6}\text{ dollars}}\right)\left(\frac{1\text{ day}}{24\text{ hr}}\right)\left(\frac{1\text{ yr}}{365\text{ days}}\right)$$

$$= \frac{114\text{ yr}}{\text{person}} = 114\text{ yr/person}$$

•
•

Of course it would be difficult to survive surrounded by all these pennies because they would cover the earth with a depth of over fifty miles.

•
•

SELF-TEST

1. Your mother expects a lot of company tonight and tells you to go to the store and buy a mole of eggs. How many eggs do you have to buy?
2. A box contains 10 mol of carbon dioxide (CO_2). How many CO_2 molecules are in the box?
3. A bag contains about 100 000 ozone (O_3) molecules. How many moles of ozone are in the bag?

ANSWERS **1.** 6×10^{23} eggs **2.** $6.0 \times 10^{24}CO_2$ molecules **3.** 2×10^{-19} mol O_3

---

*Adapted from Doris Kolb, "The Mole," *Journal of Chemical Education* 55 (Nov. 1978): 728–32.

4.3 THE MOLE AS A VOLUME

One mole of any gas occupies a volume of approximately 22.4 L under standard conditions (*standard conditions* refers to a temperature of 0°C and 1 atmosphere of pressure). For example, 6×10^{23} molecules of hydrogen occupy about 22.4 L. It was Amedeo Avogadro who first suggested in 1811 that equal volumes of gas contained equal numbers of molecules. Since his death, several methods have been used to determine how many molecules are in one mole of any substance. A few of these methods are shown in this text.

EXAMPLE 6 Under standard conditions (abbreviated STP for "standard temperature and pressure"), what volume would 2.00 mol of helium (He) occupy?

Solution Two moles of helium would occupy 44.8 L at STP.

$$(2.00 \text{ mol He})\left(\frac{22.4 \text{ L He}}{\text{mol He}}\right) = 44.8 \text{ L He}$$

EXAMPLE 7 If 20 L of nitrogen (N_2) are present in a container at STP, how many moles of nitrogen are present in the container?

Solution The container has 0.89 mol of nitrogen gas.

$$(20 \text{ L } N_2)\left(\frac{1 \text{ mol } N_2}{22.4 \text{ L } N_2}\right) = 0.89 \text{ mol } N_2$$

EXAMPLE 8 A tank contains exactly 100 atoms of argon gas (Ar) under standard conditions of temperature and pressure. What is the volume of this container?

Solution The volume of the container must be 3.72×10^{-21} L.

$$(100 \text{ Ar atoms})\left(\frac{1 \text{ mol Ar atoms}}{6.02 \times 10^{23} \text{ Ar atoms}}\right)\left(\frac{22.4 \text{ L Ar}}{1 \text{ mol Ar atoms}}\right) = 3.72 \times 10^{-21} \text{ L Ar}$$

Notice that we used 6.02×10^{23} (not just 6×10^{23}) in the preceding calculation. Do you know why?

The number of significant digits present in Avogadro's number should equal or exceed the number of significant digits in the problem data. In this last example, the factor 22.4 L has three significant digits, and the number 100 has an infinite number of significant digits. If we had used a value of Avogadro's number with less than three significant digits, our answer would not have had a number of significant digits that reflected the number of digits in the given data.

SELF-TEST

1. Under standard conditions, what volume does 0.500 mol of oxygen gas occupy?
2. A sample of hydrogen gas occupies 44.8 L under standard conditions. How many moles of hydrogen gas are present?
3. What volume is occupied by 12×10^{23} molecules of ammonia gas (NH_3) under standard conditions?

ANSWERS **1.** 11.2 L O_2 **2.** 2.00 mol H_2 **3.** 45 L NH_3

4.3.1 THE BASKETBALL AS A UNIT VOLUME, PART I*

It is relatively easy to visualize the volume occupied by 1 L of gas because a liter is about the same as a quart. Visualizing the volume occupied by 22.4 L is not as easy. You may find it useful to know that 22.4 L is exactly the volume occupied by three standard basketballs. Briefly, here are the calculations that show this. A standard basketball has a circumference of 30.0 in.

$$(30.0 \text{ in. circumference basketball})\left(\frac{2.54 \text{ cm}}{\text{in.}}\right) = 76.2 \text{ cm circumference basketball}$$

$$\text{radius of a basketball (sphere)} = \frac{\text{basketball circumference}}{2\,\pi}$$

pi = $\pi \doteq 3.14$, a math constant

$$\text{radius of a basketball (sphere)} = \frac{76.2 \text{ cm}}{(2)(3.14)}$$

$$= 12.13 \text{ cm (basketball radius)}$$

$$\text{volume of basketball (sphere)} = \frac{4}{3}\pi r^3 \leftarrow r = \text{radius}$$

$$= \left(\frac{4}{3}\right)(3.14)(12.13 \text{ cm})^3$$

$$= 7472 \text{ cm}^3 \leftarrow \text{volume of one basketball}$$

$$\left(\frac{7472 \text{ cm}^3}{\text{basketball}}\right)\left(\frac{1 \text{ L}}{1000 \text{ cm}^3}\right) = 7.472 \text{ L per basketball}$$

$$1 \text{ cm}^3 = 1 \text{ mL}$$

$$(3 \text{ basketballs})\left(\frac{7.472 \text{ L}}{\text{basketball}}\right) = 22.4 \text{ L}$$

volume of 3 basketballs and approximate volume of 1 mol of any gas

4.4 ATOMIC WEIGHT

Atoms are very small. It is impossible to place a single atom on any analytical balance and directly determine its mass (weight). If it were possible, you would find that the mass of a single atom is about 1×10^{-23} g, a figure that is somewhat clumsy to write. To help you avoid this, a carbon atom that is made up of 6 protons, 6 neutrons, and 6 electrons has been arbitrarily defined to have an atomic weight of exactly 12 atomic mass units ($12.\bar{0}$ amu). The symbol for such a carbon atom is ^{12}C (or carbon-12). It is read "carbon twelve." From this comes the definition of 1 amu. One atomic mass unit is the amount of mass that is exactly one-twelfth of the mass of a carbon atom with 6 protons, 6 neutrons, and 6 electrons; that is, 1 amu = 1/12 mass of a ^{12}C atom. All elements on the periodic table are given atomic weights in atomic mass units (amu's).

•
•

*Adapted from Fred H. Jardine, "3 Basketballs = 1 Mole of Ideal Gas at STP," *Journal of Chemical Education* 54 (Feb. 1977): 112–13.

EXAMPLE 9 A helium atom (He) has one-third of the mass of a carbon atom. What is the atomic weight of helium?

Solution The atomic weight of helium is 4 amu. One-third of 12 amu is equal to 4 amu. A mass of 4 amu is one-third of the mass of the standard ^{12}C atom.

$$\text{He mass} = \frac{1}{3}(^{12}\text{C mass}) = \frac{1}{3}(12 \text{ amu}) = 4 \text{ amu}$$

EXAMPLE 10 On some stars, carbon atoms fuse together to form atoms of magnesium (Mg). The resulting magnesium atoms have a mass that is double that of carbon. What is the atomic weight of magnesium?

Solution The atomic weight of magnesium is 24 amu.

$$(2)(\text{mass of } ^{12}\text{C}) = (2)(12 \text{ amu}) = 24 \text{ amu}$$

EXAMPLE 11 An atom of hydrogen (H) is about one-twelfth of the mass of a carbon atom. What is the atomic weight of hydrogen?

Solution The atomic weight of hydrogen is 1 amu.

$$(\tfrac{1}{12})(12 \text{ amu}) = 1 \text{ amu}$$

It is more convenient to say that the mass of an atom of hydrogen is 1 amu than to say that the mass of an atom of hydrogen is about 1×10^{-23} g.

Atomic weights of elements can be found from any periodic table. You will notice that all of the elements (including carbon) do not have whole number atomic weights. For example, the atomic weight of carbon is not 12 amu but 12.011 amu. Likewise, the atomic weight of magnesium is not 24 amu but 24.305 amu. The detailed reasons for this are explained in the isotope and binding energy sections (Sections 15.3 and 15.5) in the chapter on nuclear chemistry. For now it is sufficient for you to understand that although we defined the mass of one carbon atom made up of 6 protons, 6 neutrons, and 6 electrons to be exactly 12 amu, not all carbon atoms contain 6 neutrons. Some have more than 6 neutrons, some have less than 6 neutrons. Taken as a group, these various carbon atoms are called isotopes of carbon. All carbon isotopes contain the same number of protons (6 protons), but they contain different numbers of neutrons. The atomic weight shown for each element on a periodic table is the average atomic weight of all of the isotopes for that element based upon the isotope percentage abundance. Again, we will spend more time on this idea when we study nuclear chemistry.

4.5 MOLECULAR WEIGHT AND FORMULA WEIGHT

When two or more atoms chemically bond together, they form molecules. To find the molecular weight of a molecule, simply add the atomic weights of the atoms making up the molecule. Subscripts in chemical formulas tell us the number of atoms of each element making up the molecule.

EXAMPLE 12 Find the molecular weight of water (H_2O).

Solution One water molecule is made up of 2 atoms of hydrogen and 1 atom of oxygen (the 1 is not written as a subscript). Water has a molecular weight of 18 amu.

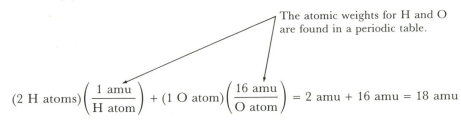

The atomic weights for H and O are found in a periodic table.

$$(2 \text{ H atoms})\left(\frac{1 \text{ amu}}{\text{H atom}}\right) + (1 \text{ O atom})\left(\frac{16 \text{ amu}}{\text{O atom}}\right) = 2 \text{ amu} + 16 \text{ amu} = 18 \text{ amu}$$

EXAMPLE 13 Find the molecular weight of sulfuric acid (H_2SO_4).

Solution The molecular weight of sulfuric acid is 98 amu. One molecule of sulfuric acid is made up of 2 atoms of hydrogen, 1 atom of sulfur, and 4 atoms of oxygen.

$$(2 \text{ H atoms})\left(\frac{1 \text{ amu}}{\text{H atom}}\right) + (1 \text{ S atom})\left(\frac{32 \text{ amu}}{\text{S atom}}\right) + (4 \text{ O atoms})\left(\frac{16 \text{ amu}}{\text{O atom}}\right)$$

$$= 2 \text{ amu} + 32 \text{ amu} + 64 \text{ amu} = 98 \text{ amu}$$

Sodium chloride is made up of positively charged sodium ions (Na^+) and negatively charged chloride ions (Cl^-). A crystal of sodium chloride is held together by the attraction between the opposite electrically charged ions (opposite charges attract). There are no discrete molecular units of NaCl present in a sample of NaCl (sodium chloride does not exist as molecules). We say that sodium chloride is an ionic substance, not a molecular substance, because NaCl is made up of ions, not molecules.

Technically speaking, it is incorrect to refer to the *molecular weight* of an ionic compound, such as NaCl, since the term *molecular weight* implies that the compound is made up of molecules. However, as a beginning student, you may not be able to recognize which substances are made up of molecules (e.g., CO) and which substances are made up of ions (e.g., NaCl). For this reason, the term *formula weight*, which can be used for molecular and ionic substances, can be very useful.

Formula weight refers to the mass of one formula unit (e.g., NaCl or CO) of either an ionic or a molecular compound. To find the formula weight of any substance (whether that substance is made up of molecules or ions), simply add the atomic weights of the atoms or the ions making up the substance just as you did to determine the molecular weight.

For example, the formula weight of NaCl is 58.5 amu. The atomic weights of Na and Cl are 23.0 and 35.5 amu, respectively, so 23.0 amu + 35.5 amu = 58.5 amu. The formula weight of CO is 28 amu. The atomic weights of C and O are 12 and 16 amu, respectively, so 12 amu + 16 amu = 28 amu.

In Example 13, we determined that the molecular weight of sulfuric acid (H_2SO_4) is 98 amu. The formula weight of sulfuric acid is also 98 amu (one formula unit of sulfuric acid is the same as one molecule of sulfuric acid, that is, H_2SO_4).

EXAMPLE 14 Find the formula weight of zinc diphosphate ($Zn_2P_2O_7$).

Solution The formula weight of $Zn_2P_2O_7$ is 305 amu. One formula unit of $Zn_2P_2O_7$ is made up of 2 atoms of Zn, 2 atoms of phosphorus, and 7 atoms of oxygen.

$$(2 \text{ Zn atoms})\left(\frac{65.4 \text{ amu}}{\text{Zn atom}}\right) + (2 \text{ P atoms})\left(\frac{31 \text{ amu}}{\text{P atom}}\right) + (7 \text{ O atoms})\left(\frac{16 \text{ amu}}{\text{O atom}}\right)$$

$$= 130.8 \text{ amu} + 62 \text{ amu} + 112 \text{ amu} = 305 \text{ amu}$$

EXAMPLE 15 Find the formula weight of aluminum sulfate ($Al_2(SO_4)_3$).

Solution

$$(2 \text{ Al atoms})\left(\frac{27 \text{ amu}}{\text{Al atom}}\right) + (3 \text{ S atoms})\left(\frac{32 \text{ amu}}{\text{S atom}}\right) + 12 \text{ O atoms}\left(\frac{16 \text{ amu}}{\text{O atom}}\right)$$

$$= 54 \text{ amu} + 96 \text{ amu} + 192 \text{ amu} = 342 \text{ amu}$$

SELF-TEST

1. An atom of ^{36}Cl has an atomic weight that is about three times that of carbon-12. What is the approximate atomic weight of ^{36}Cl?
2. What is the molecular weight of ozone (O_3)?
3. What is the formula weight of calcium phosphate ($Ca_3(PO_4)_2$)?

ANSWERS **1.** 36 amu **2.** 48 amu **3.** 310 amu

4.6 THE MOLE AS A MASS

We've defined a mole as a number (6.0221×10^{23}). We have also discussed the volume of one mole of gas molecules at STP. One mole of gas molecules (6.0221×10^{23} molecules) will occupy a volume equal to that of about three basketballs (22.4 L) under standard conditions. Obviously, all these molecules must weigh something. One mole of any element weighs its atomic or molecular weight in grams.

EXAMPLE 16 The atomic weight of helium is 4 amu. How much does 1 mol of helium weigh?

Solution The atomic weight of He is 4 amu (found in a periodic table). Therefore, 1 mol of helium has a mass of 4 g. Four grams of helium contains 6×10^{23} helium atoms and occupies a volume of about 22.4 L at STP.

EXAMPLE 17 What is the mass of 2 mol of carbon?

Solution Two moles of carbon has a mass of 24 g. The atomic weight of carbon (from a periodic table) is 12 amu. One mole of carbon has a mass of 12 g; therefore, 2 mol of carbon has a mass of 24 g.

$$(2 \text{ mol C})\left(\frac{12 \text{ g C}}{\text{mol C}}\right) = 24 \text{ g C}$$

EXAMPLE 18 How many grams does 0.50 mol of bromine (Br) weigh?

Solution The atomic weight of bromine is 80 amu. One mole of bromine weighs 80 g.

$$(0.50 \text{ mol Br})\left(\frac{80 \text{ g Br}}{\text{mol Br}}\right) = 40 \text{ g Br}$$

One mole of any molecule has a mass equal to its molecular weight in grams.

EXAMPLE 19 Oxygen (O_2) has a molecular weight (or formula weight) of 32 amu. How many grams does 1 mol of oxygen weigh?

Solution One mole of oxygen weighs 32 g. Thirty-two grams of oxygen also occupies about 22.4 L at STP and contains 6×10^{23} oxygen molecules (or O_2 formula units).

EXAMPLE 20 Find the weight in grams of 1 mol of $Al_2(SO_4)_3$.

Solution We have already determined the formula weight of this compound (see Example 15). Aluminum sulfate has a formula weight of 342 amu. One mole (6×10^{23} formula units) of aluminum sulfate must weigh 342 g.

EXAMPLE 21 How many formula units of aluminum sulfate are present in 0.500 g of aluminum sulfate?

Solution

$$(0.500 \text{ g } Al_2(SO_4)_3)\left(\frac{1 \text{ mol } Al_2(SO_4)_3}{342 \text{ g } Al_2(SO_4)_3}\right)\left(\frac{6.02 \times 10^{23}\ Al_2(SO_4)_3 \text{ formula units}}{\text{mol } Al_2(SO_4)_3}\right)$$

$$= 8.77 \times 10^{20} \text{ formula units } Al_2(SO_4)_3$$

4.6.1 HIDDEN SUGAR

EXAMPLE 22 Table sugar (cane sugar, $C_{12}H_{22}O_{11}$) has been linked to various diseases including diabetes, obesity, dental caries (tooth decay), and vitamin deficiency. People trying to restrict their daily intake of sugar for health reasons find that it is difficult because most of their average daily consumption ($\frac{1}{3}$ lb per person per day) is consumed as a food additive in processed foods. About 70% of the average daily intake of sugar is consumed this way. How many moles of sugar do you consume in processed foods each day?

Solution First, determine the molecular weight of sugar.

$$(12 \text{ C atoms})\left(\frac{12 \text{ amu}}{\text{C atom}}\right) + (22 \text{ H atoms})\left(\frac{1 \text{ amu}}{\text{H atom}}\right) + (11 \text{ O atoms})\left(\frac{16 \text{ amu}}{\text{O atom}}\right)$$

$$= 342 \text{ amu}$$

$$(0.3 \text{ lb sugar eaten daily})\left(\frac{70 \text{ lb sugar in processed foods}}{100 \text{ lb sugar eaten daily}}\right)\left(\frac{454 \text{ g}}{\text{lb}}\right)\left(\frac{1 \text{ mol sugar}}{342 \text{ g sugar}}\right)$$

$$= 0.3 \text{ mol sugar in processed foods}$$

EXAMPLE 23 Many sources of processed food sugar are easy to identify, for example, sweetened cereals (some of which are 50% sugar) and candy bars. (A candy bar has as much sugar as 1 to 3 lb of apples and is a lot easier to eat.) Unfortunately, there are many foods that we do not think of as being sweetened that are actually high in sugar content. Ketchup, for example, is 30% sugar. Find the number of molecules of sugar in a 1-oz serving of ketchup.

Solution

$$(1 \text{ oz ketchup})\left(\frac{30 \text{ oz ketchup}}{100 \text{ oz ketchup}}\right)\left(\frac{28 \text{ g}}{\text{oz}}\right)\left(\frac{1 \text{ mol sugar}}{342 \text{ g sugar}}\right)$$

$$\left(\frac{6.02 \times 10^{23} \text{ sugar molecules}}{\text{mol sugar}}\right) = 1 \times 10^{22} \text{ sugar molecules}$$

4.6.2 INTERCONVERSIONS

One mole of anything (e.g., atoms, ions, or molecules) contains 6×10^{23} particles. One mole of any gas occupies about 22.4 L at STP. (Amedeo Avogadro suggested that equal volumes of gas contain equal numbers of molecules.) However, the masses of equal volumes of gas are not necessarily the same.

As you can see in Figure 4.1, 1 mol of hydrogen contains the same number of molecules as 1 mol of oxygen. One mole of hydrogen occupies about the same volume as 1 mol of oxygen. However, 1 mol of oxygen does not weigh the same as 1 mol of hydrogen. One mole of oxygen weighs 32 g while 1 mol of hydrogen weighs only 2 g.

FIGURE 4.1 Avogadro's number. Box A contains 6×10^{23} H_2 molecules, has a volume of 22.4 L, and weighs 2 g. Box B contains 6×10^{23} O_2 molecules, has a volume of 22.4 L, and weighs 32 g.

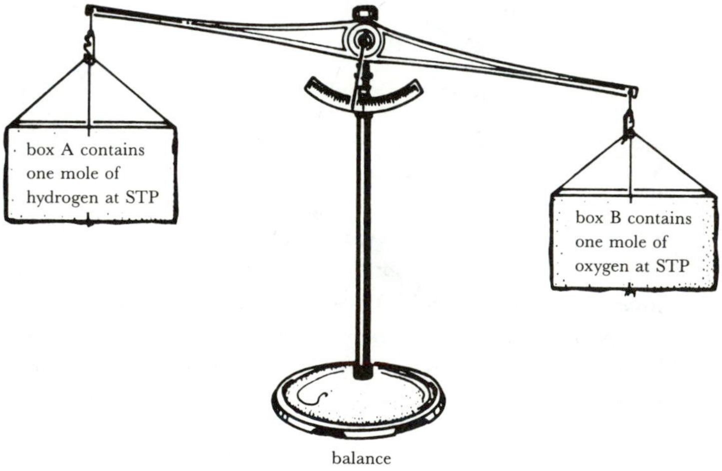
box A contains
one mole of
hydrogen at STP
box B contains
one mole of
oxygen at STP
balance

You should be able to interconvert any one of these units (grams, moles, liters, and molecules). For example, you should be able to convert grams to moles and liters to molecules. The following examples illustrate several of these conversions.

EXAMPLE 24 How many moles of oxygen molecules (O_2) are present in 64 g of oxygen?

Solution Sixty-four grams of oxygen contains 2.0 mol of oxygen. The atomic weight of oxygen (O) is 16 amu. The molecular weight of oxygen (O_2) is 32 amu. One mole of oxygen gas weighs 32 g.

$$(64 \text{ g } O_2)\left(\frac{1 \text{ mol } O_2}{32 \text{ g } O_2}\right) = 2.0 \text{ mol } O_2$$

EXAMPLE 25 Every time you take a deep breath you breathe in about 0.01 mol of oxygen (O_2). What is the weight of the oxygen you inhale?

Solution With every breath you inhale about 0.3 g of oxygen.

$$(0.01 \text{ mol } O_2)\left(\frac{32 \text{ g } O_2}{1 \text{ mol } O_2}\right) = 0.3 \text{ g } O_2$$

•
•

EXAMPLE 26 With every breath of fresh air you inhale about 2.4×10^{22} nitrogen molecules (N_2). What is the mass of the nitrogen you inhale with each breath?

Solution You inhale about 1.1 g of nitrogen with each breath. The molecular weight of nitrogen is 28 amu. One mole of nitrogen has a mass of 28 g.

$$(2.4 \times 10^{22} \text{ } N_2 \text{ molecules})\left(\frac{1 \text{ mol } N_2}{6.0 \times 10^{23} \text{ molecules}}\right)\left(\frac{28 \text{ g } N_2}{\text{mol } N_2}\right) = 1.1 \text{ g } N_2$$

•
•

EXAMPLE 27 What is the volume of 34 g of ammonia (NH_3) at STP?

Solution The volume of 34 g of NH_3 at STP is 45 L. The molecular weight of ammonia is 17 amu. One mole of ammonia weighs 17 g.

$$(34 \text{ g } NH_3)\left(\frac{1 \text{ mol } NH_3}{17 \text{ g } NH_3}\right)\left(\frac{22.4 \text{ L } NH_3}{\text{mol } NH_3}\right) = 45 \text{ L } NH_3$$

•
•

EXAMPLE 28 How many molecules of ammonia are present in a 51-g sample of ammonia?

Solution A 51-g sample of ammonia contains 1.8×10^{24} ammonia molecules.

$$(51 \text{ g } NH_3)\left(\frac{1 \text{ mol } NH_3}{17 \text{ g } NH_3}\right)\left(\frac{6.0 \times 10^{23} \text{ } NH_3 \text{ molecules}}{\text{mol } NH_3}\right) = 1.8 \times 10^{24} \text{ } NH_3 \text{ molecules}$$

•
•

EXAMPLE 29 What is the approximate mass (in grams) of one NH_3 molecule?

Solution One molecule of NH_3 has a mass of approximately 2.8×10^{-23} g.

$$(1 \text{ } NH_3 \text{ molecule})\left(\frac{1 \text{ mol } NH_3}{6.0 \times 10^{23} \text{ } NH_3 \text{ molecules}}\right)\left(\frac{17 \text{ g } NH_3}{\text{mol } NH_3}\right) = 2.8 \times 10^{-23} \text{ g } NH_3$$

•
•

EXAMPLE 30 What is the approximate volume (in liters) of 1 molecule of NH_3?

Solution The approximate volume of 1 ammonia molecule is 3.8×10^{-23} L.

$$(1 \text{ } NH_3 \text{ molecule})\left(\frac{1 \text{ mol } NH_3}{6.0 \times 10^{23} \text{ } NH_3 \text{ molecules}}\right)\left(\frac{22.4 \text{ L } NH_3}{\text{mol } NH_3}\right) = 3.8 \times 10^{-23} \text{ L } NH_3$$

•
•

EXAMPLE 31 Find the mass of 10 L of carbon dioxide (CO_2) at STP.

Solution Ten liters of carbon dioxide weighs 20 g. The molecular weight of carbon dioxide is 44 amu. One mole of carbon dioxide weighs 44 g.

$$(10 \text{ L } CO_2)\left(\frac{1 \text{ mol } CO_2}{22.4 \text{ L } CO_2}\right)\left(\frac{44 \text{ g } CO_2}{\text{mol } CO_2}\right) = 20 \text{ g } CO_2$$

EXAMPLE 32 How many molecules of sulfur dioxide (SO_2) are in 40 L of sulfur dioxide at STP?

Solution There are 1.1×10^{24} sulfur dioxide molecules in a 40-L sample of sulfur dioxide at STP.

$$(40 \text{ L } SO_2)\left(\frac{1 \text{ mol } SO_2}{22.4 \text{ L } SO_4}\right)\left(\frac{6.0 \times 10^{23} \text{ } SO_2 \text{ molecules}}{\text{mol } SO_2}\right) = 1.1 \times 10^{24} \text{ } SO_2 \text{ molecules}$$

SELF-TEST

1. How many grams of oxygen molecules are in 2 mol of oxygen (O_2)?
2. How many moles of oxygen (O_2) are in 0.3 g of O_2?
3. How many molecules of nitrogen (N_2) are in 1.1 g of nitrogen?
4. What is the mass (in grams) of 45 L of ammonia (NH_3) under standard conditions of temperature and pressure?
5. How many molecules of ammonia are in a 2.8×10^{-23}-g sample of ammonia?

ANSWERS **1.** 64 g O_2 **2.** 0.009 mol O_2 **3.** 2.4×10^{22} N_2 molecules **4.** 34 g NH_3 **5.** 1 NH_3 molecule

4.7 CHEMICAL FORMULAS

Whenever you see a chemical formula such as CO_2, you probably consider it in terms of atoms and molecules. That is, you probably say to yourself, "One molecule of CO_2 is made up of one atom of carbon and two atoms of oxygen." It is perfectly correct to think about the subscripts of a chemical formula as referring to the number of atoms making up the particular molecule. However, as we will see, additional information and insight may be gained by also thinking of the subscripts as if they represent the number of moles of each element required to make up one mole of the particular molecule.

EXAMPLE 33 From the formula CO_2, indicate how you would prepare 1 mol of carbon dioxide.

Solution If you chemically combine 1 mol of carbon with 2 mol of oxygen, you will form 1 mol of carbon dioxide.

EXAMPLE 34 Consider the formula for benzene: C_6H_6. How many moles of hydrogen and how many moles of carbon are required to make 1 mol of benzene?

Solution Six moles of hydrogen, when chemically combined with 6 mol of carbon, will produce 1 mol of benzene.

EXAMPLE 35 How many moles of oxygen (O) are needed to react with 2 mol of sulfur (S) to make 2 mol of sulfur dioxide (SO_2)?

Solution The formula (SO_2) tells us that 2 mol of oxygen reacts chemically with 1 mol of sulfur to produce 1 mol of sulfur dioxide. If we have 2 mol of sulfur, then

$$(2 \text{ mol S})\left(\frac{2 \text{ mol O}}{\text{mol S}}\right) = 4 \text{ mol O}$$

Four moles of oxygen (O) are needed to react with 2 mol of sulfur (S) to produce 2 mol of SO_2.

EXAMPLE 36 How many moles of chlorine (Cl) are needed to react with 0.33 mol of aluminum (Al) to make 0.33 mol of $AlCl_3$?

Solution The formula of aluminum chloride ($AlCl_3$) tells us that 1 mol of aluminum reacts chemically with 3 mol of chlorine. If we only have 0.33 mol aluminum, then

$$(0.33 \text{ mol Al})\left(\frac{3 \text{ mol Cl}}{1 \text{ mol Al}}\right) = 1.0 \text{ mol Cl}$$

We need 1 mol of chlorine.

EXAMPLE 37 How many moles of sodium (Na) are needed to make 4 mol of sodium carbonate (Na_2CO_3)?

Solution Eight moles of sodium are needed to make 4 mol of sodium carbonate. The formula of sodium carbonate tells us that 2 mol of sodium are required to make 1 mol of sodium carbonate.

$$\underbrace{(4 \text{ mol } Na_2CO_3)}_{\text{given}}\left(\frac{2 \text{ mol Na}}{1 \text{ mol } Na_2CO_3}\right) = 8 \text{ mol Na}$$

(given ↑ ; ↑ This conversion factor is obtained from interpreting the formula for sodium carbonate.)

SELF-TEST

1. How many atoms of Ca, P, and O (calcium, phosphorus, and oxygen) are in 2 formula units of $Ca_3(PO_4)_2$ (calcium phosphate)?
2. How many moles of Ca, P, and O are required to make 1 mol of calcium phosphate?
3. How many moles of sodium (Na) are needed to make 0.5 mol of sodium carbonate (Na_2CO_3)?

ANSWERS **1.** 6 Ca atoms; 4 P atoms; 16 O atoms **2.** 3 mol Ca; 2 mol P; 8 mol O **3.** 1 mol Na

4.7.1 EMPIRICAL FORMULAS

We will study two types of chemical formulas in the remaining parts of this chapter: empirical formulas and molecular formulas. We have been using molecular formulas so far in everything we've shown. A molecular formula tells the actual number of atoms of each element that are in a compound (molecule).

EXAMPLE 38 The molecular formula for benzene is C_6H_6. How many carbon atoms and hydrogen atoms are in 1 benzene molecule?

Solution The molecular formula of benzene tells us that in 1 molecule of benzene there are actually 6 atoms of carbon (C) and 6 atoms of hydrogen (H).

An empirical formula tells us the simplest whole number ratio (the relative number) of the atoms present in a compound.

•
•

EXAMPLE 39 Find the empirical formula of benzene. The molecular formula for benzene is C_6H_6.

Solution The molecular formula of benzene tells us that for every 6 carbon atoms there are 6 hydrogen atoms. The simplest whole number ratio (relative number) of carbon atoms to hydrogen atoms (C to H, or C:H) is one to one (1:1). This means that for every 1 carbon atom there is 1 hydrogen atom. The empirical formula for benzene is therefore C_1H_1 or simply CH. The subscript 1 usually is not written in molecular and empirical formulas.

•
•

EXAMPLE 40 The molecular formula for butane (used in butane cigarette lighters) is C_4H_{10}. How many carbon atoms and hydrogen atoms are in 1 butane molecule?

Solution In a butane molecule there are actually 4 carbon atoms and 10 hydrogen atoms.

•
•

EXAMPLE 41 Find the empirical formula of butane.

Solution The relative number (simplest whole number ratio) of carbon atoms to hydrogen atoms in butane is 2 carbons for every 5 hydrogens, that is, C:H = 4:10 = 2:5.

•
•

It is possible for the empirical formula and the molecular formula to be identical. For example, 1 molecule of carbon monoxide (CO) contains just 1 carbon atom and 1 oxygen atom. The molecular formula of carbon monoxide is CO. The simplest whole number ratio of carbon atoms to oxygen atoms is 1:1 (one to one). Therefore, the empirical formula of carbon monoxide is also CO.

•
•

EXAMPLE 42 One molecule of methane (natural gas) contains 1 atom of carbon and 4 atoms of hydrogen. What is the molecular formula for methane?

Solution The molecular formula for methane is CH_4.

•
•

EXAMPLE 43 What is the empirical formula for methane?

Solution The simplest whole number ratio (relative number) of carbon atoms to hydrogen atoms is 1 to 4 (C:H = 1:4). The empirical formula for methane is also CH_4.

•
•

Now let's see how empirical formulas are determined from a percentage composition analysis of unknown compounds.

•
•

EXAMPLE 44 An unknown compound was found to contain only the elements sulfur and oxygen. Forty percent of the weight of the compound is attributed to the sulfur content. The remaining 60.0% of the compound's mass is due to the oxygen present. From this information, calculate the empirical formula of this unknown compound containing sulfur and oxygen.

Solution Here is our general attack plan: First determine the number of moles of sulfur and the number of moles of oxygen in the unknown compound sample. Once we have these two figures, we simply insert them as subscripts into the chemical formula

$$S_{(\ )}O_{(\ )}$$

because subscripts can be interpreted as the number of moles of each atom that are required to make one mole of the compound. For example, if we find that the compound contains 1 mol of sulfur and 2 mol of oxygen, we would have the empirical formula

$$S_{(\ )}O_{(\ )} = S_{(1)}O_{(2)} = SO_2 \longleftarrow \text{empirical formula}$$

This empirical formula shows that the simplest ratio of sulfur to oxygen is 1:2, that is, there is 1 atom of sulfur for every 2 atoms of oxygen.

As a general rule, the first step in solving problems such as this is to make the assumption that you have 100 g of the unknown substance. Actually, any weight could be assumed, but the mathematics is easier when you assume that you have 100 g. If our unknown sample weighs 100 g, it contains 40.0 g of sulfur.

$$(100 \text{ g sample})\left(\frac{40.0 \text{ g S}}{100 \text{ g sample}}\right) = 40.0 \text{ g S}$$

Also, the unknown sample contains 60.0 g of oxygen.

$$(100 \text{ g sample})\left(\frac{60.0 \text{ g O}}{100 \text{ g sample}}\right) = 60.0 \text{ g O}$$

Next, convert the number of grams of sulfur and oxygen in the unknown sample to moles of sulfur and oxygen. There is 1.25 mol of sulfur in the 100-g sample.

$$(40.0 \text{ g S})\left(\frac{1.00 \text{ mol S}}{32.0 \text{ g S}}\right) = 1.25 \text{ mol S}$$

There is 3.75 mol of oxygen in the 100-g sample of unknown compound.

$$(60.0 \text{ g O})\left(\frac{100 \text{ mol O}}{16.0 \text{ g O}}\right) = 3.75 \text{ mol O}$$

Insert the number of moles of sulfur and oxygen determined in the preceding calculations as subscripts into the blanks of the formula $S_{(\ )}O_{(\ )}$, giving $S_{1.25}O_{3.75}$.

This is not the empirical formula, however. Why?

The formula $S_{1.25}O_{3.75}$ is not an empirical formula because it does not show the simplest *whole number* ratio of sulfur atoms to oxygen atoms. Somehow we have to convert the subscripts 1.25 and 3.75 into whole numbers.

Perhaps you can see that 3.75 is exactly 3 × 1.25; that is, for every 1 sulfur atom there are 3 oxygen atoms. From this we can conclude that the empirical formula is SO_3.

•
•

Performing this last step (going from nonwhole numbers, such as 1.25 and 3.75, to whole numbers, such as 1 and 3) is not always this easy and it may be helpful to introduce a general technique to do this final step. Find the smallest subscript (other than 1) and divide it into all the subscripts present, including itself. For this particular formula ($S_{1.25}O_{3.75}$), the smaller subscript is 1.25. Dividing all subscripts by the smaller subscript gives us

$$S_{\frac{1.25}{1.25}}O_{\frac{3.75}{1.25}} = S_1O_3, \text{ or simply } SO_3$$

There are times when this general approach doesn't yield the final answer. However, this approach is still useful as the first step for determining empirical formulas. We'll encounter one of these situations in another example.

We determined the empirical formula of our unknown compound to be SO_3. This formula may or may not be the molecular formula. We'll see how to determine the molecular formula from an empirical formula later. For now, because we don't know if SO_3 is the actual (molecular) formula of our unknown compound, we cannot refer to the sum of the atomic weights of all of the atoms present as the molecular weight of this compound. We will call the sum of the atomic weights of all of the atoms present in an empirical formula the formula weight. The formula weight of SO_3 is found the same way that the molecular weight is found.

$$(1 \text{ S atom})\left(\frac{32 \text{ amu}}{\text{S atom}}\right) + (3 \text{ O atoms})\left(\frac{16 \text{ amu}}{\text{O atom}}\right) = 80 \text{ amu}$$

EXAMPLE 45 An unknown compound is found to be 27% by weight carbon and 73% by weight oxygen. Find the empirical formula for this unknown compound.

Solution Assume that you have 100 g of compound. Determine the number of moles of carbon and oxygen in the 100-g sample. A 100-g sample that is 27% carbon contains 27 g of carbon.

$$(27 \text{ g C})\left(\frac{1 \text{ mol C}}{12 \text{ g C}}\right) = 2.3 \text{ mol C}$$

A 100-g sample that is 73% oxygen contains 73 g of oxygen.

$$(73 \text{ g O})\left(\frac{1 \text{ mol O}}{16 \text{ g O}}\right) = 4.6 \text{ mol O}$$

Knowing the number of moles of C and O allows us to determine the empirical formula.

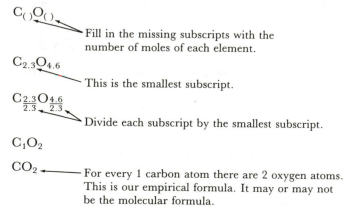

$C_{(\,)}O_{(\,)}$ ← Fill in the missing subscripts with the number of moles of each element.

$C_{2.3}O_{4.6}$ ← This is the smallest subscript.

$C_{\frac{2.3}{2.3}}O_{\frac{4.6}{2.3}}$ ← Divide each subscript by the smallest subscript.

C_1O_2

CO_2 ← For every 1 carbon atom there are 2 oxygen atoms. This is our empirical formula. It may or may not be the molecular formula.

SELF-TEST

1. Every molecule of octane contains 8 carbon atoms and 18 hydrogen atoms. What is the molecular formula for octane?
2. What is the empirical formula for octane?
3. An unknown compound is found to be 50% S and 50% O. What is the empirical formula of this compound?

ANSWERS **1.** C_8H_{18} **2.** C_4H_9 **3.** SO_2

4.7.2 MOLECULAR FORMULAS

The empirical formula only tells us the simplest whole number ratio of atoms in a molecule. The molecular formula tells us the exact number of atoms of each element in a molecule. Based upon our empirical formula of CO_2 (determined in Example 45), the molecular formula may be CO_2, C_2O_4, C_3O_6, C_4O_8, or a similar formula, all of which contain two oxygen atoms for every carbon atom. How do we know which of these possible formulas is the actual (molecular) formula? We make our decision by considering the molecular weight of the unknown compound. (In Section 8.3 we will see how molecular weights are determined.)

EXAMPLE 46 The molecular weight of the compound whose empirical formula is CO_2 is 44 amu. Determine which of the possibilities (CO_2, C_2O_4, etc.) is the molecular formula.

Solution Of the various possibilities, only CO_2 can be the molecular formula because its formula weight is the same as the given molecular weight of the unknown compound (44 amu).

$$(1\text{ C atom})\left(\frac{12\text{ amu}}{\text{C atom}}\right) + (2\text{ O atoms})\left(\frac{16\text{ amu}}{\text{O atom}}\right) = 12\text{ amu} + 32\text{ amu} = 44\text{ amu}$$

It is just a coincidence that the empirical formula and the molecular formula are the same for CO_2. This is not true in the next example.

EXAMPLE 47 An unknown compound is found to be 92.3% carbon and 7.70% hydrogen. The molecular weight of this compound is 78 amu. Find the empirical and molecular formulas of the unknown compound.

Solution The empirical formula of our unknown compound is CH. Its molecular formula is C_6H_6. Here are the calculations: A 100-g sample that is 92.3% carbon contains 92.3 g of carbon.

$$(92.3\text{ g C})\left(\frac{1.0\text{ mol C}}{12\text{ g C}}\right) = 7.7\text{ mol C}$$

A 100-g sample that is 7.70% hydrogen contains 7.70 g of H.

$$(7.70\text{ g H})\left(\frac{1.0\text{ mol H}}{1.0\text{ g H}}\right) = 7.7\text{ mol H}$$

Determine the empirical formula next.

$$C_{(\ )}H_{(\ )}$$

$$C_{7.7}H_{7.7}$$

$$C_{\frac{7.7}{7.7}}H_{\frac{7.7}{7.7}}$$

$$C_1H_1$$

$$CH$$

We now have to determine the molecular formula of the compound CH. The empirical formula only tells us the simplest ratio of atoms making up the molecule: for every 1 carbon atom there is 1 hydrogen atom. Based upon our empirical formula CH, the molecular formula could be CH, C_2H_2, C_3H_3, C_4H_4, or a similar formula. How do we know which of the various possibilities is the correct (actual, molecular) formula?

Make your decision by considering the molecular weight that was given for the unknown compound (78 amu). Of the various possibilities, only C_6H_6 has a molecular weight of 78 amu.

$$(6 \text{ C atoms})\left(\frac{12 \text{ amu}}{\text{C atom}}\right) + (6 \text{ H atoms})\left(\frac{1 \text{ amu}}{\text{H atom}}\right) = 72 \text{ amu} + 6 \text{ amu} = 78 \text{ amu}$$

There is a shortcut method for determining the molecular formula that usually works: Divide the formula weight of the empirical formula (13 amu) into the molecular weight of the compound (78 amu). The result of this division (78 amu ÷ 13 amu = 6) is multiplied by every subscript of the empirical formula to obtain the molecular formula.

$$C_{6\times1}H_{6\times1} = C_6H_6$$

•
•

EXAMPLE 48 An unknown compound with a molecular weight of 30 amu is found to be made of 40 g of carbon and 10 g of hydrogen. Find the molecular formula of this compound.

Solution

$$(40 \text{ g C})\left(\frac{1.0 \text{ mol C}}{12 \text{ g C}}\right) = 3.3 \text{ mol C}$$

$$(10 \text{ g H})\left(\frac{1.0 \text{ mol H}}{1.0 \text{ g H}}\right) = 10 \text{ mol H}$$

$C_{(\ )}H_{(\ )}$

$C_{3.3}H_{10}$

$C_{\frac{3.3}{3.3}}H_{\frac{10}{3.3}}$

C_1H_3

CH_3 ← The formula weight of CH_3 is 15 amu.

$$\frac{\text{molecular weight of compound}}{\text{formula weight of compound}} = \frac{30 \text{ amu}}{15 \text{ amu}} = 2$$

Multiply each subscript in the empirical formula by 2.

$$C_{2\times1}H_{2\times3} = C_2H_6$$

The molecular formula is C_2H_6.

•
•

EXAMPLE 49 An unknown compound is found to be composed of 53 g of aluminum (Al) and 47 g of oxygen (O). Its formula weight is 102 amu. Find the formula of one formula unit of this unknown compound.

Solution

$$(53 \text{ g Al})\left(\frac{1 \text{ mol Al}}{27 \text{ g Al}}\right) = 2.0 \text{ mol Al}$$

$$(47 \text{ g O})\left(\frac{1.0 \text{ mol O}}{16 \text{ g O}}\right) = 2.9 \text{ mol O}$$

$Al_{(\ )}O_{(\ )}$

$Al_{2.0}O_{2.9}$

$$Al_{\frac{2.0}{2.0}}O_{\frac{2.9}{2.0}}$$

$Al_1O_{1.5}$ ← Convert all subscripts to whole numbers by multiplying each subscript by 2.

Al_2O_3 ← The formula weight of this empirical formula is 102 amu.

The given formula weight (102 amu) is the same as the formula weight of the empirical formula. Therefore, the empirical formula must also be the formula of one formula unit.

•
•

SELF-TEST

1. An unknown compound is found to contain 20.0 g of sulfur and 30.0 g of oxygen. An experimental error produces an erroneous molecular weight of 160 amu. What is the empirical formula of the unknown compound?

2. What is the formula weight of the unknown compound in Problem 1 based upon its empirical formula?

3. What is the molecular formula of the unknown compound in Problem 1 based upon the incorrectly determined molecular weight?

ANSWERS **1.** SO_3 **2.** 80 amu **3.** S_2O_6

PROBLEM SET 4

The problems in Problem Set 4 parallel the examples in Chapter 4. For example, if you should have trouble working Problem 5, go back to Example 5 in this chapter to get help. The correct answers to these problems are given at the end of this book.

1. A box contains 4 mol of sulfur trioxide. How many sulfur trioxide molecules are in the box?
2. How many oxygen molecules are in a box that contains 0.8 mol of oxygen?
3. If a container is filled with 30×10^{24} molecules of hydrogen, how many moles of hydrogen are in the container?
4. Calculate how long it would take for a person to count to 6×10^{23}. Assume that a person can count ten numbers every second.
5. Imagine each of the approximately 6 billion people on earth equally dividing 6×10^{23} dollars among themselves. Each one of these 6 billion people could spend 1 million dollars an hour, 24 hours a day, throughout their lifetime. Calculate the number of years it would take for 6 billion people to spend 6×10^{23} dollars if each person individually spends dollars at the rate of 1 million dollars an hour.
6. What volume does 3 mol of helium occupy at STP?
7. If 10 L of nitrogen is in a container at STP, how many moles of nitrogen are in the container?
8. A tank contains exactly 50 atoms of argon gas at STP. What is the volume of this container?
9. A hypothetical atom has 4 times the mass of a carbon atom. What is the atomic weight of this hypothetical atom?
10. If 3 carbon atoms fuse together, what is the atomic weight of the resulting atom?
11. An atom of deuterium has a mass that is about one-sixth of the mass of a carbon atom. What is the atomic weight of deuterium?
12. Find the molecular weight of hydrogen peroxide (H_2O_2).
13. Find the molecular weight of sulfurous acid (H_2SO_3).
14. Find the formula weight of NaCl.
15. Find the formula weight of aluminum sulfite [$Al_2(SO_3)_3$].
16. The atomic weight of deuterium is 2 amu. How much does 1 mol of deuterium atoms weigh?
17. What is the mass of 3 mol of carbon?
18. How many grams of bromine does $\frac{1}{4}$ mol of bromine weigh?
19. Oxygen has a molecular weight of 32 amu. How many grams does 2 mol of oxygen weigh?
20. Find the weight in grams of 1 mol of $Al_2(SO_3)_3$.
21. How many formula units of aluminum sulfite [$Al_2(SO_3)_3$] are present in 0.500 g of aluminum sulfite?
22. A 2-oz candy bar is about 50% sugar ($C_{12}H_{22}O_{11}$). How many moles of sugar are in each 2-oz candy bar?
23. How many sugar molecules are in a 2-oz candy bar if a candy bar is about 50% sugar ($C_{12}H_{22}O_{11}$)?
24. How many moles of oxygen molecules are present in 100 g of oxygen?

25. What is the mass of 0.1 mol of oxygen?
26. What is the mass of 3×10^{21} molecules of nitrogen?
27. What volume would 50 g of ammonia occupy at STP?
28. How many molecules of ammonia are present in a 27-g sample of ammonia?
29. What is the approximate mass in grams of 3 ammonia molecules?
30. What is the approximate volume in liters of 3 molecules of ammonia?
31. Find the mass of 12 L of carbon dioxide at STP.
32. How many molecules of sulfur dioxide are present in 35 L of sulfur dioxide at STP?
33. Consider the formula of sulfur trioxide (SO_3) and indicate how you would prepare 1 mol of the compound.
34. Consider the formula for ethene (C_2H_4). How many moles of carbon and how many moles of atomic hydrogen are required to make 1 mol of ethene?
35. How many moles of oxygen (O) are needed to react with 3 mol of sulfur to make 3 mol of sulfur dioxide?
36. How many moles of chlorine (Cl) are needed to react with 4 mol of aluminum to make 4 mol of $AlCl_3$?
37. How many moles of sodium are needed to make 0.60 mol of sodium carbonate (Na_2CO_3)?
38. The molecular formula for ethane is C_2H_6. How many carbon atoms and hydrogen atoms are present in 1 ethane molecule?
39. Find the empirical formula of ethane. The molecular formula for ethane is C_2H_6.
40. The molecular formula for decane is $C_{10}H_{22}$. How many carbon atoms and hydrogen atoms are present in 1 decane molecule?
41. Find the empirical formula for decane ($C_{10}H_{22}$).
42. One molecule of propane contains 3 atoms of carbon and 8 atoms of hydrogen. What is the molecular formula for propane?
43. What is the empirical formula for propane based upon the information given in Problem 42?
44. An unknown compound was found to contain only the elements carbon and hydrogen. Eighty percent of the weight of the compound is attributed to the carbon present. The remaining 20.0% of the compound's mass is due to the hydrogen present. From this information, calculate the empirical formula of this unknown compound containing carbon and hydrogen.
45. An unknown compound is found to be 81.8% by weight carbon and 18.2% by weight hydrogen. Find the empirical formula for this unknown compound.
46. The molecular weight of a compound whose empirical formula is CO is 28 amu. Determine the molecular formula for this compound.
47. An unknown compound was found to be 80.0% carbon and 20.0% hydrogen. The molecular weight of this compound is 30 amu. Find the empirical and molecular formulas for this compound.
48. An unknown compound with a molecular weight of 80 amu is found to be made of 20 g of sulfur and 30 g of oxygen. Find the molecular formula for this compound.
49. An unknown compound is found to be composed of 106 g of aluminum and 94 g of oxygen. Its formula weight is 102 amu. Find the formula of one formula unit of this unknown compound.

5 Gas Laws

5.1 INTRODUCTION

Many students take courses (even chemistry courses) without knowing the primary purpose of the course. Students are continually confronted with a series of rules, laws, formulas, equations, and other details that can appear to be totally unrelated. What connection is there between significant digits, dimensional analysis, the metric system, calories, and the mole? And what connection is there between these things and the gas laws, which are a series of laws and dimensional analysis manipulations that predict the behavior and properties of gases under different conditions?

On the surface it is easy to suppose these topics are merely independent concepts that interest the chemist, while in reality there does exist a primary motive, a thread woven throughout these and other topics that ties these apparently independent concepts together. One such basic motive (thread) is the chemical equation. Most topics in chemistry ultimately enhance your ability to understand, interpret, and communicate by means of a balanced chemical equation. If you haven't been exposed to chemical equations yet, here is a sneak preview.

We have already mentioned that you can tell how carbon dioxide (CO_2) is made just by looking at its molecular formula. Preparing 1 mol of carbon dioxide (44 g of carbon dioxide) requires the chemical combination of 1 mol of carbon atoms (C) and 2 mol of oxygen atoms (O). This is written

This is a chemical equation.

$$\overbrace{C + O_2 \rightarrow CO_2 + 94 \text{ kcal}}$$

↑ 1 mol oxygen molecules (O_2) = 2 mol oxygen atoms (O)

and read, "One mole of carbon plus one mole of oxygen molecules (O_2) chemically combine to form one mole of carbon dioxide and 94 kcal of heat." Note that 2 mol of oxygen molecules (O_2) is made by chemically combining 2 mol of oxygen atoms (O). We know this by observing the molecular formula of oxygen (O_2).

To be able to discuss this chemical equation intelligently you need a considerable amount of background knowledge. You must know (1) nomenclature (the method for naming the elements and the compounds symbolically represented in the chemical equation); (2) the mole concept (1 mol of carbon + 1 mol of O_2 produces 1 mol of CO_2); (3) significant digits (1 mol of carbon = 1×10^1 g carbon whereas 1.0 mol of carbon = 12 g carbon); (4) math skills, such as exponential notation (1 mol of carbon contains 6×10^{23} carbon atoms); (5) metric units (1.00 mol of carbon monoxide occupies 22.4 L under standard conditions); and (6) calories (1.0 mol of carbon combines with 1.0 mol of oxygen, producing 94 kcal of heat). In later chapters we will indicate how the new material is related to interpreting the chemical equation, a universal shorthand used by scientists to communicate chemical information.

Now then, how are the gas laws related to chemistry and, in particular, to the chemical equation?

5.1.1 WHY GAS LAWS ARE IMPORTANT IN CHEMISTRY

Let's look again at our chemical equation that shows the production (formation) of carbon dioxide:

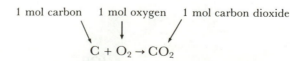

1 mol carbon ↘ 1 mol oxygen ↓ 1 mol carbon dioxide ↙

$$C + O_2 \rightarrow CO_2$$

The interpretation of this equation is as follows: "One mole of carbon atoms chemically combines with one mole of oxygen molecules (O_2) to form one mole of carbon dioxide molecules." We know the volume of 1 mol of carbon dioxide. It's about 22.4 L at 0°C and 1 atm (STP). Therefore, we can alter our interpretation slightly, as shown:

$$\underset{\text{1 mol C}}{C} + \underset{\text{1 mol } O_2}{O_2} \rightarrow \underset{\text{22.4 L } CO_2 \text{ at STP}}{CO_2}$$

Now our chemical equation tells us the volume of carbon dioxide that forms when carbon atoms (e.g., from coal) are burned. However, carbon isn't burned at 0°C, and the escaping gas (CO_2), even after it cools, is rarely at 0°C. Without gas laws we do not know the volume of a mole of carbon dioxide (or any other gas) at temperatures other than STP. That's the function of the gas laws, to tell us the volume of (and some other things) about gases at temperatures and pressures other than standard conditions.

In Chapter 4 we saw that molecular formulas could be determined from empirical formulas and molecular weights. We showed how an empirical formula is determined but not how a molecular weight is calculated. In this chapter we will show how the gas laws are used to determine the molecular weights of gases. In Chapter 8 ("Colligative Properties") we will show how to calculate the molecular weights for nongases.

5.1.2 WHAT IS ONE ATMOSPHERE OF PRESSURE?

The earth is surrounded by a layer of a gas mixture called the atmosphere. This gas is primarily composed of oxygen (20% by volume) and nitrogen (80% by volume), with small amounts of several other gases, such as water vapor, carbon dioxide, carbon monoxide, and helium. This layer of gases extends about 600 mi into space, although over 99% of its mass is confined to the first 20 mi (see Figure 5.1). All of these gas molecules in the atmosphere have a certain weight; that is, they exert a pressure upon us and upon the surface of the earth. A column of air 600 mi high with a 1-in.^{2} cross-sectional area (see Figure 5.2) would weigh about 15 lb (actually 14.7 lb). You may have heard of pressure measured in units of psi; *psi* stands for "pounds per square inch."

Water has a greater density than air, and you might expect that a 15-lb column of water

FIGURE 5.1 The earth's atmosphere

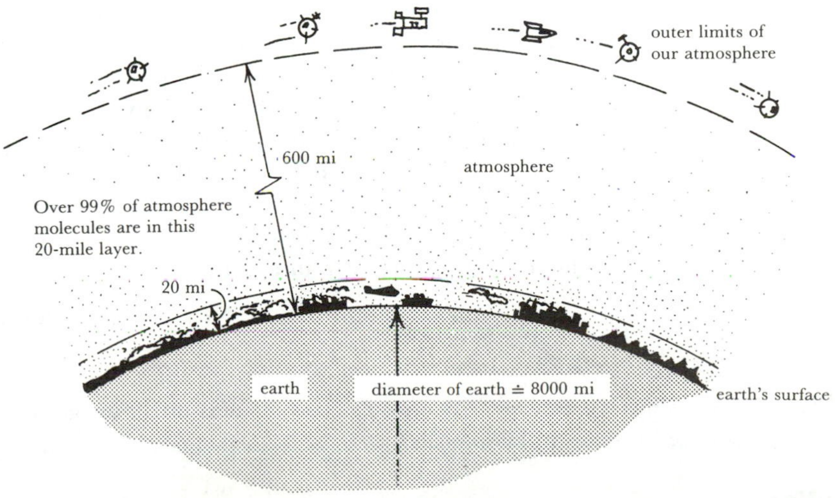

FIGURE 5.2 Relative densities of air, water, and mercury

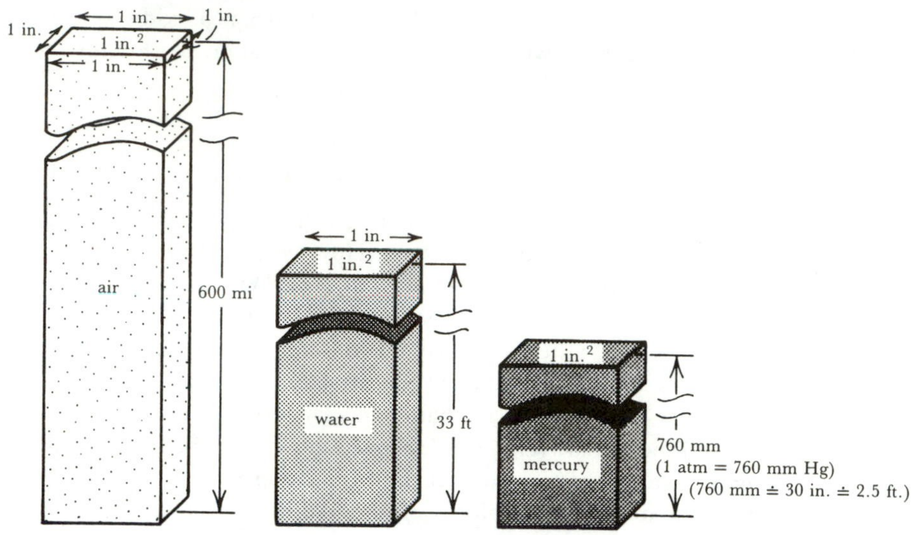

with a 1-in.^{2} cross-sectional area is much less than 600 mi high. It turns out that a column of water that is only about 33 ft high with a 1-in.^{2} cross-sectional area exerts as much pressure as an equally thick column of our atmosphere (air) that is 600 mi high (see Figure 5.2).

Mercury is even more dense than water. A column of mercury that is only 76 cm high (760 mm high) weighs the same as a column of water that is about 33 ft high or a column of air that is about 600 mi high. All three represent the same weight, the weight of our atmosphere, and the pressure that each one of these exerts is called 1 atmosphere of pressure (1 atm). We say that 760 mm Hg (millimeters of mercury) is the pressure of 1.00 atm.

5.1.3 THE PASCAL

The International Union of Pure and Applied Chemistry (IUPAC) recommends that the unit for expressing pressure should no longer be the atmosphere or millimeters of mercury (both of which are still used by many chemists). In place of the atmosphere and millimeters of mercury, the IUPAC recommends the use of the pascal (an SI unit). What is a pascal?

Before you can appreciate what a pascal is, you have to know the definition of a newton (another SI unit). A newton is an amount of pressure, or force. A normal-size apple (about 3.6 oz) exerts a force of about 1 newton (1 N) on your hand because of the earth's gravity acting on the apple. One pascal is equal to the pressure, or force, of 1 newton being exerted over an area of 1 m^2. One atmosphere is defined as 101 325 pascals (101 325 Pa), or 101.325 kilopascals (101.325 kPa). That's equivalent to the pressure exerted by about 100 000 normal-size apples in a box whose base is 1 m^2. Such a box would be about 65 m high (see Figure 5.3).

Throughout this chapter, keep in mind that the following pressures are equivalent:

$$1 \text{ atm} = 760 \text{ mm Hg} = 14.7 \text{ psi} = 101 \text{ kPa} = 33 \text{ ft } H_2O$$

EXAMPLE 1 A gas has a pressure of 2.0 atm. Express the pressure of the gas in pounds per square inch. What is the pressure in kilopascals?

Solution

$$(2.0 \text{ atm})\left(\frac{14.7 \text{ psi}}{1.00 \text{ atm}}\right) = 29 \text{ psi}$$

FIGURE 5.3 In a box with a base of 1 m² containing 100 000 apples, each 3.6-oz apple exerts a force, or pressure, of about 1 N.

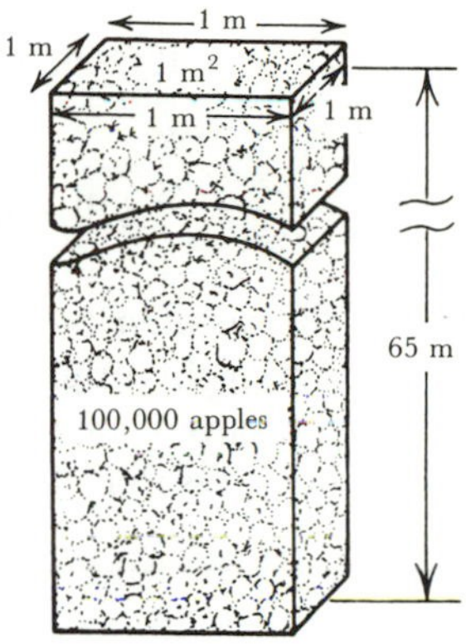

$$(2.0 \text{ atm})\left(\frac{101 \text{ kPa}}{1.00 \text{ atm}}\right) = 2.0 \times 10^2 \text{ kPa}$$

EXAMPLE 2 A gas sample has a pressure of 730 mm Hg. What is the pressure of this gas sample in atmospheres? What is its pressure in kilopascals?

Solution

$$(730 \text{ mm Hg})\left(\frac{1.00 \text{ atm}}{760 \text{ mm Hg}}\right) = 0.961 \text{ atm}$$

$$(730 \text{ mm Hg})\left(\frac{101 \text{ kPa}}{760 \text{ mm Hg}}\right) = 97.0 \text{ kPa}$$

EXAMPLE 3 A tank of oxygen is under a pressure of about 10 000 kPa. Express this pressure in millimeters of mercury.

Solution

$$(1 \times 10^4 \text{ kPa})\left(\frac{760 \text{ mm Hg}}{101 \text{ kPa}}\right) = 7 \times 10^4 \text{ mm Hg}$$

5.2 BOYLE'S LAW

Boyle's law tells us that if you increase the pressure on a gas sample, the volume of the gas sample will decrease. More precisely, the law states that at constant temperature, the volume of a fixed quantity of confined gas is inversely proportional to its pressure.

EXAMPLE 4 A sample of gas has a volume of 10 L and a pressure of 1 atm. If the pressure of the gas is increased to 2 atm, what is the new volume of the gas? Assume that the temperature remains constant while the pressure is changed.

Solution The pressure on the gas is increased (from 1 atm to 2 atm). Boyle's law predicts that the volume of the gas will decrease. Our final answer must be a volume that is less than 10 L. To make the original volume (10 L) smaller, it should be multiplied by a pressure ratio that is less than 1.

This pressure ratio equals $\frac{1}{2}$, which is less than 1.

$$(10 \text{ L gas})\left(\frac{1 \text{ atm}}{2 \text{ atm}}\right) = 5 \text{ L gas}$$

The final volume of the gas after the pressure increases from 1 atm to 2 atm is 5 L.

If we had multiplied the original volume (10 L) times a pressure ratio greater than 1 (2 atm/1 atm), we would have gotten an incorrect answer of 20 L. We knew in advance that our answer must be a volume less than 10 L, so we would have known that an answer of 20 L was incorrect.

EXAMPLE 5 A container of oxygen has a volume of 20 mL and a pressure of 10 atm. If the pressure on the oxygen gas is reduced to 5.0 atm and the temperature is kept constant, what is the new volume of the oxygen gas?

Solution The pressure on the gas is decreased (from 10 atm to 5 atm). Boyle's law predicts that a pressure decrease will cause a gas to expand (the volume of the gas will increase). The final volume (after the pressure has decreased) will be greater than the original volume (20 mL). To make the original volume (20 mL) larger, we have to multiply it by a pressure ratio that is a number greater than 1.

$$(20 \text{ mL } O_2)\left(\frac{10 \text{ atm}}{5.0 \text{ atm}}\right) = 40 \text{ mL } O_2$$

This pressure ratio equals 2 (2 is greater than 1), which, when multiplied times 20 mL, gives an answer greater than 20 mL—as we expected.

If we had multiplied the original volume (20 mL) times the pressure ratio 5.0 atm/10 atm, we would have gotten an incorrect answer of 10 mL. We knew in advance that our answer must be a volume greater than 20 mL, so we would have known that an answer of 10 mL was incorrect.

EXAMPLE 6 A tank of nitrogen has a volume of 12 L and a pressure of 760 mm Hg. Find the volume of the nitrogen when its pressure is changed to 500 mm Hg while the temperature is held constant.

Solution The pressure is decreased (from 760 mm Hg to 500 mm Hg), which causes the volume of the gas to increase (become larger than 12 L). To make the original volume (12 L) larger, multiply it by a number (pressure ratio) larger than 1.

$$(12 \text{ L } N_2)\left(\frac{760 \text{ mm Hg}}{500 \text{ mm Hg}}\right) = 18 \text{ L } N_2$$

EXAMPLE 7 A 50-L tank of ammonia has a pressure of 10 psi (pounds per square inch). Calculate the volume of the ammonia if its pressure is changed to 20 psi while its temperature remains constant.

Solution The pressure of the ammonia is increased from 10 psi to 20 psi, which causes the volume of the gas (ammonia) to decrease. Our final answer must be a volume that is less than the original volume of ammonia (50 L).

$$(50 \text{ L NH}_3)\left(\frac{10 \text{ psi}}{20 \text{ psi}}\right) = 25 \text{ L NH}_3$$

EXAMPLE 8 One hundred liters of helium at 1.00 atm and 22°C is to be placed into a tank with an internal pressure of 500 kPa. Find the volume of the helium after it is compressed into the tank when the temperature of the tank is 22°C.

Solution One atmosphere of pressure is equal to 101.3 kPa (kilopascals). The pressure is to be increased to 500 kPa, causing the gas volume to decrease. The original volume (100 L) is to be multiplied by a number (pressure ratio) less than 1 to produce a volume that is smaller than 100 L.

$$(100 \text{ L He})\left(\frac{101.3 \text{ kPa}}{500 \text{ kPa}}\right) = 20.3 \text{ L He}$$

SELF-TEST

1. A gas occupies a volume of 5.0 L at 4.0 atm. Find its volume if the pressure is changed to 2.0 atm.
2. Four hundred liters of ammonia (NH_3) at 760 mm Hg is to be compressed into a cylinder with a pressure of 3000 mm Hg at the same temperature. What is the new volume of the ammonia?
3. A gas occupies 300 mL at a pressure of 30.0 psi. Find the volume of the gas at a pressure of 15.0 psi if the temperature remains constant.

ANSWERS **1.** 10 L **2.** 101 L NH_3 **3.** 600 mL

5.2.1 SODA CAN OPENING: A PROPER METHOD

If you've ever shaken a can of soda before opening it with a can opener or popping the flip top, you know the mess that can occur when the soda sprays from the can. Some of the more experienced soda drinkers know of a technique that eliminates the mess: Tap the sides of the soda can before opening it. Tapping the sides prevents the soda from spraying out. Let's see what Boyle's law has to do with this.

When you shake a can of soda, the tiny gas bubbles created by the shaking adhere to the sides and the bottom of the can, as shown in Figure 5.4.

FIGURE 5.4 Opening a soda can. (a) Soda can before it is shaken. (b) Soda can after it is shaken. Tiny gas bubbles adhere to the inside wall of the soda can. (c) Shaken soda can after it has been opened. The gas bubbles expand as the pressure decreases. As the bubbles expand, soda is pushed out of the can.

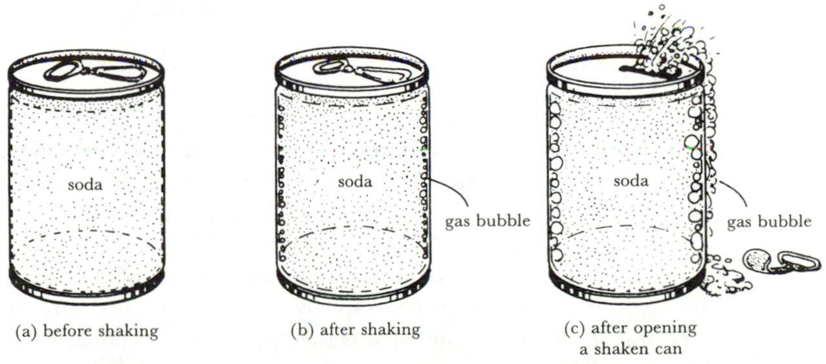

What happens to the tiny gas bubbles adhering to the sides of the soda can when the can is opened? (Hint: Soda cans are sealed under pressure.)

When a can of soda is opened, the pressure inside the can is reduced. The pressure reduction causes the gas bubbles adhering to the inside wall to expand, pushing out the soda and making the aforementioned mess.

How does tapping the sides of the soda can eliminate this mess?

Tapping the sides of the soda can causes the tiny gas bubbles to rise to the top of the can where their expansion (when the can is opened) is noticed only by a rush of gas from the can (as opposed to a rush of soda).

Buy a six-pack this weekend and perform this experiment yourself.

5.2.2 UNDERWATER SWIMMING VS. SCUBA DIVING

Recall that a column of air 600 mi high (1-in.^{2} cross-sectional area) exerts a pressure of about 1 atm. A column of water 33 ft high (1-in.^{2} cross-sectional area) will exert the same pressure, 1 atm.

•
•

EXAMPLE 9 Assume that you have a sudden urge to go deep-sea diving without any diving equipment. Being of average build, your lungs hold a total volume of about 10 L (each lung has a volume of about 5 L). You take a deep breath, jump into the water, and descend to a depth of 33 ft.

What is the volume of your lungs at this depth?

Solution Your lungs occupy a total volume of 5 L (each lung has a volume of 2.5 L) at a depth of 33 ft. The total pressure on your lungs at a depth of 33 ft is 2 atm. One atmosphere of pressure is due to the 33 ft of water, and 1 atm is caused by the 600 mi of air resting on top of the water. The initial pressure on your lungs before you dived was 1 atm. The pressure on your lungs increases from 1 atm to 2 atm after the dive. The volume of gas inside your lungs must decrease (Boyle's law).

$$(10 \text{ L lung volume})\underbrace{\left(\frac{1 \text{ atm}}{2 \text{ atm}}\right)}_{\text{pressure ratio less than 1}} = 5 \text{ L lung volume}$$

•
•

The deeper you dive, the smaller your lung volume becomes. The use of scuba (self-contained underwater breathing apparatus) gear eliminates this problem of decreasing lung volume. Scuba equipment delivers to your lungs air that is at the same pressure as your external pressure. At a depth of 33 ft, scuba equipment gives you air to breathe at a pressure of 2 atm. This keeps your lungs fully expanded (having a volume of 10 L) because the pressure inside your lungs (2 atm) is exactly equal to the pressure outside of your lungs (2 atm). Scuba regulators hold the volume of your lungs constant by changing the pressure of the air you breathe. (The predecessor of scuba equipment was the Aqua-lung invented by Jacques Cousteau.)

•
•

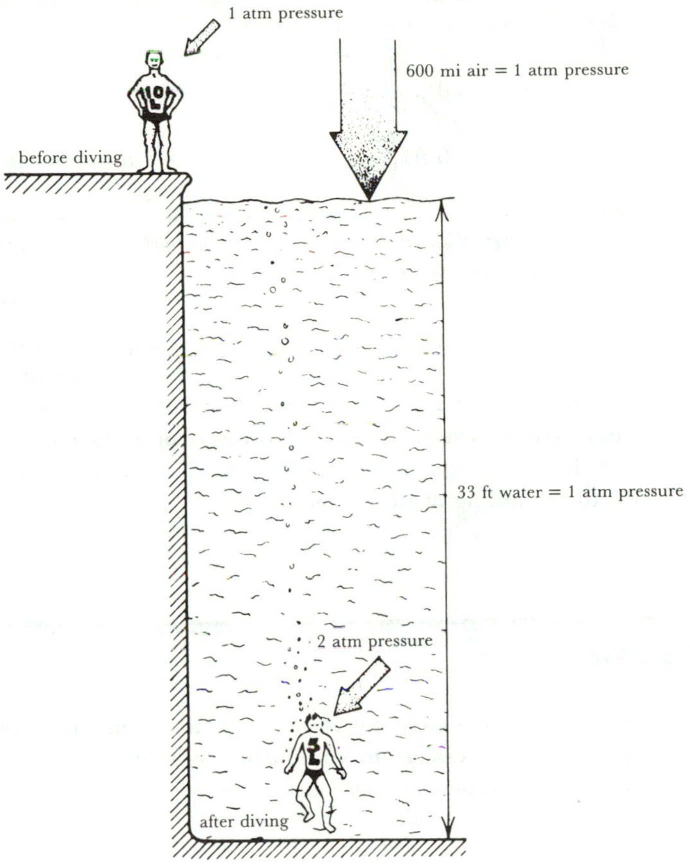

EXAMPLE 10 You are now wearing scuba gear and swimming underwater at a depth of 33 ft. You are breathing air at 2 atm and your lung volume is 10 L. Your scuba gauge indicates that your air supply is low so, to conserve air, you make a terrible (and fatal) mistake and hold your breath while you surface.

What happens to your lungs?

Solution Your lungs would probably burst and you would die. At 33 ft below the surface of the water, your lungs are under 2 atm of pressure and occupy a volume of 10 L. At the surface, the pressure is 1 atm and, if you could reach the surface while still holding your breath, your lungs would have expanded to a volume of 20 L.

$$(10 \text{ L lung volume})\left(\frac{2 \text{ atm}}{1 \text{ atm}}\right) = 20 \text{ L lung volume}$$

Of course, you would probably never reach that point because your lungs would burst long before they could expand to a volume of 20 L. This is a real problem for the almost one million scuba divers in the United States.

5.2.3 EXPLODING TEETH

The Norwegians and the British have made great oil strikes in the North Sea. To work on the scaffolding of the oil rigs, deep-sea divers must descend to great depths (1500 ft).

What is the external pressure at this depth?

The external pressure at 1500 ft is 46 atm.

$$(1500 \text{ ft})\left(\frac{1 \text{ atm}}{33 \text{ ft}}\right) = 45 \text{ atm} + 1 \text{ atm (600 mi air column)} = 46 \text{ atm}$$

Sometimes, when these divers are brought to the surface and decompressed, their teeth explode. Can you think of a possible reason for this?

Scuba regulators worn by the divers supply them with air at a pressure equal to the surrounding external pressure. Air at this pressure (46 atm at a depth of 1500 ft) is forced into spaces of the diver's teeth left by loose fillings (or improperly fitted fillings). During decompression, pressure is reduced externally causing trapped air in the tooth spaces to expand very rapidly (Boyle's law). The teeth explode. It's a very painful experience and can be quite embarrassing if you happen to be smiling at the moment.

5.3 CHARLES'S LAW

Charles's law tells us that if you increase the temperature of a gas, the volume of the gas will get larger. More precisely, the volume of a fixed quantity of confined gas is directly proportional to its absolute temperature at constant pressure.

•
•

EXAMPLE 11 The temperature of a 2.00-L sample of gas is changed from 10°C to 20°C. What will the volume of this gas be at the new temperature if the pressure is held constant?

Solution Because the temperature of the gas is increased, Charles's law predicts that the gas volume will also increase. Our answer must be a volume that is greater than the original volume (2 L). To make the original volume (2 L) larger, you must multiply it by a temperature ratio that is greater than 1:

absolute temperature has units of K

↓

$$(2.00 \text{ L gas})\left(\frac{293 \text{ K}}{283 \text{ K}}\right) = 2.07 \text{ L gas}$$

↑

This temperature ratio equals 1.04, a number greater than 1 which, when multiplied times the original volume (2.00 L), will produce an answer greater than 2 L, as expected.

If we had multiplied the original volume (2.00 L) times the temperature ratio 283 K/293 K, we would have gotten the incorrect answer 1.93 L. We knew in advance that our answer must be greater than 2.00 L, so we would have known that 1.93 L was incorrect.

•
•

EXAMPLE 12 Carbon dioxide is usually formed when gasoline is burned. If 50 kL of carbon dioxide is produced at a temperature of 1000°C and allowed to reach room temperature (25°C) without any pressure changes, what is the new volume of the carbon dioxide?

Solution Because the temperature of the carbon dioxide is decreased (from 1000°C to 25°C), we know that the volume will also decrease (Charles's law). Our answer must be a volume that is less than the original volume (50 kL). To decrease the original volume (50 kL), multiply it by a temperature ratio that is less than 1:

$$(50 \text{ kL } CO_2)\left(\frac{298 \text{ K}}{1273 \text{ K}}\right) = 12 \text{ kL } CO_2$$

EXAMPLE 13 An 800-mL sample of nitrogen is warmed from 77°F to 86°F. Find its new volume if the pressure remains constant.

Solution

$$77°F = 25°C = 298 \text{ K}$$

$$86°F = 30°C = 303 \text{ K}$$

$$(800 \text{ mL } N_2)\left(\frac{303 \text{ K}}{298 \text{ K}}\right) = 813 \text{ mL } N_2$$

↑ This temperature ratio is greater than 1, so it will increase the original volume.

SELF-TEST

1. The temperature of a 10-L sample of gas is changed from 300 K to 900 K. What is the new volume of the gas if the pressure is kept constant?
2. The temperature of a 10-L sample of gas is changed from 300 K to 100 K, while the pressure is kept constant. What is the new volume of the gas?
3. A 400-mL sample of oxygen is cooled from 50°C to −40°C. What is the new volume of the oxygen if its pressure remains constant?

ANSWERS **1.** 30 L **2.** 3.3 L **3.** 289 mL O_2

5.3.1 INFLATABLE TOYS

A few years ago, a child was playing with an inflatable toy when the cabin of the jet plane in which he was flying depressurized. The plane was flying at 30 000 ft, where temperatures are typically around −30°C and the atmospheric pressure is about 360 mm Hg.

Using Charles's law, Boyle's law, and your observations from the preceding examples, try to predict what occurred to the toy following the cooling and depressurizing of the cabin.

When the cabin cooled and depressurized, the toy expanded. Two things changed:

1. The cabin cooled. Cooling a confined gas decreases the volume of the gas.
2. The pressure of the cabin decreased. Decreasing the pressure on a confined gas increases the volume of the gas.

But which change (temperature or pressure) would have had the greater influence on the toy? You may have observed (see Example 11) that if the Celsius temperature of a gas is doubled (e.g., goes from 10°C to 20°C), the gas volume doesn't double. (The gas volume does double when the Kelvin temperature is doubled.) Therefore, doubling or halving the pressure (in atmospheres, millimeters of mercury, or any standard pressure unit) has a greater influence on a gas volume than doubling or halving the Celsius temperature. Let's look at this in more detail.

EXAMPLE 14 Assume that the pressure in the cabin remained constant and the temperature decreased from 22°C to −30°C.

What volume change would occur in a confined 200-mL sample of gas?

Solution

temp ratio = 0.82

$$(200\ \text{mL gas})\left(\frac{243\ \text{K}}{295\ \text{K}}\right) = 165\ \text{mL gas}$$

A drop of cabin temperature from 22°C to −30°C causes a volume decrease in a confined 200-mL sample of gas of only about 18%.

EXAMPLE 15 Assume that the temperature in the cabin remained constant and the pressure decreased from 760 mm Hg to 360 mm Hg.

What change in the volume of a confined 200-mL gas sample occurred?

Solution

pressure ratio = 2.11

$$(200\ \text{mL gas})\left(\frac{760\ \text{mm Hg}}{360\ \text{mm Hg}}\right) = 422\ \text{mL gas}$$

A decrease of cabin pressure from 760 mm Hg to 360 mm Hg causes a volume increase in a 200-mL gas sample of over 200%.

There are times when you cannot conveniently change just the temperature or just the pressure of a gas (such as when planes depressurize). Quite frequently, both temperature and pressure change together. We will discuss how to handle such situations later in this chapter (see Section 5.6).

5.4 GAY-LUSSAC'S LAW

Gay-Lussac's law tells us that if you increase the temperature of a gas sample, the pressure of the gas increases. More precisely, the law states that the pressure of a fixed quantity of a confined gas at constant volume is directly proportional to the absolute temperature.

EXAMPLE 16 A 20-L sample of nitrogen inside a metal container at 20°C is placed inside an oven whose temperature is 50°C. The pressure inside the container at 20°C was 1.0 atm. What is the pressure of the nitrogen after its temperature is changed?

Solution The temperature of the nitrogen is increased, so we know that the pressure of the nitrogen will also increase. Our final answer must be a pressure that is greater than the original pressure (1.0 atm). To make the original pressure (1.0 atm) larger, it should be multiplied by a temperature ratio that is greater than 1:

$$(1.0 \text{ atm})\left(\frac{323 \text{ K}}{293 \text{ K}}\right) = 1.1 \text{ atm}$$

EXAMPLE 17 A sample of gas at 1500 mm Hg inside a steel tank is cooled from 500°C to 0°C. What is the final pressure of the gas in the steel tank?

Solution Because the temperature of the gas is decreased, we know that the pressure of the gas will also decrease. Our final answer must be a pressure that is less than the original pressure (1500 mm Hg). Multiply the original pressure by a temperature ratio that is less than 1:

$$(1500 \text{ mm Hg})\left(\frac{273 \text{ K}}{773 \text{ K}}\right) = 530 \text{ mm Hg}$$

EXAMPLE 18 The temperature of a sample of gas in a steel container at 15.0 kPa is increased from −100°C to 1000°C. What is the final pressure inside the tank?

Solution The temperature of the gas is increased. The pressure of the gas will also increase. Our final answer must be a pressure that is greater than the original pressure (15.0 kPa). To make the original pressure larger, multiply it by a temperature ratio that is greater than 1:

$$(15.0 \text{ kPa})\left(\frac{1273 \text{ K}}{173 \text{ K}}\right) = 110 \text{ kPa}$$

SELF-TEST

1. A 20-L sample of nitrogen inside a metal container at 50°C is placed inside a refrigerator whose temperature is 5°C. The original pressure inside the container was 2.0 atm. What is the final pressure of the nitrogen inside the refrigerated container?
2. A sample of gas at 2000 mm Hg inside a steel tank is warmed from 10°C to 300°C. What is the final pressure of the gas in the steel tank?
3. A gas sample at 30 kPa is cooled from 800°C to 100°C. What is the final pressure inside the tank?

ANSWERS **1.** 1.7 atm **2.** 4049 mm Hg **3.** 10 kPa

5.4.1 FILLING SCUBA TANKS: A PROPER METHOD*

Scuba air tanks are reusable; when a scuba tank is empty, it can be refilled with air. The air inside a scuba tank is compressed to a pressure of 140 atm (that's more than 100 000 mm Hg, or about 2100 psi). Compressing a gas increases the temperature of the gas (Gay-Lussac's law). You may have noticed this while pumping air into your bike tires with a hand pump. The compressed air in the bike pump warms up and makes the pump itself warm to the touch. Compressing air in a scuba tank to a pressure of 140 atm can theoretically raise the tank temperature to 1000°C (1832°F). It probably wouldn't get that hot, however, because heat is lost to the surrounding air. A hot tank holds less gas than a cold tank. For example, a tank holding air at 140 atm at a temperature of 1000°C will eventually cool, lowering the pressure inside the tank.

*Adapted from E. D. Cooke, "SCUBA Diving and the Gas Laws," *Journal of Chemical Education* 50 (June 1973): 425–26.

EXAMPLE 19 *Calculate the final pressure inside a scuba tank after it cools from 1000°C to 25°C. The initial pressure in the tank was 140 atm.*

Solution The cooled tank would contain air at a pressure of about 33 atm.

$$(140 \text{ atm})\left(\frac{298 \text{ K}}{1273 \text{ K}}\right) = 32.8 \text{ atm}$$

The 33 atm of air is less than one-fourth of the tank's total capacity.

To prevent this from happening, scuba tanks are submerged in water while they are being filled with compressed air. The water helps to keep the tank cool by conducting heat away from the tank.

5.5 DALTON'S LAW

The air in your room (a container) is approximately 80% by volume nitrogen and 20% by volume oxygen. The total atmospheric pressure in your room is about 1 atm. Nitrogen is responsible for 80% of this pressure. We say that the partial pressure of the nitrogen is 0.80 atm.

$$(1 \text{ atm total pressure})\left(\frac{80 \text{ parts } N_2 \text{ pressure}}{100 \text{ parts total pressure}}\right) = 0.80 \text{ atm } N_2 \text{ pressure}$$

Oxygen is responsible for 20% of the total atmospheric pressure. We say that the partial pressure of the oxygen is 0.20 atm. Notice that the total pressure in your room (1 atm) is equal to the sum of the partial pressures of the two gases (0.80 atm N_2 and 0.20 atm O_2).

Dalton's law tells us that the total pressure of a mixture of gases (such as air) is equal to the sum of the partial pressures of the gases in the mixture.

EXAMPLE 20 A container holds three gases: oxygen, carbon dioxide, and helium. The partial pressures of the three gases are 2.0 atm, 1.5 atm, and 5.0 atm, respectively. What is the total pressure inside the container?

Solution The total pressure (P_T) of a mixture of gases is equal to the sum of their partial pressures.

$$\begin{aligned} P_T &= P_{O_2} + P_{CO_2} + P_{He} \\ &= 2.0 \text{ atm} + 1.5 \text{ atm} + 5.0 \text{ atm} \\ &= 8.5 \text{ atm} \end{aligned}$$

EXAMPLE 21 A container with two gases, helium and argon, is 40% by volume helium. Calculate the partial pressure of helium and argon. The total pressure inside the container is 2.0 atm.

Solution Forty percent of the total pressure (2 atm) is due to the helium atoms inside the container.

$$P_{He} = 40\% \text{ of } 2 \text{ atm} = 0.8 \text{ atm He}$$

The total pressure in the container (2 atm) must be equal to the sum of the partial pressures, so the partial pressure of argon must be 1.2 atm.

$$P_{He} + P_{Ar} = P_T = 2 \text{ atm}$$

$$0.8 \text{ atm} + P_{Ar} = 2 \text{ atm}$$

$$P_{Ar} = 2 \text{ atm} - 0.8 \text{ atm} = 1.2 \text{ atm Ar}$$

We could also have solved for the partial pressure of argon the same way we did for the partial pressure of helium. The mixture is 60% by volume argon, so

$$P_{Ar} = 60\% \text{ of } 2 \text{ atm} = 1.2 \text{ atm Ar}$$

Water vapor (which gets into the air when liquid water evaporates) is a gas that exerts its own partial pressure. The partial pressure of water vapor is called the vapor pressure of water. The warmer liquid water is, the more of it that evaporates. The more water that is in the air, the greater the partial pressure (vapor pressure) of the water. We say that the vapor pressure of water increases as the temperature of the water increases. Table 5.1 shows some values for the vapor pressure of water (partial pressure of water vapor) at various liquid water temperatures.

TABLE 5.1 Vapor Pressure of Water

Liquid Water Temperature in Degrees Celsius	Water Vapor Pressure in Millimeters of Mercury	Water Vapor Pressure in Kilopascals
0	4.6	0.613
10	9.2	1.23
20	17.5	2.33
30	31.8	4.24
40	55.3	7.37
50	92.5	12.3
60	149.4	19.91
70	233.7	31.15
80	355.1	47.33
90	525.8	70.08
100	760.0	101.3

EXAMPLE 22 If 50.0 L of nitrogen is collected over water (see the accompanying diagram) at 30°C when the atmospheric pressure is 760 mm Hg, what is the partial pressure of the nitrogen?

Solution The 50.0 L of collected gas is not pure nitrogen. Since nitrogen is collected over water, some water has evaporated inside the collection flask. The partial pressure of water (see Table 5.1) at 30°C is 31.8 mm Hg. When the water levels inside and outside the collection flask are equal, the total pressure inside the collection flask must be 1 atm (the same as the pressure outside the flask). If the pressure inside the collection flask were greater than 1 atm, the gas would expand and escape from the collection flask. If the pressure inside the collection flask were less than 1 atm, the partial vacuum inside the flask would draw more water into the flask. The 50.0-L space inside the collection flask remains constant (not expanding and escaping or shrinking and drawing up water), so the 50.0-L space must be at the same pressure as that of the surround-

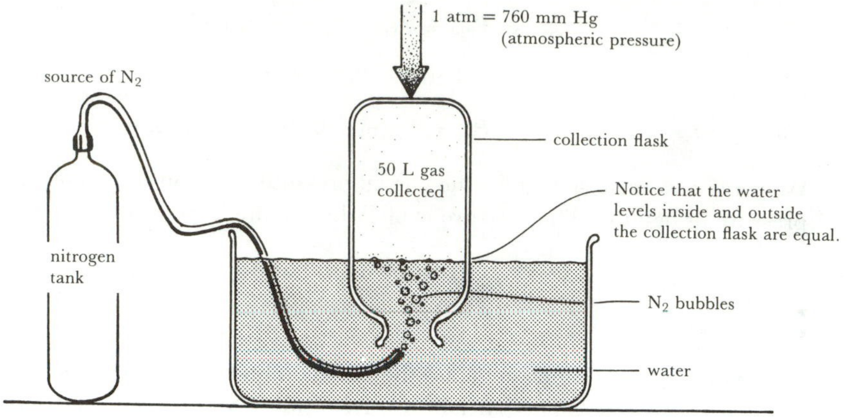

ing atmosphere. The total pressure inside the flask is 1 atm and it is made up of two gases, water vapor and nitrogen. The water vapor has a partial pressure of 31.8 mm Hg. The nitrogen must account for the rest of the total pressure of 1 atm; that is,

$$\underbrace{760.0 \text{ mm Hg}}_{\text{total pressure}} - \underbrace{31.8 \text{ mm Hg}}_{\text{partial pressure of } H_2O} = \underbrace{728.2 \text{ mm Hg}}_{\text{partial pressure of } N_2}$$

EXAMPLE 23 Thirty liters of nitrogen is collected over water at 20°C. The atmospheric pressure in the room is 99.97 kPa. What is the partial pressure of the nitrogen?

Solution The vapor pressure of the water in the flask is 2.33 kPa (see Table 5.1). The total pressure inside the flask is 99.97 kPa. The partial pressure of the nitrogen is

$$\underbrace{99.97 \text{ kPa}}_{\text{total pressure}} - \underbrace{2.33 \text{ kPa}}_{\text{partial pressure of } H_2O} = \underbrace{97.64 \text{ kPa}}_{\text{partial pressure of } N_2}$$

SELF-TEST

1. A container holds carbon monoxide and sulfur dioxide. The partial pressure of the carbon monoxide is 3 atm. The partial pressure of the sulfur dioxide is 2 atm. What is the total pressure inside the container?
2. A tank contains three gases: 25% by volume oxygen, 35% by volume nitrogen, and 40% by volume helium. The total pressure inside the tank is 10.0 atm. What is the partial pressure of each of these gases?
3. Helium is collected over water at a temperature of 50°C when the atmospheric pressure is 770 mm Hg. What is the partial pressure of the helium?

ANSWERS **1.** 5 atm **2.** $P_{O_2} = 2.5$ atm; $P_{N_2} = 3.5$ atm; $P_{He} = 4.0$ atm **3.** 677.5 mm Hg

5.5.1 CARBON DIOXIDE POISONING FROM EXCESS OXYGEN*

The air in your lungs is pretty much the same as the air that you breathe. (In this discussion we will ignore the slight differences in carbon dioxide and oxygen concentrations.) The total pres-

*Cooke, "SCUBA Diving," 425–26.

sure inside your lungs is about 1 atm, and the partial pressure of oxygen in your lungs is about 0.2 atm, the same as it is in the air surrounding your body.

Consider this problem: Wearing scuba equipment, you descend to a depth of 132 ft. What is the pressure of the air that you are breathing at this depth?

The pressure of the air that you breathe wearing scuba equipment at a depth of 132 ft underwater is 5.0 atm. The 132-ft column of water above you produces a pressure of 4.0 atm.

$$(132 \text{ ft } H_2O)\left(\frac{1.0 \text{ atm}}{33 \text{ ft } H_2O}\right) = 4.0 \text{ atm}$$

The 600-mi column of air over the surface of the water produces the fifth atmosphere of pressure.

EXAMPLE 24 What is the partial pressure of the oxygen in your lungs when you are diving at a depth of 132 ft underwater breathing air from a scuba tank that is 80% nitrogen and 20% oxygen?

Solution The partial pressure of the oxygen is 1.0 atm (20% times 5.0 atm is 1.0 atm). The partial pressure of the nitrogen is 4.0 atm.

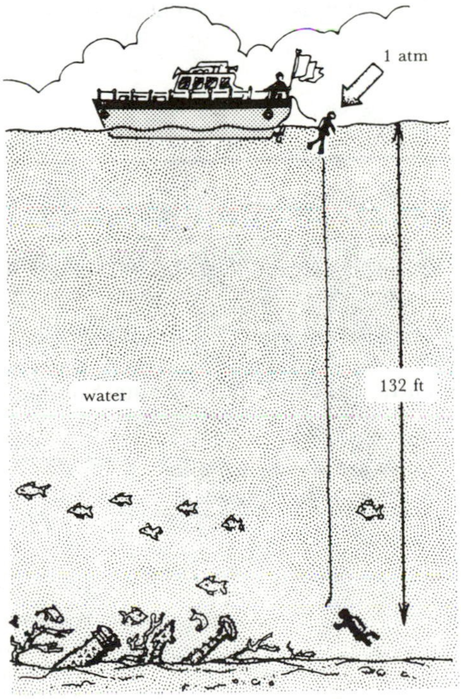

Pressure on lungs at the surface of the water is 1.0 atm.

total pressure inside lungs = 1.0 atm
partial pressure of O_2 = 0.20 atm
partial pressure of N_2 = 0.80 atm

Total pressure on lungs at a depth of 132 ft is 5.0 atm.

total pressure inside lungs = 5.0 atm
partial pressure of O_2 = 20% of 5.0 atm = 1.0 atm
partial pressure of N_2 = 80% of 5.0 atm = 4.0 atm

With such a high partial pressure of oxygen, the blood concentration of oxyhemoglobin increases. (Hemoglobin loosely combined with oxygen is called oxyhemoglobin.) This excess oxygen in the blood, oddly enough, leads to carbon dioxide poisoning. Let's discover how this happens.

Breathing accomplishes two things: (1) the intake of oxygen and (2) the removal of carbon dioxide. Normal breathing at 1 atm pressure accomplishes both of these functions. Breathing excess oxygen (oxygen at a greater partial pressure) reduces the urge to breathe (to obtain oxygen). As a result, carbon dioxide is not eliminated from the body as quickly as it should be.

How can you avoid CO_2 poisoning using your knowledge of Dalton's law?

Reduce the percentage of oxygen in the scuba air tank.

Jacques Cousteau's divers use a gas that is 2% oxygen for working at 300 ft. Can you see why?

The atmospheric pressure at a depth of 300 ft is 10.0 atm.

$$(300\ \text{ft})\left(\frac{1.0\ \text{atm}}{33\ \text{ft}}\right) = 9.0\ \text{atm}$$

$$9.0\ \text{atm (from water)} + 1.0\ \text{atm (from air)} = 10.0\ \text{atm}$$

A gas mixture that is 2% oxygen would have an oxygen partial pressure equal to 0.2 atm if the total pressure of the gas were 10.0 atm. This partial pressure (0.2 atm) is the same as the partial pressure for oxygen at the surface of the water where the total pressure is 1 atm and the oxygen is 20% of the air mixture.

The other 98% of the gas mixture used by the Cousteau divers is helium, not nitrogen. Helium is much less soluble than nitrogen. The increased solubility of nitrogen at high pressure causes the bends when divers surface. Dissolved nitrogen coming out of the blood solution causes the condition called the bends. Helium is used to prevent the bends, which are very painful and can be fatal.

5.6 COMBINED GAS LAWS

Although we have been using the gas laws separately (we have applied them only one at a time), they may be applied in combination.

•
•

EXAMPLE 25 Find the volume of a gas at 2.0 atm and 200 K if its original volume was 100 L at 1.0 atm and 400 K.

Solution

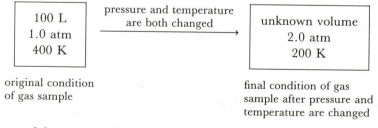

100 L 1.0 atm 400 K	pressure and temperature are both changed ⟶	unknown volume 2.0 atm 200 K
original condition of gas sample		final condition of gas sample after pressure and temperature are changed

The temperature of the gas sample is lowered from 400 K to 200 K. This will reduce the volume of the gas. The temperature ratio is 200 K/400 K. This ratio, when multiplied by the original volume (100 L), will produce a volume that is smaller than 100 L. At the same time, the pressure is being increased from 1 atm to 2 atm. This will also reduce the volume of the gas. The pressure ratio is 1.0 atm/2.0 atm. This ratio, when multiplied by the original volume (100 L), will produce a volume that is smaller than 100 L. Both ratios are multiplied times the original volume in a one-step chain multiplication, as follows:

$$(100 \text{ L gas})\left(\frac{200 \text{ K}}{400 \text{ K}}\right)\left(\frac{1.0 \text{ atm}}{2.0 \text{ atm}}\right) = 25 \text{ L gas}$$

EXAMPLE 26 One thousand liters of a gas is prepared at 700 mm Hg and 200°C. The gas is placed into a tank under high pressure. When the tank cools to 20°C, the pressure of the gas is 30.0 atm. What is the volume of the gas?

Solution The original pressure of the gas is 700 mm Hg. This pressure is increased to 2.28×10^4 mm Hg.

$$(30.0 \text{ atm})\left(\frac{760 \text{ mm Hg}}{1.00 \text{ atm}}\right) = 2.28 \times 10^4 \text{ mm Hg}$$

This pressure increase reduces the volume of the gas. The pressure ratio is 700 mm Hg/2.28×10^4 mm Hg. This ratio, when multiplied times the original volume, produces a reduced volume. The temperature of the gas decreases from 473 K (200°C) to 293 K (20°C). This cooling reduces the volume of the gas even more. The temperature ratio will therefore be 293 K/473 K. This ratio, when multiplied times the original volume, will lower that volume. Our calculation becomes

$$(1000.0 \text{ L gas})\left(\frac{700 \text{ mm Hg}}{2.28 \times 10^4 \text{ mm Hg}}\right)\left(\frac{293 \text{ K}}{473 \text{ K}}\right) = 19.0 \text{ L gas}$$

EXAMPLE 27 Let's go back to our child and his 200-mL gas sample in an inflatable toy (see Section 5.3.1) that was subjected to a temperature change from 22°C to −30°C and a pressure change from 760 mm Hg to 360 mm Hg. What is the final volume of the 200-mL gas sample after these temperature and pressure changes occur?

Solution

$$(200 \text{ mL gas})\left(\frac{760 \text{ mm Hg}}{360 \text{ mm Hg}}\right)\left(\frac{243 \text{ K}}{295 \text{ K}}\right) = 348 \text{ mL gas}$$

EXAMPLE 28 Five hundred milliliters of hydrogen are collected over water at 30°C at a pressure of 831.8 mm Hg. Find the volume of dry hydrogen collected when its pressure is reduced to 400 mm Hg.

Solution Dry hydrogen refers to hydrogen from which all water vapor (the water vapor that collected with the hydrogen inside the collection flask) has been removed. The total pressure inside the collection flask is 831.8 mm Hg, the same as the external atmospheric pressure. The partial pressure of water (water vapor pressure) is 31.8 mm Hg (from Table 5.1). The partial pressure of the hydrogen must be 800.0 mm Hg, as calculated:

$$831.8 \text{ mm Hg} - 31.8 \text{ mm Hg} = 800.0 \text{ mm Hg}$$

total pressure ↗ water vapor pressure ↑ partial pressure of hydrogen ↖

When water vapor is removed from the hydrogen–water vapor mixture, the pressure of the hydrogen gas is 800.0 mm Hg. Now that we have "dried the hydrogen," our problem can be reworded: "Find the volume of hydrogen gas that was originally 500 mL at a pressure of 800 mm Hg when the pressure is reduced to 400 mm Hg." Our pressure ratio is 800 mm Hg/400 mm Hg. This ratio, when multiplied times the original volume (500 mL), reduces the original volume:

$$(500 \text{ mL gas})\left(\frac{800 \text{ mm Hg}}{400 \text{ mm Hg}}\right)\left(\frac{1 \text{ L}}{1000 \text{ mL}}\right) = 1.00 \text{ L gas}$$

SELF-TEST

1. Ten liters of a gas have a temperature of 250 K and a pressure of 10 atm. What volume will this gas occupy if the temperature is increased to 300 K and the pressure is decreased to 8 atm?
2. Find the volume of a gas at 750 mm Hg and 30°C if its volume at 700 mm Hg and 10°C is 5.0 L.
3. Two hundred milliliters of oxygen are collected over water at 40°C and a pressure of 760 mm Hg. Find the volume of dry oxygen collected when its pressure is reduced to 700 mm Hg and its temperature is increased to 50°C.

ANSWERS **1.** 15 L gas **2.** 5.0 L gas **3.** 208 mL oxygen

5.6.1 MOLECULAR WEIGHT DETERMINATIONS

Before we show how the gas laws are used to determine molecular weights, let's look at some basic relationships between the mole concept, the volume of a gas, and the molecular weight of a gas.

EXAMPLE 29 A 35.0-g sample of gas occupies 22.4 L at STP. What is the molecular weight of this gas?

Solution This 35.0-g sample is 1.00 mol of gas. One mole of any gas occupies about 22.4 L at STP. Because this sample occupies 22.4 L at STP, it must be 1.00 mol of gas. One mole of anything has a weight equal to its molecular weight in grams. Since 1.00 mol of this gas weighs 35.0 g, its molecular weight must be 35.0 amu.

EXAMPLE 30 A 20.0-g gas sample occupies 11.2 L at STP. Find the molecular weight of this gas.

Solution The 20.0-g sample is 0.500 mol of gas. One mole of gas at STP occupies 22.4 L, so 0.500 mol occupies 11.2 L.

$$(11.2 \text{ L gas at STP})\left(\frac{1.00 \text{ mol gas}}{22.4 \text{ L gas at STP}}\right) = 0.500 \text{ mol gas}$$

To find the weight of 1.00 mol of this gas (units will be grams per mole of gas), divide the mass of the gas by the number of moles of gas.

$$\frac{20.0 \text{ g gas}}{0.500 \text{ mol gas}} = \frac{40.0 \text{ g gas}}{1.00 \text{ mol gas}}$$

That is, 40 g gas = 1 mol gas. Therefore, the molecular weight of this gas is 40.0 amu.

EXAMPLE 31 A 5.0-g sample of gas occupies 8.0 L at STP. What is the molecular weight of this gas?

Solution First, determine the number of moles of gas in the sample.

$$(8.0 \text{ L gas})\left(\frac{1.00 \text{ mol gas}}{22.4 \text{ L gas at STP}}\right) = 0.36 \text{ mol gas}$$

Next, determine the number of grams of gas per mole of gas.

$$\frac{5.0 \text{ g gas}}{0.36 \text{ mol gas}} = \frac{14 \text{ g gas}}{1.0 \text{ mol gas}}$$

That is, 14 g gas = 1.0 mol gas. The molecular weight of this unknown gas is 14 amu.

•
•

Now let's see how the gas laws are used to determine molecular weights of gases. Obviously, not every gas volume is measured at 0°C and 760 mm Hg. That would be inconvenient. The gas laws are used to calculate what volume a gas would occupy at 0°C and 760 mm Hg when its volume is measured under non-STP conditions.

•
•

EXAMPLE 32 A gas that occupies 20 L at 700 mm Hg and 20°C weighs 40 g. What is the molecular weight of this gas?

Solution First, use the gas laws to find out what volume this gas will occupy at STP. To do this we must change the temperature of the gas from 20°C to 0°C and the pressure from 700 mm Hg to 760 mm Hg. The increase in pressure (from 700 mm Hg to 760 mm Hg) will decrease the volume of our gas sample. The pressure ratio is 700 mm Hg/760 mm Hg. The decrease in temperature from 293 K (20°C = 293 K) to 273 K will reduce the volume of our gas sample. The temperature ratio is 273 K/293 K.

$$(20 \text{ L gas})\left(\frac{700 \text{ mm Hg}}{760 \text{ mm Hg}}\right)\left(\frac{273 \text{ K}}{293 \text{ K}}\right) = 17 \text{ L gas at STP}$$

Now this problem is just like the simpler problems, that is, we have 17 L of gas at STP that weighs 40 g. What is the molecular weight of this gas?

$$(17 \text{ L gas at STP})\left(\frac{1.0 \text{ mol gas}}{22.4 \text{ L gas at STP}}\right) = 0.76 \text{ mol gas}$$

$$\frac{40 \text{ g gas}}{0.76 \text{ mol gas}} = \frac{53 \text{ g gas}}{1.0 \text{ mol gas}}$$

That is, 1.0 mol gas = 53 g gas. The molecular weight of the gas is 53 amu.

•
•

EXAMPLE 33 A gas that occupies 10 L at 79.97 kPa and 30°C weighs 50 g. What is the molecular weight of this gas?

Solution First, find the volume this gas would occupy at 0°C and 101.3 kPa.

$$(10 \text{ L gas})\left(\frac{79.97 \text{ kPa}}{101.3 \text{ kPa}}\right)\left(\frac{273 \text{ K}}{303 \text{ K}}\right) = 7.1 \text{ L gas at STP}$$

Next, find how many moles of gas are present in this sample.

$$(7.1 \text{ L gas at STP})\left(\frac{1.00 \text{ mol gas}}{22.4 \text{ L gas at STP}}\right) = 0.32 \text{ mol gas}$$

Then, find the weight of 1 mol of this gas.

$$\frac{50 \text{ g gas}}{0.32 \text{ mol gas}} = \frac{1.6 \times 10^2 \text{ g gas}}{1.0 \text{ mol gas}}$$

Thus, 1 mol gas weighs approximately 160 g. The molecular weight of this gas is about 160 amu (1.6×10^2 amu).

•
•

SELF-TEST

1. A 40-g sample of gas occupies 22.4 L at STP. What is the molecular weight of this gas?
2. A 10-g sample of gas occupies 8.0 L at STP. What is the molecular weight of this gas?
3. A gas that occupies 20 L at 600 mm Hg and 30°C weighs 100 g. What is the molecular weight of this gas?

ANSWERS **1.** 40 amu **2.** 28 amu **3.** 1.6×10^2 amu

5.6.2 THE CHINOOK: A WIND THAT EATS SNOW

Each year the area around the Bow River Valley in southwestern Canada experiences temperatures that go down to −40°F (−40°F = −40°C). And almost every year, when the wind called the Chinook blows, the temperature climbs to about 50°F to 60°F. In just a matter of a few hours, this Canadian area experiences a temperature increase of 90°F to 100°F. Studying factors that cause this drastic temperature change will provide you with a review of some gas properties and some of the gas laws.

Air over the Pacific Ocean is always wet (moist) due to the continual evaporation of the ocean. The partial pressure of water in the air over the ocean depends, in part, upon the ocean's temperature. The moist air travels from the Pacific Ocean (point A in Figure 5.5) to the foot of the Rocky Mountains (point B) because air masses tend to move from west to east.

FIGURE 5.5 The Chinook

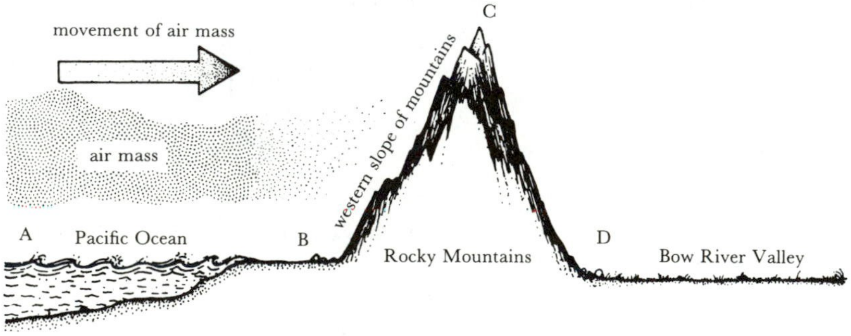

Is wet air heavier or lighter than dry air?

Most people incorrectly believe that wet air is heavier than dry air. It isn't. Dry air is heavier than wet air. Let's see why.

Air is (essentially) made up of two gases: oxygen (20%) and nitrogen (80%). The molecular weight of oxygen is 32 amu, while the molecular weight of nitrogen is 28 amu. We will approximate the mass of an average molecule of air to be about 30 amu.

What is the molecular weight of a water molecule (H_2O)?

The molecular weight of a water molecule is 18 amu, almost half the molecular weight of an air molecule. Adding water to air (more precisely, replacing some of the air molecules with water molecules) reduces the weight of the air (see Figure 5.6). Thus, wet air is lighter than dry air. The more water in the air, the lighter the air.

FIGURE 5.6 (a) Box filled with 5 air molecules weighs about 150 amu.

$$(5 \text{ air molecules})\left(\frac{30 \text{ amu}}{1 \text{ air molecule}}\right) = 150 \text{ amu}$$

(b) Box filled with 2 air molecules and 3 water molecules weighs about 114 amu.

$$(3 \text{ water molecules})\left(\frac{18 \text{ amu}}{1 \text{ water molecules}}\right) + (2 \text{ air molecules})\left(\frac{30 \text{ amu}}{1 \text{ air molecule}}\right)$$

$$= 54 \text{ amu} + 60 \text{ amu} = 114 \text{ amu}$$

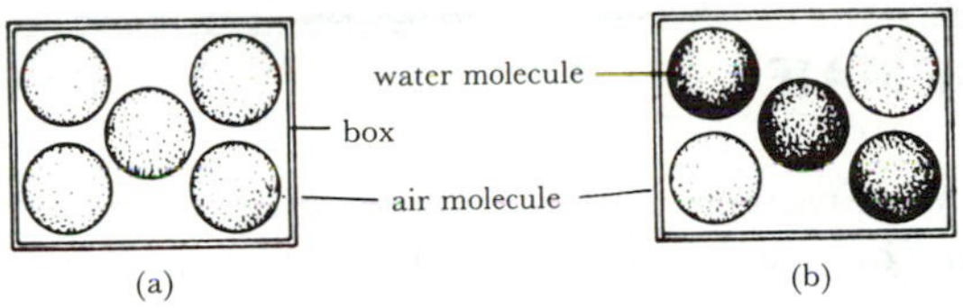

Because the wet air from the Pacific Ocean is lighter and because the Rocky Mountains are in its easterly path, this moist air mass climbs up the western slopes of the Rocky Mountains. As this air mass rises, it loses water and it rains on the western slopes.

Why does rising moist air lose water?

Temperatures drop with increasing altitude. Warm air can hold more moisture than cold air. That's why you dry your hair and your clothes with hot air. Rising air cools, it becomes saturated, and water condenses. Rain falls. The air mass becomes drier as it loses water in the form of rain.

As the air becomes drier, does its density increase or decrease?

Good, you're paying attention. The air becomes heavier as it becomes drier. The heavier dry air falls down the eastern slopes of the Rockies.

As the air falls, what happens to the pressure on this falling mass of air? Does the pressure increase, decrease, or remain the same?

The pressure on the falling gas increases as the gas approaches the ground, where the atmospheric pressure is greater. Remember that atmospheric pressure is due to a column of air, and the amount of atmosphere over point C at the top of the Rockies is less than the amount of atmosphere over point D on the ground (see Figure 5.5).

What happens to the temperature of a gas when its pressure is increased?

The temperature of a gas rises as the pressure exerted on the gas increases (Gay-Lussac's law). As the air falls down the side of the mountain, its temperature goes up 5.5°F for every thousand-foot drop. So we have warm, heavy, dry gas (air) descending upon the Bow River Valley at point D. The Chinook is this warm air (wind) moving down the Rocky Mountains at 50 mph.

Remember, it is now winter in the Bow River Valley, the ground is covered with snow, and it is quite cold (−40°F). Can you predict the effects of this dry, warm wind called the Chinook?

Since the wind is warm, the temperature of the Bow River Valley rises very rapidly. Because the wind is very dry, it evaporates and absorbs water from the melting snow. The word *Chinook* is an Indian word meaning "snow eater." The Chinook can eat a foot of snow off the ground overnight. The weather becomes warmer and the snow is cleared away. It's the sort of thing they have fantasies about in Buffalo, New York.

5.7 EQUATION OF STATE

The equation of state is a mathematical formula that relates the temperature, pressure, and volume of a gas. It is also called the ideal gas law equation.

pressure in atmospheres (→ P)
volume in liters (→ V)
moles of gas (→ n)

$$PV = nRT$$

T ← temperature in degrees Kelvin

R ↑ gas law constant, which equals 0.0821 L atm/mol K

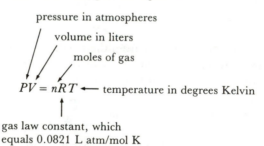

In this equation, P is the pressure of the gas in atmospheres, V is the volume of the gas in liters, T is the temperature of the gas in degrees Kelvin, n is the number of moles of gas, and R is a constant called the gas law constant. R can have several different values depending upon the units that are used with the other parts of the equation. For example, pressure can be expressed in atmospheres, millimeters of mercury, pounds per square inch, and kilopascals.

The equation of state shares a problem with all formulas (as we've mentioned in Section 2.4.2, "Formula Approach: Pros and Cons"): The formula restricts you to specific units.

EXAMPLE 34 Find the volume of 1.0 mol of a gas whose temperature is 50°C and whose pressure is 2.0 atm.

Solution Solve the equation of state for V.

$$PV = nRT$$

$$V = \frac{nRT}{P}$$

$$= \frac{(1.0\ \text{mol})\left(\dfrac{0.0821\ \text{L atm}}{1.00\ \text{mol K}}\right)(323\ \text{K})}{2.0\ \text{atm}} = 13\ \text{L}$$

EXAMPLE 35 Find the pressure of 2.0 mol of a gas whose volume is 30 L at a temperature of 10°C.

Solution Solve the ideal gas law equation for P.

$$PV = nRT$$

$$P = \frac{nRT}{V}$$

$$= \frac{(2.0\text{ mol})\left(\frac{0.0821\text{ L atm}}{1.00\text{ mol K}}\right)(283\text{ K})}{30\text{ L}} = 1.5\text{ atm}$$

•
•

EXAMPLE 36 Find the temperature of 5.0 mol of a gas whose volume is 20 L at a pressure of 380 mm Hg.

Solution Since we must use the unit *atmospheres*, convert the value 380 mm Hg to a value in the unit *atmospheres*.

$$(380\text{ mm Hg})\left(\frac{1.00\text{ atm}}{760\text{ mm Hg}}\right) = 0.500\text{ atm}$$

Now use the equation of state to find the unknown temperature.

$$PV = nRT$$

$$nRT = PV$$

$$T = \frac{PV}{nR}$$

$$= \frac{(0.500\text{ atm})(20\text{ L})}{(5.0\text{ mol})\left(\frac{0.0821\text{ L atm}}{1.00\text{ mol K}}\right)} = 24\text{ K}$$

•
•

EXAMPLE 37 A sample of gas has a temperature of 500 K, a pressure of 2.00 atm, and a volume of 20 L. How many moles of gas are present in this sample?

Solution Solve the equation of state for n, the number of moles of gas.

$$PV = nRT$$

$$nRT = PV$$

$$n = \frac{PV}{RT}$$

$$= \frac{(2.00\text{ atm})(20\text{ L})}{\left(\frac{0.0821\text{ L atm}}{1.00\text{ mol K}}\right)(500\text{ K})} = 0.97\text{ mol}$$

•
•

SELF-TEST

1. Find the temperature (in degrees Celsius) of 1.0 mol of a gas whose volume is 13 L and whose pressure is 2.0 atm.

2. Find the volume of 2.0 mol of a gas whose temperature is 10°C and whose pressure is 1.5 atm.

3. Find the pressure of 5.00 mol of a gas whose volume is 20 L at a temperature of 24 K.

4. A sample of gas has a temperature of 500 K, a pressure of 2.00 atm, and a volume of 40 L. How many moles of gas are present in this sample?

ANSWERS **1.** 44°C **2.** 31 L **3.** 0.49 atm **4.** 1.9 mol

5.7.1 THE BASKETBALL AS A UNIT VOLUME, PART II*

EXAMPLE 38 We have shown that the volume of three basketballs is 22.4 L. Use the equation of state to verify that this is also the volume occupied by 1.00 mol of gas at 0°C and 1.00 atm.

Solution Solve the equation of state for V.

$$PV = nRT$$

$$V = \frac{nRT}{P}$$

$$= \frac{(1.00 \text{ mol})\left(\dfrac{0.0821 \text{ L atm}}{1.00 \text{ mol K}}\right)(273 \text{ K})}{(1.00 \text{ atm})} = 22.4 \text{ L}$$

EXAMPLE 39 Use the equation of state to calculate the value of R for the volume of gas inside three basketballs (22.4 L) at a temperature of 0°C and a pressure of 1.00 atm.

Solution Solve the equation of state for R.

$$PV = nRT$$

$$R = \frac{PV}{nT}$$

$$= \frac{(1.00 \text{ atm})(22.4 \text{ L})}{(1.00 \text{ mol})(273 \text{ K})} = \frac{0.0821 \text{ L atm}}{1.00 \text{ mol K}}$$

PROBLEM SET 5

The problems in Problem Set 5 parallel the examples in Chapter 5. For example, if you should have trouble working Problem 5, go back to Example 5 in this chapter to get help. The correct answers to these problems are given at the end of this book.

1. A gas has a pressure of 3.0 atm. Express the pressure of the gas in pounds per square inches and in kilopascals.
2. A gas sample has a pressure of 800 mm Hg. What is the pressure of this gas sample in atmospheres and in kilopascals?
3. A tank of oxygen is under a pressure of about 4000 kPa. Express this pressure in millimeters of mercury.
4. A sample of gas has a volume of 12 L and a pressure of 1 atm. If the pressure of the gas is increased to 2 atm, what is the new volume of the gas? Assume that the temperature remains constant while the pressure is changed.
5. A container of oxygen has a volume of 30 mL and a pressure of 4 atm. If the pressure of the oxygen gas is reduced to 2 atm and the temperature is kept constant, what is the new volume of the oxygen gas?
6. A tank of nitrogen has a volume of 14 L and a pressure

*Adapted from Fred H. Jardine, "3 Basketballs = 1 Mole of Ideal Gas at STP," *Journal of Chemical Education* 54 (Feb. 1977): 112–13.

of 760 mm Hg. Find the volume of the nitrogen when its pressure is changed to 400 mm Hg while the temperature is held constant.
7. A 40-L tank of ammonia has a pressure of 8.0 psi. Calculate the volume of the ammonia if its pressure is changed to 12 psi while its temperature remains constant.
8. Two hundred liters of helium at 2.00 atm and 28°C is placed into a tank with an internal pressure of 600 kPa. Find the volume of the helium after it is compressed into the tank when the temperature of the tank is 28°C.
9. You decide to go deep-sea diving without any diving equipment. Your lungs hold a total volume of about 10 L. You jump into the water and descend to a depth of 66 ft. What is the volume of your lungs at this depth?
10. You are now wearing scuba gear and swimming underwater at a depth of 66 ft. You are breathing air at 3 atm and your lung volume is 10 L. Your scuba gauge indicates that your air supply is low so, to conserve air, you make a terrible and fatal mistake: you hold your breath while you surface. What happens to your lungs?
11. The temperature of a 4.00-L sample of gas is changed from 10°C to 20°C. What will the volume of this gas be at the new temperature if the pressure is held constant?
12. Carbon dioxide is usually formed when gasoline is burned. If 30 kL of carbon dioxide is produced at a temperature of 1000°C and allowed to reach room temperature (25°C) without any pressure changes, what is the new volume of the carbon dioxide?
13. A 600-mL sample of nitrogen is warmed from 77°F to 86°F. Find its new volume if the pressure remains constant.
14. Assume that the pressure within the cabin of a jet plane remains constant as the temperature decreases from 22°C to −30°C. What volume change occurs to a confined 400-mL sample of gas?
15. Assume that the temperature in the cabin remains constant and the pressure is reduced from 760 mm Hg to 360 mm Hg. What is the change in the volume of a confined 400-mL gas sample?
16. A 30-L sample of nitrogen inside a metal container at 20°C is placed inside an oven whose temperature is 50°C. The pressure inside the container at 20°C was 3.0 atm. What is the pressure of the nitrogen after its temperature is increased?
17. A sample of gas at 3000 mm Hg inside a steel tank is cooled from 500°C to 0°C. What is the final pressure of the gas in the steel tank?
18. The temperature of a sample of gas in a steel container at 30.0 kPa is increased from −100°C to 1000°C. What is the final pressure inside the tank?
19. Calculate the final pressure inside a scuba tank after it cools from 1000°C to 25°C. The initial pressure in the tank is 130 atm.
20. A container holds three gases: oxygen, carbon dioxide, and helium. The partial pressures of the three gases are 2 atm, 3 atm, and 4 atm, respectively. What is the total pressure inside the container?
21. A container with two gases, helium and argon, is 30% by volume helium. Calculate the partial pressure of helium and argon if the total pressure inside the container is 4.0 atm.
22. If 60.0 L of nitrogen is collected over water (see the diagram that accompanies Example 22) at 40°C when the atmospheric pressure is 760 mm Hg, what is the partial pressure of the nitrogen?
23. Eighty liters of oxygen is collected over water at 50°C. The atmospheric pressure in the room is 96.00 kPa. What is the partial pressure of the nitrogen?
24. What is the partial pressure of the oxygen in your lungs if you are diving at a depth of 165 ft underwater, breathing air from a scuba tank that is 80% nitrogen and 20% oxygen?
25. Find the volume of a gas at 2.0 atm and 200 K if its original volume was 300 L at 1.0 atm and 400 K.
26. Five hundred liters (500.0 L) of a gas are prepared at 700 mm Hg and 200°C. The gas is placed into a tank under high pressure. When the tank cools to 20°C, the pressure of the gas is 30.0 atm. What is the volume of the gas?
27. What is the final volume of a 400-mL gas sample that is subjected to a temperature change from 22°C to −30°C and a pressure change from 760 mm Hg to 360 mm Hg?
28. One liter (1.000 L) of hydrogen is collected over water at 30°C at a pressure of 831.8 mm Hg. Find the volume of dry hydrogen collected when its pressure is reduced to 400 mm Hg.
29. A 30.625-g sample of gas occupies 22.4 L at STP. What is the molecular weight of this gas?
30. A 40-g gas sample occupies 11.2 L at STP. Find the molecular weight of this gas.
31. A 12.0-g sample of gas occupies 19.2 L at STP. What is the molecular weight of this gas?
32. A gas that occupies 48.0 L at 700 mm Hg and 20°C weighs 96.0 g. What is the molecular weight of this gas?
33. A gas that occupies 4.167 L at 79.97 kPa and 30°C weighs 20.83 g. What is the molecular weight of this gas?
34. Find the volume of 2.4 mol of gas whose temperature is 50°C and whose pressure is 2.0 atm.
35. Find the pressure of 4.8 mol of a gas whose volume is 72 L at a temperature of 10°C.
36. Find the temperature of 12 mol of gas whose volume is 48 L at a pressure of 380 mm Hg.
37. A sample of gas has a temperature of 208.3 K, a pressure of 2.00 atm, and a volume of 8.333 L. How many moles of gas are present in this sample?
38. Use the equation of state to determine the volume of 2.00 mol of gas at 0°C and 1.00 atm.
39. Use the equation of state to calculate the value of R (the gas law constant) for the volume of 3.00 mol of gas (67.2 L) at a temperature of 0°C and a pressure of 1.00 atm.

6 Stoichiometry

6.1 INTRODUCTION

When two or more substances (e.g., elements, atoms, molecules, and compounds) react to form a new substance, a chemical reaction has occurred. We've already discussed the chemical reaction between carbon and oxygen, which is represented by the chemical equation

$$C + O_2 \rightarrow CO_2$$

reactants are written on the left-hand side — products are written on the right-hand side

We interpret this equation as follows: "One mole of carbon reacts with one mole of oxygen (molecules) to produce one mole of carbon dioxide." Scientists use chemical equations as a shorthand method for representing chemical reactions.

If you want to make a certain amount of a product (e.g., 50 g of carbon dioxide), you may want to calculate the amount of the reactants (carbon and oxygen) you will need. Stoichiometry calculations will give you this information. Stoichiometry calculations can also tell you the quantity of the products that will form from a given quantity of reactants.

Before you can solve a stoichiometry problem you must have a chemical equation that is balanced. The use of unbalanced chemical equations to solve stoichiometry problems is a very common student error. Many times you will have to balance a chemical equation before you can use it (interpret it) to do a stoichiometry problem. We will look at some of the general ideas of balancing chemical equations before we discuss stoichiometry problems.

6.2 CHEMICAL EQUATIONS

A chemical equation shows the reactants on the left-hand side and the products on the right-hand side. Sometimes a chemical equation will have, in parentheses, phase symbols (s, l, or g) to the right of each substance.

$$C(s) + O_2(g) \rightarrow CO_2(g)$$

These symbols let you know whether the substances are solids (s), liquids (l), or gases (g). It isn't always necessary to write phase symbols because the phases of many substances are usually obvious. For example, oxygen is usually a gas. However, there are times when phase symbols are useful. Water, for example, normally exists in both the liquid phase and the gas phase (water vapor).

To help prevent possible confusion, we will always use phase symbols in volume problems where knowledge of the phase of each substance is very important. It is also important to indicate the phase of each substance in thermodynamics problems (see Chapters 12 and 13).

•
•

EXAMPLE 1 Carbon reacts with oxygen to produce carbon dioxide. This reaction can be represented by the chemical equation

$$C(s) + O_2(g) \rightarrow CO_2(g)$$

In what phase (state) is the carbon?

Solution Carbon is in the solid state. The phase symbol (s), to the right of the carbon symbol C, tells you that carbon is a solid.

•
•

EXAMPLE 2 Hydrogen burns in oxygen to produce water. This can be represented by the chemical equation

$$2H_2(g) + O_2(g) \rightarrow 2H_2O(g)$$

Is the water that forms a solid, a liquid, or a gas?

Solution Water is produced as a gas; we know this because the phase symbol g is written to the right of H_2O.

•
•

In front of each substance in a chemical equation is a coefficient (number). If no number is written, the coefficient is assumed to be 1. The coefficients are the number of atoms, molecules, or formula units of each substance shown in the chemical equation.

•
•

EXAMPLE 3 When calcium reacts with chlorine to produce calcium chloride, the reaction is represented by the chemical equation

$$Ca + Cl_2 \rightarrow CaCl_2$$

What are the coefficients of each substance? Interpret the coefficients in this chemical equation.

Solution Since no coefficients are written, 1 is assumed to be the coefficient of each substance. We interpret this equation as follows: "One atom of calcium reacts with 1 molecule of chlorine (2 chlorine atoms) to form 1 formula unit of calcium chloride."

•
•

EXAMPLE 4 When sodium reacts with chlorine to produce sodium chloride, the reaction is represented by the chemical equation

$$2Na + Cl_2 \rightarrow 2NaCl$$

What are the coefficients of each substance? Interpret the coefficients of this chemical equation.

Solution The coefficient of sodium (Na) is 2, the coefficient of chlorine (Cl_2) is 1, and the coefficient of sodium chloride (NaCl) is 2. We can interpret this equation as follows: "Two atoms of sodium react with 1 molecule of chlorine to produce 2 formula units of sodium chloride."

•
•

6.3 BALANCING CHEMICAL EQUATIONS

A balanced chemical equation is one in which the number of atoms of all elements shown on the reagent side (left-hand side) is equal to the number of atoms of all elements shown on the product side (right-hand side). Actually, this definition is not perfectly correct because it does not mention electrical effects, which must also be balanced. However, we will not worry about these effects at this time.

•
•

EXAMPLE 5 Consider the chemical equation showing carbon reacting with oxygen to produce carbon dioxide. (Such a reaction occurs whenever a carbon-containing substance, such as gasoline, oil, or coal, is burned in air.)

$$C + O_2 \rightarrow CO_2$$

Is this chemical equation balanced? How many atoms of carbon and oxygen are on both sides of this chemical equation?

Solution Yes, this chemical equation is balanced because 1 carbon atom and 2 oxygen atoms are located on each side (the reagent side and the product side) of this chemical equation. The number of atoms of all elements shown on the reagent side (left side) is equal to the number of atoms of all elements shown on the product side (right side).

EXAMPLE 6 Consider the chemical equation showing the reaction between water and sulfur trioxide to produce sulfuric acid. (Such a reaction occurs when air polluted with SO_3 is exposed to rain to form acid rain. Acid rain is a serious worldwide problem that results in billions of dollars of damage each year.)

$$H_2O + SO_3 \rightarrow H_2SO_4$$

Is this chemical equation balanced? How many atoms of each element are there on both sides of this chemical equation?

Solution Yes, this chemical equation is balanced. There are 2 hydrogen atoms, 1 sulfur atom, and 4 oxygen atoms on both sides of this chemical equation.

EXAMPLE 7 Consider the chemical equation showing the reaction between sodium and chlorine to produce sodium chloride (common table salt).

$$Na + Cl_2 \rightarrow NaCl$$

Is this a balanced chemical equation? How many atoms of each element are shown on the reagent and the product sides in this equation?

Solution No, this chemical equation is not balanced because there are 2 chlorine atoms on the reagent side and only 1 chlorine atom on the product side.

Many times a chemical equation can be balanced by inspection or by trial and error. When balancing a chemical equation, you are only allowed to change the coefficients of the reagents and the products.

EXAMPLE 8 Balance the chemical equation showing the reaction between bromine and chlorine to produce bromine chloride. (Bromine chloride is a very effective disinfecting agent in wastewater treatment. It is even more efficient than chlorine.)

$$Br_2 + Cl_2 \rightarrow BrCl$$

↑ This chemical equation is not balanced because it shows 2 bromine atoms on the reagent side and only 1 bromine atom on the product side. Also, the chlorine atoms aren't balanced.

Solution By changing the coefficient of BrCl from 1 (it is understood that there is a 1 in front of any substance in a chemical equation if no coefficient is written) to 2, we obtain

$$Br_2 + Cl_2 \rightarrow 2BrCl$$

This is now a balanced chemical equation because there are 2 atoms of chlorine and 2 atoms of bromine on each side. Note that it would have been incorrect to balance this chemical equation by changing the subscripts in BrCl (understood to be Br_1Cl_1) from 1's to 2's, giving Br_2Cl_2. The compound Br_2Cl_2 is not the same compound as BrCl. The molecular formula BrCl indicates that 1 molecule of bromine chloride contains only 1 bromine atom and only 1 chlorine atom. We cannot balance a chemical equation by substituting for the products or reactants in the chemical equation. We can only balance a chemical equation by varying the amounts of products and reactants (i.e., by changing the coefficients of the products and reactants in the chemical equation).

•
•

EXAMPLE 9 Balance the chemical equation showing the incomplete combustion of carbon to form carbon monoxide. (Carbon monoxide is a frequent cause of automobile accidents and fatalities; see Section 9.7.1.)

$$C + O_2 \rightarrow CO$$

Solution We cannot balance this equation by changing the subscript of oxygen in CO (understood to be C_1O_1) to CO_2 because CO (carbon monoxide) and CO_2 (carbon dioxide) are two different chemicals with different physical and chemical properties. For example, CO is very toxic to humans, while CO_2 is not toxic. To balance this chemical equation, first change the coefficient of CO from 1 to 2, giving

$$C + O_2 \rightarrow 2CO$$

Now the number of oxygen atoms on both sides of this chemical equation is equal (there are 2 oxygen atoms on each side), but the number of carbon atoms on each side is not equal. There is 1 carbon atom on the reagent side and 2 carbon atoms on the product side of this chemical equation. By changing the coefficient of carbon on the reagent side from 1 to 2, we obtain

$$2C + O_2 \rightarrow 2CO$$

This is a balanced chemical equation because it has an equal number of carbon and oxygen atoms on each side.

•
•

Many chemical equations are more difficult to balance than those shown in the last two examples. Some of these chemical equations can be balanced by trial and error. Before looking at these more difficult equations, there are three hints (as opposed to rigid rules) that you may choose to observe to minimize your efforts when balancing difficult equations.

> **HINT 1:** Start to balance a chemical equation by focusing your attention on the most complicated substance present. The most complicated substance is the one made up of the greatest number of atoms. Count the number of atoms of each element making up the most complicated substance and change the coefficients on the other side of the chemical equation so that those elements are balanced.

> **HINT 2:** Treat groups of atoms (such as carbonate, CO_3^{2-}) that appear on both sides of the chemical equation as single units. The common oxygen-containing anions (negative ions), such as carbonate (CO_3^{2-}), sulfate (SO_4^{2-}), phosphate (PO_4^{3-}),

chlorate (ClO_3^-), and nitrate (NO_3^-), can usually be treated as single units. For example, don't think of $Al_2(SO_4)_3$ as containing 3 sulfur atoms and 12 oxygen atoms; rather, think of this compound as containing 3 sulfates.

HINT 3: If water is present as a product or a reagent, avoid changing the coefficient of water until the end of the equation-balancing process, and then change the coefficient of water so that the number of hydrogen atoms on both sides of the chemical equation are balanced. It is not always possible to use this third hint, but it comes in mighty handy when it can be used.

EXAMPLE 10 Balance the chemical equation

$$H_2SO_4 + Al(OH)_3 \rightarrow Al_2(SO_4)_3 + H_2O$$

Solution The most complicated compound present is $Al_2(SO_4)_3$. It contains 17 atoms, more atoms than there are in any other substance shown. Therefore, we focus our attention on aluminum sulfate, noticing that it contains 2 aluminum atoms. Change the coefficient of $Al(OH)_3$ from 1 to 2 to balance the number of aluminum atoms on both sides of this chemical equation. We now have

$$H_2SO_4 + 2Al(OH)_3 \rightarrow Al_2(SO_4)_3 + H_2O$$

As we continue to focus our attention on aluminum sulfate, we notice that it contains 3 sulfates (SO_4^{3-}). There is only 1 sulfate on the left-hand side in H_2SO_4, so we change the coefficient of H_2SO_4 from 1 to 3 to balance the number of sulfates on both sides. We now have

$$3H_2SO_4 + 2Al(OH)_3 \rightarrow Al_2(SO_4)_3 + H_2O$$

We no longer need to focus our attention on aluminum sulfate (as suggested by the first hint) because we have exhausted the information it can provide. We are following the advice given in the second hint when we treat sulfate as a single unit to make sure that both sides of this chemical equation have an equal number of sulfates. The third hint suggests that we do nothing with the water until we have first examined the coefficients of the other substances in the chemical equation. We have considered or changed the coefficients of sulfuric acid (H_2SO_4), aluminum hydroxide ($Al(OH)_3$), and aluminum sulfate ($Al_2(SO_4)_3$). Only water remains. We will change the coefficient of water to balance the hydrogen atoms on both sides of the equation. We notice that there are 12 hydrogen atoms on the left-hand side (6 in $3H_2SO_4$ and 6 in $2Al(OH)_3$). By changing the coefficient of water from 1 to 6, we balance the chemical equation. We now have

$$3H_2SO_4 + 2Al(OH)_3 \rightarrow Al_2(SO_4)_3 + 6H_2O$$

This is a balanced chemical equation because both sides have 12 H atoms, 2 Al atoms, 18 O atoms, and 3 S atoms.

EXAMPLE 11 Balance the chemical equation

$$H_3PO_4 + Ca(OH)_2 \rightarrow Ca_3(PO_4)_2 + H_2O$$

Solution The most complicated compound present is $Ca_3(PO_4)_2$ (calcium phosphate). It contains 13 atoms. Therefore, we focus our attention on calcium phosphate, noticing that it contains 3 calcium atoms. Change the coefficient of calcium hydroxide [$Ca(OH)_2$] from 1 to 3 to balance the number of calcium atoms on both sides of the chemical equation. We now have

$$H_3PO_4 + 3Ca(OH)_2 \rightarrow Ca_3(PO_4)_2 + H_2O$$

Notice that calcium phosphate contains 2 phosphates (PO_4^{3-}). Change the coefficient of H_3PO_4 (phosphoric acid) from 1 to 2 to balance the number of phosphates on both sides. We now have

$$2H_3PO_4 + 3Ca(OH)_2 \rightarrow Ca_3(PO_4)_2 + H_2O$$

We are following the advice of the second hint when we treat phosphate as a single unit. We now change the coefficient of water from 1 to 6 in order to balance the number of hydrogen atoms on both sides of the equation. We now have

$$2H_3PO_4 + 3Ca(OH)_2 \rightarrow Ca_3(PO_4)_2 + 6H_2O$$

This is a balanced chemical equation because both sides contain 12 H atoms, 2 P atoms, 14 O atoms, and 3 Ca atoms.

EXAMPLE 12 Balance the chemical equation

$$GaCl_3 + Ag_2CO_3 \rightarrow Ga_2(CO_3)_3 + AgCl$$

Solution The most complicated compound present here is $Ga_2(CO_3)_3$ (gallium carbonate). It contains 14 atoms, more atoms than there are in any other substance shown. This substance contains 2 gallium atoms. Change the coefficient of gallium chloride ($GaCl_3$) from 1 to 2 to balance the number of gallium atoms on both sides. We now have

$$2GaCl_3 + Ag_2CO_3 \rightarrow Ga_2(CO_3)_3 + AgCl$$

Gallium carbonate also contains 3 carbonates (CO_3^{2-}). Change the coefficient of silver carbonate (Ag_2CO_3) from 1 to 3 to balance the number of carbonates on both sides, giving

$$2GaCl_3 + 3Ag_2CO_3 \rightarrow Ga_2(CO_3)_3 + AgCl$$

Water is absent from this chemical reaction, so we do not follow the third hint. Notice that there are 6 silver atoms on the left-hand side of this chemical equation; therefore, we must change the coefficient of silver chloride (AgCl) from 1 to 6, giving

$$2GaCl_3 + 3Ag_2CO_3 \rightarrow Ga_2(CO_3)_3 + 6AgCl$$

This is a balanced chemical equation since each side contains 2 gallium atoms, 6 chlorine atoms, 6 silver atoms, 3 carbon atoms, and 9 oxygen atoms.

SELF-TEST Balance each of the following chemical equations:

1. $H_2 + Cl_2 \rightarrow HCl$
2. $K + Br_2 \rightarrow KBr$
3. $H_2CO_3 + Al(OH)_3 \rightarrow Al_2(CO_3)_3 + H_2O$
4. $HNO_3 + Ba(OH)_2 \rightarrow Ba(NO_3)_2 + H_2O$
5. $BaCl_2 + Na_2SO_4 \rightarrow BaSO_4 + NaCl$

ANSWERS **1.** $H_2 + Cl_2 \rightarrow 2HCl$ **2.** $2K + Br_2 \rightarrow 2KBr$
3. $3H_2CO_3 + 2Al(OH)_3 \rightarrow Al_2(CO_3)_3 + 6H_2O$
4. $2HNO_3 + Ba(OH)_2 \rightarrow Ba(NO_3)_2 + 2H_2O$
5. $BaCl_2 + Na_2SO_4 \rightarrow BaSO_4 + 2NaCl$

We said that stoichiometry calculations can tell you the quantity of the products that will form from a given quantity of reactants. Stoichiometry calculations can also tell you the quantity of reactants that will react to form a certain quantity of products. Quantity can have various units, such as moles, liters, and grams. Calculations dealing with just moles are called mole problems. Likewise, calculations dealing with just volume or mass are called volume problems and mass problems. We will treat all stoichiometry problems as a series of dimensional analysis manipulations to enhance your ability to interpret chemical equations. However, let's first see how we can interpret chemical equations on a macroscopic level (a larger scale than atoms and molecules).

6.4 CHEMICAL EQUATIONS: MACROSCOPIC INTERPRETATIONS

Like the subscripts of a molecular formula (see Section 4.7.2), the coefficients of a balanced chemical equation can be interpreted as the number of moles of the substances shown in the equation.

•
•

EXAMPLE 13 When hydrogen burns in oxygen to produce water, the reaction is represented with the balanced chemical equation

$$2H_2(g) + O_2(g) \rightarrow 2H_2O(l)$$

How many moles of hydrogen and oxygen are reacting? How many moles of water form?

Solution The coefficients of hydrogen, oxygen, and water are 2, 1, and 2, respectively. Two moles of hydrogen reacts with 1 mol of oxygen to produce 2 mol of water.

•
•

When we interpret the coefficients as the number of moles, this gives us a macroscopic (large) interpretation of the chemical reaction because we can convert moles into visible and measurable quantities of substances, such as grams and liters.

•
•

EXAMPLE 14 In the reaction between carbon and oxygen to form carbon dioxide, how many grams of carbon are reacting with 1 mol of oxygen?

$$C + O_2 \rightarrow CO_2$$ ← Always check to make sure you are working with a balanced equation.

Solution The coefficients of carbon (1) and oxygen (1) tell you that 1 mol of carbon reacts with 1 mol of oxygen. One mole of carbon is 12 g of carbon. Therefore, 12 g of carbon reacts with 1 mol of oxygen.

•
•

EXAMPLE 15 How many grams of oxygen react with 1 mol of carbon in the following reaction?

$$C + O_2 \rightarrow CO_2$$

Solution The coefficient of oxygen is 1. Therefore, 32 g of oxygen (1 mol of oxygen weighs 32 g) reacts with 1 mol of carbon.

•
•

EXAMPLE 16 How many grams of carbon dioxide are produced when 1 mol of oxygen reacts with 1 mol of carbon in the following equation?

$$C + O_2 \rightarrow CO_2$$

Solution Carbon dioxide has a coefficient of 1. One mole of carbon dioxide has a mass of 44 g. We can write the masses of the reactants and products over the balanced chemical equation, as shown:

$$\overset{12\ g}{C} + \overset{32\ g}{O_2} \rightarrow \overset{44\ g}{CO_2}$$

This equation is read (interpreted) as follows: "12 grams of carbon reacts with 32 grams of oxygen to produce 44 grams of carbon dioxide." Note that the total mass (sum) of the reactants equals the total mass of the products (12 g + 32 g = 44 g). We say that mass is conserved (we don't create or destroy mass when a chemical reaction occurs).

•
•

We also know that 1 mol of anything contains 6×10^{23} particles (atoms, ions, or molecules).

•
•

EXAMPLE 17 The following reaction is the balanced chemical equation representing the reaction between hydrogen and oxygen to produce water:

$$2H_2 + O_2 \rightarrow 2H_2O$$

Interpret this equation in terms of the number of molecules of hydrogen and of oxygen that react and the number of water molecules that form.

Solution The coefficient of oxygen is 1. One mole of oxygen contains 6×10^{23} oxygen molecules. Hydrogen's coefficient is 2. Two moles of hydrogen contains 12×10^{23} hydrogen molecules.

$$(2 \text{ mol } H_2)\left(\frac{6.0 \times 10^{23} \text{ } H_2 \text{ molecules}}{\text{mol } H_2}\right) = 12 \times 10^{23} \text{ } H_2 \text{ molecules}$$

The coefficient of water is 2. This means that 12×10^{23} water molecules form.

•
•

We can write the number of molecules of the substances over the chemical reaction:

$$\overset{12 \times 10^{23} \text{ molecules}}{2H_2} + \overset{6 \times 10^{23} \text{ molecules}}{O_2} \rightarrow \overset{12 \times 10^{23} \text{ molecules}}{2H_2O}$$

This is interpreted in the following way: "12×10^{23} hydrogen molecules react with 6×10^{23} oxygen molecules to produce 12×10^{23} water molecules."

•
•

EXAMPLE 18 The equation

$$C + O_2 \rightarrow CO_2$$

represents the chemical reaction between carbon and oxygen to produce carbon dioxide. Interpret the equation in terms of the number of molecules of each substance that is reacting and forming.

Solution Because the coefficient of each substance is 1, there are 6×10^{23} atoms of carbon reacting with 6×10^{23} molecules of oxygen to form 6×10^{23} carbon dioxide molecules. This information can be written over the chemical equation:

6×10^{23} atoms 6×10^{23} molecules 6×10^{23} molecules

$$C + O_2 \rightarrow CO_2$$

We also know that 1 mol of any gas occupies about 22.4 L at STP.

EXAMPLE 19 The equation representing the reaction between hydrogen and oxygen to produce water is

$$2H_2(g) + O_2(g) \rightarrow 2H_2O(l)$$

Interpret this equation in terms of the volume of each substance at STP.

Solution Two moles of hydrogen occupies 44.8 L.

$$(2 \text{ mol } H_2 \text{ at STP})\left(\frac{22.4 \text{ L } H_2 \text{ at STP}}{\text{mol } H_2 \text{ at STP}}\right) = 44.8 \text{ L } H_2 \text{ at STP}$$

One mole of oxygen has a volume of 22.4 L. Two moles of liquid water has a negligible volume relative to the volumes of hydrogen and oxygen. The volume of 2 mol of liquid water is 0.036 L.

$$(2 \text{ mol liquid } H_2O)\left(\frac{18 \text{ g } H_2O}{\text{mol } H_2O}\right)\left(\frac{1 \text{ mL } H_2O}{1 \text{ g } H_2O}\right)\left(\frac{1 \text{ L}}{1000 \text{ mL}}\right) = 0.036 \text{ L } H_2O$$

We can write this information over the chemical equation:

$$\overset{44.8\text{ L}}{2H_2(g)} + \overset{22.4\text{ L}}{O_2(g)} \rightarrow \overset{0.036\text{ L}}{2H_2O(l)}$$

This is read as follows: "44.8 liters of hydrogen reacts with 22.4 liters of oxygen at STP to produce 0.036 liters of liquid water."

EXAMPLE 20 The equation that represents the reaction between carbon and oxygen to produce carbon dioxide is

$$C(s) + O_2(g) \rightarrow CO_2(g)$$

Interpret this equation in terms of the volume of each substance at STP.

Solution A negligible volume of carbon (1 mol C = 12 g C $\doteq \frac{1}{3}$ oz C; carbon is a solid) reacts with 22.4 L of oxygen to produce 22.4 L of carbon dioxide. This can be written

$$\overset{\sim 0\text{ L}}{C(s)} + \overset{22.4\text{ L}}{O_2(g)} \rightarrow \overset{22.4\text{ L}}{CO_2(g)}$$

In summary, the chemical equation

$$C + O_2 \rightarrow CO_2$$

tells us that 1 mol of carbon reacts with 1 mol of oxygen and produces 1 mol of carbon dioxide. Also, we know that

- 1 mol C = 12 g C = 6×10^{23} C atoms $\doteq$ 0 L C (STP)
- 1 mol O_2 = 32 g O_2 = 6×10^{23} O_2 molecules = 22.4 L O_2 (STP)
- 1 mol CO_2 = 44 g CO_2 = 6×10^{23} CO_2 molecules = 22.4 L CO_2 (STP)

All of this information has been summarized in Table 6.1.

TABLE 6.1

	Column 1		Column 2		Column 3
Line 1	C (s)	+	O_2 (g)	→	CO_2 (g)
Line 2	1 mol C	+	1 mol O_2	→	1 mol CO_2
Line 3	12 g C	+	32 g O_2	→	44 g CO_2
Line 4	~0 L C	+	22.4 L O_2	→	22.4 L CO_2
Line 5	6×10^{23} C atoms	+	6×10^{23} O_2 molecules	→	6×10^{23} CO_2 molecules

Look at column 2 in Table 6.1. All four quantities (shown in lines 2, 3, 4, and 5) are equal to one another, that is,

$$1 \text{ mol } O_2 = 32 \text{ g } O_2 = 22.4 \text{ L } O_2 = 6 \times 10^{23} \ O_2 \text{ molecules}$$

The same is true for each of the other columns (all quantities within a given column are equal to one another).

Since each of the five lines says the same thing, we can interchange the quantities in these five lines.

EXAMPLE 21 How many grams of oxygen react with 1 mol of carbon? Refer to Table 6.1 for help.

Solution Combining quantities in the second and third lines of Table 6.1, we see that 1 mol of carbon reacts with 32 g of oxygen.

EXAMPLE 22 How many liters of carbon dioxide are produced when 12 g of carbon reacts with 32 g of oxygen? (See Table 6.1.)

Solution Interchanging the quantities in lines 3 and 4, we see that 22.4 L of carbon dioxide is produced when 32 g of oxygen reacts with 12 g of carbon.

EXAMPLE 23 How many atoms of carbon are needed to make 1 mol of carbon dioxide? (See Table 6.1.)

Solution Interchanging the information on the second and fifth lines of Table 6.1 tells us that 6×10^{23} carbon atoms must react to produce 1 mol of carbon dioxide.

EXAMPLE 24 Give an interpretation of the reaction between carbon and oxygen by interchanging quantities in lines 2, 3, and 4 in Table 6.1.

Solution One possible way of interchanging quantities in lines 2, 3, and 4 gives this interpretation: "1 mole of carbon plus 32 grams of oxygen produces 22.4 liters of carbon dioxide." Another possible interpretation using the same lines (2, 3, and 4) is "12 grams of carbon plus 22.4 liters of oxygen produces 1 mole of carbon dioxide."

•
•

EXAMPLE 25 Interchange information from lines 3, 4, and 5 in Table 6.1 to produce another interpretation of the reaction between carbon and oxygen.

Solution One possible way of interchanging information from lines 3, 4, and 5 of Table 6.1 is "6×10^{23} carbon atoms plus 22.4 liters of oxygen produces 44 grams of carbon dioxide." Another possible interpretation using the same lines (3, 4, and 5) is "12 grams of carbon plus 6×10^{23} oxygen molecules yields 22.4 liters of carbon dioxide."

•
•

EXAMPLE 26 Interchange the quantities from lines 2, 3, and 5 to produce another interpretation of the chemical equation in Table 6.1.

Solution One possibility is "1 mole of carbon plus 6×10^{23} oxygen molecules produces 44 grams of carbon dioxide." Another possibility is "12 grams of carbon plus 1 mole of oxygen yields 6×10^{23} carbon dioxide molecules."

•
•

6.4.1 SIMPLE MULTIPLE APPROACH

In addition to interchanging quantities between lines in Table 6.1, you should be able to work with simple multiples of those quantities.

•
•

EXAMPLE 27 How many moles of oxygen will react with 2 mol of carbon? (See Table 6.1.)

Solution By multiplying each quantity in the second line by 2, we get the following: "2 moles of carbon plus 2 moles of oxygen produces 2 moles of carbon dioxide." Two moles of oxygen reacts with 2 mol of carbon.

•
•

EXAMPLE 28 How many grams of carbon dioxide are produced when 6 g of carbon burns in oxygen? (See Table 6.1.)

Solution Divide each quantity in the third line of Table 6.1 by 2. The line now reads, "6 grams of carbon plus 16 grams of oxygen produces 22 grams of carbon dioxide."

•
•

EXAMPLE 29 How many grams of oxygen will react with 24 g of carbon? (See Table 6.1.)

Solution Double each of the quantities in line 3 of Table 6.1. The line now reads, "24 grams of carbon plus 64 grams of oxygen yields 88 grams of carbon dioxide." Therefore, 64 g of oxygen reacts with 24 g of carbon.

•
•

EXAMPLE 30 How many molecules of oxygen combine with 12×10^{23} atoms of carbon? (See Table 6.1.)

Solution Multiply the quantities in the fifth line by 2. This tells us that 12×10^{23} carbon atoms react with 12×10^{23} oxygen molecules.

•
•

EXAMPLE 31 How many liters of carbon dioxide will form when 11.2 L of oxygen react? (See Table 6.1.)

Solution Divide the quantities in the fourth line by 2. This tells us that 11.2 L of carbon dioxide will form when 11.2 L of oxygen is consumed.

•
•

SELF-TEST

1. Make a table similar to Table 6.1 for the reaction in which hydrogen burns in oxygen to form liquid water. Use this table to answer Questions 2 through 5.
2. How many molecules of hydrogen react with 12×10^{23} oxygen molecules?
3. How many moles of water are produced when 10 mol of oxygen is consumed?
4. How many water molecules are produced when 6×10^{23} hydrogen molecules burn?
5. How many liters of hydrogen combine with 11.2 L of oxygen?

ANSWERS

1.

$2H_2(g)$	$+ O_2(g)$	$\rightarrow 2H_2O(l)$
2 mol H_2	+ 1 mol O_2	$\rightarrow$ 2 mol H_2O
4 g H_2	+ 32 g O_2	$\rightarrow$ 36 g H_2O
44.8 L H_2	+ 22.4 L O_2	$\rightarrow$ ~0 L $H_2O(l)$
12×10^{23} H_2 molecules	+ 6×10^{23} O_2 molecules	$\rightarrow$ 12×10^{23} H_2O molecules

2. 24×10^{23} H_2 molecules **3.** 20 mol H_2O **4.** 6×10^{23} H_2O molecules
5. 22.4 L H_2

6.5 MOLE PROBLEMS

When solving stoichiometry problems, it is more efficient to use dimensional analysis than it is to construct tables such as Table 6.1. We used Table 6.1 to help you develop an appreciation for the many ways that chemical equations can be interpreted. This appreciation is absolutely necessary before you can intelligently and easily use dimensional analysis to solve stoichiometry problems.

•
•

EXAMPLE 32 Find the number of moles of oxygen that will react with 3.0 mol of hydrogen. (*Note*: $2H_2 + O_2 \rightarrow 2H_2O$.)

Solution The chemical equation for this reaction tells us that 1 mol of oxygen is consumed for every 2 mol of hydrogen consumed.

$$(3.0 \text{ mol } H_2 \text{ reacts})\left(\frac{1 \text{ mol } O_2 \text{ reacts}}{2 \text{ mol } H_2 \text{ reacts}}\right) = 1.5 \text{ mol } O_2 \text{ reacts}$$

↑ given

↑ This is our conversion factor, which is obtained from interpreting the chemical equation. If you have trouble finding conversion factors, see Section 2.3.1 for step-by-step instructions.

Our answer is 1.5 mol of oxygen reacts with 3.0 mol of hydrogen.

•
•

EXAMPLE 33 How many moles of water will form when 0.50 mol of hydrogen burns? (*Note*: $2H_2 + O_2 \rightarrow 2H_2O$.)

Solution The chemical equation tells us that 2 mol of water forms for every 2 mol of hydrogen that burns.

$$\underset{\text{given}}{(0.50\text{ mol } H_2 \text{ burns})}\underset{\text{conversion factor}}{\left(\frac{2\text{ mol } H_2O \text{ forms}}{2\text{ mol } H_2 \text{ burns}}\right)} = 0.50\text{ mol } H_2O \text{ forms}$$

The answer is 0.50 mol of water will form when 0.50 mol of hydrogen burns.

EXAMPLE 34 If you wanted to produce 20 mol of water, how many moles of oxygen would be consumed? (*Remember*: $2H_2 + O_2 \rightarrow 2H_2O$.)

Solution The chemical equation tells us that 1 mol of oxygen is consumed for every (per) 2 mol of water produced.

$$(\text{produce } 20\text{ mol } H_2O)\left(\frac{1\text{ mol } O_2 \text{ consumed}}{2\text{ mol } H_2O \text{ produced}}\right) = 10\text{ mol } O_2 \text{ consumed}$$

(per: the fraction bar)

Ten moles of oxygen is needed to produce 20 mol of water.

EXAMPLE 35 Zinc sulfide (ZnS) reacts with oxygen (O_2) to produce zinc oxide (ZnO) and sulfur dioxide (SO_2). This reaction is represented by the chemical equation

$$4ZnS + 6O_2 \rightarrow 4ZnO + 4SO_2$$

You are asked to make 10 mol of zinc oxide. How many moles of zinc sulfide and of oxygen are needed?

Solution

$$(\text{make } 10\text{ mol ZnO}\left(\frac{6\text{ mol } O_2 \text{ reacts}}{4\text{ mol ZnO made}}\right) = 15\text{ mol } O_2 \text{ reacts}$$

$$(\text{make } 10\text{ mol ZnO})\left(\frac{4\text{ mol ZnS required}}{4\text{ mol ZnO made}}\right) = 10\text{ mol ZnS required}$$

Fifteen moles of oxygen must be chemically combined with 10 mol of zinc sulfide to produce 10 mol of zinc oxide.

EXAMPLE 36 Sulfuric acid (H_2SO_4) reacts with hydrogen iodide (HI) to produce hydrogen sulfide (H_2S), iodine (I_2), and water (H_2O). This chemical reaction is represented by the chemical equation

$$H_2SO_4 + 8HI \rightarrow H_2S + 4I_2 + 4H_2O$$

You have 30 mol of HI available and all the sulfuric acid you can use. (We say that HI is the limiting reagent because it is present in limited amounts and it limits the amount of product that will form. The sulfuric acid is present in unlimited amounts. We say that sulfuric acid is present in excess.) How many moles of hydrogen sulfide and iodine can you make from 30 mol of hydrogen iodide?

Solution

$$(30\text{ mol HI reacts})\left(\frac{1\text{ mol } H_2S \text{ forms}}{8\text{ mol HI reacts}}\right) = 3.8\text{ mol } H_2S \text{ forms}$$

$$(30 \text{ mol HI reacts})\left(\frac{4 \text{ mol } I_2 \text{ forms}}{8 \text{ mol HI reacts}}\right) = 15 \text{ mol } I_2 \text{ forms}$$

We calculate that 3.8 mol of hydrogen sulfide and 15 mol of iodine can be made from 30 mol of hydrogen iodide.

•
•

SELF-TEST Nitrogen can react with oxygen to form the gas nitric oxide (NO), as represented by the chemical equation

$$N_2(g) + O_2(g) \rightarrow 2NO(g)$$

1. How many moles of nitric oxide can be produced from 4 mol of nitrogen?
2. How many moles of oxygen are needed to produce 6 mol of nitric oxide?
3. How many moles of nitrogen will react with 14 mol of oxygen?

ANSWERS **1.** 8 mol NO **2.** 3 mol O_2 **3.** 14 mol N_2

6.6 VOLUME PROBLEMS (STP CONDITIONS)

In the chemical equation

$$N_2(g) + O_2(g) \rightarrow 2NO(g)$$

1 mol of nitrogen reacts with 1 mol of oxygen to produce 2 mol of nitric oxide. Because each substance is a gas, we can write the volume over each substance in the chemical equation:

$$(2 \text{ mol NO})\left(\frac{22.4 \text{ L NO}}{\text{mol NO}}\right) = 44.8 \text{ L NO}$$

$$\overset{22.4 \text{ L}}{N_2(g)} + \overset{22.4 \text{ L}}{O_2(g)} \rightarrow \overset{44.8 \text{ L}}{2NO(g)}$$

We interpret this equation as follows: "22.4 liters of nitrogen reacts with 22.4 liters of oxygen to produce 44.8 liters of nitric oxide at STP." In the following examples, assume that all gas volumes are measured under standard temperature and pressure (STP) conditions.

•
•

EXAMPLE 37 How many liters of oxygen are needed to react with 3 L of nitrogen (at STP) according to the chemical equation

$$N_2(g) + O_2(g) \rightarrow 2NO(g)$$

Solution Write the volume of each substance over the chemical equation:

$$\overset{22.4 \text{ L}}{N_2(g)} + \overset{22.4 \text{ L}}{O_2(g)} \rightarrow \overset{44.8 \text{ L}}{2NO(g)}$$

The chemical equation tells you that 22.4 L of nitrogen is consumed for every (per) 22.4 L of oxygen consumed. This is the information we need to make the conversion factor.

$$(3 \text{ L N}_2 \text{ reacts})\left(\frac{22.4 \text{ L O}_2 \text{ reacts}}{22.4 \text{ L N}_2 \text{ reacts}}\right) = 3 \text{ L O}_2 \text{ reacts}$$

given (3 L N_2 reacts); conversion factor from interpreting the chemical equation (22.4 L O_2 reacts / 22.4 L N_2 reacts)

Three liters of oxygen is needed to react with 3 L of nitrogen.

•
•

EXAMPLE 38 How many liters of nitric oxide (NO) will form when 5.0 L of nitrogen combines with oxygen at STP according to the chemical equation

$$N_2(g) + O_2(g) \rightarrow 2NO(g)$$

Solution The chemical equation

$$\overset{22.4\text{ L}}{N_2(g)} + \overset{22.4\text{ L}}{O_2(g)} \rightarrow \overset{44.8\text{ L}}{2NO(g)}$$

tells us that 44.8 L of nitric oxide forms (is produced) for every (per) 22.4 L of nitrogen that reacts.

$$(5.0 \text{ L N}_2 \text{ reacts})\left(\frac{44.8 \text{ L NO forms}}{22.4 \text{ L N}_2 \text{ reacts}}\right) = 10 \text{ L NO forms}$$

given (5.0 L N_2 reacts); conversion factor from chemical equation interpretation (44.8 L NO forms / 22.4 L N_2 reacts)

Ten liters of NO forms when 5.0 L of nitrogen reacts with oxygen.

•
•

EXAMPLE 39 Methane reacts with oxygen to produce carbon dioxide and water. This chemical reaction is represented by the chemical equation

$$CH_4(g) + 2O_2(g) \rightarrow CO_2(g) + 2H_2O(l)$$

How many liters of methane (CH_4) react with 10 L of oxygen at STP?

Solution Write the volume of each substance over the chemical equation:

$$(2 \text{ mol O}_2)\left(\frac{22.4 \text{ L O}_2}{\text{mol O}_2}\right) = 44.8 \text{ L O}_2$$

$$\overset{22.4\text{ L}}{CH_4(g)} + \overset{44.8\text{ L}}{2O_2(g)} \rightarrow \overset{22.4\text{ L}}{CO_2(g)} + \overset{0.036\text{ L}}{2H_2O(l)}$$

$$(2 \text{ mol H}_2\text{O(l)})\left(\frac{18 \text{ g H}_2\text{O(l)}}{\text{mol H}_2\text{O(l)}}\right)\left(\frac{1 \text{ mL H}_2\text{O(l)}}{1 \text{ g H}_2\text{O(l)}}\right)\left(\frac{1 \text{ L}}{1000 \text{ mL}}\right) = 0.036 \text{ L H}_2\text{O(l)}$$

We can obtain our conversion factor by interpreting this chemical equation as follows: "22.4 L of methane reacts per (for every) 44.8 L of oxygen that reacts."

$$(10 \text{ L O}_2 \text{ reacts})\left(\frac{22.4 \text{ L CH}_4 \text{ reacts}}{44.8 \text{ L O}_2 \text{ reacts}}\right) = 5.0 \text{ L CH}_4 \text{ reacts}$$

We see that 5.0 L of methane reacts with 10 L of oxygen.

•
•

EXAMPLE 40 Some methane and oxygen is allowed to react and produce 50.0 L of carbon dioxide. How many liters of oxygen had to react to produce this much carbon dioxide? The chemical equation for this reaction is

$$CH_4(g) + 2O_2(g) \rightarrow CO_2(g) + 2H_2O(l)$$

Solution

$$\begin{array}{cccc} 22.4\text{ L} & 44.8\text{ L} & 22.4\text{ L} & 0.036\text{ L} \end{array}$$
$$CH_4(g) + 2O_2(g) \rightarrow CO_2(g) + 2H_2O(l)$$

The conversion factor is obtained from the chemical equation, which is interpreted as follows: "44.8 liters of oxygen reacts for every 22.4 liters of carbon dioxide produced."

$$(50.0\text{ L } CO_2)\left(\frac{44.8\text{ L } O_2}{22.4\text{ L } CO_2}\right) = 100\text{ L } O_2$$

Our calculation shows that 100 L of oxygen reacts to produce 50.0 L of carbon dioxide.

SELF-TEST Ammonia (NH_3) is prepared by reacting nitrogen (N_2) with hydrogen (H_2). This reaction is represented by the chemical equation

$$N_2(g) + 3H_2(g) \rightarrow 2NH_3(g)$$

1. What volume of nitrogen is required to react with 25 L of hydrogen at STP?
2. If you want to make 20 L of ammonia, how many liters of hydrogen do you need at STP?
3. If 5.0 L of nitrogen is allowed to react with hydrogen to form ammonia, how many liters of ammonia (at STP) will form?

ANSWERS **1.** 8.3 L H_2 **2.** 30 L H_2 **3.** 10 L NH_3

6.7 MASS PROBLEMS

The chemical equation

$$C + O_2 \rightarrow CO_2$$

tells us that 1 mol of carbon reacts with 1 mol of oxygen to produce 1 mol of carbon dioxide. We can write the mass of each substance over the chemical equation, as shown:

$$\begin{array}{ccc} 12\text{ g} & 32\text{ g} & 44\text{ g} \end{array}$$
$$C + O_2 \rightarrow CO_2$$

We can then reinterpret the chemical equation: "12 grams of carbon reacts with 32 grams of oxygen to produce 44 grams of carbon dioxide."

EXAMPLE 41 Carbon burns in oxygen to produce carbon dioxide. How many grams of carbon dioxide will form when 20 g of carbon is burned? The chemical equation is

$$C + O_2 \rightarrow CO_2$$

Solution Write the mass (grams) of each substance over the chemical equation:

$$\begin{matrix} 12\text{ g} & 32\text{ g} & 44\text{ g} \\ C\ + & O_2 \rightarrow & CO_2 \end{matrix}$$

The chemical equation provides us with our conversion factor: "44 grams of carbon dioxide forms for every (per) 12 grams of carbon that reacts."

$$(20\text{ g C burned})\left(\frac{44\text{ g } CO_2\text{ forms}}{12\text{ g C burned}}\right) = 73\text{ g } CO_2\text{ forms}$$

given ↑ (20 g C burned)

↑ conversion factor is an interpretation of the chemical equation

Our calculations indicate that 73 g of carbon dioxide forms when 20 g of carbon is burned in oxygen.

•
•

EXAMPLE 42 How many grams of oxygen are consumed when 4.0 g of carbon burns? The reaction is $C + O_2 \rightarrow CO_2$.

Solution The chemical equation

$$\begin{matrix} 12\text{ g} & 32\text{ g} & 44\text{ g} \\ C\ + & O_2 \rightarrow & CO_2 \end{matrix}$$

tell us that 32 g of oxygen is consumed for every 12 g of carbon burned.

$$(4.0\text{ g C burns})\left(\frac{32\text{ g } O_2\text{ consumed}}{12\text{ g C burns}}\right) = 10\text{ g } O_2\text{ consumed}$$

Ten grams of oxygen is consumed when 4.0 g of carbon burns.

•
•

EXAMPLE 43 Sodium carbonate (Na_2CO_3) reacts with calcium hydroxide [$Ca(OH)_2$] to produce sodium hydroxide (NaOH) and calcium carbonate ($CaCO_3$). This reaction is represented by the chemical equation

$$Na_2CO_3 + Ca(OH)_2 \rightarrow 2NaOH + CaCO_3$$

How many grams of sodium carbonate are required to produce 500 g of sodium hydroxide?

Solution Write the mass of each substance over the chemical equation:

$$(2\text{ mol NaOH})\left(\frac{40.0\text{ g NaOH}}{1\text{ mol NaOH}}\right) = 80.0\text{ g NaOH}$$

$$\begin{matrix} 106\text{ g} & 74\text{ g} & 80.0\text{ g} & 100\text{ g} \\ Na_2CO_3\ + & Ca(OH)_2 \rightarrow & 2NaOH\ + & CaCO_3 \end{matrix}$$

This can be interpreted to get our conversion factor: "106 grams of Na_2CO_3 is required to make 80.0 grams of NaOH."

$$(\text{produce 500 g NaOH})\left(\frac{106\text{ g } Na_2CO_3\text{ required}}{80.0\text{ g NaOH produced}}\right) = 663\text{ g } Na_2CO_3\text{ required}$$

Our calculations indicate that 663 g of sodium carbonate is required to make 500 g of sodium hydroxide.

•
•

EXAMPLE 44 We have only 35 g of sodium carbonate. How much calcium hydroxide will react with 35 g of sodium carbonate according to the following chemical equation:

$$Na_2CO_3 + Ca(OH)_2 \rightarrow 2NaOH + CaCO_3$$

Solution The chemical equation

$$\overset{106\text{ g}}{Na_2CO_3} + \overset{74\text{ g}}{Ca(OH)_2} \rightarrow \overset{80.0\text{ g}}{2NaOH} + \overset{100\text{ g}}{CaCO_3}$$

tells us that 74 g of $Ca(OH)_2$ reacts with every 106 g of Na_2CO_3 that reacts. (This is our conversion factor.)

$$(35\text{ g } Na_2CO_3\text{ reacts})\left(\frac{74\text{ g } Ca(OH)_2\text{ reacts}}{106\text{ g } Na_2CO_3\text{ reacts}}\right) = 24\text{ g } Ca(OH)_2\text{ reacts}$$

We find that 24 g of calcium hydroxide will react with the available 35 g of sodium carbonate.

•
•

EXAMPLE 45 Water forms when hydrogen burns in oxygen. This chemical reaction is represented by the chemical equation

$$2H_2 + O_2 \rightarrow 2H_2O$$

You wish to make 100 g of water. How many grams of hydrogen and oxygen are needed? If you began with equal masses of hydrogen and oxygen, which reactant would be left over (as excess, unused reagent)?

Solution Write the mass of each substance on top of the chemical equation:

$$\overset{4\text{ g}}{2H_2} + \overset{32\text{ g}}{O_2} \rightarrow \overset{36\text{ g}}{2H_2O}$$

$$(2\text{ mol } H_2)\left(\frac{2\text{ g } H_2}{\text{mol } H_2}\right) = 4\text{ g } H_2 \qquad (2\text{ mol } H_2O)\left(\frac{18\text{ g } H_2O}{\text{mol } H_2O}\right) = 36\text{ g } H_2O$$

Determine the number of grams of hydrogen and oxygen that are needed to make 100 g of water by obtaining conversion factors from the interpretations of this chemical equation.

$$(\text{produce } 100\text{ g } H_2O)\left(\frac{4\text{ g } H_2\text{ needed}}{36\text{ g } H_2O\text{ produced}}\right) = 11.1\text{ g } H_2\text{ needed}$$

$$(\text{produce } 100\text{ g } H_2O)\left(\frac{32\text{ g } O_2\text{ needed}}{36\text{ g } H_2O\text{ produced}}\right) = 88.9\text{ g } O_2\text{ needed}$$

It takes 11.1 g of hydrogen and 88.9 g of oxygen to produce 100 g of water.

If we had equal amounts of hydrogen and oxygen (e.g., 88.9 g of oxygen and 88.9 g of hydrogen), all of the oxygen would be used up and we would have some hydrogen left over. We refer to oxygen as the limiting reagent because it is the reactant that limits the amount of product produced. Hydrogen is the reagent in excess. There is more hydrogen present than would get used up. Only 11.1 g of hydrogen (of the 88.9 g of hydrogen present) would be used. If we started with 88.9 g of hydrogen and 88.9 g of oxygen, 77.8 g (88.9 – 11.1) of hydrogen would be left unreacted.

•
•

SELF-TEST Ammonia (NH_3) can react with oxygen to produce nitric oxide (NO) and water. This reaction is represented by the chemical equation

$$4NH_3 + 5O_2 \rightarrow 4NO + 6H_2O.$$

1. How many grams of oxygen will react with 20 g of ammonia?
2. How many grams of nitric oxide will form when 40 g of oxygen is consumed?
3. How many grams of ammonia would be consumed if 45 g of nitric oxide formed?

ANSWERS **1.** 47 g O_2 **2.** 30 g NO **3.** 26 g NH_3

6.7.1 LARGER SCALE INTERPRETATIONS: TONS VS. GRAMS

When interpreting a chemical equation, it is possible to use any mass unit (e.g., grams, tons, or pounds). We have been using grams because the word *mole* is short for *gram-molecular weight*, which is defined as the molecular weight in grams. Grams are a convenient unit for chemists. However, we could use the unit *ton-molecular weight*, which would be the molecular weight in tons. If we use ton-molecular weight, the chemical equation representing the burning of carbon in oxygen

$$C + O_2 \rightarrow CO_2$$

would be interpreted as follows: "12 tons of carbon reacts with 32 tons of oxygen to produce 44 tons of carbon dioxide." This interpretation may have more meaning to people working in an electric power plant that burns coal (carbon) than to you. For every 12 tons of carbon burned (coal is about 70% to 90% carbon), 44 tons of carbon dioxide is released into the atmosphere.

EXAMPLE 46 A company wishes to make 50 tons of nitrogen dioxide (NO_2) by the chemical reaction that is represented by the following chemical equation:

$$N_2 + 2O_2 \rightarrow 2NO_2$$

How many tons of oxygen will be used?

Solution Write the mass in tons over each substance in the chemical equation (using ton-molecular weights):

$$\overset{28 \text{ tons}}{N_2} + \overset{64 \text{ tons}}{2O_2} \rightarrow \overset{92 \text{ tons}}{2NO_2}$$

There is 28 g of N_2 in 1 mol (gram-molecular weight) of N_2.
There is 28 tons of N_2 in 1 ton-molecular weight of N_2.

This chemical equation is interpreted as follows: "64 tons of oxygen is required per 92 tons of nitrogen dioxide produced."

$$(\text{produce 50 tons } NO_2)\left(\frac{64 \text{ tons } O_2 \text{ required}}{92 \text{ tons } NO_2 \text{ produced}}\right) = 35 \text{ tons } O_2 \text{ required}$$

It would take 35 tons of oxygen to make 50 tons of NO_2.

EXAMPLE 47 You want to react 35 lb of nitrogen. How many pounds of oxygen are needed? (*Remember*: $N_2 + 2O_2 \rightarrow 2NO_2$.)

Solution Write the number of pounds over each substance in the chemical equation:

$$\overset{28\text{ lb}}{N_2} + \overset{64\text{ lb}}{2O_2} \rightarrow \overset{92\text{ lb}}{2NO_2}$$

There is 28 lb of N_2 in 1.0 pound-molecular weight of N_2.

This equation tells us that 64 lb of oxygen is needed for every 28 lb of nitrogen that reacts.

$$(\text{react 35 lb } N_2)\left(\frac{64\text{ lb } O_2\text{ reacts}}{28\text{ lb } N_2\text{ reacts}}\right) = 80\text{ lb } O_2\text{ reacts}$$

Eighty pounds of oxygen reacts with 35 lb of nitrogen.

•
•

EXAMPLE 48 How many ounces of nitrogen dioxide will form when 15 ounces of nitrogen is allowed to react with oxygen? The reaction that takes place is represented by $N_2 + 2O_2 \rightarrow 2NO_2$.

Solution

$$\overset{28\text{ oz}}{N_2} + \overset{64\text{ oz}}{2O_2} \rightarrow \overset{92\text{ oz}}{2NO_2}$$

$$(15\text{ oz } N_2\text{ reacts})\left(\frac{92\text{ oz } NO_2\text{ forms}}{28\text{ oz } N_2\text{ reacts}}\right) = 49\text{ oz } NO_2\text{ forms}$$

When 15 oz of N_2 is allowed to react, 4.9 oz of NO_2 forms.

•
•

SELF-TEST Carbon monoxide will burn in oxygen to produce carbon dioxide. This chemical reaction is represented by

$$2CO + O_2 \rightarrow 2CO_2$$

1. How many pounds of CO_2 will form when 15 lb of carbon monoxide is burned in oxygen?
2. How many tons of oxygen are required to produce 30 tons of carbon dioxide?
3. How many ounces of CO will burn in 6.5 oz of oxygen?

ANSWERS **1.** 24 lb CO_2 **2.** 11 tons O_2 **3.** 11 oz CO

6.8 HEAT PROBLEMS

When heat is given off during a chemical reaction, the reaction is said to be exothermic. Carbon burns with oxygen to produce carbon dioxide and heat. This exothermic chemical reaction can be represented by the equation

$$C + O_2 \rightarrow CO_2 + 94\text{ kcal}$$

This chemical equation can be interpreted in several ways, one of which is "12 grams of carbon combines with 32 grams of oxygen to produce 44 grams of carbon dioxide plus 94 kilocalories (94 000 calories) of heat."

•
•

EXAMPLE 49 How many grams of carbon must be burned to produce 50 kcal of heat? (*Remember*: $C + O_2 \rightarrow CO_2 + 94$ kcal.)

Solution Write the mass (in grams) of each substance in the chemical equation:

$$\begin{array}{cccc} 12\text{ g} & 32\text{ g} & 44\text{ g} & \\ C + & O_2 \rightarrow & CO_2 + & 94\text{ kcal} \end{array}$$

This equation tells us that 12 g of carbon is burned for every 94 kcal of heat produced. This is our conversion factor. The 94 kcal produced is treated just like one of the products formed.

$$(\text{produce 50 kcal heat})\left(\frac{12\text{ g C burned}}{94\text{ kcal heat produced}}\right) = 6.4\text{ g C burned}$$

You need to burn 6.4 g of carbon to produce 50 kcal of heat.

•
•

EXAMPLE 50 How much heat will be produced if the amount of oxygen consumed in the preceding reaction is 20 g?

Solution

$$(20\text{ g } O_2\text{ consumed})\left(\frac{94\text{ kcal produced}}{32\text{ g } O_2\text{ consumed}}\right) = 59\text{ kcal produced}$$

When 20 g of oxygen is consumed, 59 kcal of heat is given off.

•
•

EXAMPLE 51 Hydrogen burns in oxygen to produce water and heat. This chemical reaction is represented by the chemical equation

$$2H_2 + O_2 \rightarrow 2H_2O + \underset{\text{569 kilojoules}}{569\text{ kJ}}$$

How many grams of hydrogen would you have to burn if you wanted to produce 800 kJ of heat?

Solution

$$\begin{array}{cccc} 4.00\text{ g} & 32\text{ g} & 36\text{ g} & \\ 2H_2 + & O_2 \rightarrow & 2H_2O + & 569\text{ kJ} \end{array}$$

$$(\text{produce 800 kJ})\left(\frac{4.00\text{ g } H_2\text{ burned}}{569\text{ kJ produced}}\right) = 5.62\text{ g } H_2\text{ burned}$$

You would have to burn 5.62 g of hydrogen to produce 800 kJ of heat.

•
•

EXAMPLE 52 If 50 g of hydrogen burned, how much heat would be given off?

Solution

$$(50\text{ g } H_2\text{ burned})\left(\frac{569\text{ kJ given off}}{4.0\text{ g } H_2\text{ burned}}\right) = 7.1 \times 10^3\text{ kJ heat given off}$$

About 7100 kJ of heat would be given off when 50 g of H_2 burns.

•
•

SELF-TEST Manganese dioxide (MnO_2) reacts with hydrochloric acid (HCl) to produce manganese dichloride ($MnCl_2$), chlorine (Cl_2), and water (H_2O). This chemical reaction can be represented by the chemical equation

$$MnO_2 + 4HCl \rightarrow MnCl_2 + Cl_2 + 2H_2O + 200\text{ kcal}$$

1. How many grams of MnO_2 are needed to produce 45 kcal of heat?
2. How many grams of HCl are needed to produce 40 kcal of heat?
3. How much heat will be given off when 30 g of chlorine are formed?

ANSWERS **1.** 20 g MnO_2 **2.** 29 g HCl **3.** 85 kcal heat

6.9 COMBINATION PROBLEMS

We will now practice solving problems in which different quantity units are used together. For example, you may be given grams of hydrogen and asked to find liters of oxygen. Assume that all gas volumes are measured at STP.

•
•

EXAMPLE 53 Hydrogen burns in oxygen to produce water and heat according to the chemical reaction

$$2H_2 + O_2 \rightarrow 2H_2O + 569\ \text{kJ}$$

How many kilojoules of heat will be given off if 4 mol of oxygen is consumed?

Solution

$$2H_2 + \underset{}{\overset{1\ \text{mol of}\ O_2}{O_2}} \rightarrow 2H_2O + 569\ \text{kJ}$$

$$(4\ \text{mol}\ O_2)\left(\frac{569\ \text{kJ}}{1\ \text{mol}\ O_2}\right) = 2 \times 10^3\ \text{kJ}$$

•
•

EXAMPLE 54 How many moles of water are produced when 30 L of hydrogen at STP is burned? Use the equation $2H_2 + O_2 \rightarrow 2H_2O + 136$ kcal.

Solution Write the volume of hydrogen and the number of moles of water over the chemical equation:

$$\overset{44.8\ \text{L}\ H_2}{2H_2} + O_2 \rightarrow \overset{2\ \text{mol}\ H_2O}{2H_2O} + 136\ \text{kcal}$$

$$(30\ \text{L}\ H_2\ \text{at STP})\left(\frac{2\ \text{mol}\ H_2O}{44.8\ \text{L}\ H_2\ \text{at STP}}\right) = 1.3\ \text{mol}\ H_2O$$

When 30 L of hydrogen is burned, 1.3 mol of water forms.

•
•

EXAMPLE 55 How many liters of oxygen will be consumed at STP when 30 g of hydrogen is burned? Use the equation $2H_2 + O_2 \rightarrow 2H_2O + 136$ kcal.

Solution Write the volume of oxygen and the mass of hydrogen over the chemical equation:

$$\overset{4.0\ \text{g}}{2H_2} + \overset{22.4\ \text{L}\ O_2}{O_2} \rightarrow 2H_2O + 136\ \text{kcal} \quad \longleftarrow \text{volume of 1 mol of oxygen}$$

$$(2\ \text{mol}\ H_2)\left(\frac{2\ \text{g}\ H_2}{\text{mol}\ H_2}\right) = 4.0\ \text{g}\ H_2$$

This chemical equation tells us (is interpreted as saying) that 22.4 L of oxygen is consumed for every 4.0 g of hydrogen burned. This is our conversion factor.

$$\underset{\text{given}}{(30\ \text{g}\ H_2\ \text{burns})}\underset{\text{conversion factor}}{\left(\frac{22.4\ \text{L}\ O_2\ \text{consumed at STP}}{4.0\ \text{g}\ H_2\ \text{burns}}\right)} = 1.7 \times 10^2\ \text{L}\ O_2\ \text{consumed at STP}$$

About 170 L of oxygen at STP is consumed when 30 g of hydrogen burns.

•
•

EXAMPLE 56 How many molecules of water are produced when 300 kcal of heat are given off? (*Remember*: $2H_2 + O_2 \rightarrow 2H_2O + 136$ kcal.)

Solution

$$\overset{12.0 \times 10^{23} \text{ molecules of } H_2O}{\downarrow}$$

$$2H_2 + O_2 \rightarrow 2H_2O + 136 \text{ kcal}$$

$$(300 \text{ kcal})\left(\frac{12.0 \times 10^{23} \text{ } H_2O \text{ molecules}}{136 \text{ kcal}}\right) = 2.65 \times 10^{24} \text{ } H_2O \text{ molecules}$$

Our calculations indicate that 2.65×10^{24} water molecules are produced along with 300 kcal of heat.

•
•

EXAMPLE 57 How many molecules of oxygen will combine with 2 L of hydrogen at STP? (*Remember*: $2H_2 + O_2 \rightarrow 2H_2O + 136$ kcal.)

Solution Write the volume of hydrogen and the number of moles of oxygen over the chemical equation:

$$\overset{44.8 \text{ L } H_2}{\downarrow} \quad \overset{6 \times 10^{23} \text{ } O_2 \text{ molecules}}{\downarrow}$$

$$2H_2 + O_2 \rightarrow 2H_2O + 136 \text{ kcal}$$

Then, use the conversion factor to determine the answer:

$$(2 \text{ L } H_2 \text{ at STP})\left(\frac{6 \times 10^{23} \text{ } O_2 \text{ molecules}}{44.8 \text{ L } H_2 \text{ at STP}}\right) = 3 \times 10^{22} \text{ } O_2 \text{ molecules}$$

Our calculations show that 3×10^{22} oxygen molecules will react with 2 L (measured at STP) of hydrogen.

•
•

EXAMPLE 58 How many grams of oxygen are consumed when 500 molecules of hydrogen are burned? (*Remember*: $2H_2 + O_2 \rightarrow 2H_2O + 136$ kcal.)

Solution Write the number of hydrogen molecules and the mass of oxygen over the chemical equation:

$$\overset{12.0 \times 10^{23} \text{ } H_2 \text{ molecules}}{\searrow} \quad \overset{32 \text{ g } O_2}{\downarrow}$$

$$2H_2 + O_2 \rightarrow 2H_2O + 136 \text{ kcal}$$

Then, use the conversion factor to determine the answer:

$$(500 \text{ } H_2 \text{ molecules})\left(\frac{32 \text{ g } O_2}{12.0 \times 10^{23} \text{ } H_2 \text{ molecules}}\right) = 1.33 \times 10^{-20} \text{ g } O_2$$

When 500 molecules of hydrogen are burned, 1.33×10^{-20} g of oxygen are consumed.

•
•

SELF-TEST Carbon burns in oxygen to produce carbon dioxide plus heat according to the following chemical equation:

$$C + O_2 \rightarrow CO_2 + 393 \text{ kJ}$$

1. How much heat is produced when 4.0 g of carbon is burned?
2. How many liters of carbon dioxide at STP are produced when 3.0 mol of carbon is burned?
3. How many atoms of carbon must be burned to produce 1 cal of heat?

ANSWERS **1.** 1.3×10^2 kJ **2.** 67 L CO_2 **3.** 6×10^{18} C atoms

6.9.1 NON-STP VOLUME PROBLEMS

Not all gases form and react at 0°C and 1 atm of pressure (STP). It is necessary to use gas laws to determine the volumes of such gases.

EXAMPLE 59 Hydrogen is burned in oxygen at a temperature of 1000°C. What volume of oxygen is consumed when 5.0 g of hydrogen is burned? (*Remember*: $2H_2 + O_2 \rightarrow 2H_2O + 136$ kcal.)

Solution

$$\underset{4.0\text{ g }H_2}{\overset{\downarrow}{2H_2}} + \underset{22.4\text{ L }O_2\text{ at STP}}{\overset{\downarrow}{O_2}} \rightarrow 2H_2O + 136\text{ kcal}$$

$$(5.0\text{ g }H_2)\left(\frac{22.4\text{ L }O_2\text{ at STP}}{4.0\text{ g }H_2}\right) = 28\text{ L }O_2\text{ at STP}$$

Under STP conditions, 28 L of oxygen is consumed when 5.0 g of hydrogen is burned. (If the volume of oxygen consumed were measured at STP, it would equal 28 L.) At 1000°C (1273 K) the volume of oxygen consumed will be larger (Charles's law).

$$(28\text{ L }O_2\text{ consumed at STP})\left(\frac{1273\text{ K}}{273\text{ K at STP}}\right) = 1.3 \times 10^2\text{ L }O_2\text{ consumed}$$

About 130 L of oxygen reacts with 5.0 g of hydrogen at 1000°C.

EXAMPLE 60 What volume of hydrogen is consumed at 1000°C when 50 kcal of heat is given off? (*Remember*: $2H_2 + O_2 \rightarrow 2H_2O + 136$ kcal.)

Solution

$$\underset{44.8\text{ L }H_2}{\overset{\downarrow}{2H_2}} + O_2 \rightarrow 2H_2O + 136\text{ kcal}$$

$$(50\text{ kcal})\left(\frac{44.8\text{ L }H_2\text{ consumed at STP}}{136\text{ kcal}}\right) = 16\text{ L }H_2\text{ consumed at STP}$$

Find the volume of hydrogen consumed at 1000°C.

$$(16\text{ L }H_2\text{ consumed at STP})\left(\frac{1273\text{ K}}{273\text{ K at STP}}\right) = 75\text{ L }H_2\text{ consumed}$$

At 1000°C, 75 L of hydrogen is consumed when 50 kcal of heat is produced.

6.9.2 LIMITING REAGENT PROBLEMS

Sometimes one of two reagents (reactants) is present in excess (some of it will be left over when the reaction is complete). The other reactant (the one not in excess) is called the limiting reagent. The limiting reagent is always used up. The limiting reagent limits the amount of product that can form. Many students find this the most difficult type of stoichiometry problem, so before proceeding, let's consider a limiting reagent analogy.

We live in a society in which most marriages are made up of just one man and just one woman. We can write the "chemical reaction" for a marriage as

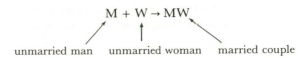

$$M + W \rightarrow MW$$

unmarried man — unmarried woman — married couple

where M is the symbol for an unmarried man, W for an unmarried woman, and MW for a married couple.

Now let's place five women and three men inside a room and tell them to get married. The women are present in excess. After the marriages take place, there will be two women in excess (left over, unmarried, unreacted). The men are the limiting reagents. All of the men will be married. The number of men determines how much product (number of married couples) will form. There are three men, which limits the number of marriages to three.

Solving limiting reagent problems usually requires two steps: (1) Determine which reagent is the limiting reagent, and (2) work with the limiting reagent to determine the amount of product that will form. Let's illustrate these two steps with the marriage analogy.

Step 1 Determine which reagent is present in excess and which reagent is the limiting reagent. M (men) is the limiting reagent. W (women) is present in excess.

Step 2 Use the limiting reagent to calculate the amount of product formed. The three men will produce three married couples.

If we don't use the limiting reagent to calculate the amount of product, we will get an incorrect answer. For example, if we had tried to figure out the number of married couples using five women (ignoring the number of men available), we would have obtained the incorrect answer of five married couples.

•
•

EXAMPLE 61 Silicon reacts with oxygen to produce silicon dioxide. The following chemical equation represents this reaction:

$$Si + O_2 \rightarrow SiO_2$$

How many grams of silicon dioxide (SiO_2) will form when 10.0 g of silicon (Si) is mixed with 11.0 g of oxygen?

Solution **Step 1** First we have to determine which reagent (Si or O_2) is the limiting reagent. To do this, pick either of the two reagents (at random) and see how much of the remaining reagent is needed to react with the one you picked. For example, let's pick oxygen (as opposed to silicon) and determine how many grams of silicon are required to react with 11.0 g of oxygen.

$$\overset{28.0\text{ g}}{Si} + \overset{32.0\text{ g}}{O_2} \rightarrow \overset{60.0\text{ g}}{SiO_2}$$

$$(11.0\text{ g } O_2\text{ available})\left(\frac{28.0\text{ g Si needed}}{32.0\text{ g } O_2\text{ available}}\right) = 9.60\text{ g Si needed}$$

This calculation tells us that 9.60 g of silicon will react with 11.0 g of oxygen. We were given 10.0 g of silicon and only 9.60 g of silicon is needed, so there will be some silicon left over (0.4 g left over, in excess). Oxygen is totally used up; that is, oxygen is the limiting reagent.

Before proceeding to the second step of this two-step solution, let's see what would have happened had we picked (at random) silicon. Determine how many grams of oxygen are needed to react with 10.0 g of silicon.

$$\begin{array}{ccc} 28.0\text{ g} & 32.0\text{ g} & 60.0\text{ g} \\ \text{Si} + & \text{O}_2 \rightarrow & \text{SiO}_2 \end{array}$$

$$(10.0\text{ g Si available})\left(\frac{32.0\text{ g O}_2\text{ needed}}{28.0\text{ g Si available}}\right) = 11.4\text{ g O}_2\text{ needed}$$

This calculation tells us that 10.0 g of silicon requires 11.4 g of oxygen to complete the reaction. But we were only given 11.0 g of oxygen, so there isn't enough oxygen present to react with all 10.0 g of silicon. Again, we conclude that oxygen is the limiting reagent and silicon is present in excess.

Step 2 Use the limiting reagent (oxygen) to determine how much product SiO_2 will form.

$$\begin{array}{ccc} 28.0\text{ g} & 32.0\text{ g} & 60.0\text{ g} \\ \text{Si} + & \text{O}_2 \rightarrow & \text{SiO}_2 \end{array}$$

$$(11.0\text{ g O}_2\text{ reacts})\left(\frac{60.0\text{ g SiO}_2\text{ forms}}{32.0\text{ g O}_2\text{ reacts}}\right) = 20.6\text{ g SiO}_2\text{ forms}$$

Mixing 10.0 g of silicon with 11.0 g of oxygen produces 20.6 g of SiO_2 with 0.4 g of silicon left over.

EXAMPLE 62 A mixture of 35 g of hydrogen and 270 g of oxygen is allowed to react chemically to form water. The chemical reaction is represented by the chemical equation

$$2\text{H}_2 + \text{O}_2 \rightarrow 2\text{H}_2\text{O}$$

How many grams of water will form?

Solution We cannot assume that all 35 g of hydrogen will react with all 270 g of oxygen. First find which substance is present in excess and which substance is the limiting reagent. Let's randomly pick oxygen (as opposed to hydrogen) and determine how much hydrogen will react with 270 g of oxygen.

$$\begin{array}{ccc} 4.0\text{ g} & 32\text{ g} & 36\text{ g} \\ 2\text{H}_2 + & \text{O}_2 \rightarrow & 2\text{H}_2\text{O} \end{array}$$

$$(270\text{ g O}_2\text{ available})\left(\frac{4.0\text{ g H}_2\text{ needed}}{32\text{ g O}_2\text{ available}}\right) = 34\text{ g H}_2\text{ needed}$$

Our calculations show that 34 g of hydrogen will react with 270 g of oxygen to complete the reaction. We were given 35 g of hydrogen, but only 34 g of it will react. So hydrogen is present in excess (1 g of it will be left over after the reaction). Oxygen is the limiting reagent. All 270 g of the oxygen will be used up. Oxygen will limit (determine) the amount of water (product) that forms. Working with the limiting reagent (270 g oxygen) to determine the amount of water that forms, we have

$$(270\text{ g O}_2\text{ reacts})\left(\frac{36\text{ g H}_2\text{O forms}}{32\text{ g O}_2\text{ reacts}}\right) = 3.0 \times 10^2\text{ g H}_2\text{O forms}$$

About 300 g of water will form.

EXAMPLE 63 Nitrogen and oxygen can react to form nitric oxide as represented by the chemical equation

$$N_2(g) + O_2(g) \rightarrow 2NO(g)$$

A mixture of 40 g of nitrogen and 45 g of oxygen is allowed to react chemically to produce nitric oxide. How many liters of nitric oxide (at STP) are formed?

Solution First determine the limiting reagent by determining how much oxygen is required to combine with 40 g of nitrogen.

$$\overset{28\text{ g}}{N_2(g)} + \overset{32\text{ g}}{O_2(g)} \rightarrow \overset{44.8\text{ L}}{2NO(g)}$$

$$(40\text{ g } N_2 \text{ available})\left(\frac{32\text{ g } O_2 \text{ needed}}{28\text{ g } N_2 \text{ available}}\right) = 46\text{ g } O_2 \text{ needed}$$

We see that 46 g of oxygen is required to combine with 40 g of nitrogen. But we were only given 45 g of oxygen. There isn't enough oxygen to react with all 40 g of the nitrogen. Some nitrogen will be left unreacted. Nitrogen is present in excess. Oxygen is the limiting reagent. Work with the limiting reagent (oxygen) to determine the volume of nitric oxide (product) that forms.

$$(45\text{ g } O_2 \text{ reacts})\left(\frac{44.8\text{ L NO forms}}{32\text{ g } O_2 \text{ reacts}}\right) = 63\text{ L NO forms}$$

Sixty-three liters of nitric oxide at STP forms when 45 g of oxygen is allowed to react with 40 g of nitrogen. Some nitrogen is left unreacted when the reaction is complete.

•
•

SELF-TEST Ten grams of carbon is allowed to react with 30 g of oxygen. This reaction is represented by the chemical equation

$$C + O_2 \rightarrow CO_2 + 94\text{ kcal}$$

1. Which reactant is the limiting reagent?
2. How many kcal of heat are given off?
3. What volume of carbon dioxide will form at STP?
4. What volume of carbon dioxide will form if the temperature of the reaction is 500 K?

ANSWERS **1.** C **2.** 78 kcal **3.** 19 L CO_2 **4.** 35 L CO_2

PROBLEM SET 6

The problems in Problem Set 6 parallel the examples in Chapter 6. For example, if you should have trouble working Problem 5, go back to Example 5 in this chapter to get help. The correct answers to these problems are given at the end of this book.

1. Potassium reacts with chlorine to produce potassium chloride. This reaction can be represented by the chemical equation

$$2K(s) + Cl_2(g) \rightarrow 2KCl(s)$$

In what phase (state) is the potassium?

2. In the reaction shown in Problem 1, in what state does the chlorine exist?
3. In the reaction shown in Problem 1, what are the coefficients of each substance? Interpret the coefficients in this chemical equation.
4. The reaction between hydrogen and oxygen to form water is represented by the balanced chemical equation

$$2H_2 + O_2 \rightarrow 2H_2O$$

What are the coefficients of each substance in this equation? Interpret the coefficients in this chemical equation.

5. Consider the chemical equation

$$H_2 + O_2 \rightarrow H_2O_2$$

Is this chemical equation balanced?

6. Consider the chemical equation

$$2Al + 3S \rightarrow Al_2S_3$$

Is this chemical equation balanced?

7. Consider the chemical equation

$$H_2O + NO_2 \rightarrow HNO_3$$

Is this chemical equation balanced?

8. Balance the chemical equation

$$H_2 + F_2 \rightarrow HF$$

9. Balance the chemical equation

$$C + H_2 \rightarrow C_2H_4$$

10. Balance the chemical equation

$$H_2CO_3 + Al(OH)_3 \rightarrow Al_2(CO_3)_3 + H_2O$$

11. Balance the chemical equation

$$H_3PO_3 + Ba(OH)_2 \rightarrow Ba_3(PO_3)_2 + H_2O$$

12. Balance the chemical equation

$$AlBr_3 + Na_2CO_3 \rightarrow Al_2(CO_3)_3 + NaBr$$

13. When sulfur burns in oxygen to produce sulfur trioxide, the reaction is represented by the balanced chemical equation

$$2S(s) + 3O_2(g) \rightarrow 2SO_3(g)$$

How many moles of sulfur and oxygen react together according to this equation? How many moles of sulfur trioxide form?

Refer to the chemical equation in Problem 13 when working Problems 14–21.

14. In the reaction between sulfur and oxygen, how many grams of sulfur react with 3 mol of oxygen?
15. When sulfur reacts with oxygen, how many grams of oxygen react with 2 mol of sulfur?
16. How many grams of sulfur trioxide form when 2 mol of sulfur reacts with 3 mol of oxygen?
17. How many atoms of sulfur react to form 2 mol of sulfur trioxide?
18. How many molecules of oxygen react with 2 mol of sulfur?
19. What volume of oxygen at STP is consumed when 2 mol of sulfur is consumed?
20. What volume of sulfur trioxide at STP is produced when 2 mol of sulfur burns in 3 mol of oxygen?
21. Construct a table similar to Table 6.1 giving the various ways you can interpret the balanced chemical equation that represents the reaction of sulfur with oxygen to produce sulfur trioxide.

Refer to the table you constructed for Problem 21 when working Problems 22–31.

22. How many liters of sulfur trioxide form when 64 g of sulfur burns?
23. How many sulfur atoms react with 67.2 L of oxygen?
24. Give an interpretation of the reaction between sulfur and oxygen to produce sulfur trioxide by interchanging quantities in lines 2, 3, and 4.
25. Interchange information from lines 3, 4, and 5 to produce another interpretation of the chemical equation.
26. Interchange the quantities from lines 2, 3, and 5 to produce another interpretation of the chemical equation.
27. How many moles of oxygen will react with 4 mol of sulfur?
28. How many grams of sulfur trioxide are produced when 32 g of sulfur burns in oxygen?
29. How many grams of oxygen will react with 128 g of sulfur?
30. How many molecules of oxygen combine with 6×10^{23} atoms of sulfur?
31. How many liters of sulfur trioxide will form when 134.4 L of oxygen reacts?
32. Find the number of moles of oxygen that will react with 7.0 mol of hydrogen according to the chemical equation

$$2H_2 + O_2 \rightarrow 2H_2O$$

33. How many moles of water will form when 4 mol of hydrogen burns according to the reaction shown in Problem 32?
34. If you produced 30 mol of water, how many moles of oxygen would be consumed according to the reaction in Problem 32?
35. Zinc sulfide reacts with oxygen to produce zinc oxide and sulfur dioxide as represented by

$$4ZnS + 6O_2 \rightarrow 4ZnO + 4SO_2$$

How many moles of zinc sulfide are needed to make 8 mol of ZnO?

36. Sulfuric acid reacts with hydrogen iodide to produce hydrogen sulfide, iodine, and water, as represented by

$$H_2SO_4 + 8HI \rightarrow H_2S + 4I_2 + 4H_2O$$

How many moles of hydrogen sulfide can you make from 14 mol of hydrogen iodide?

37. How many liters of oxygen are needed to react with 4 L of nitrogen at STP according to the balanced chemical equation

$$N_2(g) + O_2(g) \rightarrow 2NO(g)$$

38. How many liters of nitric oxide (NO) will form when 6.0 L of nitrogen combines with oxygen at STP according to the equation shown in Problem 37?
39. Methane reacts with oxygen to produce carbon dioxide and water. This is represented by the balanced chemical equation

$$CH_4(g) + 2O_2(g) \rightarrow CO_2(g) + 2H_2O(l)$$

How many liters of methane react with 8.0 L of oxygen at STP?

40. Some methane and oxygen are allowed to react and they produce 30.0 L of carbon dioxide. How many liters of oxygen had to react to produce this much carbon dioxide according to the balanced chemical equation shown in Problem 39?
41. Carbon burns in oxygen to produce carbon dioxide. How many grams of carbon dioxide will form when 15 g of carbon is burned? (*Remember*: $C + O_2 \rightarrow CO_2$.)
42. How many grams of oxygen are used when 6.0 g of carbon burns according to the balanced chemical reaction shown in Problem 41?
43. Sodium carbonate reacts with calcium hydroxide to produce sodium hydroxide and calcium carbonate, as represented by the balanced chemical equation

$$Na_2CO_3 + Ca(OH)_2 \rightarrow 2NaOH + CaCO_3$$

How many grams of sodium carbonate are required to produce 300 g of sodium hydroxide?

44. How much calcium hydroxide will react with 30.6 g of sodium carbonate according to the reaction shown in Problem 43?
45. You wish to make 87.5 g of water. How many grams of hydrogen are needed according to the reaction shown in Problem 32?
46. A company wishes to make 43.8 tons of nitrogen dioxide according to the balanced chemical equation

$$N_2 + 2O_2 \rightarrow 2NO_2$$

How many tons of oxygen will be used?

47. How many pounds of oxygen are needed to react with 28 lb of nitrogen according to the equation shown in Problem 46?
48. How many ounces of nitrogen dioxide will form when 12 oz of nitrogen is allowed to react with oxygen according to the reaction shown in Problem 46?
49. How many grams of carbon must be burned to produce 40 kcal of heat according to the reaction

$$C + O_2 \rightarrow CO_2 + 94 \text{ kcal}$$

50. How much heat will be produced if 16 g of oxygen is consumed according to the equation shown in Problem 49?
51. How many grams of hydrogen would you have to burn if you wanted to produce 640 kJ of heat according to the reaction

$$2H_2 + O_2 \rightarrow 2H_2O + 569 \text{ kJ}$$

Refer to the balanced chemical equation shown in Problem 51 when working Problems 52–55.

52. If 40 g of hydrogen burned, how much heat would be given off?
53. How many kilojoules of heat would be given off if 3.2 mol of oxygen was consumed?
54. How many moles of water are produced when 24 L of hydrogen at STP is burned?
55. How many liters of oxygen are consumed when 24 g of hydrogen is burned?
56. How many molecules of water are produced when 240 kcal of heat is given off according to the chemical equation

$$2H_2 + O_2 \rightarrow 2H_2O + 136 \text{ kcal}$$

Refer to the balanced chemical equation in Problem 56 when working Problems 57–60.

57. How many molecules of oxygen will combine with 1.6 L of hydrogen?
58. How many grams of oxygen are consumed when 400 molecules of hydrogen is burned?
59. Hydrogen is burned in oxygen at a temperature of 1000°C. What volume of oxygen is consumed when 4.0 g of hydrogen is burned?
60. What volume of hydrogen is consumed at 1000°C when 40 kcal of heat is given off?
61. Silicon reacts with oxygen to produce silicon dioxide. How many grams of silicon dioxide will form when 20.0 g of silicon is mixed with 20.0 g of oxygen? (*Hint*: $Si + O_2 \rightarrow SiO_2$.)
62. A mixture of 40 g of hydrogen and 300 g of oxygen is allowed to react as shown in Problem 56. How many grams of water form?
63. A mixture of 50 g of nitrogen and 55 g of oxygen reacts according to the balanced equation

$$N_2(g) + O_2(g) \rightarrow 2NO(g)$$

How many liters of nitric oxide at STP form?

7 Concentration Units

7.1 INTRODUCTION

Many students struggle to learn concentration units and memorize concentration unit formulas without ever fully understanding the material and its importance. One reason most students have trouble with concentration units is that they don't understand the mole concept. Become familiar with the mole concept! (See Chapter 4.) It will lessen your burden as you go through this chapter.

Another reason many students have trouble grasping concentration unit material is their lifelong use of memorized formulas to obtain numerical answers. We will solve all the problems in this chapter using intuition, logic, definitions, and dimensional analysis. It's really easier than it may sound. The most difficult part of this work may be for you to keep the definitions straight in your mind.

Why bother to learn concentration units? There are several reasons. It is through the use of concentration units that molecular weights can be determined. Molecular weights are of fundamental importance in chemistry. You have been using the molecular weight concept throughout most of this course.

A knowledge of concentration units is also necessary when working with equilibrium problems, a major topic in general chemistry. You must know concentration units to solve problems involving osmotic pressure, boiling point elevation, and freezing point depression.

Concentration units are important for practical reasons as well. If you ever work in a laboratory (e.g., as a student, a nurse, a technician, or a scientist), you will be working with many chemicals dissolved in water. Knowing the concentration units of the chemicals that you are using is obviously important.

It is important for you to know the difference between a solute, a solvent, and a solution. If a teaspoon of sugar is dissolved in water, sugar is the solute (the dissolved substance is called the solute), water is the solvent (the substance that dissolves the solute is called the solvent), and the combination of the two substances (sugar dissolved in water) is called the solution (see Figure 7.1).

FIGURE 7.1 Preparing a solution by dissolving a solute in a solvent

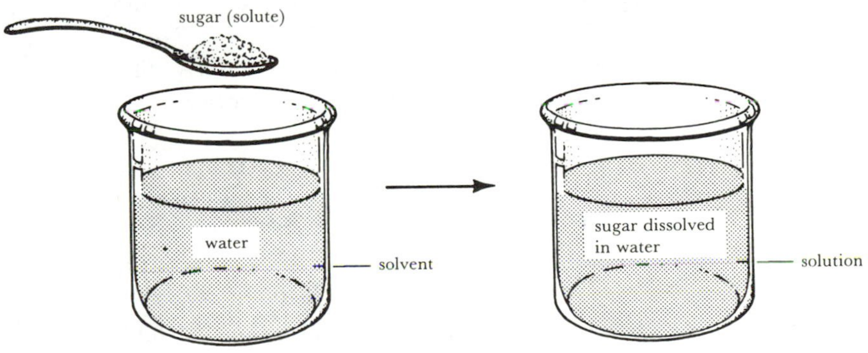

EXAMPLE 1 A pound of salt is dissolved in 20 gallons of water. Identify the solute, the solvent, and the solution.

Solution The salt is the solute. Salt is being dissolved in water and the dissolved substance is called the solute. Water is the solvent. The substance (water) that dissolves the solute (salt) is called the solvent. The resulting combination of the solute and solvent (salt dissolved in water) is called the solution.

EXAMPLE 2 Fifty milliliters of alcohol is mixed with 50 mL of water. Identify the solute, the solvent, and the solution.

Solution When two substances are mixed together to form a solution, the substance that is present in the greater amount is usually called the solvent. When two substances forming a solution are present in about equal amounts, the solvent and solute distinctions are arbitrary. In this problem we can choose either the alcohol or the water to be the solvent. The remaining liquid will be the solute. If water is called the solvent, then alcohol is the solute. If alcohol is called the solvent, then water is the solute. Both ways are valid.

The purpose of the next few sections is to convince you that you already know, intuitively, the basic ideas about concentration units. We will consider four concentration units: percent by weight, mole fraction, molarity, and molality.

7.2 PERCENT BY WEIGHT

Percent by weight is easy to learn because you already know 99% of this concept. Before applying percent by weight to chemical problems, first consider the "body" problems in the next section.

7.2.1 BODY ANALOGIES

If you weigh 100 lb and 20 lb of that weight is fat, what percent by weight is the fat in your body?

You probably figured out, almost intuitively, that 20% of your body weight is fat. This may seem to be a trivial calculation, but let's review how the 20% value is determined. (For more details, see Chapter 3.)

The fraction of your body weight that is fat.

$$\%\text{ by wt fat} = \left(\frac{\text{wt of fat}}{\text{total body wt}}\right)\left(\frac{100}{\text{cent.}}\right) = \left(\frac{20\text{ lb fat}}{100\text{ lb body}}\right)\left(\frac{100}{\text{cent.}}\right) = 20\%\text{ by wt fat}$$

The total body weight includes the weight of the fat.

Assume that, from your neck down, your body weight is 130 lb. Your head weighs 20 lb. What is the percent by weight of your head?

Your head is 13% of the weight of your body.

$$\% \text{ by wt head} = \left(\frac{\text{wt head}}{\text{wt whole body}}\right)\left(\frac{100}{\text{cent.}}\right) = \left(\frac{20 \text{ lb head}}{150 \text{ lb body}}\right)\left(\frac{100}{\text{cent.}}\right) = 13\% \text{ by wt head}$$

The fraction of your body weight that is your head.

7.2.2 CHEMICAL APPLICATIONS

EXAMPLE 3 Consider a solution that is made up of 5.0 g of hydrochloric acid (HCl) and 10 g of water (H_2O). What is the percent by weight of the hydrochloric acid?

Solution The percent by weight of the hydrochloric acid is 33%.

$$\% \text{ by wt HCl} = \left(\frac{\text{wt HCl}}{\text{wt soln}}\right)\left(\frac{100}{\text{cent.}}\right) = \left(\frac{5.0 \text{ g HCl}}{15 \text{ g soln}}\right)\left(\frac{100}{\text{cent.}}\right) = 33\% \text{ by wt HCl}$$

The fraction of the solution that is HCl.

EXAMPLE 4 A solution is made up of three different substances: 20 g of ethanol (drinking alcohol), 29 g of water, and 1 g of sodium chloride (NaCl, table salt). What is the percent by weight of the alcohol?

Solution The percent by weight of the ethanol is 40%.

$$\% \text{ by wt alcohol} = \left(\frac{\text{wt alcohol}}{\text{wt soln}}\right)\left(\frac{100}{\text{cent.}}\right)$$

$$= \left(\frac{20 \text{ g alcohol}}{20 \text{ g alcohol} + 29 \text{ g } H_2O + 1 \text{ g NaCl}}\right)\left(\frac{100}{\text{cent.}}\right)$$

$$= \left(\frac{20 \text{ g alcohol}}{50 \text{ g soln}}\right)\left(\frac{100}{\text{cent.}}\right) = 40\% \text{ by wt alcohol}$$

EXAMPLE 5 *A solution of sulfuric acid that is 50.0% by weight sulfuric acid is half sulfuric acid and half water. If you have 1000 g of this solution, how many grams of sulfuric acid are present?*

Solution There is 500 g of sulfuric acid in 1000 g of a sulfuric acid solution that is 50.0% by weight sulfuric acid. Fifty percent by weight sulfuric acid means that there are 50.0 grams of sulfuric acid for every 100 grams of solution.

$$(1000 \text{ g soln})\left(\frac{50.0 \text{ grams } H_2SO_4}{100 \text{ grams soln}}\right) = 500 \text{ g } H_2SO_4$$

or

$$(1000 \text{ g soln})\left(\frac{50.0 \text{ g } H_2SO_4}{100 \text{ g soln}}\right) = 500 \text{ g } H_2SO_4$$

EXAMPLE 6 A solution is prepared by mixing together alcohol, water, and sugar. The total weight of the final solution is 350 g. Ten percent of this solution is alcohol. How many grams of alcohol are in the solution?

Solution

$$(350 \text{ g soln})\left(\frac{10 \text{ g alcohol}}{100 \text{ g soln}}\right) = 35 \text{ g alcohol}$$

SELF-TEST

1. A solution is made up of 45 g of water and 5.0 g of salt. What is the percent by weight of the salt?
2. A solution is made up of three liquids: 40 g of liquid A, 30 g of liquid B, and 50 g of liquid C. What is the percent by weight of liquid C?
3. A solution is made up of alcohol and water. Find the weight of alcohol in 500 g of this solution if the solution is 25.0% by weight alcohol.

ANSWERS **1.** 10% by wt NaCl **2.** 42% by wt liquid C **3.** 125 g alcohol

7.3 MOLE FRACTION

You will be happy to learn that you also know most of the basic ideas associated with the second concentration unit, mole fraction.

7.3.1 MARBLE AND CLASSROOM ANALOGIES

If you have two blue marbles and three red marbles, what fraction of the marbles are blue?

The fraction of the marbles that are blue is $\frac{2}{5}$, or 0.4. Here's how this fraction is calculated:

$$\text{blue marble fraction} = \frac{\text{number of blue marbles}}{\text{total number of marbles}} = \frac{2}{5} = 0.4$$

Consider a class made up of 3 girls and 2 boys. What fraction of the class is female?

The female fraction of the class is $\frac{3}{5}$, or 0.6.

$$\text{female fraction} = \frac{\text{number of girls}}{\text{total number of students}} = \frac{3}{5} = 0.6$$

You should realize that the male fraction is equal to 1 minus the female fraction (1.0 − 0.6 = 0.4). Likewise, the red marble fraction equals 1 minus the blue marble fraction (1.0 −

0.4 = 0.6). This is because the fraction of blue marbles plus the fraction of red marbles must equal 1. The male fraction plus the female fraction must also add up to 1.

7.3.2 CHEMICAL APPLICATIONS

The mole fraction concept is very similar to the blue marble fraction or the female fraction concept.

EXAMPLE 7 If you have a solution made up of 2 mol of alcohol (CH_3CH_2OH) and 4 mol of water, what is the mole fraction of the alcohol?

Solution The mole fraction of alcohol is 0.3, or $\frac{1}{3}$.

$$\text{mole fraction of alcohol} = \frac{\text{mol alcohol}}{\text{total mol}} = \frac{2}{6} = \frac{1}{3} = 0.3$$

Our answer can be abbreviated $X_{\text{alcohol}} = 0.3$.

EXAMPLE 8 What is the mole fraction of water in the solution in Example 7?

Solution The mole fraction of water is 0.7. There are two methods that can be used to obtain this answer.

First Method:

$$X_{H_2O} = \frac{\text{mol } H_2O}{\text{total mol}} = \frac{4}{6} = \frac{2}{3} = 0.7$$

Second Method:

$$X_{H_2O} = 1 - X_{\text{alcohol}} = 1 - \frac{1}{3} = \frac{2}{3} = 0.7$$

EXAMPLE 9 A solution is made up of 10 mol of water, 1.0 mol of sodium hydroxide (NaOH), and 2.0 mol of sugar. What is the mole fraction of the sodium hydroxide?

Solution The mole fraction of the sodium hydroxide is $\frac{1}{13}$, or 0.077.

$$X_{NaOH} = \frac{\text{mol NaOH}}{\text{total mol}} = \frac{1.0}{13} = 0.077$$

EXAMPLE 10 A solution is made up of 65.0 g of H_2O, 2.0 g of NaCl, and 1.0 g of HNO_3. What is the mole fraction of the sodium chloride (NaCl)?

Solution First, calculate the number of moles of each substance present in the solution.

$$(65.0 \text{ g } H_2O)\left(\frac{1.00 \text{ mol } H_2O}{18.0 \text{ g } H_2O}\right) = 3.61 \text{ mol } H_2O$$

$$(2.0 \text{ g NaCl})\left(\frac{1.00 \text{ mol NaCl}}{58.5 \text{ g NaCl}}\right) = 0.034 \text{ mol NaCl}$$

$$(1.0 \text{ g HNO}_3)\left(\frac{1.0 \text{ mol HNO}_3}{63 \text{ g HNO}_3}\right) = 0.016 \text{ mol HNO}_3$$

The total number of moles present is 3.61 mol + 0.034 mol + 0.016 mol = 3.66 mol.

$$X_{NaCl} = \frac{\text{mol NaCl}}{\text{total mol}} = \frac{0.034}{3.66} = 0.0093$$

SELF-TEST

1. A solution is made up of 10 mol of alcohol and 15 mol of water. What is the mole fraction of the alcohol?
2. What is the mole fraction of water in the preceding problem?
3. A solution is made up of 5.0 g of NaCl, 50 g of water, and 2.0 g of NaOH. Find the mole fraction of the NaOH.

ANSWERS **1.** 0.40 **2.** 0.60 **3.** 0.017

7.4 MOLARITY

Before going into the last two concentration units (molarity and molality), let's make up a concentration unit that will give you a better grasp of the concepts involved with these last two units.

7.4.1 SPOONFULS-PER-CUP ANALOGY

The unit we will make up is the spoonfuls-per-cup (spc) unit of concentration. As you will see, spc, molarity, molality, and normality all tell the amount of solute present per unit volume of something.

Consider two cups of coffee. The first cup contains 1 spoonful of sugar and the second cup contains 2 spoonfuls of sugar. Which cup has the higher sugar concentration?

The sugar in the second cup is more concentrated than the sugar in the first cup. The second cup has more solute (sugar) per cup (a unit volume) of solution.

What is the sugar concentration of each cup of coffee?

The concentration of sugar in the first cup is 1 spc of sugar (1 spoonful of sugar per cup). The concentration of sugar in the second cup of coffee is 2 spc of sugar.

Notice that the unit *spoonfuls per cup* is made up of two parts, one divided by the other. The spc concentration unit tells us how much solute (spoonfuls of sugar) there is in (per) a specific volume of solution (1 cup of coffee). Sugar is the solute and the coffee solution is the solvent.

If you have 1 qt of coffee that has a concentration of 3 spc of sugar, how much sugar is present in the quart of coffee?

There are 12 spoonfuls of sugar in the quart of coffee.

$$(1 \text{ qt coffee})\left(\frac{4 \text{ cups}}{\text{qt}}\right)\left(\frac{3 \text{ spoonfuls sugar}}{\text{cup coffee}}\right) = 12 \text{ spoonfuls sugar}$$

7.4.2 CHEMICAL APPLICATIONS

Molarity is a concentration unit. Molarity tells the number of moles of solute that are dissolved in 1 L of a solution. Molarity is similar to spoonfuls per cup. With spoonfuls per cup, the amount of solute is measured in spoonfuls and the amount of solution is measured in cups. With molarity, the amount of solute is measured in moles and the amount of solution is measured in liters. The spc unit of concentration is read: "spoonfuls of solute per cup of solution." The unit of concentration called molarity is read: "moles of solute per liter of solution."

•
•

EXAMPLE 11 *Consider a 1-L solution of sodium hydroxide (NaOH). Two moles of sodium hydroxide is dissolved in this 1-L solution, which is made of water and NaOH. What is the molarity of the sodium hydroxide in this solution?*

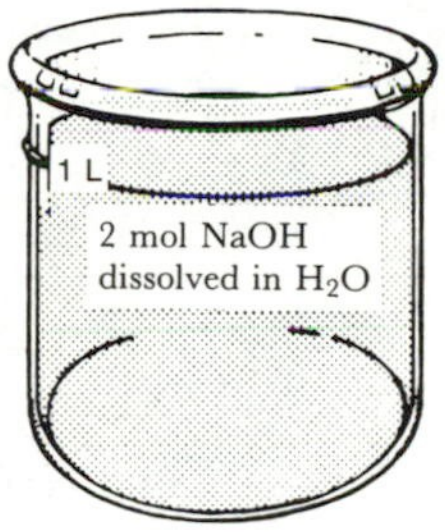

Note that the total volume of H_2O plus NaOH is exactly 1 L. The volume of just the H_2O in this solution is less than 1 L.

Solution The molarity of the sodium hydroxide is 2. There is 2 mol of solute (sodium hydroxide) per liter of solution.

$$\text{molarity of NaOH} = \text{moles of solute per liter of solution} = \frac{\text{mol solute}}{\text{L soln}} \quad (\text{per})$$

$$= \frac{2 \text{ mol NaOH}}{1 \text{ L soln}} = 2 \text{ M NaOH}$$

A solution that has a sodium hydroxide molarity of 2 is said to be a 2-molar sodium hydroxide solution. This is abbreviated 2 M NaOH. These three ways of expressing molarity are used interchangeably. You should be familiar with each way.

•
•

EXAMPLE 12 Two moles of sodium hydroxide is dissolved in 1 L of water. More water is added until the solution volume increases to 4 L. What is the molarity (concentration) of this sodium hydroxide solution? Express your answer three different ways.

Solution The molarity of this NaOH solution is 0.5. This is a 0.5-molar NaOH solution. The solution is 0.5 M NaOH.

$$\text{NaOH molarity} = \frac{\text{mol solute}}{\text{L soln}} = \frac{2 \text{ mol NaOH}}{4 \text{ L soln}} = 0.5 \text{ M NaOH}$$

•
•

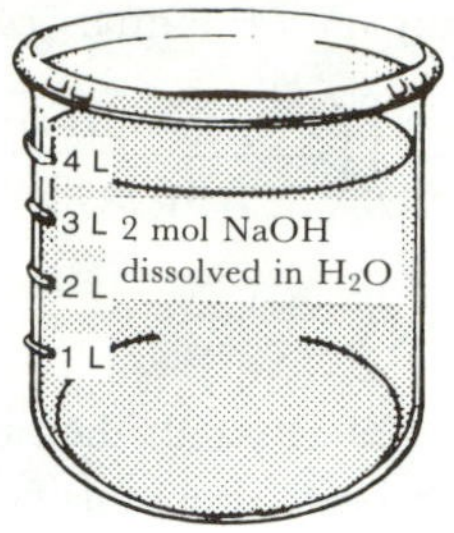

← The total volume of NaOH plus H_2O is 4 L.

EXAMPLE 13 Five hundred milliliters of solution contains 2 mol of sodium hydroxide. What is the molarity of this solution?

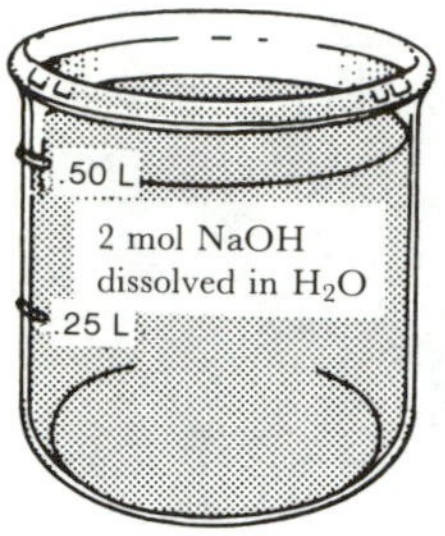

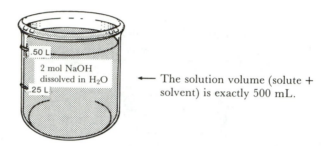

← The solution volume (solute + solvent) is exactly 500 mL.

Solution The solution is 4 M NaOH.

$$\text{NaOH molarity} = \frac{\text{mol solute}}{\text{L soln}} = \frac{2 \text{ mol NaOH}}{0.500 \text{ L soln}} = 4 \text{ M NaOH}$$

EXAMPLE 14 Place 40 g of sodium hydroxide in 500 mL of water and stir until the sodium hydroxide dissolves. Then increase the volume of this solution to 1 L. What is the molarity (concentration) of this sodium hydroxide solution?

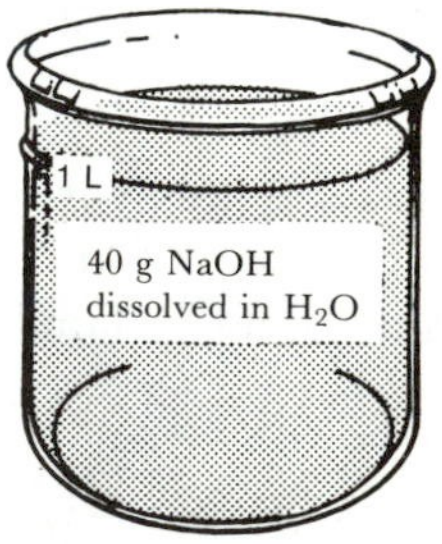

Solution The solution is 1 M NaOH. First find the number of moles of NaOH that are dissolved in the water, then calculate the molarity.

$$(40 \text{ g NaOH})\left(\frac{1.0 \text{ mol NaOH}}{40 \text{ g NaOH}}\right) = 1.0 \text{ mol NaOH}$$

$$\text{molarity} = \frac{\text{mol solute}}{\text{L soln}} = \frac{1.0 \text{ mol NaOH}}{1 \text{ L soln}} = 1 \text{ M NaOH}$$

EXAMPLE 15 If 80 g of sodium hydroxide is present in 500 mL of solution, what is the molarity (concentration) of the sodium hydroxide in this solution?

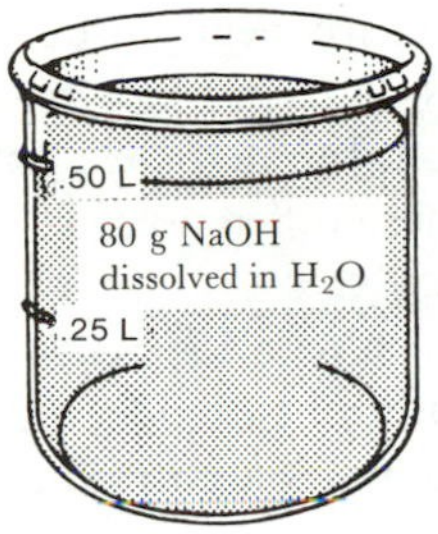

Solution This solution is 4.0 M NaOH. Eighty grams of sodium hydroxide is equal to 2 mol of sodium hydroxide, as shown:

$$(80 \text{ g NaOH})\left(\frac{1.0 \text{ mol NaOH}}{40 \text{ g NaOH}}\right) = 2.0 \text{ mol NaOH}$$

$$\text{molarity} = \frac{\text{mol solute}}{\text{L soln}} = \frac{2.0 \text{ mol NaOH}}{0.500 \text{ L soln}} = 4.0 \text{ M NaOH}$$

EXAMPLE 16 Place 27 g of sodium hydroxide into a beaker and add enough water to bring the total volume up to 700 mL. What is the concentration (molarity) of this sodium hydroxide solution?

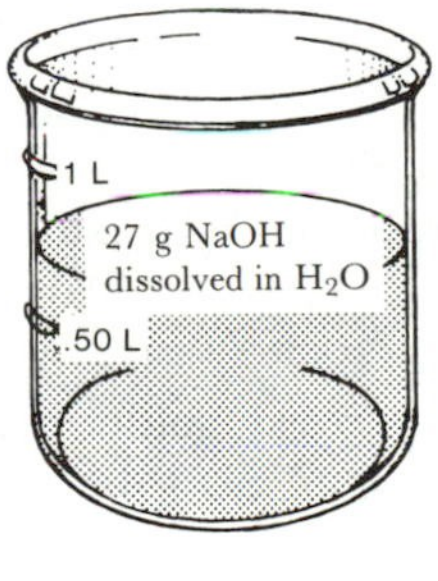

Solution This solution is 0.97 M NaOH.

$$(27 \text{ g NaOH})\left(\frac{1.0 \text{ mol NaOH}}{40 \text{ g NaOH}}\right) = 0.68 \text{ mol NaOH}$$

$$\text{molarity} = \frac{\text{mol solute}}{\text{L soln}} = \frac{0.68 \text{ mol NaOH}}{0.700 \text{ L soln}} = 0.97 \text{ M NaOH}$$

SELF-TEST

1. A 10-L solution contains 5.0 mol of sodium chloride (NaCl). What is the molarity of this solution?
2. A 500-mL solution contains 3.0 mol of sugar ($C_6H_{12}O_6$). What is the molarity of the sugar?
3. Place 30 g of sodium hydroxide (NaOH) into a beaker and add enough water to bring the total volume up to 850 mL. What is the molarity of this sodium hydroxide solution?

ANSWERS **1.** 0.50 M NaCl **2.** 6.0 M $C_6H_{12}O_6$ **3.** 0.88 M NaOH

7.5 MOLALITY

Molality is another frequently used concentration unit. Molality tells us the number of moles of solute that are dissolved in 1 kg (1000 g) of solvent. This molality definition is very similar to the molarity definition, but there are three important differences that we will discuss in detail. Before looking at these differences, let's first consider some molality examples.

•
•

EXAMPLE 17 A solution of sodium hydroxide is prepared by mixing 2 mol of NaOH with 1000 g of water. What is the molality (concentration) of this solution?

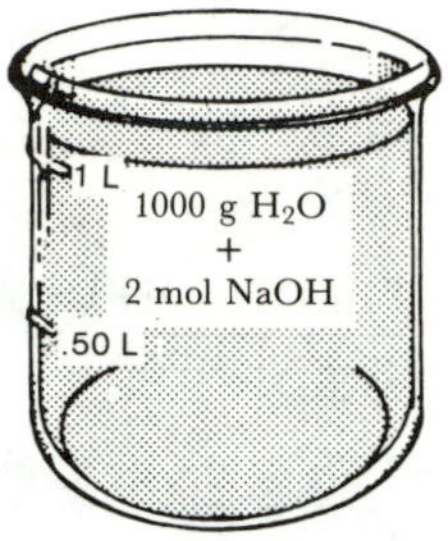

← Note that the total volume of NaOH + H_2O is more than 1 L.

Solution The molality of this sodium hydroxide solution is 2.

$$\text{molality} = \text{moles of solute per kilogram of solvent}$$

$$= \frac{\text{mol solute}}{\text{kg solvent}} \;(\text{per}) = \frac{2 \text{ mol NaOH}}{1.000 \text{ kg solvent}} = 2 \text{ m NaOH}$$

A solution with a sodium hydroxide molality equal to 2 is written 2 m NaOH.

•
•

EXAMPLE 18 Another sodium hydroxide solution is prepared by adding 1 mol of sodium hydroxide to 500 g of water. Find the molality of this solution.

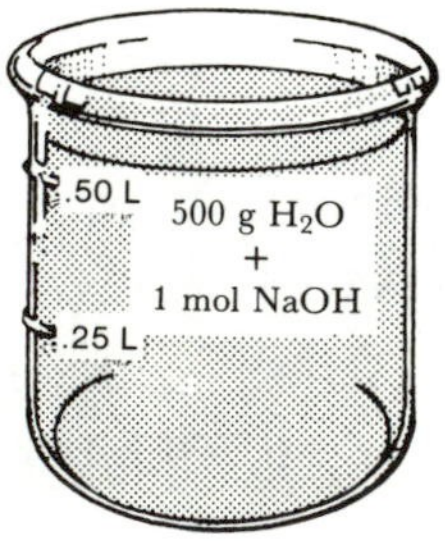

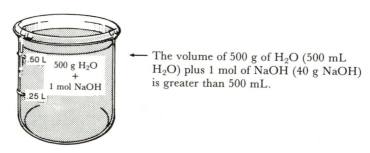

← The volume of 500 g of H_2O (500 mL H_2O) plus 1 mol of NaOH (40 g NaOH) is greater than 500 mL.

Solution This is a 2 m solution of sodium hydroxide.

$$\text{molality} = \frac{\text{mol solute}}{\text{kg solvent}} = \frac{1 \text{ mol NaOH}}{0.500 \text{ kg } H_2O} = 2 \text{ m NaOH}$$

•
•

EXAMPLE 19 If 80 g of NaOH is mixed with 2000 g of water, what is the sodium hydroxide molality of the resulting solution?

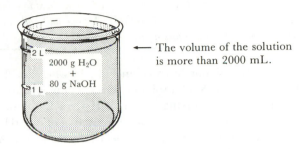

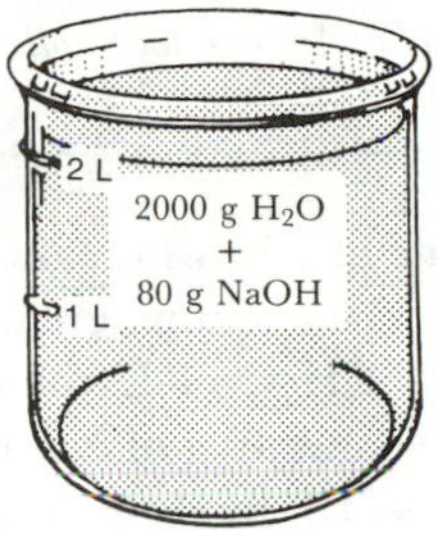

Solution This solution is 1.0 m sodium hydroxide. First, find out how many moles of sodium hydroxide are present in 80 g of NaOH. Then, calculate the molality.

$$(80 \text{ g NaOH})\left(\frac{1.0 \text{ mol NaOH}}{40 \text{ g NaOH}}\right) = 2.0 \text{ mol NaOH}$$

$$\text{molality} = \frac{\text{mol solute}}{\text{kg solvent}} = \frac{2.0 \text{ mol NaOH}}{2.000 \text{ kg H}_2\text{O}} = 1.0 \text{ m NaOH}$$

EXAMPLE 20 Find the molality of a sodium hydroxide solution prepared by mixing 15 g of NaOH with 600 g of H_2O.

Solution This solution is 0.63 m NaOH.

$$(15 \text{ g NaOH})\left(\frac{1.0 \text{ mol NaOH}}{40 \text{ g NaOH}}\right) = 0.38 \text{ mol NaOH}$$

$$\text{molality} = \frac{\text{mol solute}}{\text{kg solvent}} = \frac{0.38 \text{ mol NaOH}}{0.600 \text{ kg H}_2\text{O}} = 0.63 \text{ m NaOH}$$

SELF-TEST

1. A solution of sodium hydroxide is prepared by mixing 2 mol of NaOH with 5000 g of water. What is the molality of this solution?
2. Mix 160 g of NaOH with 3000 g of water. What is the molality of this solution?
3. Find the molality of a sodium hydroxide solution prepared by mixing 3.0 g of NaOH with 400 mL of water.

ANSWERS **1.** 0.4 m NaOH **2.** 1.33 m NaOH **3.** 0.19 m NaOH

7.6 MOLARITY VS. MOLALITY

In this section we will discuss three important differences between molarity and molality.

First Difference:

To keep things simple, let's compare a one molal sodium hydroxide solution with a one molar sodium hydroxide solution (i.e., 1 m NaOH vs. 1 M NaOH). Assume that both are made by mixing NaOH with water. Which of these two solutions occupies a greater volume?

A 1 M NaOH solution has a volume of 1 L, whereas a 1 m NaOH solution has a volume greater than 1 L. A 1 m NaOH solution has a volume greater than 1 L because a 1 m solution of NaOH is prepared by mixing 1000 g of H_2O (1-L volume) with 1 mol of NaOH (40 g of NaOH). The total volume of 1 L of water plus 40 g of NaOH is greater than 1 L. A 1 M solution of NaOH has a volume exactly equal to 1 L by definition. A 1 M solution is made by mixing 1 mol of solute (NaOH) with enough water so that the total volume of the solution is exactly 1 L.

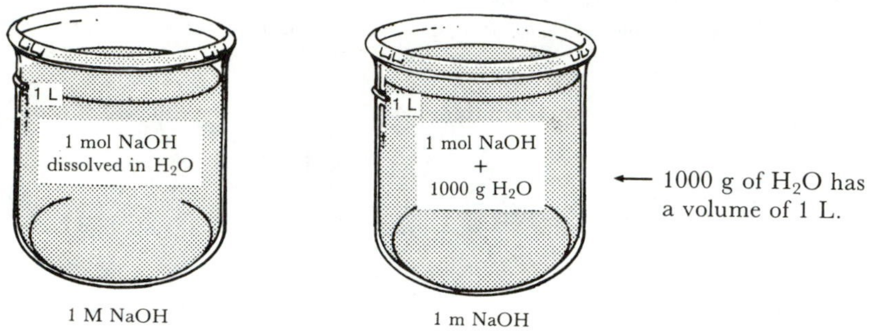

Second Difference:

Assume that you would like to have a friend prepare two solutions for you. The first solution is to be 1 m NaOH and the second solution is to be 1 M NaOH. Write out the directions for preparing these two solutions.

The mechanical steps required to prepare a 1 m NaOH solution are different from the mechanical steps required to prepare a 1 M NaOH solution.

To make a 1 m solution of NaOH, weigh out 1000 g of water, and then weigh out 40 g of NaOH. Mix the 1000 g of H_2O with the 40 g of NaOH in a beaker. The result of mixing the water and the NaOH together is a solution with a concentration of 1 m NaOH.

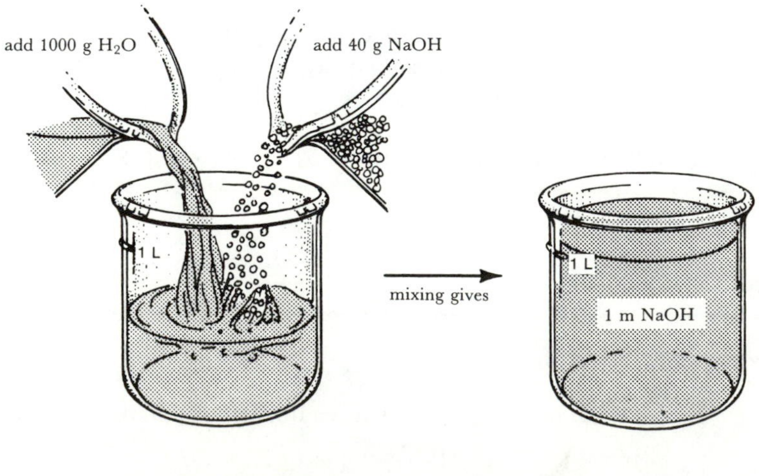

To prepare a 1 M solution of NaOH, first add 40 g of NaOH (1 mol of NaOH) to an empty beaker. Then carefully add just enough water (about 500 mL) to dissolve the NaOH. After the NaOH has dissolved, carefully add more water to the beaker until the total volume of the solution is exactly 1 L.

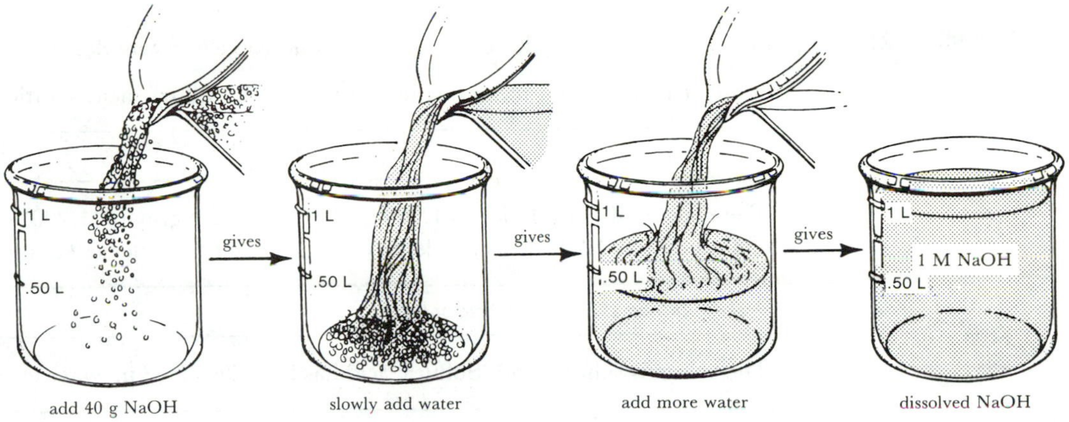

These steps produce a 1-L solution of 1 M NaOH.

Third Difference:

In the preceding discussion we assumed that water is the solvent. One thousand grams of water has a volume of 1 L. As a result, a 1 M NaOH solution and a 1 m NaOH solution have about the same total volume and about the same NaOH concentration. One of them is a little more concentrated than the other. Which solution is more concentrated?

The 1 M NaOH solution would be more concentrated because it has 1 mol of NaOH dissolved in a total solution volume of 1 L. The 1 m solution has 1 mol of NaOH dissolved in a total solution volume that is slightly more than 1 L.

The third difference exists because the solvent doesn't have to be water. The solvent can be alcohol, benzene, mercury, or any other liquid that will dissolve the solute. And, although 1000 g of water has a volume of 1 L, 1000 g of mercury has a volume of about 74 mL, or 0.074 L. If you were to make a solution using mercury as your solvent, a 0.01 m solution would have a volume of about 75 mL, whereas a 0.01 M solution would have a volume of exactly 1 L. When mercury is used as a solvent, the concentration difference between a 0.01 m solution and a 0.01 M solution is very large. A 1 m solution is much more concentrated than a 1 M solution.

7.7 REVIEW AND INTERRELATIONSHIPS

The following Review Example ties together the four concentration units (percent by weight, mole fraction, molarity, and molality) with which we have been working, along with related vocabulary:

-
-

REVIEW EXAMPLE A 20.0% by weight solution of sulfuric acid (H_2SO_4) has a specific gravity of 1.200. Find the molarity, molality, and mole fraction of this sulfuric acid solution.

We can conveniently establish the parallels between this problem and its multiple solutions with the problems in Problem Set 7 by dividing this problem into three separate examples.

•

•

EXAMPLE 21 Find the molarity of the solution discussed in the Review Example.

Solution First of all, it is important to know the density of this sulfuric acid solution.

What is the density of this solution?

The density of this solution is 1.200 g/cc. The specific gravity (1.200) indicates that the sulfuric acid solution is 1.200 times the density of pure water. The density of pure water is $1.0\bar{0}$ g/cc.

What is the solute in this sulfuric acid solution?

The solute is sulfuric acid. Sulfuric acid has been dissolved in water to produce the solution.

What is the solvent?

The solvent is water. The solute sulfuric acid was dissolved in the solvent water to produce the solution.

For reasons that will be clear later, let's assume that we have 1.000 L of this sulfuric acid solution. What is the weight of 1 L of this solution?

One liter of this solution weighs 1200 g. Each cubic centimeter (cc) or milliliter (mL) weighs 1.200 g, so a liter (1000 mL) has a weight of 1200 g.

$$\left(\frac{1.200 \text{ g soln}}{\text{mL soln}}\right)\left(\frac{1000 \text{ mL soln}}{\text{L soln}}\right) = \frac{1200 \text{ g soln}}{\text{L soln}}$$

How many grams of sulfuric acid (solute) are present in 1 L of this sulfuric acid solution?

The weight of the sulfuric acid in 1 L of this solution is 240 g. The solution has a weight of 1200 g and 20.0% of that weight is sulfuric acid. Twenty percent of 1200 g is 240 g. A solution that is 20.0% by weight sulfuric acid has 20.0 grams of sulfuric acid for every 100 grams of solution. Thus

$$(1200 \text{ g soln})\left(\frac{20.0 \text{ grams } H_2SO_4}{100 \text{ grams soln}}\right) = 240 \text{ g } H_2SO_4$$

What is the weight of the water (solvent) making up this sulfuric acid solution?

The water has a weight of 960 g. The entire solution (made up of sulfuric acid and water) weighs 1200 g. Of this total weight, 240 g is sulfuric acid. That leaves 960 g (1200 g − 240 g), which must be the weight of water.

$$\begin{array}{rl} 1200 \text{ g} & [\text{total weight of } H_2O + H_2SO_4 \text{ (solution)}] \\ -\ 240 \text{ g} & [\text{weight of } H_2SO_4 \text{ (solute)}] \\ \hline 960 \text{ g} & [\text{weight of } H_2O \text{ (solvent)}] \end{array}$$

How many moles of sulfuric acid are present in 1 L of this solution?

There are 2.45 mol of sulfuric acid present. One liter of this sulfuric acid solution contains 240 g H_2SO_4. This is 2.45 mol of H_2SO_4, as shown:

$$(240 \text{ g } H_2SO_4)\left(\frac{1.00 \text{ mol } H_2SO_4}{98.1 \text{ g } H_2SO_4}\right) = 2.45 \text{ mol } H_2SO_4$$

What is the molarity of this sulfuric acid solution?

The solution is 2.45 M H_2SO_4. We assumed that we were working with 1 L of solution. We calculated that 1 L of this solution contains 2.45 mol of sulfuric acid. Molarity is defined as the number of moles of solute per liter of solution.

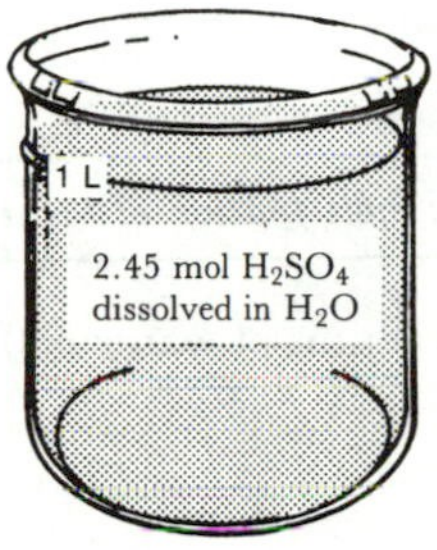

$$\text{molarity} = \frac{\text{mol solute}}{\text{L soln}} = \frac{2.45 \text{ mol } H_2SO_4}{1.000 \text{ L soln}} = 2.45 \text{ M } H_2SO_4$$

Can you explain why we assumed that we had 1 L of solution?

Molarity is defined as the number of moles of solute per liter of solution. We assumed we had 1 L of solution to keep our calculations simple. Division by 1 is easy.

•
•

EXAMPLE 22 What is the molality of the solution discussed in Example 21?

Solution The molality of this sulfuric acid solution is 2.55. The solvent is water and it has a mass of 960 g, or 0.960 kg.

$$(960 \text{ g } H_2O)\left(\frac{1 \text{ kg}}{1000 \text{ g}}\right) = 0.960 \text{ kg } H_2O$$

$$\text{molality} = \frac{\text{mol solute}}{\text{kg solvent}} = \frac{2.45 \text{ mol } H_2SO_4}{0.960 \text{ kg } H_2O} = 2.55 \text{ m } H_2SO_4$$

•
•

EXAMPLE 23 What is the mole fraction of the sulfuric acid solution discussed in Example 21?

Solution The mole fraction of the sulfuric acid solution is 0.0439. To obtain this number, first calculate the number of moles of water present in the solution. Remember that the weight of the water is 960 g.

$$(960 \text{ g } H_2O)\left(\frac{1.00 \text{ mol } H_2O}{18.0 \text{ g } H_2O}\right) = 53.3 \text{ mol } H_2O$$

We have already determined that the number of moles of sulfuric acid in our solution is 2.45. The total number of moles (2.45 mol of sulfuric acid plus 53.3 mol of water) is 55.8.

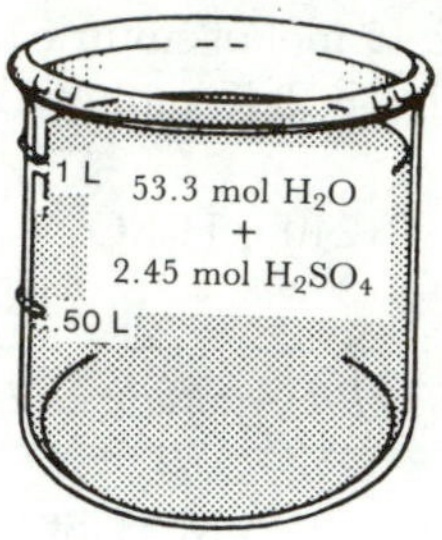

$$\text{mole fraction } H_2SO_4 = \frac{\text{mol } H_2SO_4}{\text{total mol}} = \frac{2.45}{55.8} = 0.0439$$

Finally, what is the mole fraction of the water in this sulfuric acid solution?

The mole fraction of water is 0.956. There are two ways to calculate this value.

First Method:

$$X_{H_2O} = \frac{\text{mol } H_2O}{\text{total mol}} = \frac{53.3}{55.8} = 0.956$$

Second Method:

$$X_{H_2O} = (1 - X_{H_2SO_4}) = (1 - 0.0439) = 0.957$$

We can summarize our results as follows:

- Solution: sulfuric acid dissolved in water
- Solute: sulfuric acid (H_2SO_4)
- Solvent: water (H_2O)
- Volume of solution: 1.000 L
- Specific gravity: 1.200
- Density of solution: 1.200 g/mL
- Mass of solution: 1200 g
- Percent by weight of sulfuric acid: 20.0%
- Mass of solute: 240 g sulfuric acid
- Mass of solvent: 960 g water
- Moles of solute: 2.45 mol sulfuric acid
- Mole fraction sulfuric acid: $X_{H_2SO_4} = 0.0439$
- Mole fraction water: $X_{H_2O} = 0.957$
- Molarity of sulfuric acid: 2.45 M H_2SO_4
- Molality of sulfuric acid: 2.55 m H_2SO_4

All of this information was obtained from just two numbers. You were given a 20.0% by weight solution of sulfuric acid with a specific gravity of 1.20. Because so many ideas, calculations, and vocabulary words are tied together in this one example, it would be wise for you to master this problem.

Try to obtain all of the summarized information shown by yourself. The first time you try will be the most difficult, and you may have to cheat a few times by looking back at the solutions to some of the parts. That is all right. Keep trying until you can reproduce all of the summarized information without cheating.

Now that you think you have this material mastered, let's try to redo the same Review Example by assuming a total solution volume of 200 mL (instead of 1 L).

•
•

EXAMPLE 24 A 20.0% by weight solution of sulfuric acid has a specific gravity of 1.200. Find the molarity of this solution assuming a total solution volume of 200 mL.

Solution First, determine the weight of this solution. The solution must weigh 240 g.

$$(200 \text{ mL soln})\left(\frac{1.200 \text{ g soln}}{\text{mL soln}}\right) = 240 \text{ g soln}$$

Next, determine the number of grams of sulfuric acid (solute) present in this 200-mL solution of sulfuric acid.

$$(240 \text{ g soln})\left(\frac{20.0 \text{ parts } H_2SO_4}{100 \text{ parts soln}}\right) = 48.0 \text{ g } H_2SO_4$$

Now, calculate the weight of the water (solvent) making up this sulfuric acid solution.

$$\begin{array}{rl} 240 \text{ g} & [\text{total weight of } H_2O + H_2SO_4 \text{ (solution)}] \\ -\ 48.0 \text{ g} & [\text{weight of } H_2SO_4 \text{ (solute)}] \\ \hline 192 \text{ g} & [\text{weight of } H_2O \text{ (solvent)}] \end{array}$$

Then, calculate the number of moles of sulfuric acid present in 200 mL of this solution.

$$(48.0 \text{ g } H_2SO_4)\left(\frac{1.00 \text{ mol } H_2SO_4}{98.1 \text{ g } H_2SO_4}\right) = 0.489 \text{ mol } H_2SO_4$$

The molarity of this solution is defined as the number of moles of solute (H_2SO_4) per liter of solution.

$$(200 \text{ mL soln})\left(\frac{1 \text{ L}}{1000 \text{ mL}}\right) = 0.200 \text{ L soln}$$

$$\text{molarity} = \frac{0.490 \text{ mol } H_2SO_4}{0.200 \text{ L soln}} = 2.45 \text{ M } H_2SO_4$$

•
•

Compare this answer (molarity) with the answer obtained in Example 21. Can you explain why these two answers are the same?

The molarity of a solution depends only upon the percent by weight of the solution and not upon the volume of the solution. One liter of a sulfuric acid solution that is 20.0% by weight sulfuric acid has the same concentration as 200 mL of the same solution.

•
•

EXAMPLE 25 What is the molality of the sulfuric acid solution discussed in Example 24?

Solution Based upon our discussion in Example 24, you should know in advance that the molality of this sulfuric acid solution is the same as that calculated in Example 22 because both solutions are 20.0% by weight sulfuric acid. One liter of a sulfuric acid solution that is 20.0% by weight sulfuric acid has the same molality as 200 mL of the same solution (2.55 m H_2SO_4). Even one drop of this solution has the same concentration as the concentration of 1 L of the same solution.

Molality is defined as the number of moles of solute per kilogram of solvent. We have shown in Example 24 that in 200 mL of a 20.0% by weight solution of sulfuric acid there are 0.489 mol of sulfuric acid (solute) and 192 g of water (solvent).

$$(192 \text{ g } H_2O)\left(\frac{1 \text{ kg } H_2O}{1000 \text{ g } H_2O}\right) = 0.192 \text{ kg } H_2O$$

$$\text{molality} = \frac{\text{mol solute}}{\text{kg } H_2O} = \frac{0.489 \text{ mol } H_2SO_4}{0.192 \text{ kg } H_2O} = 2.55 \text{ m } H_2SO_4$$

EXAMPLE 26 What is the mole fraction of the sulfuric acid solution discussed in Example 25?

Solution For the reasons discussed in Example 25, we know that the mole fraction of the sulfuric acid solution must be the same as the mole fraction of the sulfuric acid solution in Example 23. Here are the calculations.

We showed in Example 24 that this 200-mL solution is made up of 0.489 mol of H_2SO_4 and 192 g of H_2O. Find the number of moles of water present in this 200-mL solution.

$$(192 \text{ g } H_2O)\left(\frac{1.00 \text{ mol } H_2O}{18.0 \text{ g } H_2O}\right) = 10.7 \text{ mol } H_2O$$

The total number of moles (0.489 mol of H_2SO_4 plus 10.7 mol of H_2O) is 11.2.

$$\text{mol fraction } H_2SO_4 = \frac{\text{mol } H_2SO_4}{\text{total mol}} = \frac{0.489}{11.2} = 0.0437$$

The slight difference between the mole fraction of sulfuric acid that we obtain here (0.0437) and the mole fraction of sulfuric acid that we obtained in Example 23 (0.0439) is caused when we round off each intermediate answer that we calculate prior to the calculation of the final answer. Such differences usually disappear when problems are worked out in one step, that is, in one chain calculation (see Section 2.5, Example 30).

SELF-TEST

1. A 10.0% by weight solution of a hypothetical acid HX has a specific gravity of 1.09. Find the molarity of this solution. The molecular weight of HX is 169.87 amu.

2. Find the molality of the HX solution discussed in Problem 1.

3. Find the mole fraction of HX in the solution described in Problem 1.

ANSWERS **1.** 0.642 M HX **2.** 0.654 m HX **3.** 0.0117

PROBLEM SET 7

The problems in Problem Set 7 parallel the examples in Chapter 7. For example, if you should have trouble working Problem 5, refer back to Example 5 in this chapter to get help. The correct answers to these problems are given at the end of this book.

1. A pound of honey is dissolved in 20 gallons of water. Identify the solute, the solvent, and the solution.
2. Equal volumes of liquid A and liquid B are mixed, forming a solution that is 50% A and 50% B. Which is the solvent and which is the solute?
3. Consider a solution made up of 3.0 g of hydrochloric acid and 10 g of water. What is the percent by weight of the hydrochloric acid?
4. A solution is prepared by mixing 10 g of liquid A with 15 g of liquid B and 20 g of liquid D. What is the percent by weight of liquid D?
5. A solution of sulfuric acid that is 40.0% by weight sulfuric acid is prepared. How many grams of sulfuric acid are present in 800 g of this solution?
6. A solution is prepared by mixing together some A, B, and C. The total weight of the final solution is 570 g. The solution is 12.0% by weight A. How many grams of A are present in this solution?

7. If you have a solution made up of 1.0 mol of propanol and 3.0 mol of water, what is the mole fraction of the propanol?
8. What is the mole fraction of water in the solution discussed in Problem 7?
9. A solution is made up of 4.0 mol of A, 5.0 mol of B, and 6.0 mol of C. What is the mole fraction of A?
10. A solution is made up of 60.0 g of H_2O, 1.0 g of NaCl, and 0.5 g of HNO_3. What is the mole fraction of NaCl?
11. Consider a 1-L solution of sodium hydroxide. Three moles of sodium hydroxide is dissolved in this 1-L solution, which is made of water and sodium hydroxide. What is the molarity of the sodium hydroxide in this solution?
12. Four moles of sodium hydroxide is dissolved in 1 L of water. More water is added until the solution volume increases to 4 L. What is the molarity of this sodium hydroxide solution? Express your answer three different ways.
13. Five hundred milliliters of solution contains 6.0 mol of sodium hydroxide. What is the molarity of this solution?
14. Place 20 g of sodium hydroxide in 500 mL of water and stir until the sodium hydroxide dissolves. Then increase the volume of this solution to 1 L. What is the molarity of this sodium hydroxide solution?
15. If 40 g of sodium hydroxide is present in 500 mL of solution, what is the molarity of the sodium hydroxide in this solution?
16. Place 21.6 g of sodium hydroxide into a beaker and add enough water to bring the total volume up to 560 mL. What is the molarity of this sodium hydroxide solution?
17. A solution of sodium hydroxide is prepared by mixing 4 mol of NaOH with 1000 g of H_2O. What is the molality of this solution?
18. Another sodium hydroxide solution is prepared by adding 1.00 mol of sodium hydroxide to 400 g of water. Find the molality of this solution.
19. If 64 g of NaOH is mixed with 1600 g of H_2O, what is the NaOH molality of the resulting solution?
20. Find the molality of a sodium hydroxide solution prepared by mixing 12 g of NaOH with 480 g of water.
21. A 5.000% by weight solution of sulfuric acid has a specific gravity of 1.033. Find the molarity of this sulfuric acid solution. Assume that the volume of this solution is 1.000 L.

Refer to the solution described in Problem 21 when working Problems 22–26.

22. What is the molality of the solution?
23. What is the mole fraction of sulfuric acid in the solution?
24. Assume that the volume of the solution is 300 mL. Recalculate the molarity of this sulfuric acid solution.
25. Assume that the volume of the solution is 400 mL. Recalculate the molality of this sulfuric acid solution.
26. Assume that the volume of the solution is 150 mL. Recalculate the mole fraction of sulfuric acid in this solution.

8 Colligative Properties

8.1 INTRODUCTION

A colligative property of a solution is a property that depends upon the number of moles of solute that are dissolved per unit weight of solvent. Pure water will freeze at 0°C. The addition of solute (e.g., salt) to the water produces a solution with a freezing point less than 0°C. The addition of more solute (salt) lowers the freezing point of the solution even more.

Is the freezing point (fp) of water a colligative property?

Yes, the freezing point of water is a colligative property. The freezing point of water depends upon the weight (moles) of the solute dissolved in a specific weight of solvent. The more solute (salt) added to a fixed weight of water, the lower the freezing point of the solution.

The boiling point (bp) of water also depends upon the number of moles of solute dissolved per unit weight of solvent. The boiling point of water is also a colligative property.

Colligative properties are used in chemistry to determine the molecular weights of unknown compounds. Understanding colligative properties will also help you understand several everyday phenomena, such as candy making, cloudy ice cubes, frozen hot-water pipes, pickle making, food preservation with salt, and energy generation from the ocean, all of which are discussed later in this chapter.

8.2 FREEZING POINT DEPRESSION AND BOILING POINT ELEVATION

One mole (6.02×10^{23}) of dissolved solute particles will lower the freezing point of 1 kg of water by 1.86°C. (A solute particle is either a molecule, an atom, or an ion.) One mole of dissolved solute particles will raise the boiling point of 1 kg of water by 0.51°C.

•
•

EXAMPLE 1 If 1.00 mol of sucrose (common table sugar, $C_{12}H_{22}O_{11}$) is dissolved in 1.00 kg of water, what are the freezing and boiling points of this sugar solution?

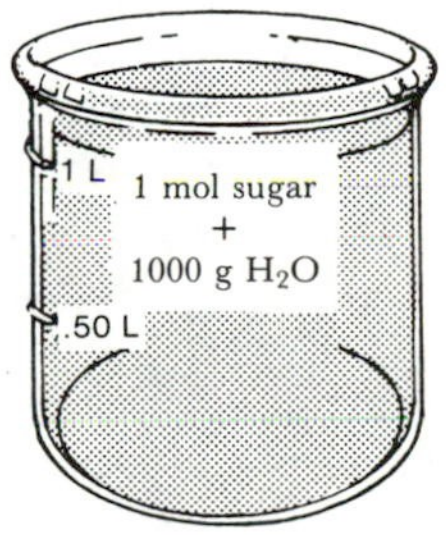

Solution The freezing point of this solution is −1.86°C. The boiling point of this sugar solution is 100.51°C. One mole of sugar molecules (dissolved solute particles) lowers the freezing point of 1 kg of water by 1.86°C and raises the boiling point of 1 kg of water by 0.51°C.

$$(1.00 \text{ mol sugar in } 1.00 \text{ kg } H_2O)\left(\frac{1.86°\text{C fp depression}}{1.00 \text{ mol sugar in } 1.00 \text{ kg } H_2O}\right) = 186°\text{C fp depression}$$

(1.00 mol sugar in 1.00 kg H_2O) ↑ given; the fraction ↑ conversion factor

$$\text{freezing point} = (0.00°\text{C} - 1.86°\text{C}) = -1.86°\text{C}$$

- 0.00°C: normal fp of H_2O
- 1.86°C: fp depression due to 1 mol of solute particles in 1 kg of H_2O
- −1.86°C: fp of 1 kg of H_2O with 1 mol of sugar dissolved in it.

$$(1.00 \text{ mol sugar in } 1.00 \text{ kg } H_2O)\left(\frac{0.51°\text{C bp elevation}}{1.00 \text{ mol sugar in } 1.00 \text{ kg } H_2O}\right) = 0.51°\text{C bp elevation}$$

$$\text{boiling point} = (100.00°\text{C} + 0.51°\text{C}) = 100.51°\text{C}$$

- 100.00°C: normal bp of H_2O
- 0.51°C: bp elevation caused by 1 mol of sugar
- 100.51°C: bp of 1 kg H_2O with 1 mol of sugar

•
•

EXAMPLE 2 What is the concentration (molality) of the sugar solution in Example 1?

Solution The solution is 1.00 m sugar.

$$\text{molality} = \frac{\text{mol sugar}}{\text{kg solvent}} = \frac{1.00 \text{ mol sugar}}{1.00 \text{ kg } H_2O} = 1.00 \text{ m sugar}$$

(the fraction bar in $\frac{\text{mol sugar}}{\text{kg solvent}}$ ← per)

•
•

EXAMPLE 3 If 0.500 mol of sugar (sucrose, $C_{12}H_{22}O_{11}$) is dissolved in 250 g of water, what is the molality of this sugar solution and what are the freezing and boiling points of this solution?

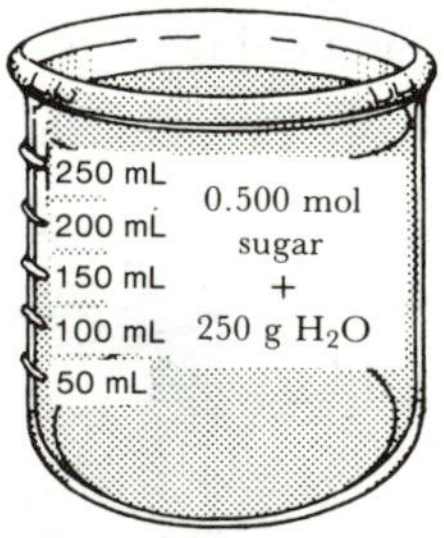

Solution The molality of this sugar solution is 2.00, that is, there are 2.00 mol of sugar (solute) for every kilogram of water (solvent).

$$\text{molality} = \frac{\text{mol solute}}{\text{kg solvent}} = \frac{0.500 \text{ mol sugar}}{0.250 \text{ kg } H_2O} = 2.00 \text{ m sugar}$$

The freezing point of this solution is −3.72°C, and the boiling point is 101.02°C.

$$(2.00 \text{ mol particles in } 1.00 \text{ kg } H_2O)\left(\frac{1.86°\text{C fp depression}}{1.00 \text{ mol particles in } 1.00 \text{ kg } H_2O}\right)$$

↑ sugar molecules

$$= 3.72°\text{C fp depression}$$

$$\text{freezing point} = (0.00°\text{C} - 3.72°\text{C}) = -3.72°\text{C}$$

$$(2.00 \text{ mol particles in } 1.00 \text{ kg } H_2O)\left(\frac{0.51°\text{C bp elevation}}{1.00 \text{ mol particles in } 1.00 \text{ kg } H_2O}\right)$$

↑ sugar molecules

$$= 1.02°\text{C bp elevation}$$

$$\text{boiling point} = (100.00°\text{C} + 1.02°\text{C}) = 101.02°\text{C}$$

We can now restate our earlier definition of freezing point depression (lowering) and boiling point elevation (raising). For a 1 m solution of solute particles dissolved in water, the freezing point is depressed by 1.86°C, and the boiling point is elevated by 0.51°C.

•
•

EXAMPLE 4 Find the freezing and boiling points of a solution made by adding 342 g of sugar (sucrose) to 2.00 kg of water. The molecular weight of sugar is 342 amu.

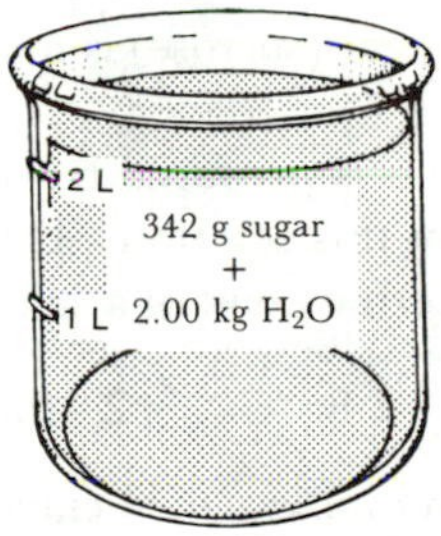

Solution The freezing point of this solution is −0.93°C. Its boiling point is 100.26°C.

$$(342 \text{ g sugar})\left(\frac{1.00 \text{ mol sugar}}{342 \text{ g sugar}}\right) = 1.00 \text{ mol sugar}$$

$$\text{molality} = \frac{\text{mol solute}}{\text{kg solvent}} = \frac{1.00 \text{ mol sugar}}{2.00 \text{ kg } H_2O} = 0.500 \text{ m sugar}$$

$$(0.500 \text{ mol sugar in } 1.00 \text{ kg } H_2O)\left(\frac{1.86°\text{C fp depression}}{1.00 \text{ mol sugar in } 1.00 \text{ kg } H_2O}\right) = 0.93°\text{C fp depression}$$

$$\text{freezing point} = (0.00°\text{C} - 0.93°\text{C}) = -0.93°\text{C}$$

$$(0.500 \text{ mol sugar in } 1.00 \text{ kg } H_2O)\left(\frac{0.51°\text{C bp elevation}}{1.00 \text{ mol sugar in } 1.00 \text{ kg } H_2O}\right) = 0.26°\text{C bp elevation}$$

$$\text{boiling point} = (100.00°\text{C} + 0.26°\text{C}) = 100.26°\text{C}$$

•
•

EXAMPLE 5 Find the freezing and boiling points of a solution made by adding 85.5 g of sugar to 125 g of water.

Solution The freezing point of this solution is −3.72°C. Its boiling point is 101.02°C.

$$(85.5 \text{ g sugar})\left(\frac{1.00 \text{ mol sugar}}{342 \text{ g sugar}}\right) = 0.250 \text{ mol sugar}$$

$$\text{molality} = \frac{\text{mol solute}}{\text{kg solvent}} = \frac{0.250 \text{ mol sugar}}{0.125 \text{ kg } H_2O} = 2.00 \text{ m sugar}$$

$$(2.00 \text{ mol particles in } 1.00 \text{ kg } H_2O)\left(\frac{1.86°C \text{ fp depression}}{1.00 \text{ mol particles in } 1.00 \text{ kg } H_2O}\right)$$

↑ sugar molecules

$$= 3.72°C \text{ fp depression}$$

$$\text{freezing point} = (0.00°C - 3.72°C) = -3.72°C$$

$$(2.00 \text{ mol particles in } 1.00 \text{ kg } H_2O)\left(\frac{0.51°C \text{ bp elevation}}{1.00 \text{ mol particles in } 1.00 \text{ kg } H_2O}\right)$$

$$= 1.02°C \text{ bp elevation}$$

$$\text{boiling point} = (100.00°C + 1.02°C) = 101.02°C$$

SELF-TEST

1. 2.00 moles of sugar (sucrose) is mixed with 2.00 kg of water. What is the freezing point of this solution?
2. A sugar solution is prepared by mixing 171 g of sucrose with 1000 g of water. What is the boiling point of this solution? The molecular weight of sucrose is 342 amu.
3. What is the freezing point of a solution made by adding 478.8 g of sucrose to 700.0 g of water?

ANSWERS **1.** −1.86°C **2.** 100.26°C **3.** −3.72°C

Sugar doesn't ionize (dissociate, split up into ions) when it is dissolved in water. Sugar is not an electrolyte. When 1 mol of sugar is dissolved in water, 1 mol of sugar molecules (particles) can be found in the water. A different situation occurs when salts (electrolytes), such as sodium chloride, are added to water. When 1 mol of NaCl is dissolved in water, 2 mol of ions (dissolved particles) are formed when the NaCl dissociates (ionizes, splits up, breaks up).

It may sound strange, but there are no molecules of NaCl in an NaCl solution. Perhaps an analogy will help you understand this more clearly.

Pretend you place 10 married couples into a certain room that causes all of the couples to divorce. The divorce rate is 100% for all couples who enter this room. If you place 10 married couples into the room, how many divorced men are found in the room?

There are 10 divorced men in the room. If all 10 couples divorce (a couple is made up of 1 man and 1 woman), 10 divorced men are produced. The divorce "reaction" can be represented with the following "equation":

$$\text{MW} \xrightarrow{\text{100\% divorce rate}} \text{M} + \text{W}$$

where MW represents a married couple, M represents a divorced man, and W represents a divorced woman.

How many divorced women are found in the room?

Since you found 10 divorced men, there must also be 10 divorced women.

How many married couples are left in the room?

No married couples are left in the room. They are all divorced. They no longer exist as couples but as individual men and women.

And so it is with NaCl, which completely breaks up (divorces, ionizes, splits up, dissociates) when placed into water. If you put 1 mol of NaCl into 1 L of water, you will find 1 mol of chloride ions (Cl^-) and 1 mol of sodium ions (Na^+), but you will not find any NaCl molecules or formula units.

$$NaCl \xrightarrow{100\% \text{ dissociation}} Na^+ + Cl^-$$

Assume that you dissolve 1 mol of NaCl in water by mixing 1 mol of NaCl with 1000 g of water. How many moles of particles are in this solution?

There are 2 mol of particles (ions) in this 1 m sodium chloride solution: 1 mol of sodium ions plus 1 mol of chloride ions. There are 0 mol of NaCl in this 1 m NaCl solution because all of the NaCl particles have split up into sodium ions and chloride ions.

$$(1 \text{ mol NaCl in 1 kg } H_2O)\left(\frac{2 \text{ mol ions}}{1 \text{ mol NaCl}}\right) = 2 \text{ mol ions in 1 kg } H_2O$$

EXAMPLE 6 What are the freezing and boiling points of a 1.00 m solution of NaCl?

Solution The freezing point of a 1.00 m NaCl solution is −3.72°C. The boiling point of this solution is 101.02°C.

$$\underset{\substack{\uparrow \\ Na^+ \text{ and } Cl^- \text{ ions}}}{(2.00 \text{ mol particles in 1.00 kg } H_2O)}\left(\frac{1.86°C \text{ fp depression}}{1.00 \text{ mol particles in 1.00 kg } H_2O}\right)$$

$$= 3.72°C \text{ fp depression}$$

$$\text{freezing point} = (0.00°C - 3.72°C) = -3.72°C$$

$$(2.00 \text{ mol particles in 1.00 kg } H_2O)\left(\frac{0.51°C \text{ bp elevation}}{1.00 \text{ mol particles in 1.00 kg } H_2O}\right)$$

$$= 1.02°C \text{ bp elevation}$$

$$\text{boiling point} = (100.00°C + 1.02°C) = 101.02°C$$

EXAMPLE 7 Assume that $Sc(NO_3)_3$ (scandium nitrate) is a strong electrolyte (strong electrolytes completely—100%—dissociate into ions when placed in water). What are the freezing and boiling points of a solution made by mixing 1.00 mol of $Sc(NO_3)_3$ with 1000 g of water?

Solution The freezing point of a 1.00 m $Sc(NO_3)_3$ solution is −7.44°C. Its boiling point is 102.04°C. One mole of $Sc(NO_3)_3$ splits up into 4 mol of particles (1 mol of scandium ion and 3 mol of nitrate ion). There are no molecules or formula units of scandium nitrate in a scandium nitrate solution.

$$Sc(NO_3)_3 \xrightarrow{100\%} Sc^{3+} + 3NO_3^-$$

$$(1.00 \text{ mol } Sc(NO_3)_3 \text{ in 1.000 kg } H_2O)\left(\frac{4 \text{ mol ions}}{1 \text{ mol } Sc(NO_3)_3}\right) = 4.00 \text{ mol ions in 1 kg } H_2O$$

One mole of particles lowers the freezing point of 1000 g of water by 1.86°C. Four moles of particles lower the freezing point of 1000 g of water by 4 times 1.86°C, that is, by 7.44°C.

$$(4.00 \text{ mol particles in } 1.00 \text{ kg } H_2O)\left(\frac{1.86°\text{C fp depression}}{1.00 \text{ mol particles in } 1.00 \text{ kg } H_2O}\right)$$

↑ Sc^{3+} and NO_3^- ions

$$= 7.44°\text{C fp depression}$$

$$\text{freezing point} = (0.00°\text{C} - 7.44°\text{C}) = -7.44°\text{C}$$

$$(4.00 \text{ mol ions in } 1.00 \text{ kg } H_2O)\left(\frac{0.51°\text{C bp elevation}}{1.00 \text{ mol ions in } 1.00 \text{ kg } H_2O}\right) = 2.04°\text{C bp elevation}$$

$$\text{boiling point} = (100.00°\text{C} + 2.04°\text{C}) = 102.04°\text{C}$$

Sulfuric acid dissociates into hydrogen ions and hydrogen sulfate (HSO_4^-) ions, as represented by the following equation:

$$H_2SO_4 \xrightarrow{100\%} H^+ + HSO_4^-$$

This equation indicates that every mole of sulfuric acid present will dissociate into 2 mol of particles (1 mol of H^+ ions and 1 mol of HSO_4^- ions).

•
•

EXAMPLE 8 What are the freezing and boiling points of a 20.0% by weight solution of sulfuric acid whose specific gravity is 1.200?

Solution This H_2SO_4 solution has the same percent by weight and specific gravity as the sulfuric acid solution that we played with at the end of the last section. There, we calculated the molality of this solution to be 2.55 m H_2SO_4. If you have trouble obtaining this answer for the molality, go back to the last chapter and review this material.

H^+ and HSO_4^- ↙

$$(2.55 \text{ mol } H_2SO_4 \text{ in } 1 \text{ kg } H_2O)\left(\frac{2 \text{ mol ions}}{1 \text{ mol } H_2SO_4}\right) = 5.10 \text{ mol ions in } 1 \text{ kg } H_2O$$

$$(5.10 \text{ mol particles in } 1 \text{ kg } H_2O)\left(\frac{1.86°\text{C fp depression}}{1.00 \text{ mol particles in } 1.00 \text{ kg } H_2O}\right)$$

↑ H^+ and HSO_4^- ions

$$= 9.49°\text{C fp depression}$$

$$\text{freezing point} = (0.00°\text{C} - 9.49°\text{C}) = -9.49°\text{C}$$

$$(5.10 \text{ mol particles in } 1 \text{ kg } H_2O)\left(\frac{0.51°\text{C bp elevation}}{1.00 \text{ mol particles in } 1.00 \text{ kg } H_2O}\right)$$

$$= 2.60°\text{C bp elevation}$$

$$\text{boiling point} = (100.00°\text{C} + 2.60°\text{C}) = 102.60°\text{C}$$

•
•

These boiling and freezing point values are ideal. In reality, solutions that are this concentrated (2.55 m H_2SO_4) are not as likely to fully dissociate (achieve 100% dissociation) as are solutions with a lower solute concentration (e.g., 0.01 m NaCl). This idea will be discussed in greater detail in Section 11.2, "Effects of HCl and Aspirin on the Stomach."

•
•

SELF-TEST

1. Place 1.00 mole of NaCl into a beaker. Add 500 g of water to the beaker and stir it until the sodium chloride has dissolved. What is the freezing point of this solution?
2. A mixture of 500 g of water and 0.500 mol of $AlCl_3$ is prepared. What is the boiling point of this solution?
3. What is the freezing point of a solution prepared by dissolving 100 g of sulfuric acid in 200 g of water?

ANSWERS **1.** −7.44°C **2.** 102.04°C **3.** −18.96°C

8.2.1 BOILING SOLVENTS VS. BOILING SOLUTIONS

We're going to do a mental experiment that may give you a better understanding of boiling point elevation.

Imagine placing a quart (liter) of water on your stove, putting a thermometer into the water, and heating the water to make it boil. Continue boiling the water until half of it has evaporated. Describe the thermometer readings taken during this experiment.

As the water warms, the thermometer shows the temperature of the water increasing from room temperature (around 22°C) to the temperature of boiling water (around 100°C). These temperature readings are approximate because the temperature of boiling water depends upon atmospheric pressure and water purity. The temperature of the boiling water stays constant at about 100°C the entire time during which 500 mL of water boils away (evaporates while the water boils).

Now, in your mind, prepare a solution by mixing 1 L of pure water with 1 mol of sugar (sucrose). This is a 1 m sugar solution (1 mol of solute dissolved in 1 kg of solvent). Place this sugar solution on your stove and heat it to make it boil. Continue boiling the solution until half of the water has boiled away. Assume that the atmospheric pressure is 760 mm Hg. Describe in detail the thermometer readings taken during this experiment.

The temperature of the sugar solution rises from room temperature (around 22°C) to 100.51°C, the boiling point of a 1 m solution.

As the solution continues to boil, the boiling point does not remain constant. The boiling point of the sugar solution increases. The temperature of the sugar solution rises as the boiling process continues. Why?

As water boils away, the remaining sugar solution becomes more concentrated. The concentration of the sugar increases as the amount of solvent (water) decreases (by boiling away).

•
•

EXAMPLE 9 What will be the temperature of a 1.00 m boiling sugar solution when half of the water (500 mL) boils away? The initial boiling point was 100.51°C.

Solution The resulting sugar solution will have a boiling point of 101.02°C. We evaporate 500 mL of water, but we still have 1.00 mol of sugar dissolved in the remaining 500 mL of water. The concentration of the sugar increases to 2.00 m. None of the sugar is lost when the water boils away.

$$\text{molality} = \frac{\text{mol solute}}{\text{kg solvent}} = \frac{1.00 \text{ mol sugar}}{0.500 \text{ kg } H_2O} = 2.00 \text{ m sugar}$$

$$(2.00 \text{ mol sugar in } 1.00 \text{ kg } H_2O)\left(\frac{0.51°\text{C bp elevation}}{1.00 \text{ mol sugar in } 1.00 \text{ kg } H_2O}\right) = 1.02°\text{C bp elevation}$$

$$\text{boiling point} = (100.00°\text{C} + 1.02°\text{C}) = 101.02°\text{C}$$

As more water boils away, the sugar solution becomes more concentrated, and the temperature of the boiling sugar solution rises.

•
•

8.2.2 MAKING CANDY

You can make your own syrup and candy by applying boiling point elevation and concentration unit concepts. To make candy, boil a sugar solution until its boiling point reaches the correct temperature (which can be found in many cookbooks). The main difference between syrup, soft candy, and hard candy is the sugar concentration of each (see Figure 8.1). The temperature of a boiling sugar solution is a measure of its sugar concentration. When the proper temperature (concentration of sugar) is reached, stop boiling and slowly cool the solution to room temperature.

FIGURE 8.1 The main difference between the pancake syrup and the candy is their boiling point.

•
•

EXAMPLE 10 Soft ball candy is prepared from a boiling sugar solution that has a boiling point of 234°F. Find the molality of this sugar solution.

Solution The molality of a sugar solution with a boiling point of 234°F is 24. To obtain this value, first convert the boiling point from degrees Fahrenheit to degrees centigrade.

$$\text{deg C} = \frac{5°\text{C}}{9°\text{F}}(\text{deg F} - 32°\text{F})$$

$$= \frac{5°\text{C}}{9°\text{F}}(234°\text{F} - 32°\text{F}) = 112°\text{C}$$

The soft ball sugar solution has a boiling point of 112°C, which is 12°C above the boiling point of pure water. The sugar present in the sugar solution has elevated the normal boiling point of water by 12°C. The molality of the sugar solution can be found from its boiling point elevation.

$$(12°\text{C bp elevation})\left(\frac{1.00 \text{ m sugar soln}}{0.51°\text{C bp elevation}}\right) = 24 \text{ m sugar soln}$$

EXAMPLE 11 How many grams of sugar must be dissolved in 1.00 L of water (1.00 kg of water) to make a 24 m sugar solution? The molecular weight of sugar (sucrose) is 342 amu.

Solution A 24 m sugar solution has about 8200 g of sugar dissolved in 1.00 kg of water.

$$\left(\frac{24 \text{ mol sugar}}{\text{L water}}\right)\left(\frac{342 \text{ g sugar}}{\text{mol sugar}}\right) = \frac{8.2 \times 10^3 \text{ g sugar}}{\text{L water}}$$

EXAMPLE 12 Hard ball candy is prepared from a boiling sugar solution that has a boiling point of 250°F. Find the number of grams of sugar that must be dissolved in 1.00 L of water to make a sugar solution with a boiling point of 250°F.

Solution You must dissolve about 14 000 g of sugar in 1.00 L of water (1.00 kg of water) to have a sugar solution that boils at 250°F. To obtain this answer, first determine the boiling point temperature of the sugar solution in degrees Celsius.

$$\text{deg C} = \frac{5°\text{C}}{9°\text{F}}(\text{deg F} - 32°\text{F}) = \frac{5°\text{C}}{9°\text{F}}(250°\text{F} - 32°\text{F}) = 121°\text{C}$$

Now calculate the molality of a sugar solution with a boiling point of 121°C. The solution has a boiling point elevation of 21°C (121°C − 100°C = 21°C).

$$(21°\text{C bp elevation})\left(\frac{1.0 \text{ m sugar soln}}{0.51°\text{C bp elevation}}\right) = 41 \text{ m sugar soln}$$

Finally, determine the number of grams of sugar present in a 41 m sugar solution.

$$\left(\frac{41 \text{ mol sugar}}{\text{kg } H_2O}\right)\left(\frac{342 \text{ g sugar}}{\text{mol sugar}}\right) = \frac{1.4 \times 10^4 \text{ g sugar}}{\text{kg } H_2O}$$

Preparing a sugar solution that contains 14 000 g of sugar for every 1000 g (1 kg) of water is not possible at room temperature. Hot water can dissolve more sugar than cold water can, and it is easier to concentrate sugar by boiling away water than by dissolving more sugar into the hot water. The explanation is actually more complicated than this discussion may lead you to believe. We have been forming supersaturated sugar solutions.

EXAMPLE 13 It is possible to dissolve even more than 14 000 g of sugar in 1 L of water by raising the boiling point of the sugar solution above 250°F. As the sugar concentration increases (as the boiling point increases), the mixtures that result are referred to as hard ball candy, very hard ball candy, light crack candy, and hard crack candy. The boiling point of a sugar solution known as hard crack candy is 300°F. How many grams of sugar are dissolved in 1.000 L of water if the sugar solution has a boiling point of 300°F?

Solution A sugar solution with a boiling point of 300°F has a molality of 94 and contains about 33 000 g of sugar for every 1000 g (1.000 L) of water.

$$300°F = 149°C$$

$$\text{bp elevation} = 49°C$$

$$(49°C \text{ bp elevation})\left(\frac{1.00 \text{ m sugar soln}}{0.51°C \text{ bp elevation}}\right) = 96 \text{ m sugar soln}$$

$$\left(\frac{96 \text{ mol sugar}}{\text{kg } H_2O}\right)\left(\frac{342 \text{ g sugar}}{\text{mol sugar}}\right) = \frac{3.3 \times 10^4 \text{ g sugar}}{\text{kg } H_2O}$$

8.2.3 WHY CAR RADIATOR ANTIFREEZE IS DILUTED

Antifreeze is mixed with a car's radiator water to prevent the water from freezing in the winter. Water expands when it freezes, and the expansion can crack a radiator. Antifreeze is added to lower the freezing point of the radiator water.

EXAMPLE 14 A common antifreeze is ethylene glycol (1,2-ethanediol), which is sold under various trade names, such as Prestone antifreeze and Xerex antifreeze. This antifreeze has a molecular weight of 62.0 amu and a density of 1.110 g/cm^3. A 50-50 mixture of antifreeze and water (i.e., 50% by weight antifreeze) is usually recommended to protect cars in the winter. What is the freezing point of a solution made by mixing 1.000 L of antifreeze with 1.000 L of water?

Solution The freezing point of a 50-50 mixture of antifreeze and water is −33°C. Here are the calculations:

$$(1000 \text{ mL antifreeze})\left(\frac{1.110 \text{ g antifreeze}}{\text{mL antifreeze}}\right) = 1110 \text{ g antifreeze}$$

$$(1110 \text{ g antifreeze})\left(\frac{1.00 \text{ mol antifreeze}}{62.0 \text{ g antifreeze}}\right) = 17.9 \text{ mol antifreeze} \leftarrow \text{solute}$$

$$\text{mass of water} = 1.000 \text{ kg } H_2O \leftarrow \text{solvent}$$

$$\text{molality} = \frac{\text{mol solute}}{\text{kg solvent}} = \frac{17.9 \text{ mol antifreeze}}{1.000 \text{ kg } H_2O} = 17.9 \text{ m antifreeze}$$

$$\text{freezing point depression} = (17.9 \text{ mol antifreeze in } 1.00 \text{ kg } H_2O) \times \left(\frac{1.86°C \text{ fp depression}}{1.00 \text{ mol antifreeze in } 1.00 \text{ kg } H_2O}\right) = 33.3°C$$

$$\text{freezing point} = (0°C - 33.3°C) = -33.3°C$$

A 50-50 mixture of antifreeze and water has a freezing point of −33°C, so why not use 100% pure antifreeze to protect your car radiator? Pure 1,2-ethanediol (antifreeze) has a freezing point of −11.5°C.

Can you explain why a 50-50 mixture has a freezing point of −33°C while pure antifreeze freezes at −11.5°C?

It is not the low freezing point of antifreeze that gives a water-antifreeze mixture a low freezing point. Antifreeze lowers the freezing point of water in the same way any solute does, that is, by 1.86°C for every mole of solute particles added. Salt (NaCl) could be used if you were only concerned with preventing radiator water from freezing. But salt and other solutes have disadvantages. For example, they may accelerate radiator corrosion.

8.2.4 WHY ANTIFREEZE IS USED IN THE SUMMER

The purpose of a car radiator is to help prevent the car engine from overheating. Water circulates through and cools the car engine. This water becomes hot as it cools the engine and must be passed through the cooling coils of the radiator. A fan forces air through the cooling coils, which helps cool the water before it circulates through the engine again.

In hot weather an engine can overheat more easily, especially with the extra strain of the car air conditioner. The engine can become so hot that is causes the radiator water to boil. If the radiator water boils away, the engine (having lost its cooling water) becomes even hotter and can be seriously damaged.

It is usually recommended that you leave antifreeze in your car radiator all year long (even in the hot summer). What advantage does this have?

In Example 14, we showed that a 50-50 mixture of water and antifreeze is 17.9 m antifreeze. The antifreeze not only depresses the freezing point of water, but it also elevates the boiling point.

$$(17.9 \text{ mol antifreeze in } 1.00 \text{ kg } H_2O)\left(\frac{0.51°\text{C bp elevation}}{1.00 \text{ mol antifreeze in } 1.00 \text{ kg } H_2O}\right)$$

$$= 9.1°\text{C bp elevation}$$

$$\text{boiling point} = (100.0°\text{C} + 9.1°\text{C}) = 109.1°\text{C}$$

$$109.1°\text{C} = 228.4°\text{F}$$

If we increase the boiling point of the cooling water, the water will not boil away if its temperature reaches 212°F (the normal boiling point of water). Your car engine remains protected even at water temperatures slightly above the normal boiling point of water.

8.2.5 WHY HOT-WATER PIPES FREEZE BEFORE COLD-WATER PIPES

Houses without cellars or basements usually have a crawl space between the ground and the first floor for servicing plumbing. People living in houses with crawl spaces may find their water pipes frozen after a very cold winter night. To their surprise, it is usually the hot-water pipe that is frozen and not the cold-water pipe. You might expect, since the cold-water pipe is colder than the hot-water pipe, that the cold-water pipe would freeze first. Let's see if we can explain why this doesn't occur.

Can you dissolve more gas in hot water or in cold water?

More gas can be dissolved in cold water than in hot water. Most students answering this question for the first time give the wrong answers, probably because they know that hot water will dissolve more solid solutes (like salt and sugar) than cold water. However, colder solvents dissolve more gas than warmer solvents. One reason soda pop is refrigerated is to keep it from losing its carbonation and going flat. Warm soda pop goes flat faster than cold soda pop.

You've probably observed a related phenomenon when heating water to the boiling point. Before boiling occurs, many little gas bubbles can be seen forming at the bottom of the pan. The heat from the stove degasses the water. The bubbles you see are dissolved air coming out of solution.

When water is heated in a hot-water heater, the water is degassed. Dissolved air escapes from cold water as it is heated in a hot-water heater. Water in a hot-water pipe (supplied by the hot-water heater) contains less dissolved gas than water in a cold-water pipe.

Now can you explain why the hot-water pipe freezes before the cold-water pipe freezes?

The freezing point of water decreases as the concentration of solute increases. A solute can be a solid, like sugar, or it can be another liquid, like antifreeze. It can even be a gas, like air. When air dissolves in water, the freezing point of the water is lowered. Even though water in a cold-water pipe may reach the outside temperature faster than water in a hot-water pipe, the freezing point of the cold water is lower than the freezing point of the hot water.

There's another factor that must be considered. Just as the boiling point of a sugar solution increases while the sugar solution boils, the freezing point of water inside a cold-water pipe continues to decrease as the freezing process continues. As water in contact with the sides of the water pipe freezes, gas is lost from this layer of ice to the unfrozen water in the center of the pipe. The center water becomes more concentrated with gas (solute), and its freezing point decreases further. It is possible to have a pipe with frozen water touching the sides of the pipe and unfrozen water in the center (see Figure 8.2).

FIGURE 8.2 Partially frozen pipe

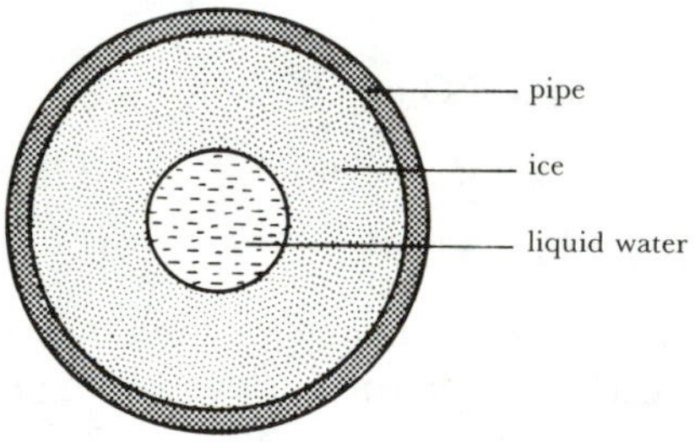

Another factor to consider is that the pressure inside the pipe increases as water freezes. Why does it increase?

Water expands when it freezes, and the expansion increases the pressure inside a closed container (such as a pipe). You have probably observed this if you have ever frozen a liquid in a closed bottle. The frozen water will either push the top off the bottle or cause the bottle to crack.

An increase in pressure lowers the freezing point of water. The freezing point of water in a cold-water pipe is lowered for two reasons. The freezing point of water decreases when solute is added to the water and when pressure is applied to the water. You will understand why pressure lowers the freezing point of water after studying Le Chatelier's principle in Chapter 9.

Have you ever noticed that the center of an ice cube is the last part to freeze? Also, there are usually air bubbles in the center of the ice cube (making the center appear cloudy) while the outer layer of the ice cube is clear. Can you explain why these phenomena are observed?

Gases are usually forced from the outer layer of an ice cube (which freezes first) to the center of the cube. The increase in the gas concentration in the center of an ice cube keeps the center unfrozen for a longer period of time. Eventually the center portion also freezes and loses its dissolved air, which appears as a cloudy part of the fully frozen ice cube (see Figure 8.3).

FIGURE 8.3 Ice cubes. Note that trapped bubbles in the center of the fully frozen ice cube give a cloudy appearance.

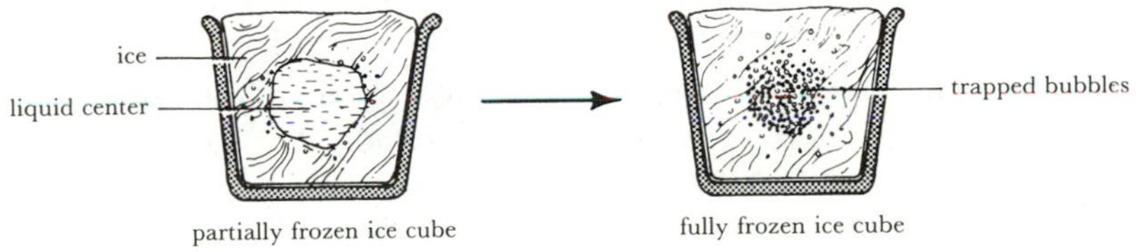

8.3 DETERMINING MOLECULAR WEIGHTS

The concept of molecular weight is used throughout all courses in chemistry. If you've ever wondered how molecular weights are determined, here's your opportunity to find out.

There are several ways to determine the molecular weight of a substance. We will illustrate the freezing point depression method. In each of the following problems we will make the assumption that the solute doesn't ionize (dissociate). We would check for ionization by measuring the electrical conductivity of our solution: If our solution strongly conducts an electric current, then the solute must be dissociated into ions. In a water solution, ions conduct electricity. Molecules don't.

•
•

EXAMPLE 15 *If 50.0 g of an unknown solute is mixed with 1000 g of water and the resulting solution has a freezing point of −1.86°C, what is the molecular weight of the unknown solute? Treat this problem as a brain teaser: Try to determine the molecular weight without using a pencil and paper. Just think about it for a while.*

The unknown solute has a molecular weight of 50.0 amu. If you didn't get this answer, let's go over the logic that is necessary to solve this brain teaser.

If you think about it, you should know the number of moles of solute that were added to the 1000 g of water and produced the −1.86°C freezing point depression. Think about that for a moment or two.

One mole of unknown solute was added to the 1000 g of water. How do we know that? We know it because 1.00 mol of solute added to 1000 g of water will lower the freezing point of water by 1.86°C. Since the freezing point of our solution is −1.86°C, then its freezing point depression is 1.86°C. A solution with a freezing point depression of 1.86°C is a 1.00 m solution (1.00 mol of solute dissolved in 1000 g of water).

$$\underset{\text{given}}{(1.86^\circ\text{C fp depression})}\underset{\text{conversion factor}}{\left(\frac{1.00 \text{ m solute soln}}{1.86^\circ\text{C fp depression}}\right)} = 1.00 \text{ m solute soln}$$

We added 50.0 g of unknown solute to 1000 g of water. Therefore, 50.0 g of solute must be 1.00 mol of solute. The gram-molecular weight of the solute is 50.0 g. The molecular weight of the solute is 50.0 amu.

Let's try another problem to see if you are getting the hang of this.

•
•

EXAMPLE 16 *If 20 g of an unknown solute is dissolved in 1000 g of water and the solution has a freezing point of −0.93°C, what is the molecular weight of the solute? Again, try doing this in your head. If you have trouble, look over Example 15 before giving up and reading the solution given.*

Solution The molecular weight of the solute is 40 amu. You can get this answer if you can determine the number of moles of solute that were added to the water.

How many moles of solute were added to the water?

We added 0.50 mol of solute to 1.000 kg of water. A full mole of solute would have lowered the freezing point of 1000 g of water by 1.86°C. But our solute lowered the freezing point by only half of this amount, that is, 0.93°C (0.93 is half of 1.86).

$$\underbrace{(0.93°\text{C fp depression})}_{\text{given}}\underbrace{\left(\frac{1.00 \text{ m solute soln}}{1.86°\text{C fp depression}}\right)}_{\text{conversion factor}} = 0.50 \text{ m solute soln}$$

We know that we added 20 g of solute. Therefore, 20 g of solute is 0.50 mol of solute. A full mole of solute would weigh 40 g.

$$\frac{20 \text{ g solute/1 kg } H_2O}{0.50 \text{ mol solute/1 kg } H_2O} = \frac{40 \text{ g solute}}{\text{mol solute}}$$

$$1.00 \text{ mol solute} = 40 \text{ g solute}$$

$$\text{molecular weight of solute} = 40 \text{ amu}$$

The next examples are more difficult and you'll need a pencil and paper to work them out. However, the logic used to solve these next examples is the same as that used in the preceding two examples.

It may help you to break up these problems into three steps:

Step 1 Determine how many moles of solute are present in 1 kg of solvent.

Step 2 Determine how many grams of solute are present in 1 kg of solvent.

Step 3 Determine the molecular weight of the solute by setting the results of Step 1 equal to the results of Step 2.

•
•

EXAMPLE 17 Find the molecular weight of a solute if 10.0 g of this solute, added to 500 g of water, produces a solution with a freezing point of −2.00°C.

Solution **Step 1** Determine the number of moles of solute in 1 kg of solvent (i.e., determine the molality of the solute).

$$(2.00°\text{C fp depression})\left(\frac{1.00 \text{ m solute soln}}{1.86°\text{C fp depression}}\right) = 1.08 \text{ m solute soln}$$

The result 1.08 m solute soln means 1.08 mol solute/1.00 kg solvent.

Step 2 Determine the number of grams of solute present in 1 kg of solvent.

$$\left(\frac{10.0 \text{ g solute}}{500 \text{ g solvent}}\right)\left(\frac{2}{2}\right) = \frac{20.0 \text{ g solute}}{1000 \text{ g solvent}}$$

$\frac{2}{2} = 1$, and you can multiply anything by 1 without changing its value.

We have completed Step 2; that is, there is 20.0 g of solute in 1 kg of solvent.

There is an alternative way of calculating the results required by the second step. Let's look at this approach:

Step 2 (alternative approach)

Determine the number of grams of solute present in 1 kg of solvent.

$$(500 \text{ g solvent})\left(\frac{1 \text{ kg}}{1000 \text{ g}}\right) = 0.50 \text{ kg solvent}$$

$$\frac{10.0 \text{ g solute}}{0.500 \text{ kg solvent}} = \frac{20.0 \text{ g solute}}{\text{kg solvent}}$$

Although this alternative approach may appear to be more complex than the first approach, you won't always be given information that can be so easily manipulated. You'll find that the alternative approach to the second step is more universally applicable.

Step 3 Set the results of Steps 1 and 2 equal to each other to determine the molecular weight of the unknown solute.

$$\frac{20.0 \text{ g solute}}{\text{kg } H_2O} = \frac{1.08 \text{ mol solute}}{\text{kg } H_2O}$$

$$20.0 \text{ g solute} = 1.08 \text{ mol solute}$$

$$\frac{20.0 \text{ g solute}}{1.08 \text{ mol solute}} = \frac{18.5 \text{ g solute}}{\text{mol solute}}$$

$$1.00 \text{ mol solute} = 18.5 \text{ g solute}$$

$$\text{molecular weight of solute} = 18.5 \text{ amu}$$

•
•

EXAMPLE 18 Find the molecular weight of a solute if 12.0 g of this solute, added to 280 g of water, produces a solution with a freezing point of −1.27°C.

Solution **Step 1** Determine the number of moles of solute in 1.000 kg of solvent (i.e., determine the molality of the solute).

$$(1.27°\text{C fp depression})\left(\frac{1.00 \text{ m solute soln}}{1.86°\text{C fp depression}}\right) = 0.683 \text{ m solute soln}$$

The result 0.683 m solute soln means 0.683 mol solute/1.000 kg of solvent.

Step 2 Determine the number of grams of solute present in 1 kg of solvent.

$$(280 \text{ g solvent})\left(\frac{1 \text{ kg}}{1000 \text{ g}}\right) = 0.280 \text{ kg solvent}$$

$$\frac{12.0 \text{ g solute}}{0.280 \text{ kg solvent}} = \frac{42.9 \text{ g solute}}{\text{kg solvent}}$$

There are 42.9 g of solute dissolved in 1.000 kg of water.

Step 3 Set the results of Steps 1 and 2 equal to each other to determine the molecular weight of the unknown solute.

$$\frac{42.9 \text{ g solute}}{\text{kg solvent}} = \frac{0.683 \text{ mol solute}}{\text{kg solvent}}$$

$$42.9 \text{ g solute} = 0.683 \text{ mol solute}$$

$$\frac{42.9 \text{ g solute}}{0.683 \text{ mol solute}} = \frac{62.8 \text{ g solute}}{\text{mol solute}}$$

$$1.00 \text{ mol solute} = 62.8 \text{ g solute}$$

$$\text{molecular weight of solute} = 62.8 \text{ amu}$$

SELF-TEST

1. If 20 g of an unknown solute is mixed with 1000 g of water and the resulting solution has a freezing point of −3.62°C, what is the molecular weight of the unknown solute?
2. If 30 g of a solute is dissolved in 500 g of water and the solution has a freezing point of −0.93°C, what is the molecular weight of the solute?
3. Find the molecular weight of a solute if 5.00 g of this solute, added to 250 g of water, produces a solution with a freezing point of −2.00°C.

ANSWERS **1.** 10 amu **2.** 120 amu **3.** 18.5 amu

8.4 OSMOSIS AND OSMOTIC PRESSURE

Osmosis is the diffusion of a solvent (frequently water) through a semipermeable membrane. A semipermeable membrane is a porous (pore- or hole-containing), sheetlike material, such as skin or cellophane. Particles (e.g., molecules and ions) smaller than the pore diameter can pass through the membrane, while larger particles cannot (see Figure 8.4).

If pure water is placed on one side of a semipermeable membrane and a sugar solution is placed on the other side of the membrane, water molecules will travel through the membrane from either side. However, water molecules tend to travel more frequently from the pure water (solvent) side to the sugar solution side because the water molecule concentration is greater on the pure water side. The greater the water molecule concentration, the greater the number of water molecule-membrane collisions per second.

The sugar solution liquid level begins to rise as water molecules from the solvent side enter the solution side faster than water molecules from the solution side enter the solvent side (see Figure 8.5).

FIGURE 8.4 Osmosis

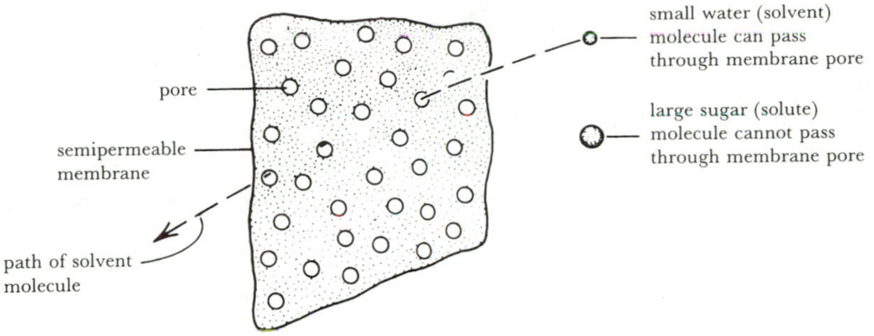

You can stop the sugar solution from rising by exerting a pressure against the surface of the sugar solution. The minimum pressure that is used to prevent the solution from rising (to stop osmosis from happening) is called the osmotic pressure of the solution.

Osmosis is responsible for the following common phenomena:

1. Grass dies when watered with salt water because the salt water draws water away from the grass cells.
2. Swimming in a freshwater lake makes your body waterlogged and increases your need to urinate because your more concentrated body solution draws in water from the fresh water.
3. Cucumbers wrinkle, lose water, and shrink when placed in brine because brine is a concentrated saltwater solution that has a higher solute concentration than the juice within the cucumber.
4. Prunes swell in water because water passes through the prune skin (membrane) into the more concentrated solution of prune juice within the prune.
5. You get thirsty and your skin wrinkles when you swim in the ocean because the salty ocean water is more concentrated than your body fluids, and water passes through your skin, a membrane, into the ocean.
6. Water rises 600 ft from the bottom of a root system to the leaves near the top of a 300-ft redwood tree because the leaves contain more dissolved solute than ground water.

FIGURE 8.5 Osmosis in a U-tube

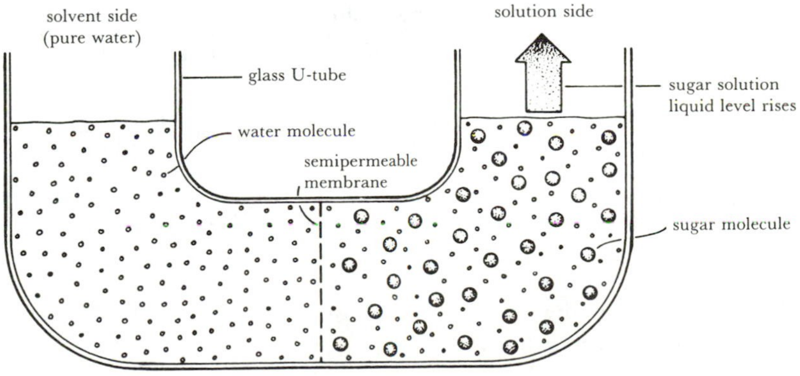

7. Salt and sugar are used to preserve many foods because bacteria, which cause food to spoil, die when water is drawn out through their cell membranes.
8. The solute concentration in intravenous solutions administered to hospital patients must be the same as that in blood plasma because patients would die if their blood cells were made to swell and burst or shrink and wrinkle by coming into contact with a solution with a different solute concentration.
9. Many saltwater fish die when placed in fresh water and vice versa because osmosis causes the solute concentrations of the fish body tissue to increase or decrease.

Osmotic pressure is a colligative property. Like freezing point depression and boiling point elevation, osmotic pressure is proportional to solute molality. A solution exerts an osmotic pressure of 24.5 atm for every mole of solute dissolved in 1 kg of solvent at 25°C.

EXAMPLE 19 Calculate the osmotic pressure of a solution prepared by dissolving 0.500 mol of sugar ($C_{12}H_{22}O_{11}$) in 1.00 kg of water at 25°C.

Solution This sugar solution has an osmotic pressure of 12.3 atm.

for every

$$(0.500 \text{ mol sugar in } 1.00 \text{ kg } H_2O)\left(\frac{24.5 \text{ atm osmotic pressure}}{1.00 \text{ mol sugar in } 1.00 \text{ kg } H_2O}\right)$$

$$= 12.3 \text{ atm osmotic pressure}$$

If this were the sugar solution in the U-tube shown in Figure 8.5, a pressure of 12.3 atm exerted upon the 0.5 m sugar solution would be required to stop osmosis. That's the same pressure that would be exerted on your body if you were about 400 ft underwater (see Section 5.2.3), and this is the amount of pressure required to raise water 400 ft high from a root system to the uppermost leaves on a tree.

EXAMPLE 20 Calculate the osmotic pressure of a solution prepared by dissolving 0.100 mol of sugar in 600 g of water at 25°C.

Solution The sugar solution would have an osmotic pressure of 4.09 atm. To obtain this answer, first calculate the molality of the sugar solution.

$$\text{molality} = \frac{\text{mol solute}}{\text{kg solvent}} = \frac{0.100 \text{ mol sugar}}{0.600 \text{ kg } H_2O} = 0.167 \text{ m sugar}$$

$$(0.167 \text{ mol sugar in } 1.00 \text{ kg } H_2O)\left(\frac{24.5 \text{ atm osmotic pressure}}{1.00 \text{ mol sugar in } 1.00 \text{ kg } H_2O}\right)$$

$$= 4.09 \text{ atm osmotic pressure}$$

EXAMPLE 21 Calculate the osmotic pressure of a solution prepared by dissolving 8.55 g of sugar ($C_{12}H_{22}O_{11}$) in 125 g of water at 25°C.

Solution The osmotic pressure of this sugar solution is 4.90 atm. To obtain this answer, first determine the number of moles of sugar present, and then determine the sugar solution molality.

$$(8.55 \text{ g sugar})\left(\frac{1.00 \text{ mol sugar}}{342 \text{ g sugar}}\right) = 0.0250 \text{ mol sugar}$$

$$\text{molality} = \frac{\text{mol solute}}{\text{kg solvent}} = \frac{0.0250 \text{ mol sugar}}{0.125 \text{ kg } H_2O} = 0.200 \text{ m sugar}$$

$$(0.200 \text{ mol sugar in } 1.00 \text{ kg } H_2O)\left(\frac{24.5 \text{ atm osmotic pressure}}{1.00 \text{ mol sugar in } 1.00 \text{ kg } H_2O}\right)$$

$$= 4.90 \text{ atm osmotic pressure}$$

SELF-TEST

1. Calculate the osmotic pressure of a solution prepared by dissolving 0.250 mol of sugar in 1.00 kg of water at 25°C.
2. Calculate the osmotic pressure of a solution prepared by dissolving 0.0500 mol of sugar in 300 g of water at 25°C.
3. Calculate the osmotic pressure of a solution prepared by dissolving 5.00 g of sugar in 125 g of water at 25°C.

ANSWERS **1.** 6.13 atm **2.** 4.08 atm **3.** 2.87 atm

8.4.1 MOLECULAR WEIGHTS OF LARGE MOLECULES

Osmotic pressure measurements are used frequently to estimate the molecular weights of very large molecules, such as proteins. We will use the three steps shown in Example 17 to solve the following examples:

EXAMPLE 22 An aqueous solution of an unknown protein is prepared by dissolving 30.00 g of the protein in 1.000 kg of water at 25°C. The osmotic pressure of this solution is found to be 18.62 mm Hg. What is the approximate molecular weight of this protein?

Solution

Step 1 Determine the molality of the protein solution.

$$(18.62 \text{ mm Hg osmotic pressure})\left(\frac{1.000 \text{ atm}}{760.0 \text{ mm Hg}}\right)\left(\frac{1.00 \text{ m protein soln}}{24.5 \text{ atm osmotic pressure}}\right)$$

$$= 0.00100 \text{ m protein soln}$$

Step 2 Determine the number of grams of protein present in 1 kg of solvent.

$$\frac{30.00 \text{ g protein}}{1.000 \text{ kg solvent}}$$

Step 3 Set the results of Steps 1 and 2 equal to each other to determine the molecular weight of the protein.

$$\frac{30.00 \text{ g protein}}{\text{kg solvent}} = \frac{0.00100 \text{ mol protein}}{\text{kg solvent}}$$

$$30.00 \text{ g protein} = 0.00100 \text{ mol protein}$$

$$\frac{30.00 \text{ g protein}}{0.00100 \text{ mol protein}} = \frac{3.000 \times 10^4 \text{ g protein}}{\text{mol protein}}$$

The molecular weight of the unknown protein is about 30 000 amu.

EXAMPLE 23 An aqueous solution of an unknown polymer is prepared by dissolving 20.00 g of the unknown polymer in 500.0 g of water at 25°C. The osmotic pressure of this solution is 16.45 mm Hg. What is the approximate molecular weight of the polymer?

Solution **Step 1** Determine the molality of the polymer solution.

$$(16.45 \text{ mm Hg osmotic pressure})\left(\frac{1.000 \text{ atm}}{760.0 \text{ mm Hg}}\right)\left(\frac{1.00 \text{ m polymer}}{24.5 \text{ atm osmotic pressure}}\right)$$

$$= 8.83 \times 10^{-4} \text{ m polymer}$$

Step 2 Determine the number of grams of solute present in 1 kg of solvent.

$$\frac{20.00 \text{ g polymer}}{0.5 \text{ kg } H_2O} = \frac{40.0 \text{ g polymer}}{1.00 \text{ kg } H_2O}$$

Step 3 Set the results of Steps 1 and 2 equal to each other to determine the molecular weight of the unknown solute.

$$\frac{40.0 \text{ g polymer}}{1.00 \text{ kg water}} = \frac{8.83 \times 10^{-4} \text{ mol polymer}}{1.00 \text{ kg } H_2O}$$

$$40.0 \text{ g polymer} = 8.83 \times 10^{-4} \text{ mol polymer}$$

$$\frac{40.0 \text{ g polymer}}{8.83 \times 10^{-4} \text{ mol polymer}} = \frac{4.53 \times 10^{4} \text{ g polymer}}{\text{mol polymer}}$$

The molecular weight of the polymer is approximately 45 300 amu.

8.4.2 ELECTRIC POWER PLANTS FUELED BY OSMOSIS

There are many places in the United States where fresh water drains into an ocean. If salty ocean water and fresh water are piped through a container in which they are separated only by a semipermeable membrane, fresh water passes through the membrane to the ocean water side. Mechanical power can be obtained as the pressurized saltwater solution rises and is directed through fins driving a motor that, in turn, drives an electric generator. In Canada, where this process is being studied, it has been estimated that if only 10% of the total freshwater drainage were used in this way with only a 20% efficiency, the electrical power generated would be equal to that of five current nuclear power plants (about 4 billion watts of power).

The energy that theoretically can be obtained by this process is approximately equal to the energy released by the same volume of fresh water falling from a 500-ft waterfall. By comparison, Niagara Falls (a hydroelectric power source shared by New York and Canada) is only 167 ft high.

PROBLEM SET 8

The problems in Problem Set 8 parallel the examples in Chapter 8. For example, if you should have trouble working Problem 5, go back to Example 5 in this chapter to get help. The correct answers to these problems are given at the end of this book.

1. If 2.00 mol of sucrose is dissolved in 2.00 kg of water, what are the freezing and boiling points of this sugar solution?
2. What is the molality of the sugar solution in Problem 1?
3. If 0.250 mol of sugar is dissolved in 250 g of water, what

is the molality of this sugar solution, and what are the freezing and boiling points of this solution?

4. Find the freezing and boiling points of a solution made by adding 342 g of sugar ($C_{12}H_{22}O_{11}$) to 4.0 kg of water. The molecular weight of sugar is 342 amu.
5. Find the freezing and boiling points of a solution made by adding 68.4 g of sugar to 100 g of water.
6. What are the freezing and boiling points of a 2.00 m NaCl solution?
7. Assume that $Sc(NO_3)_3$ dissociates into ions 100% when placed in water. What are the freezing and boiling points of a solution made by mixing 0.500 mol of $Sc(NO_3)_3$ with 1000 g of water?
8. What are the freezing and the boiling points of a 20.0% by weight solution of sulfuric acid solution whose specific gravity is 1.200?
9. A solution is made with 1000 g of water and has a boiling point of 100.51°C. What will be the temperature of this boiling sugar solution when 400 mL of water boils away?
10. Find the molality of a sugar solution that has a boiling point of 110.0°C.
11. How many grams of sugar must be dissolved in 1.000 L of water to make a 20 m sugar solution?
12. Find the number of grams of sugar that must be dissolved in 1.000 L of water to make a sugar solution with a boiling point of 260°F.
13. Find the number of grams of sugar that must be dissolved in 1.000 L of water to make a sugar solution with a boiling point of 290°F.
14. What is the freezing point of a solution made by mixing 10.00 L of antifreeze (1.110 g/mL antifreeze, 62.0 g/mol antifreeze) with 10.00 L of water?
15. If 40 g of an unknown solute is mixed with 800 g of water and the resulting solution has a freezing point of −3.72°C, what is the molecular weight of the unknown solute?
16. If 30 g of an unknown solute is dissolved in 1000 g of water and the solution has a freezing point of −0.93°C, what is the molecular weight of the solute?
17. Find the molecular weight of a solute if 8.00 g of this solute, added to 400 g of water, produces a solution with a freezing point of −2.00°C.
18. Find the molecular weight of a solute if 9.60 g of this solute, added to 224 g of water, produces a solution with a freezing point of −1.27°C.
19. Calculate the osmotic pressure of a solution prepared by dissolving 0.400 mol of sugar in 800 g of water at 25°C.
20. Calculate the osmotic pressure of a solution prepared by dissolving 0.150 mol of sugar in 900 g of water at 25°C.
21. Calculate the osmotic pressure of a solution prepared by dissolving 6.84 g of sugar ($C_{12}H_{22}O_{11}$) in 100 g of water at 25°C.
22. An aqueous solution of an unknown protein is prepared by dissolving 24.00 g of the protein in 800 g of water at 25°C. The osmotic pressure of this solution is found to be 18.62 mm Hg. What is the approximate molecular weight of this protein?
23. An aqueous solution of an unknown polymer is prepared by dissolving 16.00 g of the unknown polymer in 400.0 g of water at 25°C. The osmotic pressure of this solution is 2.1645×10^{-2} atm. Find the approximate molecular weight of the polymer.

9

Equilibrium and Le Chatelier's Principle

9.1 INTRODUCTION TO EQUILIBRIUM

Up until now we have been interpreting chemical equations such as

$$A + B \rightarrow C + D$$

by saying, "One mole of A reacts with one mole of B to produce one mole of C and one mole of D." For many reactions, however, once products form, the products begin reacting with themselves, re-forming the reactants. After C and D form in the preceding reaction, they begin to react together to produce A and B.

The chemical equation representing the chemical preparation of ammonia (NH_3) is

$$N_2 + 3H_2 \rightarrow 2NH_3$$

In this reaction, as the ammonia concentration increases, ammonia molecules begin colliding with each other, producing nitrogen and hydrogen. To represent this two-way reaction more accurately, double arrows are usually written:

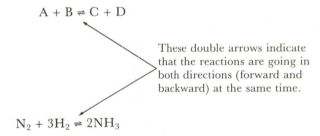

$$A + B \rightleftharpoons C + D$$

These double arrows indicate that the reactions are going in both directions (forward and backward) at the same time.

$$N_2 + 3H_2 \rightleftharpoons 2NH_3$$

When the rate (speed) of the forward reaction (nitrogen reacting with hydrogen) exactly equals the rate of the reverse (backward) reaction (ammonia molecules reacting together), we say that the system (chemical reaction) is at equilibrium.

When a system is at equilibrium (when the forward rate equals the reverse rate), the concentrations of the products and the reactants do not change. At equilibrium, the concentrations of all substances (products and reactants) remain constant. However, the forward and reverse reactions continue to occur. We say that equilibrium is a dynamic (active) situation.

To illustrate the dynamic nature of equilibrium, consider the equilibrium between solid sodium chloride and its ions, sodium (Na^+) and chloride (Cl^-).

$$NaCl(s) \rightleftharpoons Na^+(aq) + Cl^-(aq)$$

This equilibrium exists when a crystal of sodium chloride is placed into a saturated solution of sodium chloride. A saturated solution of NaCl is a solution that cannot increase its concentration of sodium chloride solute. A saturated solution is a solution that has dissolved the maximum concentration of a solute that is possible.

Is the crystal of sodium chloride dissolving?

Most students incorrectly answer that the sodium chloride crystal is not dissolving because the solution is saturated and cannot hold any more sodium and chloride ions. Although it is true that the solution cannot increase its concentration of sodium and chloride ions, the sodium chloride crystal is dissolving. However, just as quickly as the NaCl crystal is dissolving, sodium ions and chloride ions already in solution are reacting together to form solid sodium chloride, which precipitates back onto the crystal. The forward reaction (dissolving of the sodium chloride crys-

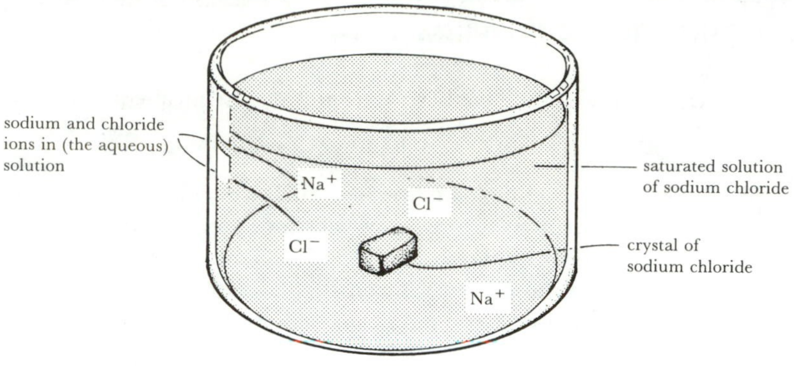

tal) and the reverse reaction (sodium ions reacting with chloride ions) are occurring at equal rates.

$$NaCl \rightleftharpoons Na^+ + Cl^-$$

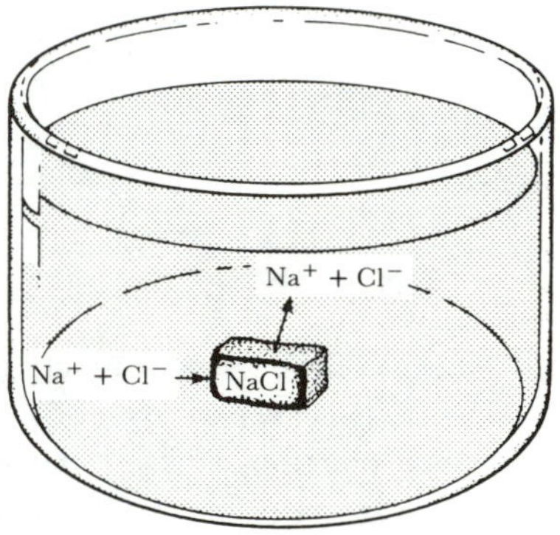

To a casual observer it appears that the sodium chloride crystal is not dissolving into sodium and chloride ions, because these ions are coming out of solution (precipitation) just as quickly as they are going into solution (dissolving). It is possible to demonstrate that NaCl is dissolving and precipitating at the same time. Chip off a corner of the sodium chloride crystal and remove it from the saturated solution. The broken corner will "grow back" while the dimensions of the cube (height, width, and length) decrease. Crystals tend to form in perfect geometric shapes (e.g., cubes). The corner grows back and the dimensions of the cube decrease as sodium and chloride ions dissolve from the sides and edges of the cube and precipitate out at the chipped-off corner. The study of equilibrium will give us further insight into the interpretation of chemical equations and lay the foundation for understanding other areas of chemistry, such as chemical thermodynamics and electrochemistry.

9.2 EQUILIBRIUM CONSTANTS

You will be happy to learn that there is only one formula (the equilibrium constant formula) that you have to memorize in order to do the equilibrium problems in this book. Despite this, many students find that more time is required to master equilibrium problems than is needed to master other types of problems. Although you will have less to memorize, a greater amount of thinking is usually required to solve equilibrium problems.

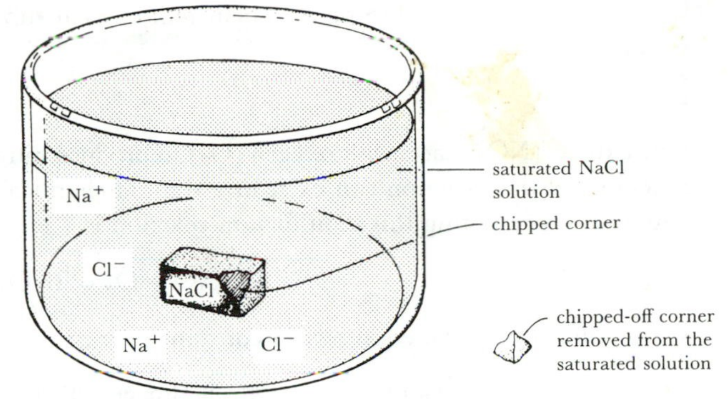

For the general equilibrium equation

$$aA + bB \rightleftharpoons cC + dD$$ ← chemical equilibrium equation

These are the numerical coefficients of the balanced chemical equation.

there exists an equilibrium constant, K:

$$K = \frac{[C]^c[D]^d}{[A]^a[B]^b}$$

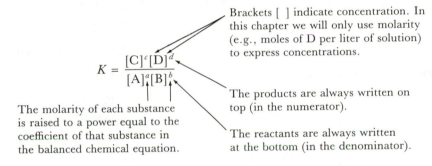

Brackets [ ] indicate concentration. In this chapter we will only use molarity (e.g., moles of D per liter of solution) to express concentrations.

The products are always written on top (in the numerator).

The reactants are always written at the bottom (in the denominator).

The molarity of each substance is raised to a power equal to the coefficient of that substance in the balanced chemical equation.

This is the equilibrium constant formula. It is used to determine equilibrium constant expressions for chemical equilibrium reactions. In the numerator (top), the molarity of C raised to the *c* power is multiplied times the molarity of D raised to the *d* power. In the denominator (bottom), the molarity of A raised to the *a* power is multiplied times the molarity of B raised to the *b* power.

•
•

EXAMPLE 1 Hydrogen (H_2) reacts with carbon dioxide (CO_2) to produce water (H_2O) and carbon monoxide (CO). The products and reactants of this reaction can exist in a state of chemical equilibrium. The balanced chemical equation representing this equilibrium reaction is

$$H_2(g) + CO_2(g) \rightleftharpoons H_2O(g) + CO(g)$$

Write and discuss the equilibrium constant expression for this chemical equilibrium reaction.

Solution The equilibrium constant expression for this equilibrium reaction is

$$K = \frac{[H_2O][CO]}{[H_2][CO_2]}$$

← products on top
← reactants at the bottom

For this equilibrium, there exists an equilibrium constant, K, which is equal to the product of the water and carbon monoxide molarities divided by the product of the hydrogen and carbon dioxide molarities. Each concentration (molarity) is raised to the first power (the exponent is 1) because 1 is understood to be the coefficient of each substance in the balanced chemical equilibrium equation.

•
•

EXAMPLE 2 Nitric oxide (NO) reacts with oxygen (O_2) to produce nitrogen dioxide (NO_2). The products and reactants of this reaction can exist in a state of chemical equilibrium. The balanced chemical equation representing this equilibrium reaction is

$$2NO + O_2 \rightleftharpoons 2NO_2$$

Write and discuss the equilibrium constant expression for this reaction.

Solution The equilibrium constant expression for this equilibrium is

$$K = \frac{[NO_2]^2}{[NO]^2[O_2]}$$

For this equilibrium reaction, there exists an equilibrium constant, K, which is equal to the molarity of NO_2 squared divided by the product of the molarity of NO squared times the oxygen molarity.

•
•

EXAMPLE 3 Consider the hypothetical equilibrium that is represented by

$$3A + 2B \rightleftharpoons C + 4D$$

Write the equilibrium constant expression for this chemical equilibrium equation.

Solution The equilibrium constant expression for this hypothetical equilibrium is

$$K = \frac{[C][D]^4}{[A]^3[B]^2}$$

•
•

9.2.1 CALCULATING *K*

Values of equilibrium constants K can be calculated if the molarity of each substance in equilibrium is known.

•
•

EXAMPLE 4 A 1.0-L box contains 1.0 mol of hydrogen, 10 mol of carbon dioxide, 2.0 mol of water vapor, and 8.0 mol of carbon monoxide. The temperature inside this box is 990°C. The equilibrium that exists between the substances in the box is represented by

$$H_2(g) + CO_2(g) \rightleftharpoons H_2O(g) + CO(g)$$

Calculate the value of the equilibrium constant K for this equilibrium.

Solution **Step 1** Write the equilibrium constant expression for the given equilibrium reaction.

$$K = \frac{[H_2O][CO]}{[H_2][CO_2]}$$

Step 2 Calculate the concentration (moles per liter) of water, carbon monoxide, hydrogen, and carbon dioxide.

$$[H_2O] = \frac{2.0 \text{ mol } H_2O}{1.0 \text{ L}} = 2.0 \text{ M}$$ ← M stands for molarity (moles of solute per liter solution).

$$[CO] = \frac{8.0 \text{ mol CO}}{1.0 \text{ L}} = 8.0 \text{ M}$$

$$[H_2] = \frac{1.0 \text{ mol } H_2}{1.0 \text{ L}} = 1.0 \text{ M}$$

$$[CO_2] = \frac{10 \text{ mol } CO_2}{1.0 \text{ L}} = 10 \text{ M}$$

Step 3 Substitute these concentration values into the equilibrium constant expression and calculate the numerical value for K.

$$K = \frac{(2.0 \text{ M})(8.0 \text{ M})}{(1.0 \text{ M})(10 \text{ M})} = \frac{16}{10} = 1.6$$

EXAMPLE 5 A 2.0-L box contains 2.0 mol of hydrogen, 20 mol of carbon dioxide, 4.0 mol of water vapor, and 16 mol of carbon monoxide. The temperature of this box is 990°C. The chemical equilibrium that exists between these substances is represented by

$$H_2 + CO_2 \rightleftharpoons H_2O(g) + CO$$

Calculate the value of the equilibrium constant for this equilibrium.

Solution **Step 1** Write the equilibrium constant expression for this equilibrium reaction.

$$K = \frac{[H_2O][CO]}{[H_2][CO_2]}$$

Step 2 Calculate the concentration (moles per liter) of the substances in the box.

$$[H_2O] = \frac{4.0 \text{ mol } H_2O}{2.0 \text{ L}} = 2.0 \text{ M}$$

$$[CO] = \frac{16 \text{ mol CO}}{2.0 \text{ L}} = 8.0 \text{ M}$$

$$[H_2] = \frac{2.0 \text{ mol } H_2}{2.0 \text{ L}} = 1.0 \text{ M}$$

$$[CO_2] = \frac{20 \text{ mol } CO_2}{2.0 \text{ L}} = 10 \text{ M}$$

Step 3 Substitute these concentrations into the equilibrium constant expression.

$$K = \frac{(2.0 \text{ M})(8.0 \text{ M})}{(1.0 \text{ M})(10 \text{ M})} = 1.6$$

EXAMPLE 6 A 500-mL container holds 0.50 mol of PCl_3 (phosphorus trichloride), 1.0 mol of Cl_2, and 6.5 mol of PCl_5 (phosphorus pentachloride) in equilibrium. What is the equilibrium constant for this equilibrium reaction? The equilibrium reaction is represented by the equation

$$PCl_3(g) + Cl_2(g) \rightleftharpoons PCl_5(g)$$

Solution **Step 1** Write the equilibrium constant expression for this equilibrium reaction.

$$K = \frac{[PCl_5]}{[PCl_3][Cl_2]}$$

Step 2 Calculate the concentrations of the substances in the container.

$$[PCl_5] = \frac{6.5 \text{ mol } PCl_5}{0.500 \text{ L}} = 13 \text{ M}$$

$$[PCl_3] = \frac{0.50 \text{ mol } PCl_3}{0.500 \text{ L}} = 1.0 \text{ M}$$

$$[Cl_2] = \frac{1.0 \text{ mol } Cl_2}{0.500 \text{ L}} = 2.0 \text{ M}$$

Step 3 Substitute these concentrations into the equilibrium constant expression.

$$K = \frac{(13 \text{ M})}{(1.0 \text{ M})(2.0 \text{ M})} = \frac{6.5}{\text{M}} = 6.5 \text{ M}^{-1}$$

9.2.2 UNITS OF *K*

Earlier in this book (see Section 2.4.1, Example 23) we discussed the importance of identifying substances when using units of quantity to make your calculations easier to follow. Unfortunately, this advice doesn't work well in K calculations. For example, if we use the information given in Example 6, here's what happens when we identify substances in our units of quantity:

$$K = \frac{[PCl_5]}{[PCl_3][Cl_2]} = \frac{\left(\dfrac{13 \text{ mol } PCl_5}{1.0 \text{ L soln}}\right)}{\left(\dfrac{1.0 \text{ mol } PCl_3}{1.0 \text{ L soln}}\right)\left(\dfrac{2.0 \text{ mol } Cl_2}{1.0 \text{ L soln}}\right)}$$

$$= \frac{6.5 \text{ mol } PCl_5 \text{ L soln}}{\text{mol } PCl_3 \text{ mol } Cl_2}$$

This difficulty doesn't occur in advanced chemistry courses because something called *activity* rather than *molarity* is used to calculate equilibrium constants. Activity values have no units. Activities are unitless numbers, just like specific gravity (see Section 3.3). Therefore, when activity is used to express the concentration of products and reactants, K doesn't have a unit.

Many textbooks often omit units in equilibrium constant values. We will follow the same practice in the remaining pages. We can justify (rationalize?) this practice by making believe our molarity values are approximate values for unitless activity values. This turns out to be quite close to the truth: Molarity values are approximations of activity values.

9.2.3 MAGNITUDE OF *K*

The magnitude (size) of K is very important. The magnitude of K tells us the extent to which a chemical reaction occurs (i.e., how much of the reactants are used up). We will designate three possible arbitrary magnitudes for K:

- K greater than 10
- K less than 0.1
- K about equal to 1.

If K is greater than 10, usually most of the reactants react. (There are exceptions, but we will not worry about them now.) If the products and reactants were confined to a box, we would find, at equilibrium, predominantly products in the box and very few reactants.

If K is less than 0.1, usually very few products form. Most of the original reactants are left unreacted. If the products and reactants were confined to a box, we would find, at equilibrium, predominantly reactants in the box and very few products.

If K is about equal to 1, usually there is about a 50-50 mixture of reactants and products; that is, about half of the reactants will react to form products. If the products and reactants were confined to a box, we would find, at equilibrium, about equal amounts of reactants and products inside the box.

For the hypothetical equilibrium between R (reactants) and P (products)

$$R \rightleftharpoons P$$

the three possibilities are illustrated in Figure 9.1.

FIGURE 9.1 Magnitude of K

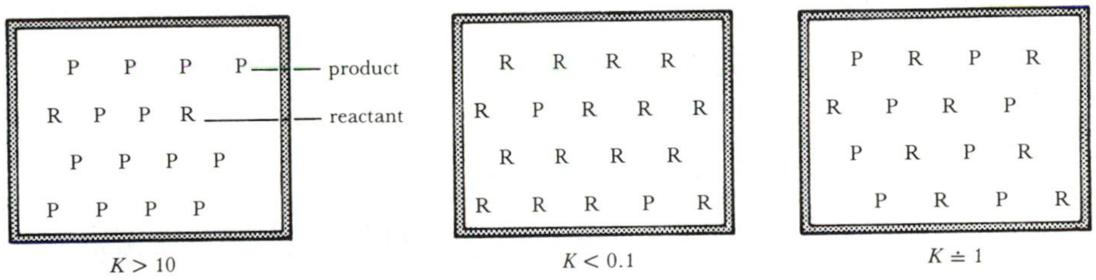

EXAMPLE 7 Consider the hypothetical chemical reaction

$$R \rightleftharpoons P$$

Ten moles of R is placed into a 1.00-L container, and the reaction occurs. After equilibrium is reached (when R turns into P just as quickly as P turns into R), it is found that there is 0.0100 mol of R and 9.99 mol of P in the container. Determine the equilibrium constant for this reaction and interpret the magnitude of the equilibrium constant.

Solution The equilibrium constant expression for this equilibrium reaction is

$$K = \frac{[P]}{[R]}$$

At equilibrium, the concentrations of R and P are

$$[R] = \frac{0.0100 \text{ mol R}}{1.00 \text{ L}} = 0.0100 \text{ M}$$

$$[P] = \frac{9.99 \text{ mol P}}{1.00 \text{ L}} = 9.99 \text{ M}$$

Substituting these concentrations into the equilibrium constant expression gives us the value of K.

$$K = \frac{[P]}{[R]} = \frac{9.99}{0.0100} = 999$$

The forward reaction ($R \rightarrow P$) occurs to a greater extent. We know that most of R is consumed and that large amounts of P form because K is greater than 10. Since only 0.0100 mol of R is left from the original 10.0 mol of R, about 99.9% of the original reactants have decomposed (reacted). The only way K can be greater than 10 is for the numerator (top) of the equilibrium constant expression to be much larger than the denominator. For K to be large, usually there must be much more product than reactant at equilibrium.

EXAMPLE 8 Consider the hypothetical reaction

$$B \rightleftharpoons A$$

This time, 10.0 mol of B is placed into a 2.00-L container, and the reaction is allowed to occur until equilibrium exists. (We say that the reaction is allowed to reach equilibrium.) At equilibrium, there is 9.99 mol of B and 0.0100 mol of A in the container. Determine the equilibrium constant for this reaction and interpret the magnitude of the equilibrium constant.

Solution The equilibrium constant expression for this equilibrium is

$$K = \frac{[A]}{[B]}$$

At equilibrium, the concentrations of A and B are

$$[A] = \frac{0.0100 \text{ mol A}}{2.00 \text{ L}} = 0.00500 \text{ M}$$

$$[B] = \frac{9.99 \text{ mol B}}{2.00 \text{ L}} = 5.00 \text{ M}$$

Substituting these concentrations into the equilibrium constant expression gives us $K = 0.001\,00$.

$$K = \frac{[A]}{[B]} = \frac{0.00500}{5.00} = 0.001\,00$$

This small value of K (less than 0.1) indicates that very little B reacted to form A at equilibrium. The extent of this chemical reaction is small. About 0.01% of the original reactant (B) reacts to form the product (A). The only way K can be less than 0.1 is for the denominator (reactant concentration) to be much larger than the numerator (product concentration) in the equilibrium constant expression. A small K usually means that there must be much more reactant than product.

EXAMPLE 9 Twenty moles of D is placed into a 5.0-L container, and the reaction is allowed to reach equilibrium.

$$D \rightleftharpoons E$$

At equilibrium, the container holds 10 mol of D and 10 mol of E. Determine the equilibrium constant for this reaction and interpret its magnitude.

Solution The equilibrium constant expression for this equilibrium reaction is

$$K = \frac{[E]}{[D]}$$

At equilibrium, the concentrations of E and D are

$$[E] = \frac{10 \text{ mol E}}{5.0 \text{ L}} = 2.0 \text{ M}$$

$$[D] = \frac{10 \text{ mol D}}{5.0 \text{ L}} = 2.0 \text{ M}$$

Substituting these concentrations into the equilibrium constant expression gives us $K = 1.0$.

$$K = \frac{[E]}{[D]} = \frac{2.0}{2.0} = 1.0$$

When K is about equal to 1, about half of the reactants react at equilibrium to form products. The only way for K to be about equal to 1 is for the numerator and denominator of the equilibrium constant expression to be approximately equal, that is, there must be about equal amounts of product and reactant.

SELF-TEST

1. A 2.0-L box contains 6.0 mol of water vapor, 10.0 mol of carbon monoxide, 4.0 mol of hydrogen, and 7.0 mol of carbon dioxide. All of these substances are in chemical equilibrium, as represented by

$$H_2(g) + CO_2(g) \rightleftharpoons H_2O(g) + CO(g)$$

Calculate the value of the equilibrium constant for this reaction.

2. Interpret the extent of the forward reaction in Problem 1 by examining the magnitude of the equilibrium constant.

3. Consider the hypothetical equilibrium that is represented by

$$3A + 2B \rightleftharpoons C + 2D$$

A 1.00-L box contains 1.0 mol of A, 2.0 mol of B, 3.0 mol of C, and 3.0 mol of D at equilibrium. What is the equilibrium constant for this reaction?

ANSWERS **1.** 2.1 **2.** About half of the hydrogen and the carbon dioxide is used up. **3.** 6.8

9.3 INTRODUCTION TO LE CHATELIER'S PRINCIPLE

Being able to understand and apply Le Chatelier's principle is of fundamental importance in any study of chemical equilibrium. Le Chatelier's principle can be stated in many different ways. Here are three working definitions of it:

1. A system in equilibrium tends to relieve itself of applied stress.
2. A system in equilibrium tends to adapt (respond) to its surroundings.
3. A system in equilibrium tends to do the opposite of whatever is done to it.

We will illustrate the application of Le Chatelier's principle with three types of stress: (1) concentration, (2) temperature, and (3) pressure.

9.4 CONCENTRATION STRESS

9.4.1. TEARING EYES

The moisture in your eyes has a specific *p*H (hydrogen ion concentration). If strong acid (a quantity of H^+ ions) accidentally gets into your eyes, the hydrogen ion concentration in your eyes increases. When this happens, your eyes tear.

What function do the tears have? What does the tearing accomplish?

The tears dilute (decrease) the acid (H^+ ions) concentration. The tears are your eye's attempt to bring the hydrogen ion concentration back to normal. Your eyes are relieving themselves of the applied stress of the increase in hydrogen ion concentration. They are adapting (responding) to surrounding conditions. Your eyes do not remain passive. They do the opposite of what happens to them; that is, when the hydrogen ion concentration is increased, your eyes *attempt* to decrease the hydrogen ion concentration.

The word *attempt* is very important here. A system in equilibrium isn't always successful at relieving itself of applied stress. Tears effectively dilute small amounts of acid in your eyes, but they are useless when the acid quantity is too large. However, the tendency to bring relief is present.

9.4.2 CHEMICAL EQUILIBRIUM EXAMPLES

Let's turn our attention to an inanimate (nonliving) system. Let's see how Le Chatelier's principle applies to a simple chemical equilibrium.

•
•

EXAMPLE 10 A box contains PCl_5, PCl_3, and Cl_2 at equilibrium, as shown. After this system reaches equilibrium, more chlorine (Cl_2) is added to the box (the chlorine concentration is increased). Generally speaking, what will this system in equilibrium try to do?

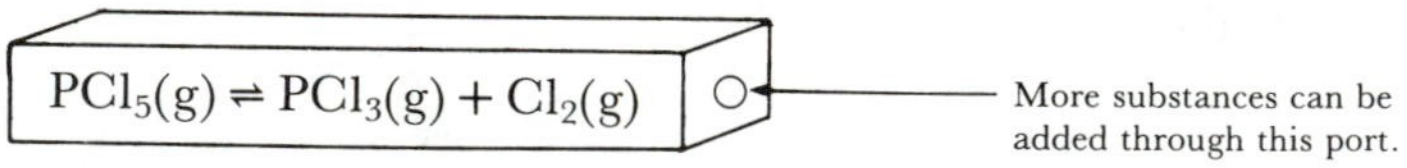
$PCl_5(g) \rightleftharpoons PCl_3(g) + Cl_2(g)$
More substances can be added through this port.

Solution The system will try to relieve itself of the increased chlorine concentration (stress) by decreasing the chlorine concentration. The system will try to do the opposite of what was done to it; that is, the chlorine concentration of the system was increased and the system will try (tend) to decrease the chlorine concentration.

•
•

EXAMPLE 11 How can the preceding system in equilibrium (the chemical equilibrium itself) decrease the concentration of chlorine in the box?

Solution The system can decrease the chlorine concentration by increasing the rate of the backward reaction. When this happens, we say that the equilibrium shifts to the left.

$$\xleftarrow{\text{shifts to the left}}$$
$$PCl_5 \rightleftharpoons PCl_3 + Cl_2$$

When the equilibrium shifts to the left, some of the added chlorine reacts with PCl_3 to form additional amounts of PCl_5. When PCl_3 reacts with Cl_2, both the Cl_2 and the PCl_3 concentrations decrease. The concentration of PCl_5 increases because PCl_5 forms when Cl_2 and PCl_3 combine.

•
•

EXAMPLE 12 What would happen if some PCl_3 were removed from the box? In general, what would the system tend to do?

Solution If PCl_3 were removed from the box, the system would try (tend) to relieve itself of this applied stress by producing more PCl_3. The system would tend to do the opposite of what was done to it; that is, it would produce more PCl_3 because PCl_3 was removed.

•
•

EXAMPLE 13 How can this equilibrium system produce more PCl_3?

Solution This system can produce more PCl_3 by shifting the equilibrium to the right (increasing the rate of the forward reaction).

$$\xrightarrow{\text{shifts to the right}}$$
$$PCl_5 \rightleftharpoons PCl_3 + Cl_2$$

More PCl_5 decomposes, producing additional amounts of PCl_3 and Cl_2, when the forward rate increases. When the equilibrium shifts to the right, the concentration of PCl_5 decreases while the concentrations of both PCl_3 and Cl_2 increase.

•
•

EXAMPLE 14 A box contains nitrogen (N_2), oxygen (O_2), and nitric oxide (NO) in equilibrium. If the oxygen concentration is increased, which way will the equilibrium shift?

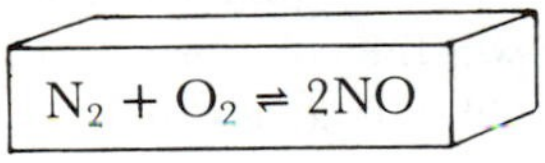

Solution The equilibrium will shift to the right to reduce the oxygen concentration.

$$\xrightarrow{\text{shifts to the right}}$$
$$N_2 + O_2 \rightleftharpoons 2NO$$

When the equilibrium shifts to the right, nitrogen reacts with the increased oxygen concentration, decreasing the oxygen concentration.

•
•

EXAMPLE 15 If the nitric oxide concentration in the last example is increased, which way will the equilibrium shift?

Solution The equilibrium will shift to the left to reduce the nitric oxide concentration.

$$\xleftarrow{\text{shifts to the left}}$$
$$N_2 + O_2 \rightleftharpoons 2NO$$

The system in equilibrium tries to do the opposite of what is done to it. The system's nitric oxide concentration is increased, so the system tries to decrease the nitric oxide concentration by increasing the rate of the reverse (backward) reaction.

EXAMPLE 16 If the concentration of nitric oxide is decreased, which way will the equilibrium shift?

Solution The equilibrium will shift to the right to produce more nitric oxide.

$$\xrightarrow{\text{shifts to the right}}$$
$$N_2 + O_2 \rightleftharpoons 2NO$$

EXAMPLE 17 If the concentration of oxygen is decreased, what does the equilibrium do?

Solution The equilibrium will shift to the left to increase the oxygen concentration. The concentration of nitrogen also increases when the equilibrium shifts to the left.

$$\xleftarrow{\text{shifts to the left}}$$
$$N_2 + O_2 \rightleftharpoons 2NO$$

SELF-TEST Some carbon monoxide (CO), water (H_2O), carbon dioxide (CO_2), and hydrogen (H_2) are placed into a container and allowed to come to equilibrium.

$$CO(g) + H_2O(g) \rightleftharpoons CO_2(g) + H_2(g)$$

1. Which way will the equilibrium shift if the concentration of CO is increased?
2. Which way will the equilibrium shift if the concentration of water is decreased?
3. Which way will the equilibrium shift if the concentration of CO_2 is increased?
4. Which way will the equilibrium shift if the concentration of hydrogen is decreased?

ANSWERS **1.** right **2.** left **3.** left **4.** right

9.5 TEMPERATURE STRESS

9.5.1 WINTER IN A NUDIST COLONY

Assume that (1) it is winter, (2) the outside temperature is below zero, and (3) you feel like running around in the snow without the benefit of clothing. Also assume that (4) the police don't interrupt your endeavors and (5) you have understanding neighbors.

What does your body do to adapt to the stress of the cold weather?

No, it doesn't blush. Your body begins to shiver. Shivering is a series of rapid muscle contractions that burn as many calories as hard physical exertion. Shivering produces heat that warms you. Your body tries to relieve itself of the applied stress (cold, heat loss). Your body tends to do the opposite (produce more heat) of what was done to it. Death will result if you stay out too long, because it is always possible to overwhelm an equilibrium system. Remember, equilibrium systems are not always successful at relieving applied stress. They just have a built-in tendency to relieve themselves.

On hot summer days your body tends to perspire. Evaporating perspiration cools your body. Your body adapts to the hot surroundings. Your body does the opposite of what is done to it; that is, if its temperature is raised, it tries to lower its temperature.

9.5.2 CHEMICAL EQUILIBRIUM EXAMPLES

Now let's subject an inanimate system, a simple chemical equilibrium, to a temperature change and see how Le Chatelier's principle is applied.

•
•

EXAMPLE 18 A box contains nitrogen (N_2), hydrogen (H_2), and ammonia (NH_3) at equilibrium.

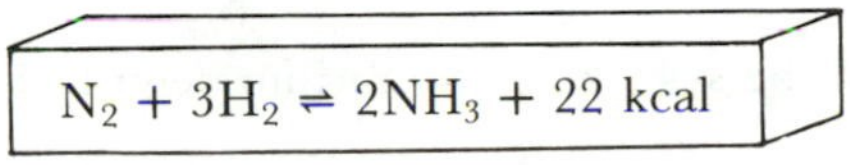
$N_2 + 3H_2 \rightleftharpoons 2NH_3 + 22$ kcal

The forward reaction is exothermic (it produces heat). Twenty-two thousand calories of heat is produced for every 2 mol (34 g) of ammonia (NH_3) produced. In general, according to Le Chatelier's principle, what will happen if this system in equilibrium is cooled down?

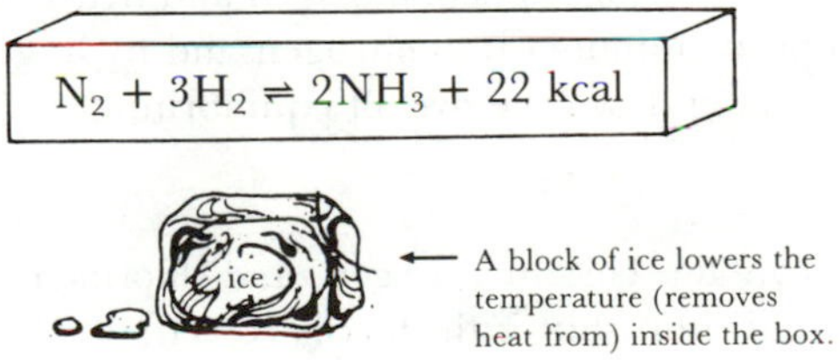
$N_2 + 3H_2 \rightleftharpoons 2NH_3 + 22$ kcal
ice
A block of ice lowers the temperature (removes heat from) inside the box.

Solution The equilibrium system will try to adapt to its surroundings. If its temperature is lowered, it will try to do the opposite, that is, raise its temperature. It will try to relieve itself of the applied stress (cold) by producing more heat.

•
•

EXAMPLE 19 How can the system described in Example 18 produce more heat?

Solution The system in equilibrium can produce more heat by shifting to the right.

$$\xrightarrow{\text{shifts to the right}}$$
$$N_2 + 3H_2 \rightleftharpoons 2NH_3 + 22 \text{ kcal}$$

When this equilibrium shifts to the right, more nitrogen and hydrogen are consumed, more ammonia and heat are produced. As ammonia is produced, heat is produced. Both ammonia and heat are products of the forward reaction.

•
•

EXAMPLE 20 Generally speaking, what would the system described in Example 19 do if it were warmed up (if calories were added to the box)?

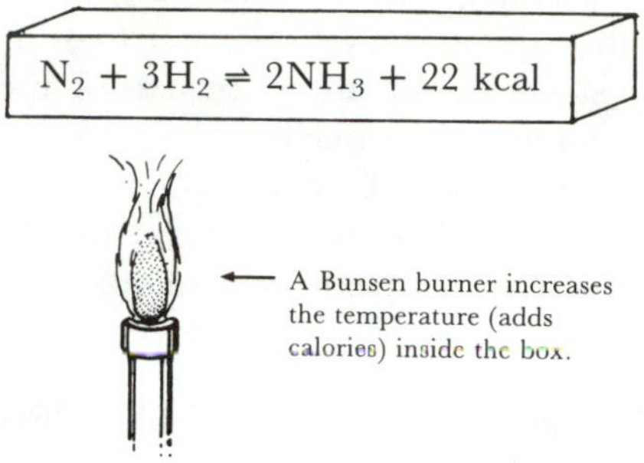

$N_2 + 3H_2 \rightleftharpoons 2NH_3 + 22$ kcal
A Bunsen burner increases the temperature (adds calories) inside the box.

Solution This equilibrium system would try to lower its temperature (cool down, use up calories). The system tends to do the opposite of what is done to it. It tries to relieve itself of the applied stress (heat) by consuming the added heat.

•
•

EXAMPLE 21 How can the system in equilibrium described in Example 20 cool down itself?

Solution This system can cool down itself by shifting to the left.

$$\xleftarrow{\text{shifts to the left}}$$
$$N_2 + 3H_2 \rightleftharpoons 2NH_3 + 22 \text{ kcal}$$

As the rate of the reverse reaction increases, both ammonia and heat (calories) are consumed and additional amounts of nitrogen and hydrogen form. Energy (heat, calories) is required to decompose ammonia into nitrogen and hydrogen. When ammonia decomposes, it consumes heat, cooling down the overall equilibrium.

•
•

EXAMPLE 22 A box contains chlorine, oxygen, and chlorine monoxide in equilibrium. In which direction will the equilibrium shift if the box is cooled?

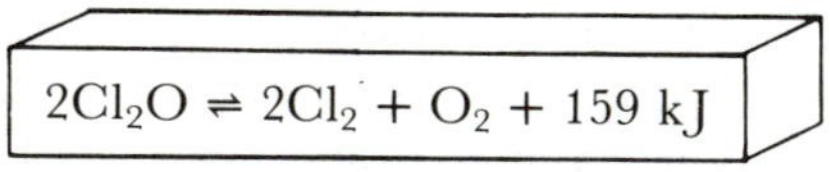

Solution Cooling the box (removing heat) will cause the equilibrium to shift in the direction that produces heat. The equilibrium will shift to the right.

•
•

EXAMPLE 23 If heat is added to the box described in Example 22, what will happen to the concentration of the chlorine monoxide?

Solution Adding heat to the box causes the equilibrium to shift in the direction that removes heat, that is, to the left. The concentration of the chlorine monoxide increases when the equilibrium shifts to the left.

•
•

EXAMPLE 24 Consider the equilibrium described in Example 23. If chlorine monoxide is added to the box, will the equilibrium give off or absorb heat?

Solution If the chlorine monoxide concentration is increased, the equilibrium will shift to the right to decrease the chlorine monoxide concentration. When the equilibrium shifts to the right, heat is produced (given off).

•
•

EXAMPLE 25 If the preceding equilibrium is warmed, what happens to the oxygen concentration?

Solution Warming this equilibrium causes the equilibrium to shift in the direction that consumes heat, that is, to the left. When this equilibrium shifts to the left, the oxygen concentration decreases.

•
•

SELF-TEST Some nitrogen, oxygen, and nitric oxide are placed into a container and allowed to reach equilibrium.

$$2NO \rightleftharpoons N_2 + O_2 + 43 \text{ kcal}$$

1. Which way does the equilibrium shift if the temperature of this system is increased?
2. Is heat given off or absorbed if nitrogen is removed from the container?
3. What happens to the nitric oxide (NO) concentration if the temperature of the system is lowered?

ANSWERS **1.** left **2.** given off **3.** it will decrease

9.6 PRESSURE STRESS

9.6.1 FLAT TIRES

If your car has a flat tire (low pressure inside the tire), how do you increase the pressure inside the tire?

Usually the pressure inside a tire is increased by adding molecules of air (gas molecules). The addition of gas molecules (air) to the inside of a tire increases the pressure inside the tire. If the tire has too much pressure, you can reduce the pressure by removing gas molecules (air) from inside the tire.

9.6.2 CHEMICAL EQUILIBRIUM EXAMPLES

•
•

EXAMPLE 26 The substances PCl_5, PCl_3, and Cl_2 are placed into an airtight container with a flexible wall. If the pressure inside the box is decreased (e.g., by pulling outward on the flexible wall, creating a partial vacuum), what will this system in equilibrium tend to do? Assume that the temperature is kept constant.

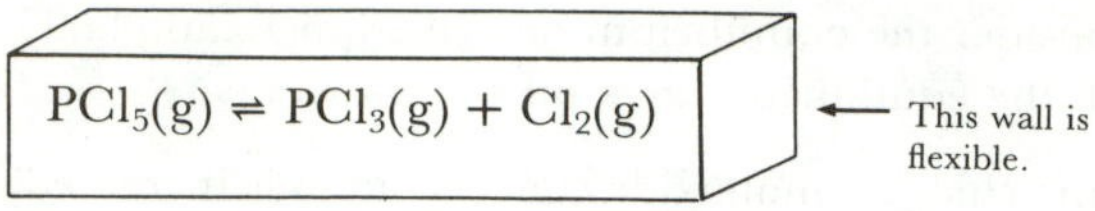

Solution The equilibrium will tend to undo the stress (reduced pressure). When the pressure of the system is reduced, the system tends to increase the pressure inside the container.

•
•

EXAMPLE 27 How can the equilibrium system increase the pressure inside the container?

Solution The system can increase the pressure inside the container by shifting the equilibrium to the right.

$$\xrightarrow{\text{shifts to the right}}$$
$$PCl_5(g) \rightleftharpoons PCl_3(g) + Cl_2(g)$$

When the equilibrium shifts to the right, PCl_5 is consumed (reacts) and additional amounts of $PCl_3(g)$ and $Cl_2(g)$ form. For every one gas molecule of PCl_5 that decomposes, two gas molecules (one gas molecule of PCl_3 and one gas molecule of Cl_2) are produced. There is a net gain of one gas molecule (one gas molecule is decomposed and two are formed). When the equilibrium shifts to the right, the number of gas molecules inside the container increases. As gas molecules are added to a container (e.g., a car tire), the pressure inside the container increases.

•
•

EXAMPLE 28 If the pressure inside the container described in Example 27 is increased, what will the system in equilibrium try to do?

Solution The system will try to decrease its pressure. It tries to do the opposite of what is done to it. It tries to relieve itself of the applied stress (increased pressure).

•
•

EXAMPLE 29 How can the system described in Example 28 reduce its pressure?

Solution The system can reduce its pressure by shifting to the left.

$$\xleftarrow{\text{shifts to the left}}$$
$$PCl_5(g) \rightleftharpoons PCl_3(g) + Cl_2(g)$$

For every two gas molecules (one PCl_3 and one Cl_2) that are used up (combined, reacted), only one gas molecule (PCl_5) forms. There is a net loss of one gas molecule every time one PCl_3 molecule reacts with one Cl_2 molecule. As gas molecules are removed from a container (e.g., a car tire), the pressure inside the container decreases.

•
•

EXAMPLE 30 Some nitrogen, hydrogen, and ammonia are placed into a container and allowed to come to equilibrium. Which way will the equilibrium shift if the pressure inside the container is reduced while the temperature is kept constant?

Solution The equilibrium will shift to the left. This shift to the left will increase the number of gas molecules inside the container. If the number of gas molecules inside the container increases, the pressure inside the container increases.

•
•

EXAMPLE 31 Some nitric oxide, oxygen, and nitrogen dioxide are placed into a container and allowed to reach equilibrium. Which way will the equilibrium shift if the pressure inside the container is increased while the temperature remains constant?

$$2NO(g) + O_2(g) \rightleftharpoons 2NO_2(g)$$

Solution The equilibrium will shift to the right. This shift will decrease the number of gas molecules inside the container. Decreasing the number of gas molecules inside the container decreases the pressure inside the container.

•
•

EXAMPLE 32 Some hydrogen, chlorine, and hydrogen chloride are introduced into a container and allowed to come to equilibrium. If the pressure inside this container is increased, which way will the equilibrium shift? Assume that the temperature is held constant.

$$H_2(g) + Cl_2(g) \rightleftharpoons 2HCl(g)$$

Solution The equilibrium will not shift to the left or to the right. A shift in either direction will not decrease the number of gas molecules inside the container. When one gas molecule of hydrogen combines with one gas molecule of chlorine (a total of two gas molecules), two gas molecules of hydrogen chloride form. When two gas molecules of hydrogen chloride react, they form one gas molecule of hydrogen and one gas molecule of chlorine (two gas molecules form). There is no net decrease in the total number of gas molecules when either the forward or the reverse reaction occurs. Le Chatelier's principle says that equilibrium systems tend to relieve themselves of applied stress. They don't always succeed.

•
•

EXAMPLE 33 Some carbon monoxide and water are placed into a container where they react to form carbon dioxide and hydrogen. The reaction is allowed to reach equilibrium. If the pressure inside this container is decreased while the temperature is held constant, which way will the equilibrium shift?

$$CO(g) + H_2O(g) \rightleftharpoons CO_2(g) + H_2(g)$$

Solution The equilibrium will not shift to the left or to the right. A shift in either direction will not increase the number of gas molecules inside the container.

•
•

EXAMPLE 34 Some carbon, oxygen, and carbon dioxide are placed into a container and allowed to reach chemical equilibrium. Which way will the equilibrium shift if the pressure inside the container is decreased while the temperature is held constant?

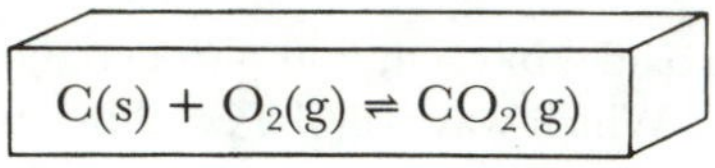

Solution The equilibrium will not shift to the left or to the right. Carbon is a solid, not a gas. The pressure inside a container is increased and decreased by changing the number of gas molecules in the container. (If you want to increase the pressure inside your car tires, you do not add pieces of a solid, such as carbon, to the inside of the tire. You add gas molecules.) In the equilibrium

$$C(s) + O_2(g) \rightleftharpoons CO_2(g)$$

one atom of solid carbon reacts with one gas molecule of oxygen (a total of one gas molecule reacting) to form one gas molecule of carbon dioxide. There is no net gain in the number of gas molecules inside the container when carbon reacts with oxygen to produce carbon dioxide.

•
•

EXAMPLE 35 Some hydrogen, oxygen, and water are placed into a container and allowed to reach equilibrium. If the pressure inside this box is decreased, which way will the equilibrium shift? Assume that the temperature remains constant.

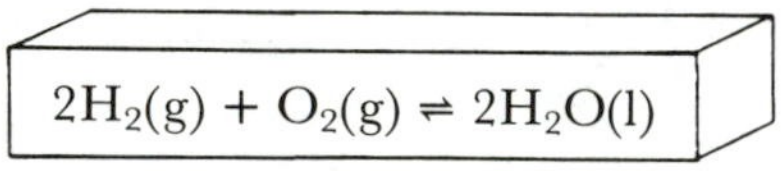

Solution The equilibrium will shift to the left to increase the number of gas molecules inside the container. When two liquid molecules of water react to form two gas molecules of hydrogen and one gas molecule of oxygen, there is a net gain of three gas molecules. Increasing the number of gas molecules inside this container will increase the pressure inside the container.

•
•

EXAMPLE 36 Some PCl_5 is placed into an empty container and allowed to decompose into PCl_3 and Cl_2. When all three substances are in equilibrium

$$PCl_5(g) \rightleftharpoons PCl_3(g) + Cl_2(g)$$

the concentration of chlorine is increased. Discuss the equilibrium shift and the pressure change that will occur.

Solution The equilibrium will shift to the left to decrease the chlorine concentration. Shifting to the left causes the pressure inside the container to decrease.

•
•

EXAMPLE 37 Some ammonia, nitrogen, and hydrogen are placed into a container and allowed to reach equilibrium. If the pressure inside the container is increased while the temperature is kept constant, in which direction will the equilibrium shift and what heat effects will occur?

$$N_2(g) + 3H_2(g) \rightleftharpoons 2NH_3(g) + 22 \text{ kcal}$$

Solution The equilibrium will shift to the right to decrease the number of gas molecules inside the container. Decreasing the number of gas molecules inside the container decreases the pressure. Heat is given off when this equilibrium shifts to the right.

•
•

EXAMPLE 38 Consider the equilibrium discussed in Example 37. Discuss the equilibrium shift, the pressure changes, and the heat effects that will occur if nitrogen is added to the container after equilibrium is reached.

Solution The equilibrium will shift to the right to decrease the nitrogen concentration. A shift to the right will decrease the number of gas molecules (and thus the pressure) inside the container. Heat will be given off when the equilibrium shifts to the right.

•
•

SELF-TEST

1. Some N_2O_4 and NO_2 are placed into a container and allowed to reach equilibrium.

$$N_2O_4(g) \rightleftharpoons 2NO_2(g)$$

Which way will the equilibrium shift if the pressure inside the container is increased while the temperature is held constant?

2. Some H_2, I_2, and HI are placed into a container at a high temperature and allowed to come to equilibrium.

$$H_2(g) + I_2(g) \rightleftharpoons 2HI(g)$$

Which way will the equilibrium shift if the pressure inside the container is decreased while the temperature is held constant?

3. Some nitrogen, hydrogen, and ammonia are placed into a container at a high temperature and allowed to reach equilibrium.

$$3H_2(g) + N_2(g) \rightleftharpoons 2NH_3(g) + 22 \text{ kcal}$$

Discuss the equilibrium shift, the pressure change, the heat effects, and the hydrogen concentration change that will occur if the nitrogen concentration is increased.

ANSWERS **1.** left **2.** no shift in either direction
3. equilibrium shifts to the right, pressure decreases, heat is given off, hydrogen concentration decreases

9.7 APPLICATIONS OF LE CHATELIER'S PRINCIPLE

9.7.1 CARBON MONOXIDE POISONING

As you may already know, the hemoglobin in your blood carries oxygen (O_2) from your lungs to the various parts of your body and brain. Unfortunately, carbon monoxide (CO) has a stronger affinity for hemoglobin than does oxygen. This means that if you breathe air containing oxygen and carbon monoxide, it is the carbon monoxide that will be picked up by the hemoglobin (in preference to the oxygen).

When hemoglobin is bonded to carbon monoxide (instead of oxygen), the amount of oxygen that your body receives is reduced. If carbon monoxide reaches a level of 40% in the blood, death usually follows. The brain begins to deteriorate irreversibly when it is deprived of sufficient oxygen for three or four minutes (damaged brain cells cannot repair themselves or regenerate). Therefore, it is important to eliminate carbon monoxide from the hemoglobin as quickly as possible.

Take a look at this equilibrium:

$$\text{hemoglobin:}O_2 + CO \rightleftharpoons \text{hemoglobin:}CO + O_2$$

Hemoglobin:O_2 represents a hemoglobin molecule bonded to oxygen, and hemoglobin:CO represents a hemoglobin molecule bonded to carbon monoxide. Carbon monoxide victims are usually treated with exposure to fresh air (which contains 20% oxygen and essentially no carbon monoxide). The oxygen in the air drives the equilibrium shown to the left in accordance with Le Chatelier's principle.

How do you suppose you could speed up the process of eliminating carbon monoxide from your body? (How could you make the equilibrium shift to the left even faster?)

Breathing in pure oxygen (100% oxygen) would increase the oxygen concentration in the equilibrium. It would also increase the rate of the reverse reaction (which increases the concentration of hemoglobin:O_2).

How could you speed up the process of eliminating carbon monoxide from your body even more quickly than by breathing pure oxygen? (What would shift the equilibrium to the left faster than using 100% oxygen?)

If you got the correct answer on your own, you're pretty sharp. Increasing the oxygen concentration beyond 100% may sound impossible, but here's how it's done.

If you put 32 g of oxygen into a box with a volume of 22.4 L, the pressure inside the box would be 1 atm. Remember that 1 mol of oxygen (32 g of O_2) occupies 22.4 L at standard conditions (1 atm, 0°C). If you put 64 g of oxygen (2 mol of O_2) into the same size box (22.4 L), you have doubled the concentration of the oxygen. There are twice the number of moles of oxygen gas per liter of volume. The pressure of 64 g of oxygen in a 22.4-L box is 2 atm.

In many hospitals around the United States, physicians have been making use of this principle. They are using hyperbaric oxygen chambers (*hyper* means "greater than" and *baric* means "pressure") that contain pure oxygen at 3 atm of pressure. Hundreds of thousands of carbon monoxide victims have been treated in these chambers. (Carbon monoxide is responsible for more deaths in the United States each year than any other type of poison.) The length of the treatment required to reduce the amount of carbon monoxide in the blood by one-half of its original concentration is 5.5 hr when the victim is exposed to fresh air, 1.5 hr when pure oxygen is inhaled at 1 atm of pressure, and only 23 min using oxygen at 3 atm of pressure.

You might be interested in knowing (or horrified, if you are a smoker) that even low levels of carbon monoxide in your blood (which result from cigarette smoking) can impair your performance on standardized exams and your motor performance (e.g., driving skills). No matter how much carbon monoxide you have in your system, its presence means that a certain fraction of your body's hemoglobin is not available to carry oxygen because it is busy carrying carbon monoxide. This makes your heart pump faster to ensure that your body gets the oxygen it needs. If your heart is forced to work harder (because of higher carbon monoxide levels or an overweight condition), you are more prone to heart disease and heart attacks.

9.7.2 RUBBER BAND DEMONSTRATION*

Here is an interesting (and less horrifying) demonstration of Le Chatelier's principle. A 500-g weight is suspended from a hook with a rubber band. The rubber band is heated with hot air from a hair dryer (see Figure 9.2).

FIGURE 9.2 Rubber band demonstration of Le Chatelier's principle

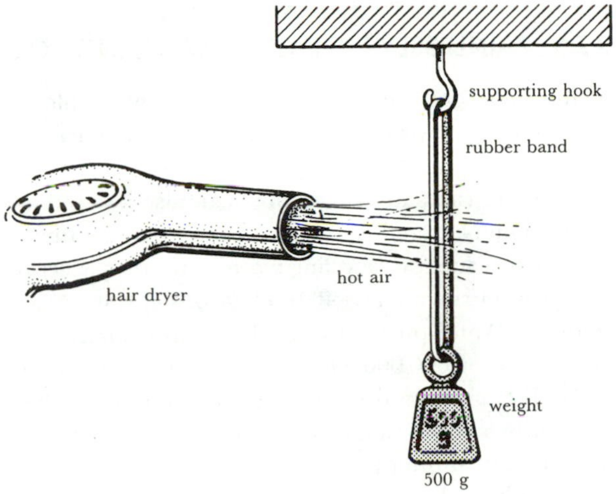

What will happen to the 500-g weight when the rubber band is heated with hot air? Will the weight drop to a lower position, rise to a higher position (closer to the supporting hook), or just sit there?

Most students guess that the weight will sink and are surprised when they see this demonstration and observe the weight rising. You could have predicted the results had you analyzed a certain property of rubber bands: they warm up when stretched (and cool down when contracted). If you did not know about this property, take a rubber band, stretch it quickly, and then touch the stretched rubber band to your cheek. It should feel warm to you. Keep the rubber band stretched out until it has had time to cool down to room temperature and no longer feels warm when touched to your cheek. Then let the rubber band contract quickly and again touch it to your cheek. It should feel cooler this time. This can be represented by the equilibrium shown in Figure 9.3.

FIGURE 9.3 Rubber band equilibrium

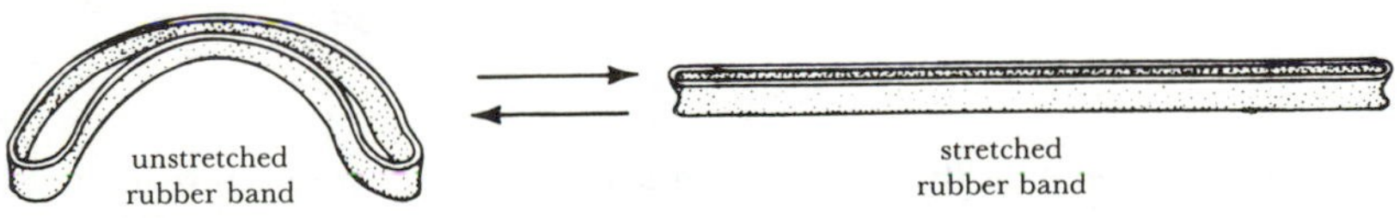

*Adapted from Ronald A. DeLorenzo, "Le Chatelier's Principle and a Rubber Band," *Journal of Chemical Education* 50 (Feb. 1973):124.

Le Chatelier's principle predicts that, if we add heat to the equilibrium in Figure 9.3, the equilibrium will shift to the left to relieve itself of the applied stress (i.e., the added heat). The reaction absorbs heat when it shifts to the left; the reaction gives off heat when it shifts to the right. When the heat of a hair dryer is applied to a stretched rubber band, the rubber band can relieve itself of the applied stress (can absorb the added heat) by contracting. (Obviously, you could heat the rubber band so much as to cause it to melt. This would make the weight fall to a lower level, perhaps on your foot!)

9.7.3 PAINLESSLY PURIFYING SWIMMING POOL WATER*

Swimming pool water is usually treated with chlorine to kill bacteria. Unfortunately, chlorine burns the eyes. Also,it is an unpleasant experience to accidently drink chlorinated swimming pool water.

Silver (actually silver ions) can also be used as a bactericide (an agent that destroys bacteria). The National Aeronautics and Space Administration (NASA) developed the use of silver as a bactericide for recycling wastewater in spacecraft, such as the Space Shuttle. A teaspoon of silver can purify about 1 400 000 gallons of water. Silver is a much more effective bactericide than chlorine. Water purified with silver ions is tasteless at this low silver ion concentration (10 parts of silver per 1 000 000 000 parts of water). Drinking such water is harmless to people and animals. To maintain the correct concentration of silver ion in a swimming pool, a small amount of slightly soluble silver salt, such as silver bromide (AgBr), can be placed in the filtering system of a swimming pool.

Let's see how Le Chatelier's principle can illuminate this use of silver as a bactericide. Consider a very small swimming pool (a 500-mL beaker) filled with a saturated solution of AgBr. A crystal of AgBr is placed at the bottom of the beaker. In this saturated (but very dilute) solution of AgBr, an equilibrium exists between the silver bromide crystal (solid) at the bottom of the beaker and the silver and bromide ions in solution.

$$AgBr(s) \rightleftharpoons Ag^+(ag) + Br^-(ag)$$

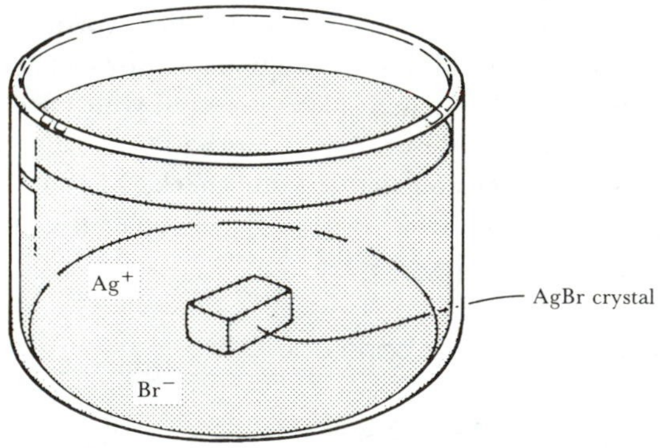

Remember that this equilibrium is dynamic: the forward reaction (dissolving) is occurring at the same rate as the reverse reaction (precipitation).

*Adapted from Robert C. Plumb, "Sparkling Pure Water in a Swimming Pool," *Journal of Chemical Education* 53 (Mar. 1976):189.

As silver ion is used up in the swimming pool, the concentration of the silver ion decreases. What can the equilibrium system do to increase the silver ion concentration to make it fairly constant?

The equilibrium shifts to the right, that is, more AgBr dissolves to increase the silver ion concentration as silver ion is used up in the swimming pool.

$$\xrightarrow{\text{shifts to the right}}$$
$$AgBr(s) \rightleftharpoons Ag^+(ag) + Br^-(ag)$$

Acting as a bactericide, the silver ion concentration decreases. This decrease in concentration is a stress to which the system in equilibrium responds by increasing the rate of the forward reaction (AgBr dissolves). When AgBr dissolves, the concentration of silver ion in solution increases.

PROBLEM SET 9

The problems in Problem Set 9 parallel the examples in Chapter 9. For example, if you should have trouble working Problem 5, go back to Example 5 in this chapter to get help. The correct answers to these problems are given at the end of this book.

1. Water (H_2O) and carbon monoxide (CO) react together to produce hydrogen (H_2) and carbon dioxide (CO_2). The products and reactants of this reaction can exist in a state of chemical equilibrium. Write a balanced chemical equation for this equilibrium reaction and then write its equilibrium constant expression.
2. Nitrogen dioxide (NO_2) decomposes into nitric oxide (NO) and oxygen (O_2). Write a balanced chemical equation representing this equilibrium. Write the equilibrium constant expression for this reaction.
3. Consider the hypothetical equilibrium represented by

$$2A + 3B \rightleftharpoons 4C + D$$

   Write the equilibrium constant expression for this chemical equilibrium equation.
4. A 1.0-L box contains 2.0 mol of hydrogen, 20 mol of carbon dioxide, 4.0 mol of water vapor, and 16.0 mol of carbon monoxide. The temperature of this box is 990°C. The equilibrium that exists between the substances in the box is represented by

$$H_2O(g) + CO(g) \rightleftharpoons H_2(g) + CO_2(g)$$

   Calculate the value of the equilibrium constant K for this equilibrium.
5. A 4.0-L box contains 4.0 mol of hydrogen, 40 mol of carbon dioxide, 8.0 mol of water vapor, and 32 mol of carbon monoxide. The temperature of this box is 990°C. The chemical equilibrium that exists between these substances is represented by

$$H_2O(g) + CO(g) \rightleftharpoons H_2(g) + CO_2(g)$$

   Calculate the value of the equilibrium constant for this equilibrium.
6. A 400-mL container holds 0.40 mol of PCl_3, 0.80 mol of Cl_2, and 5.2 mol of PCl_5 in equilibrium. What is the equilibrium constant for this equilibrium reaction? The equilibrium reaction is represented by the equation

$$PCl_5(g) \rightleftharpoons PCl_3(g) + Cl_2(g)$$

7. Consider the hypothetical chemical reaction A $\rightleftharpoons$ B. Twenty moles of A is placed into a 2.00-L container, and the reaction occurs. After equilibrium is reached, it is found that there are 0.0200 mol of A and 19.98 mol of B in the container. Determine the equilibrium constant for this reaction and interpret the magnitude of the equilibrium constant.
8. Consider the hypothetical reaction G $\rightleftharpoons$ H. We place 8.00 mol of G into a 1.60-L container and allow the reaction to occur until equilibrium exists. At equilibrium we find that there are 7.99 mol of G and 0.008 00 mol of H. Determine the equilibrium constant for this reaction and interpret the magnitude of the equilibrium constant.
9. Forty moles of X is placed into a 10.0-L container, and the reaction X $\rightleftharpoons$ Y is allowed to reach equilibrium. At equilibrium, the container holds 20 mol of X and 20 mol of Y. Determine the equilibrium constant for this reaction and interpret its magnitude.

A box contains PCl_3, PCl_5, and Cl_2 at equilibrium, as shown:

$$PCl_5(g) \rightleftharpoons PCl_3(g) + Cl_2(g)$$

After this system reaches equilibrium, more PCl_3 is added to the box. Refer to this equilibrium when working Problems 10–13.

10. Generally speaking, what will this equilibrium system try to do?
11. What does the system do when more PCl_3 is added?
12. What happens if some Cl_2 is removed from the system? In general, what is the tendency of the system?
13. Specifically, what does the system do if Cl_2 is removed?

A box contains nitrogen, oxygen, and nitric oxide in equilibrium, as shown:

$$N_2 + O_2 \rightleftharpoons 2NO$$

Refer to this equilibrium when answering Problems 14–17.

14. If the oxygen concentration is decreased, which way will the equilibrium shift?
15. If the nitric oxide concentration is decreased in the system, which way will the equilibrium shift?
16. If the concentration of nitrogen is increased in the system, which way will the equilibrium shift?
17. If the concentration of oxygen is increased in the system, which way will the equilibrium shift?

A box contains the hypothetical chemicals A, B, C, and D at equilibrium, as shown:

$$2A + 3B \rightleftharpoons 4C + 5D + 10 \text{ kcal}$$

Refer to this equilibrium when working Problems 18–21.

18. In general, according to Le Chatelier's principle, what will happen if this system in equilibrium is cooled down?
19. Specifically, what does the system do when it is cooled?
20. Generally speaking, what happens to the system if it is warmed up?
21. Specifically, what does the system do if it is warmed up?

A box contains the hypothetical chemicals M and N in equilibrium, as shown:

$$M \rightleftharpoons N + 40 \text{ kJ}$$

Refer to this equilibrium when working Problems 22–25.

22. In what direction will this equilibrium shift if the box is cooled?
23. If heat is added to the system, what happens to the concentration of M?
24. If M is added to the system, does the system give off or absorb heat?
25. If the system is warmed, what happens to the concentration of N?

The hypothetical chemicals R, S, and T are placed into an airtight container with a flexible wall.

$$2R(g) + S(g) \rightleftharpoons 2T(g)$$

Refer to this equilibrium when working Problems 26–29.

26. If the pressure inside the box is decreased, what, in general, does this system try to do?
27. Specifically, what does the system do when the pressure inside the box is decreased?
28. If the pressure inside the container is increased, in general, what does the system try to do?
29. Specifically, what does the system do when the pressure inside the box is increased?
30. Some nitrogen, hydrogen, and ammonia are placed into a container and allowed to come to equilibrium, as shown:

$$N_2(g) + 3H_2(g) \rightleftharpoons 2NH_3(g)$$

Which way will the equilibrium shift if the pressure inside the container is increased while the temperature is kept constant?

31. Some nitric oxide, oxygen, and nitrogen dioxide are placed into a container and allowed to reach equilibrium, as shown:

$$2NO(g) + O_2(g) \rightleftharpoons 2NO_2(g)$$

Which way will the equilibrium shift if the pressure inside the container is decreased while the temperature remains constant?

32. Some hypothetical chemicals R, S, T, and U are introduced into a container and allowed to come to equilibrium, as shown:

$$R(g) + 3S(g) \rightleftharpoons 2T(g) + 2U(g)$$

If the pressure inside this container is decreased, which way will the equilibrium shift if the temperature is held constant?

33. Some hypothetical chemicals E, F, G, and H are introduced into a container and allowed to come to equilibrium, as shown:

$$3E(s) + 2F(g) \rightleftharpoons 2G(g) + H(l)$$

If the pressure inside this container is increased while the temperature is held constant, which way will the equilibrium shift?

34. Some sulfur, oxygen, and sulfur dioxide are placed into a container and allowed to reach chemical equilibrium, as shown:

$$S(s) + O_2(g) \rightleftharpoons SO_2(g)$$

In which direction will the equilibrium shift if the pressure inside the container is increased while the temperature is held constant?

35. Some hydrogen, oxygen, and water are placed into a container and allowed to reach equilibrium, as shown:

$$2H_2(g) + O_2(g) \rightleftharpoons 2H_2O(l)$$

If the pressure inside this box is increased, which way will the equilibrium shift?

36. Some PCl_5 is placed into an empty container and allowed to decompose into PCl_3 and Cl_2. When all three substances are in equilibrium

$$PCl_5(g) \rightleftharpoons PCl_3(g) + Cl_2(g)$$

the concentration of chlorine is decreased. Discuss the equilibrium shift and the pressure change that will occur.

37. Ammonia, nitrogen, and hydrogen are placed into a container and allowed to reach equilibrium, as shown:

$$N_2(g) + 3H_2(g) \rightleftharpoons 2NH_3(g) + 22 \text{ kcal}$$

If the pressure inside the container is decreased while the temperature is kept constant, in which direction will the equilibrium shift and what heat effects will occur?

38. Consider the system described in Problem 37. Discuss the equilibrium shift, the pressure changes, and the heat effects that will follow if nitrogen is removed from the container after equilibrium is reached.

10

Equilibrium Calculations Part I

10.1 INTRODUCTION: ANALOGIES THAT SEPARATE THE MATH FROM THE CHEMISTRY*

Many freshmen chemistry students are uneasy when they try to solve chemical equilibrium problems. It is common to hear students say that they are having trouble in this area because they are weak in math. The purpose of this chapter is to prove to you that many of the mathematical skills needed to solve chemical equilibrium problems are really quite simple, especially when they are isolated from the chemistry. The marriage analogies discussed in this chapter will (hopefully) convince you that the math isn't difficult. These analogies will give you additional insight into the chemical nature of equilibrium problems. Keep in mind, however, that you will need to put in more time than usual to solve equilibrium problems with confidence.

Before reading this chapter you should understand molarity and Le Chatelier's principle, be familiar with the chemical equilibrium constant formula, and be able to calculate logarithms and antilogarithms. You will gain some more experience with logs and antilogs in Section 10.2, "*p* Concept," before we continue our equilibrium discussion. The *p* concept is frequently used in equilibrium problems.

10.2 *p* CONCEPT

The lowercase letter *p* stands for "negative log" (−log).

•
•

EXAMPLE 1 There is a number, N, that equals 100. Find pN.

Solution

$$pN = -\log N = -\log 100 = -2$$

The log of 100 is equal to 2; the *p* of 100 is equal to the negative log of 100, −2. (See Chapter 1 for a review of logs and antilogs.)

•
•

EXAMPLE 2 If a number, W, equals 1000, find pW.

Solution

$$pW = -\log W = -\log 1000 = -3$$

The log of 1000 is equal to 3; the *p* of 1000 is equal to the negative log of 1000, −3.

•
•

EXAMPLE 3 A number, U, is equal to 20. Find pU.

Solution

$$\begin{aligned} pU &= -\log U = -\log 20 \\ &= -[\log(2 \times 10^1)] \\ &= -(\log 2 + \log 10^1) \\ &= -(0.3 + 1) \\ &= -(1.3) \\ pU &= -1.3 \end{aligned}$$

•
•

*Adapted from Ronald A. DeLorenzo, "Chemical Equilibrium: Analogies That Separate the Mathematics from the Chemistry," *Journal of Chemical Evaluation* 54 (Nov. 1977):676.

EXAMPLE 4 A number, H, is equal to 0.0001. Find pH.

Solution

$$pH = -\log H = -\log 0.0001$$
$$= -\log 10^{-4}$$
$$= -(-4)$$
$$= +4$$
$$pH = 4$$

The log of 0.0001 is equal to −4; the p of 0.0001 is equal to the negative log of 0.0001, +4.

EXAMPLE 5 A number, R, is equal to 0.000 032. Find pR.

Solution

$$pR = -\log R = -\log 0.000\,032$$
$$= -\log(3.2 \times 10^{-5})$$
$$= -(\log 3.2 + \log 10^{-5})$$
$$= -[(0.5) + (-5)]$$
$$= -(-4.5)$$
$$= +4.5$$
$$pR = 4.5$$

The log of 0.000 032 is equal to −4.5; the negative log of 0.000 032 is equal to 4.5.

EXAMPLE 6 If $\log X = 3$, what is the value of X?

Solution We learned in Chapter 1 that the value of X is equal to the antilog of 3 and that antilog 3 = $10^3 = 1000$.

EXAMPLE 7 If $pX = 3$, what is the value of X?

Solution This antilog problem is worked just like the one in Example 6 except that now we are dealing with negative logs. If $pX = 3$, $X = 10^{-pX} = 10^{-3} = 0.001$. You can check your answer by working backward; that is, if $X = 0.001$, find pX. If you do this, you will see that $pX = 3$.

EXAMPLE 8 Given $pD = 2$, find D.

Solution

$$D = 10^{-pD} = 10^{-2} = 0.01$$

EXAMPLE 9 Given $pK = 5.7$, find K.

Solution

$$K = 10^{-pK} = 10^{-5.7} = 10^{-6} \times 10^{0.3}$$
$$10^{0.3} = \text{antilog } 0.3 = 2.0$$
$$K = 2.0 \times 10^{-6}$$

EXAMPLE 10 Given $pS = 3.6$, find S.

Solution

$$S = 10^{-pS} = 10^{-3.6} = 10^{-4} \times 10^{0.4}$$

$$10^{0.4} = \text{antilog}\, 0.4 = 2.5$$

$$S = 2.5 \times 10^{-4}$$

•
•

SELF-TEST

1. Given $N = 10\,000$, find pN.
2. Given $M = 0.016$, find pM.
3. Given $pX = 10$, find X.
4. Given $pA = 6.4$, find A.

ANSWERS **1.** -4 **2.** 1.8 **3.** 10^{-10} **4.** 4×10^{-7}

10.2.1 *p*H

There are many times that a chemist, when talking about the hydrogen ion molarity of an aqueous (water) solution, will use the term *pH*. *p*H means "$-\log[H^+]$." Recall that [ ] (brackets) means "moles per liter," that is, molarity.

In general, the *p*H of a solution can be thought of as being in one of three ranges:

- *p*H is greater than 7
- *p*H is less than 7
- *p*H is equal to 7

When the *p*H of a water solution at 25°C is equal to 7, the solution is said to be neutral (i.e., it has equal concentrations of hydrogen ion and hydroxide ion). When the *p*H of a water solution is greater than 7 at 25°C, the solution is said to be basic. If the *p*H is less than 7, the solution is considered acidic. Unless otherwise specified, assume that the temperature of all solutions we discuss is 25°C. To find the *p*H of a solution, simply find the logarithm of the solution's hydrogen ion molarity and then change the sign of the logarithm (i.e., multiply the logarithm times -1).

•
•

EXAMPLE 11 A solution has a hydrogen ion concentration of 0.001 mol of H^+ per liter of solution. Find the *p*H of this solution. Also, specify whether the solution is acidic, basic, or neutral.

Solution Simply take the log of 0.001

$$\log 0.001 = \log 10^{-3} = -3$$

and then multiply the logarithm by -1.

$$(-1)(-3) = +3$$

These two steps (taking the logarithm and then multiplying the logarithm by -1) are usually combined into one.

$$[H^+] = 0.001 = 10^{-3}$$

$$pH = -\log[H^+] = -\log 10^{-3} = -(-3) = +3$$

This solution is acidic because its *p*H is less than 7.

•
•

EXAMPLE 12 A solution has a hydrogen ion molarity of 0.000 000 01. Find the pH of this solution and indicate whether the solution is acidic, basic, or neutral.

Solution This solution has a pH of 8, and it is basic. Whenever the pH is greater than 7, the solution is basic.

$$[H^+] = 0.000\,000\,01 = 10^{-8}$$

$$pH = -\log[H^+] = -\log 10^{-8} = -(-8) = +8$$

EXAMPLE 13 A solution has a hydrogen ion molarity equal to 0.002. Find the pH of this solution. Is the solution acidic, basic, or neutral?

Solution This solution has a pH of 2.7.

$$[H^+] = 0.002 = 2 \times 10^{-3}$$

$$pH = -\log(2 \times 10^{-3}) = -(\log 2 \times \log 10^{-3})$$

$$= -[(0.3) + (-3)] = -[0.3 - 3] = -[-2.7] = 2.7$$

This solution is acidic because the pH is less than 7.

EXAMPLE 14 Find the pH of a solution whose hydrogen ion concentration is 6.4×10^{-6} mol of hydrogen ion per liter of solution.

Solution The solution has a pH of 5.2.

$$[H^+] = 6.4 \times 10^{-6}$$

$$pH = -\log(6.4 \times 10^{-6}) = -[\log 6.4 + \log 10^{-6}]$$

$$= -[(0.8) + (-6)] = -[-5.2] = +5.2$$

If $[H^+] = 10^{-7}$, then $pH = 7$. Conversely, if $pH = 7$, then $[H^+] = 10^{-pH} = 10^{-7}$.

EXAMPLE 15 Find the hydrogen ion molarity of a solution that has a pH of 8.

Solution

$$pH = 8$$

$$[H^+] = 10^{-pH} = 10^{-8}$$

EXAMPLE 16 Find the hydrogen ion concentration of a solution that has a pH of 7.3.

Solution

$$pH = 7.3$$

$$[H^+] = 10^{-7.3} = 10^{-8} \times 10^{0.7}$$

$$10^{0.7} = \text{antilog } 0.7 = 5$$

$$[H^+] = 5 \times 10^{-8}$$

EXAMPLE 17 Find the hydrogen ion concentration of a solution with a pH of 4.9.

Solution

$$pH = 4.9$$

$$[H^+] = 10^{-4.9} = 10^{-5} \times 10^{0.1}$$

$$10^{0.1} = \text{antilog}\, 0.1 = 1.3$$

$$[H^+] = 1.3 \times 10^{-5}$$

EXAMPLE 18 A solution has a pH equal to 4.35. Find the molarity of the hydrogen ion dissolved in this solution.

Solution

$$pH = 4.35$$

$$[H^+] = 10^{-4.35} = 10^{-5} \times 10^{0.65}$$

$$10^{0.65} = \text{antilog}\, 0.65 = 4.5$$

$$[H^+] = 4.5 \times 10^{-5}$$

SELF-TEST

1. A solution has a hydrogen ion concentration of 1×10^{-7}. Find the pH of this solution and determine whether the solution is acidic, basic, or neutral.
2. Find the pH of a solution whose hydrogen ion concentration is equal to 0.000 000 000 3. Is this solution acidic, basic, or neutral?
3. Find the hydrogen ion concentration of a solution whose pH is equal to 8.5.
4. Find the hydrogen ion concentration of a solution whose pH is equal to 2.35.

ANSWERS **1.** pH = 7; neutral **2.** pH = 9.5; basic **3.** 3.2×10^{-9} **4.** 4.5×10^{-3}

10.2.2 OTHER *p*'s

The p concept is not restricted to just hydrogen ion concentrations. Chemists also use terms such as pOH (the negative log of the hydroxide ion molarity) and pK ($-\log K$, where K is the equilibrium constant).

EXAMPLE 19 The equilibrium constant for a reaction is found to be 0.000 000 02. What is the pK of this reaction?

Solution

$$K = 0.000\,000\,02 = 2 \times 10^{-8}$$

$$pK = -\log K = -\log(2 \times 10^{-8}) = -(-7.7) = +7.7$$

EXAMPLE 20 The hydroxide ion molarity of a solution is found to be 0.0005. What is the pOH of this solution?

Solution

$$pOH = -\log[OH^-] = -\log(5 \times 10^{-4}) = -(-3.3) = +3.3$$

EXAMPLE 21 For the hypothetical reaction

$$3A + B \rightleftharpoons 2C + D$$

$pK = 4$. Find the equilibrium constant for this reaction.

Solution

$$pK = 4$$

$$K = 10^{-pK} = 10^{-4} = 1 \times 10^{-4}$$

Therefore,

$$1 \times 10^{-4} = \frac{[C]^2[D]}{[A]^3[B]}$$

•
•

EXAMPLE 22 The pOH of a solution is 7.7. What is the hydroxide molarity of this solution?

Solution

$$pOH = 7.7$$

$$[OH^-] = 10^{-7.7} = 10^{-8} \times 10^{0.3}$$

$$10^{0.3} = \text{antilog } 0.3$$

$$[OH^-] = 2 \times 10^{-8}$$

•
•

SELF-TEST

1. A solution has a hydroxide concentration (molarity) of 0.000 000 01 mol of OH^- per liter of solution. Find the pOH of this solution.
2. A solution has a pOH of 7.3. What is the hydroxide ion molarity?
3. A chemical reaction has an equilibrium constant equal to 1×10^{-3}. Find the pK for this reaction.
4. The pK for a chemical reaction is 4.35. What is the value of the equilibrium constant for this reaction?

ANSWERS **1.** 8 **2.** 5×10^{-8} **3.** 3 **4.** 4.5×10^{-5}

10.3 *p*H OF STRONG ACID SOLUTIONS

10.3.1 MARRIAGE ANALOGY

Consider a small town with 20 people, all of whom are married. Thus, there are 10 married couples. A local law forbids anybody from moving into or out of this town. Also assume that each of the married people does not enjoy being married. This means that within a year all of these married people will be divorced. The divorce rate for this town is 100%. There is no tendency for any of the divorced people to remarry.

The problem is this: Initially we have 10 married couples. How many divorced men will be found in this town after one year?

There will be 10 divorced men in town. If all 10 couples divorce (a couple is made up of 1 man and 1 woman), 10 divorced men will be produced. The divorce "reaction" can be illustrated with the "equation"

$$MW \xrightarrow{100\% \text{ divorce}} M + W$$

where MW represents a married couple, M represents a divorced man, and W represents a divorced woman. Notice the single arrow going from left to right. This equation does *not* represent an equilibrium because there is no tendency for the divorced people (products) to reunite (forming reactant MW).

How many divorced women will be found after 1 year?

Because you found 10 divorced men, there must also be 10 divorced women. Ten couples are made up of 10 men and 10 women. If all 10 couples break up, 10 divorced men and 10 divorced women will be produced. The important point for you to grasp is that the number of divorced men must always be equal to the number of divorced women. This may sound like a simple idea, but this is the same level of difficulty found in many equilibrium problems that confuse students. Remember, one of the goals of this chapter is to show you that the logic and mathematics of many equilibrium problems are simple when isolated from the chemistry.

Going back to the analogy, how many married couples will be left after 1 year?

No married couples are left after 1 year if all of them divorce. The answer is 0.

This discussion can be summarized by writing the initial and final values of M, W, and MW over the "chemical equilibrium" in the following manner:

initial number of MW / number of MW divorced / final number of MW	initial number of M / number of M formed / final number of M	initial number of W / number of W formed / final number of W
10 − 10 = 0	0 + 10 = 10	0 + 10 = 10

$$MW \xrightarrow{100\%\ \text{divorce}} M + W$$

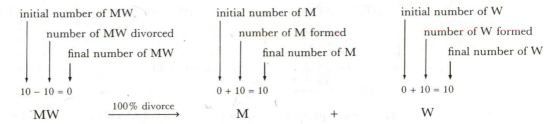

10.3.2 STRONG ACID PROBLEMS

Let's examine a chemistry problem that uses the same logic as the preceding analogy.

EXAMPLE 23 Find the *p*H of a 0.1 M HCl solution. Hydrochloric acid (HCl) is a strong acid.

Solution To simplify our discussion we are going to assume we have 1 L of the 0.1 M HCl solution. The chemical equation that represents the dissociation of hydrochloric acid is

$$HCl \xrightarrow{100\%\ \text{dissociation}} H^+ + Cl^-$$

Notice the use of a single arrow (as opposed to the equilibrium double arrows) because hydrochloric acid is a strong acid. Strong acids are defined in this book as those that dissociate (break up, ionize, split up, divorce) 100% in dilute (0.1 M or less) aqueous (water) solutions. All strong acids, strong bases, and strong electrolytes share this property (they completely dissociate in dilute aqueous solutions).

There is 0.1 mol of HCl dissolved in 1 L of solution (0.1 M HCl), and all of the HCl dissociates. What is the concentration (molarity) of the hydrogen ion?

The hydrogen ion concentration is 0.1 mol of hydrogen ion per liter of solution. Hydrogen is bonded to chlorine (a chemical bond) to form HCl in much the same way that M is bonded to W (a matrimonial bond) to form married couple MW. If 10 couples (MW) break their marital bonds, 10 divorced men and 10 divorced women are the result. Likewise, if 0.1 mol of HCl dissociates (ionizes, splits apart), 0.1 mol of H^+ will form.

How many moles of Cl^- will form?

If 0.1 mol of hydrogen ion forms, then 0.1 mol of chloride ion will also form. The number of hydrogen ions and chloride ions formed from the dissociation of HCl must always be the same (just as the number of divorced men and divorced women formed from the dissociation of MW must always be the same).

What is the concentration of HCl in a 0.1 M HCl solution?

The concentration of HCl in a 0.1 M HCl solution is 0 mol of HCl per liter of solution. If all of the HCl breaks up, then there are no HCl molecules left. (If all 10 couples break up, there are no married couples left.) It may be difficult to understand that there are no molecules of HCl in an HCl solution, but this is similar to saying that there are no married couples in a town in which the divorce rate is 100%.

We can summarize our discussion by writing the initial and final concentrations (molarities) of HCl, H^+, and Cl^- over the chemical equation, as follows:

initial [HCl]	initial [H^+]	initial [Cl^-]
[HCl] that dissociate	[H^+] that form	[Cl^-] that form
final [HCl]	final [H^+]	final [Cl^-]
$0.1 - 0.1 = 0.0$	$0.0 + 0.1 = 0.1$	$0.0 + 0.1 = 0.1$

$$\mathrm{HCl} \xrightarrow{100\%\ \text{dissociation}} \mathrm{H^+} + \mathrm{Cl^-}$$

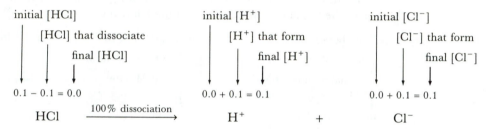

The original problem asked for the *p*H of a 0.1 M HCl solution. In 1 L of this 0.1 M HCl solution, there are 0 mol of HCl, 0.1 mol of hydrogen ion, and 0.1 mol of chloride ion.

What is the pH *of this solution?*

The *p*H of this solution is 1. Since the hydrogen ion concentration is 0.1 mol of hydrogen ion per liter of solution, then

$$\begin{aligned} pH &= -\log[H^+] = -\log 0.1 \\ &= -\log(1 \times 10^{-1}) = -(\log 1 + \log 10^{-1}) = -[(0) + (-1)] \\ &= -(-1) = +1 \end{aligned}$$

EXAMPLE 24 Find the *p*H of a 0.01 M HNO_3 solution. Nitric acid (HNO_3) is a strong acid.

Solution The *p*H of this solution is equal to 2. Nitric acid is a strong acid and therefore completely (100%) dissociates in aqueous solutions.

$$\mathrm{HNO_3} \xrightarrow{100\%} \mathrm{H^+} + \mathrm{NO_3^-}$$

In 1 L of this 0.01 M HNO_3 solution, there are 0 mol of HNO_3, 0.01 mol of hydrogen ion, and 0.01 mol of nitrate ion. We can summarize what happens by writing the initial and final concentrations (molarities) of nitric acid, hydrogen ion, and nitrate ion over the chemical equation representing this reaction.

$$\underset{\mathrm{HNO_3}}{0.01 - 0.01 = 0} \xrightarrow{100\%} \underset{\mathrm{H^+}}{0.00 + 0.01 = 0.01} + \underset{\mathrm{NO_3^-}}{0.00 + 0.01 = 0.01}$$

$$pH = -\log[H^+] = -\log 0.01 = -\log 10^{-2} = -(-2) = 2$$

EXAMPLE 25 Find the pH of a 0.02 M $HClO_3$ solution. Chloric acid ($HClO_3$) is a strong acid.

Solution The pH of this solution is 1.7. Chloric acid is a strong acid and therefore completely dissociates in water. In 1 L of this 0.02 M $HClO_3$ solution, there are 0 mol of chloric acid, 0.02 mol of hydrogen ion, and 0.02 mol of chlorate ion.

$$\underset{HClO_3}{0.02 - 0.02 = 0} \xrightarrow{100\%} \underset{H^+}{0.00 + 0.02 = 0.02} + \underset{ClO_3^-}{0.00 + 0.02 = 0.02}$$

$$pH = -\log[H^+] = -\log(0.02)$$
$$= -\log(2 \times 10^{-2}) = -[\log(2) + \log(10^{-2})] = -[(0.3) + (-2)] = 1.7$$

SELF-TEST

1. Find the pH of a 0.000 01 M HNO_3 solution.
2. Find the pH of a 1.0 M $HClO_3$ solution.
3. Find the pH of a 0.000 05 M HCl solution.

ANSWERS **1.** 5 **2.** 0 **3.** 4.3

10.4 DISSOCIATION EQUILIBRIUM CONSTANT AND PERCENTAGE OF DISSOCIATION

10.4.1 MARRIAGE ANALOGY

Consider another small town similar to the one in the previous analogy except that this second town has a lower divorce rate. Initially there are 10 married couples. After several years (when equilibrium is reached), 4 divorced men are found. (Assume that nobody has moved into or out of the town during this time.)

What percentage of the marriages have broken up?

You probably figured out in your head, almost intuitively, that 40% of the marriages ended in divorce. This may seem to be a trivial calculation, but let's examine exactly how the 40% figure is determined.

There are 4 divorced men, implying that 4 marriages broke up. There were initially 10 marriages and 4 of them broke up, so the fraction that broke up is 4 out of 10, $\frac{4}{10}$, or 0.4. Fractions can be converted to percentages, mechanically, by moving the decimal point in the fraction two places to the right. Thus, 0.4 is 40% (100% times 0.4 equals 40%).

Look at a slightly different way of solving this same problem. The percentage of broken marriages can be defined by the formula

The number of divorces equals the number of divorced men found.

$$\text{percentage of divorces} = \frac{\text{number of divorces}}{\text{initial number of marriages}} \times 100\%$$

$\frac{100}{\text{centum}}$

$$= \frac{4}{10} \times 100\% = 40\%$$

To push the analogy even further, consider writing an equilibrium expression such as

$$MW \rightleftharpoons M + W$$

where MW stands for a married couple, M stands for a divorced man, and W stands for a divorced woman. According to this equilibrium equation, divorced men and women can remarry (the double arrows imply this).

What are the initial values for MW, M, and W in this town?

MW is initially 10 and M and W are both initially 0. (Initially there were 10 married couples and no divorced people.)

Now, what are the values for MW, M, and W after equilibrium is reached?

At equilibrium, MW is 6 (4 of the 10 couples divorced, 10 – 4 = 6), M equals 4, and W equals 4 (for every divorced man there must be a divorced woman). Finding 4 divorced men implies that there must also be 4 divorced women.

initial number of MW / number of MW divorced / final number of MW	initial number of M / number of M formed / final number of M	initial number of W / number of W formed / final number of W
10 – 4 = 6	0 + 4 = 4	0 + 4 = 4

$$MW \rightleftharpoons M + W$$

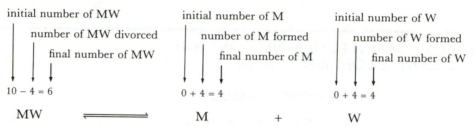

We can define a marriage equilibrium constant, K_m, as

$$K_m = \frac{(M)(W)}{(MW)} \quad \begin{matrix} \leftarrow \text{products} \\ \leftarrow \text{reactants} \end{matrix}$$

Calculate the value of K_m.

K_m is equal to $2.\bar{6}$ (2.666 666 666 . . .). The equilibrium values for M and W are both 4 ($4.\bar{0}$ implied), and the equilibrium value of MW is 6.

$$K_m = \frac{(M)(W)}{(MW)} = \frac{(4)(4)}{(6)} = 2.\bar{6}$$

10.4.2 DISSOCIATION PROBLEMS

Now let's apply all of these ideas to a chemical equilibrium problem. We are going to show how to find the value of the equilibrium constant K of the imaginary weak acid HA, where A^- is the anion (negative ion) part of the acid. (HA dissociates to form H^+ and A^-.) Then we will show how to determine the percentage dissociation of this weak acid.

•
•

EXAMPLE 26 Find the equilibrium constant K of a weak acid, HA, if 1 L of a 0.1 M HA solution has a pH of 4.0. The equilibrium is represented by

$$HA \rightleftharpoons H^+ + A^-$$

Solution First find the concentration (molarity) of the hydrogen ion in this solution. This is determined from the pH of the solution.

$$[H^+] = 10^{-pH} = 10^{-4} = 1 \times 10^{-4}$$

What is the concentration of the anion A^- at equilibrium?

Hopefully, you calculated an anion molarity of 1×10^{-4}. Why are the hydrogen ion and anion molarities the same? (Both are 1×10^{-4}.) Consider the equilibrium for the dissociation of HA and the equilibrium for the dissociation of married couples MW:

$$HA \rightleftharpoons H^+ + A^-$$

$$MW \rightleftharpoons M + W$$

Remember that finding 4 divorced men implies that there are 4 divorced women around. Likewise, when HA breaks up (ionizes, dissociates, divorces, splits up), it produces hydrogen ions (H^+) and anions (A^-) in equal amounts.

What is the concentration of undissociated HA at equilibrium?

This question is a little tricky. The best answer is that the HA concentration is 0.1. Recall that in the marriage analogy we started with 10 couples, found 4 divorced men, and concluded that the number of couples left was 10 − 4 = 6. In the present problem, we started with 0.1 mol of HA, found 1×10^{-4} (0.0001) mol of H^+, and concluded that the amount of HA left is 0.1 − 0.0001 = 0.1 mol HA. Recall from our discussion of significant figures and accuracy (Section 1.5) that the answer cannot be more accurate than the least accurate number in the subtraction.

If the original problem had stated that the HA molarity was 0.1000 (as opposed to 0.1), then we could have said that the equilibrium concentration of HA is 0.1000 − 0.0001 = 0.0999. However, to simplify our math, we might want to estimate 0.0999 as being approximately equal to 0.1 because it is easier to divide by 0.1 than by 0.0999. The error introduced by this approximation is very small (see Section 1.4, "Calculators and Estimation Techniques").

We can summarize our calculations and information by writing the initial and equilibrium concentrations (molarities) of HA, H^+, and A^- over the chemical equilibrium, as follows:

initial [HA]	initial [H⁺]	initial [A⁻]
[HA] dissociated	[H⁺] formed	[A⁻] formed
equilibrium [HA]	equilibrium [H⁺]	equilibrium [A⁻]
$0.1 - 10^{-4} = 0.1$	$0 + 10^{-4} = 10^{-4}$	$0 + 10^{-4} = 10^{-4}$

$$HA \rightleftharpoons H^+ + A^-$$

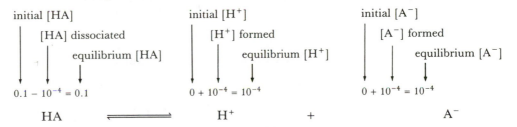

(Actually, the initial hydrogen ion concentration in water is 10^{-7}, not 0. However, $10^{-7} + 10^{-4}$ still equals 10^{-4}; that is,

$$\begin{array}{r} 0.0000001 \\ +0.0001 \\ \hline 0.0001 \end{array}$$

because of the rules that were discussed in Section 1.5.3, "Accuracy vs. Significant Digits.")

Now that we have calculated the equilibrium values for HA, H^+, and A^-, what is the value of the equilibrium constant K *for the dissociation of acid HA?*

The equilibrium constant for the dissociation of HA is 1×10^{-7}.

$$K = \frac{[H^+][A^-]}{[HA]} = \frac{(10^{-4})(10^{-4})}{(10^{-1})} = 10^{-7}$$

EXAMPLE 27 Find the pK for the dissociation of HA in Example 26.

Solution

$$pK = -\log K = -\log 10^{-7} = -(-7) = +7 = 7$$

EXAMPLE 28 What is the percentage of dissociation for the acid HA in the preceding example?

Solution The acid is 0.1% dissociated. To calculate this value we use the same sort of logic that we used to calculate the percentage of marriages that dissociated (broke up). Using the second of the two approaches shown in the previous marriage analogy,

$$\text{percentage divorces} = \frac{\text{number divorces}}{\text{initial number married couples}} \times 100\%$$

we obtain

The number of moles of HA dissociated equal the number of moles of H^+ formed when HA dissociates.

$$\text{percentage dissociation of HA} = \frac{\text{number mol HA dissociated}}{\text{initial number mol HA}} \times 100\%$$

$$= \frac{[H^+]\text{ at equilibrium}}{\text{initial [HA]}} \times 100\%$$

These are equal because the solution volume is 1 L.

$$= \frac{(10^{-4})}{(0.1)} \times 100\% = \frac{(10^{-4})}{(10^{-1})}(10^2)\% = 0.1\%$$

EXAMPLE 29 Why is the value 0.1% a reasonable answer in Example 28?

Solution K is very small (1×10^{-7}), so you would not expect the reaction

$$HA \rightleftharpoons H^+ + A^-$$

to proceed very far to the right. You would expect that a solution of HA would not dissociate appreciably, which is what we observed (i.e., the percent dissociation was 0.1%). When K is small, the amount of products formed is also small.

EXAMPLE 30 Find the equilibrium constant K of a weak acid, HB, if 1 L of 0.1 M HB solution has a pH of 3. The equilibrium is represented by

$$HB \rightleftharpoons H^+ + B^-$$

Solution First, find the hydrogen ion molarity of this solution.

$$[H^+] = 10^{-pH} = 10^{-3}$$

Then, determine the equilibrium concentrations of HB, H^+, and B^-.

$$\underset{HB}{\overset{0.1 - 10^{-3} = 0.1}{}} \rightleftharpoons \underset{H^+}{\overset{0 + 10^{-3} = 10^{-3}}{}} + \underset{B^-}{\overset{0 + 10^{-3} = 10^{-3}}{}}$$

Finally, substitute these equilibrium concentrations into the equilibrium constant expression for the dissociation of HB.

$$K = \frac{[H^+][B^-]}{[HB]} = \frac{(10^{-3})(10^{-3})}{(10^{-1})} = 10^{-5}$$

EXAMPLE 31 Find the pK for the dissociation of HB in Example 30.

Solution

$$pK = -\log K = -\log 10^{-5} = -(-5) = +5 = 5$$

EXAMPLE 32 Find the percentage of dissociation of the weak acid HB discussed in the previous problem.

Solution

$$\text{percentage dissociation of HB} = \frac{[HB] \text{ dissociated}}{\text{initial } [HB]} \times 100\%$$

$$= \frac{[H^+] \text{ at equilibrium}}{\text{initial } [HB]} \times 100\%$$

$$= \frac{(10^{-3})}{(10^{-1})} \times 100\% = 1\%$$

EXAMPLE 33 Find the equilibrium constant K of a weak acid, HC, if 1.0 L of a 0.20 M HC solution has a pH of 3.40.

Solution First, determine the hydrogen ion molarity of this solution.

$$[H^+] = 10^{-pH} = 10^{-3.4} = 10^{-4} \times 10^{0.60} = 4.0 \times 10^{-4}$$

Then, determine the equilibrium concentrations.

$$\underset{HC}{\overset{0.20 - (4.0 \times 10^{-4}) = 0.20}{}} \rightleftharpoons \underset{H^+}{\overset{0 + (4.0 \times 10^{-4}) = 4.0 \times 10^{-4}}{}} + \underset{C^-}{\overset{0 + (4.0 \times 10^{-4}) = 4.0 \times 10^{-4}}{}}$$

Finally, calculate K by inserting the equilibrium concentrations into the equilibrium constant expression.

$$K = \frac{[H^+][C^-]}{[HC]} = \frac{(4.0 \times 10^{-4})(4.0 \times 10^{-4})}{(2.0 \times 10^{-1})} = 8.0 \times 10^{-7}$$

EXAMPLE 34 Find the pK for the dissociation of HC discussed in Example 33.

Solution

$$pK = -\log K = -\log(8.0 \times 10^{-7})$$

$$= -(\log 8.0 + \log 10^{-7}) = -[(0.90) + (-7)] = 6.10$$

EXAMPLE 35 Find the percentage of dissociation of HC.

Solution

$$\text{percentage dissociation of HC} = \frac{[\text{HC}]\text{ dissociated}}{\text{initial }[\text{HC}]} \times 100\%$$

$$= \frac{[\text{H}^+]\text{ at equilibrium}}{\text{initial }[\text{HC}]} \times 100\%$$

$$= \frac{(4.0 \times 10^{-4})}{(2.0 \times 10^{-1})} \times 100\% = 0.20\%$$

SELF-TEST

1. Find the pK of a weak acid, HD, if 1 L of a 0.1 M HD solution has a pH of 3.
2. Find the percentage of dissociation of acid HW if 1 L of 1.0 M HW has a pH of 4.
3. Find the equilibrium constant K for the weak acid HV if 1.0 L of 0.30 M HV has a pH of 4.70.

ANSWERS **1.** 5 **2.** 0.01% **3.** 1.3×10^{-9}

Consider a slightly different type of problem in which the equilibrium constant is given.

EXAMPLE 36 Find the hydrogen ion concentration of a 0.1 M HA solution. HA has an equilibrium constant equal to 1×10^{-5}. The equilibrium equation showing the dissociation of this acid is

$$\text{HA} \rightleftharpoons \text{H}^+ + \text{A}^-$$

Solution First, write the equilibrium constant expression showing the relationship between K, the hydrogen ion concentration, the anion concentration, and the acid concentration.

$$K = \frac{[\text{H}^+][\text{A}^-]}{[\text{HA}]}$$

We don't know the hydrogen ion concentration, so we must set it equal to x. Let

$$x = [\text{H}^+]$$

If x *is the value of the hydrogen ion concentration, what is the anion (A^-) concentration?*

The anion concentration is the same as the hydrogen ion concentration. Both are equal to x. If you don't understand this, recall our discussion about married couples. Whenever a couple MW splits up, it forms equal numbers of M's and W's. If initially there are 100 married couples (MW's), and later 5 divorced men are found, there must also be 5 divorced women. If x divorced men are found, there must also be x divorced women. The two (M and W) are always equal to each other, whether they are expressed as numbers (5's) or as unknown values (x's).

What is the value of the HA concentration at equilibrium?

It is $(0.1 - x)$. If you didn't get this answer, think back to the initial 100 married couples that produced 5 divorced men. How many married couples were left after the 5 divorced men were formed? There were $100 - 5 = 95$ married couples left. The chemistry works the same way. If we start with 0.1 mol of HA, and some of it splits up to produce x mol of hydrogen ion, then there is $0.1 - x$ mol of HA left.

We can summarize our discussion by writing the initial and equilibrium concentrations over the chemical equilibrium equation.

initial [HA] — [HA] dissociated — equilibrium [HA]: $0.1 - x = (0.1 - x)$

initial $[H^+]$ — $[H^+]$ formed — equilibrium $[H^+]$: $0 + x = x$

initial $[A^-]$ — $[A^-]$ formed — equilibrium $[A^-]$: $0 + x = x$

$$HA \rightleftharpoons H^+ + A^-$$

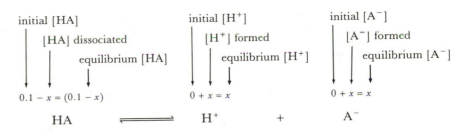

Now that we have equilibrium concentration values for each substance, substitute these values into the equilibrium constant expression.

$$K = \frac{[H^+][A^-]}{[HA]} = \frac{(x)(x)}{(0.1 - x)}$$

Do you remember from a previous problem that (0.1 – 0.0001) is essentially 0.1? There were two reasons given, one of which was that if you were to subtract 0.0001 (a very small number, almost insignificant compared with 0.1) from 0.1, you would not be decreasing the value of 0.1 enough to affect the accuracy of your answer. This is similar to subtracting 1 cent from 1 million dollars. After taking a penny away from a million dollars you could still claim to have a million dollars left (even though you actually have $999,999.99) because a penny is very small (almost insignificant) when compared with a million dollars. In this problem we are going to subtract x from 0.1 $(0.1 - x)$ and say that the answer is 0.1; that is, $0.1 - x = 0.1$ because x is so small compared to 0.1.

How do we know that x *is a very small number?*

We know that x is a very small number because K for the dissociation of HA is 1×10^{-5}. Because K is small, the reaction doesn't proceed very far to the right (little product forms). Any K value less than 0.1 is considered small. This means that few hydrogen ions form from the dissociation of HA because very few HA molecules split up. Because we let x be the number of moles of hydrogen ion in 1 L of solution that are produced by the HA dissociation, and because very few hydrogen ions form from the HA dissociation, x must be a very small number. If x is a very small number, then $0.1 - x$ will be approximately 0.1, just as when a penny is taken away from a million dollars approximately a million dollars is left.

Now our equilibrium constant expression can be solved for the hydrogen ion concentration x.

$$10^{-5} = \frac{(x)(x)}{(0.1 - x)} = \frac{(x)(x)}{(0.1)} = \frac{(x)(x)}{10^{-1}} = \frac{x^2}{10^{-1}}$$

$$x^2 = (10^{-1})(10^{-5}) = 10^{-6}$$

$$x = 10^{-3} = [H^+]$$

•
•

We've solved the original problem, which asked for the hydrogen ion concentration. Do you know what the pH *of this solution is?*

The pH of this solution is 3.

$$[H^+] = x = 10^{-3}$$

$$pH = -\log 10^{-3} = -(-3) = 3$$

Now let's see how well you understand the preceding material. What is the concentration of the anion A^- in the solution?

The anion concentration is 1×10^{-3}, the same as the hydrogen ion concentration, because both were set equal to x at the beginning of our solution to this problem.

•
•

EXAMPLE 37 Find the pH of a 0.1 M HB solution. HB is an acid with an equilibrium constant equal to 1×10^{-3}. The dissociation of HB is represented by

$$HB \rightleftharpoons H^+ + B^-$$

Solution First, write the equilibrium constant expression for this dissociation.

$$K = \frac{[H^+][B^-]}{[HB]}$$

Let

$$x = [H^+]$$

Then,

$$[B^-] = x \text{ as well}$$

and

$$[HB] = (0.1 - x) = 0.1$$

$$\underset{HB}{0.1 - x = 0.1} \quad \rightleftharpoons \quad \underset{H^+}{0 + x = x} \quad + \quad \underset{B^-}{0 + x = x}$$

Substitute these concentration values into the equilibrium constant expression.

$$10^{-3} = \frac{(x)(x)}{(0.1)} = \frac{x^2}{10^{-1}}$$

$$x^2 = 10^{-4}$$

$$x = 10^{-2} = [H^+]$$

$$pH = 2$$

•
•

EXAMPLE 38 Find the percentage of dissociation that occurs in the preceding 0.1 M HB solution.

Solution

$$\text{percentage dissociation of HB} = \frac{[\text{HB}] \text{ dissociated}}{\text{initial } [\text{HB}]} \times 100\%$$

$$= \frac{[\text{H}^+] \text{ at equilibrium}}{\text{initial } [\text{HB}]} \times 100\%$$

$$= \frac{10^{-2}}{10^{-1}} \times 100\% = 10\%$$

Note: Since all data contained one significant digit, it would be more correct to express the answer as $1 \times 10^1\%$.

•
•

EXAMPLE 39 Find the *p*H of a 1.0 M HC solution. HC is an acid with an equilibrium dissociation constant equal to 1×10^{-4}. The dissociation of HC is represented by

$$HC \rightleftharpoons H^+ + C^-$$

Solution First, write the equilibrium constant expression for the dissociation of HC.

$$K = \frac{[H^+][C^-]}{[HC]}$$

Let

$$x = [H^+] = [C^-]$$

$$\underset{1.0 - x = 1.0}{HC} \rightleftharpoons \underset{0 + x = x}{H^+} + \underset{0 + x = x}{C^-}$$

$$10^{-4} = \frac{(x)(x)}{(1.0 - x)} = \frac{x^2}{1.0} = x^2$$

$$x = 10^{-2} = [H^+]$$

$$pH = 2$$

•
•

EXAMPLE 40 Find the percentage of dissociation that occurs in the preceding 1.0 M HC solution.

Solution

$$\text{percentage dissociation of HC} = \frac{[HC] \text{ dissociated}}{\text{initial } [HC]} \times 100\%$$

$$= \frac{10^{-2}}{1.0} \times 100\% = 1\%$$

•
•

SELF-TEST

1. Find the *p*H of a 0.1 M HD solution. HD has an equilibrium constant equal to 1×10^{-5}.
2. Find the molarity of the anion B^- in a 0.1 M HB solution. HB has an equilibrium constant equal to 1×10^{-3}.
3. Find the percentage of dissociation of a 0.1 M HD solution. HD has an equilibrium constant equal to 1×10^{-5}.

ANSWERS **1.** 3 **2.** 10^{-2} M B^- **3.** 1%

10.5 COMMON ION EFFECT

10.5.1 MARRIAGE ANALOGY

We've saved the most difficult concept for last. Consider two cities, Divorcia, Nevada (called "Sin City of the World"), and Puria, Georgia (where, they say, people are "as pure as the driven snow"). We will make the assumption that within a year every marriage in Divorcia ends in divorce, while very few marriages in Puria break up. Puria is a city with 1000 married couples (total population: 2000 people). We don't know the number of divorced men walking the streets of Puria, so we'll set that number equal to x. Remember that even though we don't know the value of x, we do know that x is a very small number (because few people get divorces in Puria). Therefore, $1000 - x = 1000$ in the same way that 1 million dollars minus 1 penny is 1 million dollars.

If there were 10 000 married couples in Divorcia last year, we know that today there must be 10 000 divorced men in Divorcia (likewise, there must also be 10 000 divorced women). Equally important, we know that there are no married couples presently living in Divorcia. If a man and woman happen to get married in Divorcia, within a year they split up. Those that split up don't remarry.

What happens if 100 married couples move from Divorcia to Puria? How many divorced men are in Puria a year after this move occurs?

There are approximately 100 divorced men in Puria. There are exactly $(100 + x)$ divorced men in the city of Puria.

Can you explain why there are exactly 100 + x *divorced men in Puria and why 100 +* x *is approximately equal to 100?*

Puria has 1000 married couples, very few of which get divorced, producing x divorced men. All 100 couples from Divorcia will be divorced within a year, producing another 100 divorced men. Divorced men come from two sources, some from Puria couples who get a divorce and some from Divorcia couples who get a divorce. The sum of these two sources is $100 + x$. Because we stated that x is very small, the total number of divorced men in Puria is about 100.

Here is the most important point: The total number of divorced men in Puria is always (approximately) equal to the number of married couples who move into Puria from Divorcia. Think about that for a moment. It doesn't matter how many married couples there are in Puria. The most important indicator of the number of divorced men in Puria is the number of married couples (MW) that have moved into Puria from Divorcia.

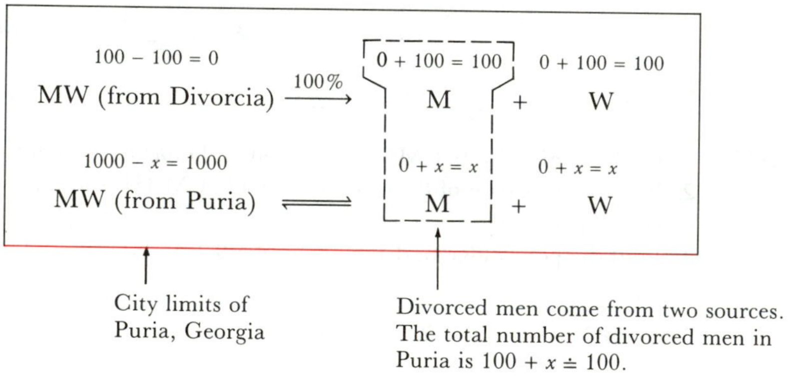

If you can follow this logic, then you can follow the logic necessary to solve chemistry problems that involve the common ion effect. Let's try one now.

10.5.2 COMMON ION PROBLEMS

EXAMPLE 41 In Example 36, we found that the hydrogen ion molarity is 1×10^{-3} in a 0.1 M HA solution. The equilibrium constant for HA is 1×10^{-5}. This time we want to find the hydrogen ion concentration of a 0.1 M HA solution that also contains 1 M NaA. (Na^+ is the sodium ion and A^- is the anion found in the acid HA.) Assume that we have a total solution volume of 1 L.

Solution Our 1 L of solution has, initially (before dissociation occurs), 0.1 mol of HA and 1 mol of NaA. HA is a weak acid (weak electrolyte, doesn't dissociate very much), and the salt NaA is a strong electrolyte (dissociates 100%).

How do we know this?

HA is a weak acid because the value of its equilibrium constant K is very small (10^{-5}). Remember that any K less than 0.1 is very small. NaA, on the other hand, is a very strong electrolyte. Why? We know NaA is a strong electrolyte (even though a K value for NaA isn't given) because NaA is a sodium salt and just about every sodium salt is a strong electrolyte.

The two chemical equations showing the dissociation of the weak acid HA and the dissociation of the strong electrolyte NaA are

$$HA \rightleftharpoons H^+ + A^-$$

$$NaA \xrightarrow{100\%} Na^+ + A^-$$

Notice that the first equation is written with two arrows, showing an equilibrium between products and reactants, while the second equation is shown with just one arrow, indicating that it fully dissociates and that Na^+ and A^- do not tend to recombine.

The salt NaA is very similar to the MW couples from Divorcia. Both completely break up. If 100 couples from Divorcia move into Puria, then 100 divorced men will be found in Puria. If 1 mol of NaA is added to water, there will be no NaA present in the solution. All of the NaA will dissociate. If 1 mol of NaA is added to water, we will find 1 mol of Na^+ ions and 1 mol of A^- ions in that solution. If you add 1 mol of NaA to a solution, you are, in effect, adding 1 mol of A^- ions to that solution. (Similarly, adding 100 MW couples from Divorcia to Puria is, in effect, equivalent to adding 100 divorced men to the city of Puria.)

What is the total number of moles of A^- in the solution made up of 0.1 M HA and 1 M NaA? (Remember that we assume that the volume of our solution is 1 L.)

There is approximately 1 mol of A^- in the solution. The exact number of moles of A^- is $1 + x$, which equals approximately 1. One mole of A^- comes from the breakup of 1 mol of NaA. There is a second source of A^- from the dissociation of HA. But HA doesn't dissociate very much because it has such a small value for its equilibrium constant K. HA is a weak electrolyte. If we let x be the number of moles per liter that dissociate, then the amount of A^- produced by the dissociation of HA will also be x. (If x couples divorce, x divorced men are produced.) Because x is a small number, $1 + x$ is about equal to 1 (just as a penny plus a million dollars is about a million dollars).

We can represent this situation as shown in the accompanying diagram:

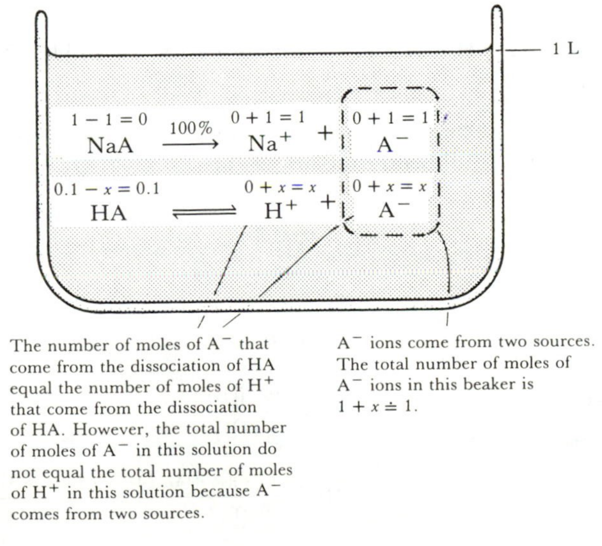
1 L
1 − 1 = 0
NaA
100%
0 + 1 = 1
Na⁺
0 + 1 = 1
A⁻
0.1 − x = 0.1
HA
0 + x = x
H⁺
0 + x = x
A⁻
The number of moles of A⁻ that come from the dissociation of HA equal the number of moles of H⁺ that come from the dissociation of HA. However, the total number of moles of A⁻ in this solution do not equal the total number of moles of H⁺ in this solution because A⁻ comes from two sources.
A⁻ ions come from two sources. The total number of moles of A⁻ ions in this beaker is 1 + x ≐ 1.

Set up the equilibrium constant expression and insert into that expression the values for K *and the concentrations of HA, A^-, and H^+.*

$$K = \frac{[H^+][A^-]}{[HA]}$$

$$10^{-5} = \frac{(x)(x+1)}{(0.1-x)}$$

We said that x represents the number of moles of HA that dissociate, so x is also equal to the number of moles of H^+ that form, and $(0.1 - x)$ is the number of moles of HA left undissociated. Since $1 + x = 1$ and $0.1 - x = 0.1$, then

$$10^{-5} = \frac{(x)(x+1)}{(0.1-x)} = \frac{(x)(1)}{(0.1)} = \frac{x}{10^{-1}}$$

$$x = 10^{-6} = [H^+] = [A^-] \text{ (from the dissociation of HA)}$$

The number of moles of hydrogen ion and anion A^- that are the result of the dissociation of HA is 1×10^{-6}. It is important for you to realize that the concentrations of the hydrogen ion and the A^- ion in this solution are *not* 1×10^{-6} mol per liter of solution.

What is the concentration of A^-?

The concentration of A^- is approximately equal to 1. Remember that we added 1 mol of NaA (which is the same as saying we added 1 mol of A^-) to our HA solution. The total number of moles of A^- is $1 + 10^{-6} = 1$.

What do you think the pH *of this solution is?*

You probably calculated a *p*H of 6, which is a pretty good approximation, but this is not the exact value of the *p*H. The *p*H is slightly less than 6. Why? Remember that pure (neutral) water at 25°C has a *p*H of 7 (which means that the hydrogen ion concentration is 1×10^{-7}). To a liter of this pure water we have added some HA, which dissociated to produce 1×10^{-6} additional moles of hydrogen ion. The total hydrogen ion concentration is closer to the sum $(1 \times 10^{-6}) + (1 \times 10^{-7}) = 1.1 \times 10^{-6}$.

•
•

EXAMPLE 42 Find the *p*H of a solution that is 0.1 M HB and 0.1 M NaB. The equilibrium constant for HB is 1×10^{-3}.

Solution The chemical equations representing the reactions occurring in the solution, along with the initial and equilibrium molarities, are

$$\underset{0.1 - x = 0.1}{HB} \rightleftharpoons \underset{0 + x = x}{H^+} + \underset{0 + x = x}{B^-}$$

$$\underset{0.1 - 0.1 = 0}{NaB} \xrightarrow{100\%} \underset{0.0 + 0.1 = 0.1}{Na^+} + \underset{0.0 + 0.1 = 0.1}{B^-}$$

$$K = \frac{[H^+][B^-]}{[HB]}$$

$$10^{-3} = \frac{(x)(x+0.1)}{(0.1-x)} = \frac{(x)(x+0.1)}{(0.1)} = \frac{(x)(0.1)}{(0.1)} = x$$

$$x = 10^{-3} = [H^+]$$

$$pH = 3$$

EXAMPLE 43 Find the pH of a solution that is 1 M HC and 1 M NaC. The equilibrium constant for HC is 10^{-5}.

Solution

$$\underset{HC}{\overset{1-x=1}{}} \rightleftharpoons \underset{H^+}{\overset{0+x=x}{}} + \underset{C^-}{\overset{0+x=x}{}}$$

$$\underset{NaC}{\overset{1-1=0}{}} \xrightarrow{100\%} \underset{Na^+}{\overset{0+1=1}{}} + \underset{C^-}{\overset{0+1=1}{}}$$

$$K = \frac{[H^+][C^-]}{[HC]}$$

$$10^{-5} = \frac{(x)(1+x)}{(1)} = \frac{(x)(1)}{(1)} = x$$

$$x = 10^{-5} = [H^+]$$

$$pH = 5$$

SELF-TEST

1. Find the pH of a solution that is 0.1 M HA and 0.2 M NaA. The equilibrium constant for HA is 1×10^{-5}.
2. Find the pH of a solution that is 0.1 M HB and 0.2 M NaB. The equilibrium constant for HB is 1×10^{-6}.
3. Find the pH of a solution that is 0.03 M HC and 0.2 M NaC. The equilibrium constant for HC is 1×10^{-6}.

ANSWERS **1.** 5.3 **2.** 6.3 **3.** 6.8

10.5.3 LE CHATELIER'S PRINCIPLE AND THE COMMON ION EFFECT

In combination, Examples 36 and 41 are a good illustration of Le Chatelier's principle. In Example 36 we found that the hydrogen ion concentration produced by the dissociation of acid HA

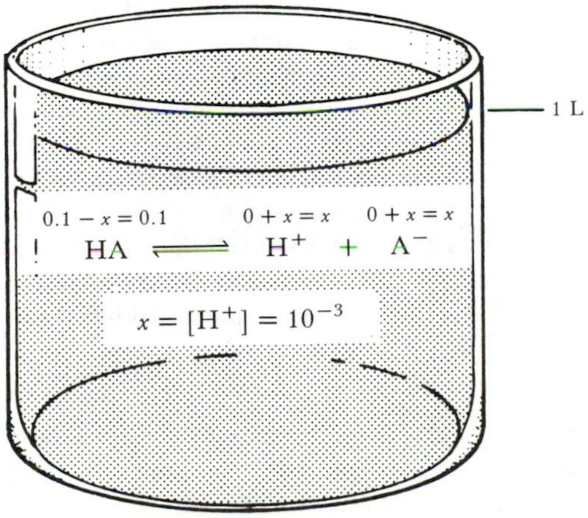

summary of Example 36

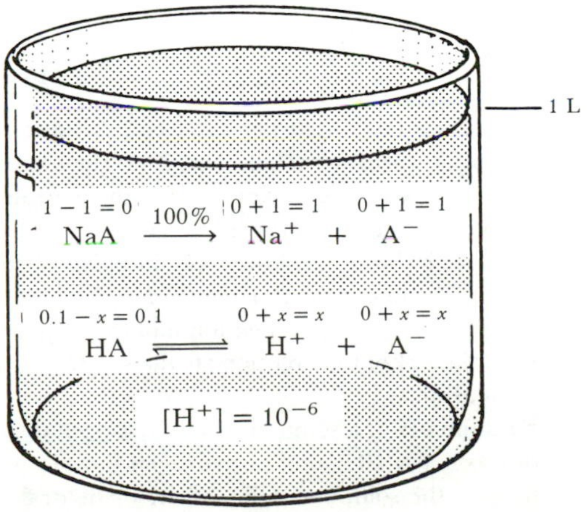

summary of Example 41

is 1×10^{-3}. In Example 41 we added some A^- (in the form of NaA) to the HA solution, that is, we increased the concentration of A^- in the system in equilibrium. We increased the concentration of A^- by adding NaA, which completely dissociates into Na^+ and A^-.

What does Le Chatelier's principle predict will happen if the concentration of A^- is increased (in the left-hand beaker)?

Le Chatelier's principle predicts that the equilibrium will shift to the left because a system in equilibrium tends to relieve itself of any applied stress. The stress applied to the equilibrium in the left-hand beaker is the addition of anion A^-. By shifting to the left, the equilibrium reduces the concentration of A^-. The equilibrium can only shift to the left by combining hydrogen ions already present in the solution with the added A^- ion (hydrogen ions are used up). If this does occur, our calculations will show that the hydrogen ion concentration from the HA dissociation decreases after NaA is added to a solution of HA. We observed this in Example 41 (in the right-hand beaker). The hydrogen ion concentration decreased from 1×10^{-3} (in the left-hand beaker) to 1×10^{-6} (in the right-hand beaker). The hydrogen ion concentration produced from the HA dissociation was reduced by a factor of 1000 (it became 1000 times smaller after NaA was added to the HA solution).

PROBLEM SET 10

The problems in Problem Set 10 parallel the examples in Chapter 10. For example, if you should have trouble working Problem 5, go back to Example 5 in this chapter for help. The correct answers to these problems are given at the end of this book.

1. There is a number, N, that equals 10. Find pN.
2. Given that a number, W, equals 10 000, find pW.
3. A number, U, is equal to 32. Find pU.
4. A number, H, is equal to 0.000 01. Find pH.
5. A number, R, is equal to 0.000 32. Find pR.
6. If $\log X = 4$, what is the value of X?
7. If $pX = 4$, what is the value of X?
8. If $pD = 3$, find D.
9. If $pK = 5.6$, find K.
10. If $pS = 3.7$, find S.
11. A solution has a hydrogen ion concentration of 0.0001 mol of H^+ per liter of solution. Find the pH of this solution. Is this solution acidic, basic, or neutral?
12. A solution has a hydrogen ion molarity of 0.000 000 1. Find the pH of this solution and indicate whether the solution is acidic, basic, or neutral.
13. A solution has a hydrogen ion molarity equal to 0.0002. Find the pH of this solution. Is the solution acidic, basic, or neutral?
14. Find the pH of a solution whose hydrogen ion concentration is 6.4×10^{-8} mol of hydrogen ion per liter of solution. Is the solution acidic, basic, or neutral?
15. Find the hydrogen ion molarity of a solution that has a pH equal to 9.
16. Find the hydrogen ion concentration of a solution that has a pH of 8.3.
17. Find the hydrogen ion concentration of a solution with a pH of 6.9.
18. A solution has a pH equal to 3.35. Find the molarity of the hydrogen ion dissolved in this solution.
19. The equilibrium constant for a reaction is found to be 0.000 000 002. What is the pK of this reaction?
20. The hydroxide ion molarity of a solution is found to be 0.000 050. What is the pOH of this solution?
21. The pK for a hypothetical reaction is equal to 6. Find the equilibrium constant for this reaction.
22. The pOH of a solution is 5.7. What is the hydroxide molarity of this solution?
23. Find the pH of a 0.01 M HCl solution.
24. Find the pH of a 0.001 M HNO_3 solution.
25. Find the pH of a 0.2 M $HClO_3$ solution.
26. Find the equilibrium constant K of a weak acid, HA, if 1 L of a 0.1 M HA solution has a pH of 5.0. The equilibrium is represented by

$$HA \rightleftharpoons H^+ + A^-$$

27. Find the pK for the dissociation of HA in Problem 26.
28. What is the percentage of dissociation for the acid HA in Problem 26?

29. Is the value for the percentage of dissociation of HA that you obtained in Problem 28 a reasonable value? Why?
30. Find the equilibrium constant K of a weak acid, HB, if 1 L of 0.1 M HB has a pH of 2. The equilibrium is represented by

$$HB \rightleftharpoons H^+ + B^-$$

31. Find the pK for the dissociation of HB in Problem 30.
32. Find the percentage of dissociation of the weak acid HB discussed in Problem 30.
33. Find the equilibrium constant K of a weak acid, HC, if 1 L of a 0.20 M HC solution has a pH of 4.4.
34. Find the pK for the dissociation of the acid HC discussed in Problem 33.
35. Find the percentage of dissociation of HC discussed in Problem 33.
36. Find the hydrogen ion concentration of a 0.1 M HA solution, in which HA has an equilibrium constant equal to 1×10^{-3}. The equilibrium equation showing the dissociation of this acid is

$$HA \rightleftharpoons H^+ + A^-$$

37. Find the pH of a 0.1 M HB solution. HB is an acid with an equilibrium constant equal to 1×10^{-5}. The dissociation of HB is represented by

$$HB \rightleftharpoons H^+ + B^-$$

38. Find the percentage of dissociation of HB that occurs in Problem 37.
39. Find the pH of a 1.0 M HC solution. HC is an acid with an equilibrium dissociation constant equal to 1×10^{-2}. The dissociation of HC is represented by

$$HC \rightleftharpoons H^+ + C^-$$

40. Find the percentage of dissociation of HC that occurs in Problem 39.
41. Find the hydrogen ion concentration of a 0.1 M HA solution that also contains 1 M NaA. The equilibrium constant for HA is 10^{-7}.
42. Find the pH of a solution that is 0.1 M HB and 0.1 M NaB. The equilibrium constant for HB is 1×10^{-5}.
43. Find the pH of a solution that is 1 M HC and 1 M NaC. The equilibrium constant for HC is 10^{-3}.

11

Equilibrium Calculations

Part II: A Review

11.1 INTRODUCTION

In this chapter we will review what you have learned about equilibrium. If you feel that you have mastered (or at least feel somewhat comfortable with) *p*H, percentage of ionization, equilibrium calculations, Le Chatelier's principle, and the common ion effect (many students never feel comfortable with the common ion effect), then this chapter will benefit you by tying together these ideas and reviewing some of the problem-solving techniques that you may have found difficult.

11.2 EFFECTS OF HCl AND ASPIRIN ON THE STOMACH*

11.2.1 STOMACH ACID, PART I

On June 6, 1822, Alexis St. Martin accidentally shot himself in the stomach while cleaning his shotgun in Fort Mackinac, Michigan. Dr. William Beaumont was the attending physician who stitched the wound. The edges of the wound eventually healed but did not knit. A hole remained in St. Martin's stomach, covered by a flap of skin that could be easily lifted for a firsthand view of what was happening inside a human stomach. Dr. Beaumont attached a piece of meat to the end of a string and lowered it through the hole in St. Martin's stomach. He was able to observe the churning, kneading, and oozing of juices that followed. The juices were removed by suction through a rubber tube and found to be, among other things, 0.1 M HCl.

What is the p*H of this solution?*

The *p*H of your stomach juices (0.1 M HCl) is 1.0. HCl is a strong acid, so it is, for all practical purposes, fully dissociated (ionized). Therefore, a solution that is 0.1 M HCl contains no HCl molecules because each of the HCl molecules has ionized (broken up) into H^+ ions and Cl^- ions.

	HCl	**H^+**	**Cl^-**
initial concentrations:	0.1 M	0.0 M	0.0 M
final concentrations:	0.0 M	0.1 M	0.1 M

If this is unclear, read Section 10.3.1, "Marriage Analogy," where this problem is worked out in greater detail.

The problem now is to find the *p*H of a solution whose hydrogen ion concentration is 0.1 m.

$$[H^+] = 0.1 = 1 \times 10^{-1}$$

$$pH = -\log[H^+] = -\log(1 \times 10^{-1})$$

$$= -(\log 1 + \log 10^{-1}) = -(0 - 1) = -(-1) = +1$$

11.2.2 TWO-PART MYSTERY

Some people are bothered when they realize that stomach acid is HCl because hydrochloric acid is a strong acid and capable of dissolving a wide variety of metals, including zinc and iron. HCl can dissolve all of the metals listed above hydrogen in the electrochemical series. Hydrochloric

*Adapted from Ronald A. DeLorenzo, "The Effects of HCl and Aspirin on the Stomach: An Equilibrium Review," *Journal of Chemical Education* 54 (May 1977):306.

acid helps digest food proteins, but not the proteins in the stomach's own cells. Why doesn't the hydrochloric acid attack the stomach? This is the first part (and the easiest part to solve) of our two-part mystery.

We know that aspirin (acetylsalicylic acid) is a weak acid with an equilibrium constant of approximately 1×10^{-4} (see Figure 11.1). Yet, when most people take two aspirin, they lose 1 to 2 mL of blood from the stomach wall. Some people lose $\frac{1}{2}$ qt or more of blood and become anemic. The second part of our mystery is to explain why a weak acid like aspirin can attack your stomach.

FIGURE 11.1 Aspirin (acetylsalicylic acid)

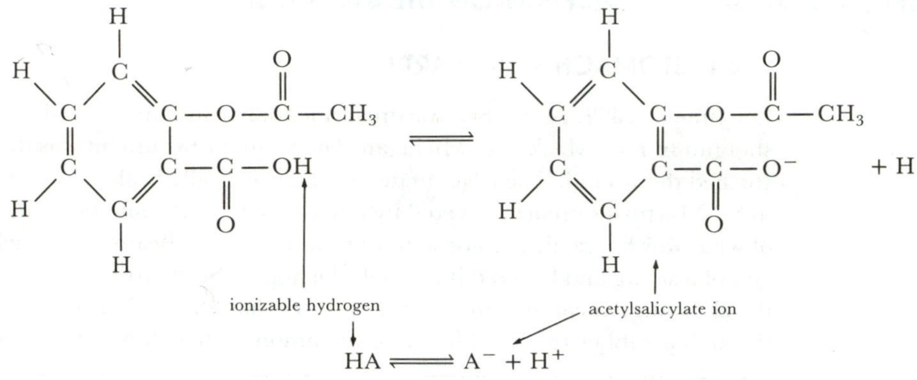

By the way, how would you know if you did lose a significant amount of blood after taking aspirin?

Your stools would turn black if you had a significant amount of internal bleeding. Simply feeling run down is a symptom of several other problems, as well as of anemia.

The full question (mystery) before us is this: Why does aspirin, a weak acid, dissolve (attack, harm) your stomach, while hydrochloric acid, a very strong acid, does not dissolve your stomach?

Dr. Horace W. Davenport of the University of Michigan Medical School has demonstrated that the stomach is protected from hydrochloric acid by thin outer membrane cells containing fat (lipids). Hydrochloric acid of a known concentration was placed inside a dog's stomach, and the *p*H of this acid solution was measured. The *p*H meter reading agreed with the calculated *p*H value based on the concentration of the known acid. In other words, if the hydrochloric acid concentration was 0.1 M, then the *p*H meter should have read *p*H = 1.0, which it did.

The dog's stomach was then washed with detergent (detergents dissolve fats and grease). After the dog's stomach was washed, the experiment was repeated; that is, hydrochloric acid of a known concentration was placed inside the washed stomach and the *p*H of this acid solution was measured. This time the acid attacked the unprotected stomach, and the *p*H meter reading changed. Let's assume that 0.1 M HCl was added to the washed stomach.

Would you expect the pH *meter reading to be higher or lower than 1.0?*

If you thought that the *p*H meter reading would be less than 1.0, you probably made a common error in your reasoning. If the *p*H meter reading dropped, that would mean that the solution was becoming more acidic. The protective lipid layer had been removed with detergent, so the hydrochloric acid attacked the stomach (the hydrochloric acid was used up). A decrease in the hydrogen ion concentration caused the pH of the solution to increase. A solution whose hydrogen ion concentration is 1×10^{-1} has a pH of 1.0. A weaker acidic solution with a hydrogen ion concentration of 1×10^{-2} has a pH of 2.0.

This discussion of the dissolving power of detergents on grease focuses on an important principle of chemistry. Some people state the principle in this way: "Like dissolves like." This means that polar solvents dissolve polar solutes. For example, water is a polar solvent, and it dissolves sodium chloride, which is a polar solute. Similarly, nonpolar solvents dissolve nonpolar solutes. For example, carbon tetrachloride (CCl_4) is a nonpolar solvent, and it easily dissolves iodine, a nonpolar solute.

What factors cause chemical bonds to be polar?

Polar bonds are the result of unequal sharing of electrons between bonded atoms. You should realize that a molecule may have polar bonds and still be a nonpolar molecule. In carbon tetrachloride, carbon and chlorine have different electronegativities and don't share electrons equally.

Carbon tetrachloride has polar bonds, but it is a nonpolar molecule. Why?

The polar bonds in carbon tetrachloride are geometrically arranged in such a manner as to cancel each other, and the CCl_4 molecule does not act as a dipole.

Aspirin can exist in both an essentially nonpolar form (HA) or an ionized form (H^+ and A^-), where H^+ represents the dissociated hydrogen ion, and A^-, the dissociated anion (acetylsalicylate ion). The ionization of aspirin is shown in Figure 11.1.

The lipid (fat) barrier lining the stomach is nonpolar. This nonpolar lipid layer will not dissolve ionic substances, such as hydrogen ions, from dissociated hydrochloric acid. This answers the first part of our mystery. At this point you may think you have the complete mystery solved, but it gets a little tricky, so hang on for a while.

Do you think that aspirin, HA, will exist in an ionized form ($H^+ + A^-$) or an un-ionized form (HA) when an aspirin tablet is placed into a quart (liter) of water?

You probably guessed that the aspirin will remain in the un-ionized form HA because of its very small ionization constant (1×10^{-4}). However, this is not the correct answer. It turns out that about two-thirds of the aspirin will be ionized. Weak electrolytes can dissociate significantly in very dilute solutions. Even "insoluble" salts will dissolve (if enough water is added) to produce very dilute solutions. Why? There are two ways to look at this: using arm-waving arguments (qualitatively) and using calculations (quantitatively).

11.2.3 QUALITATIVE ARGUMENTS: PERCENTAGE OF DISSOCIATION IN DILUTE SOLUTIONS, PART I

Consider the equilibrium

$$HA \rightleftharpoons H^+ + A^-$$

If you add water to this equilibrium, you will be diluting the hydrogen ion (H^+) concentration as well as the anion (A^-) concentration. You will not be diluting the HA (aspirin) concentration. Why?

HA exists as an undissolved solid, sitting at the bottom of the solution. The concentration of a solid is a constant. Ten grams of solid HA has the same concentration as 10 tons of HA. Both are 100% pure; neither is more concentrated than the other.

According to Le Chatelier's principle, if you decrease the concentration of the hydrogen ion or the anion (and we have done both by adding water), the equilibrium will shift to the right to increase the concentrations of H^+ and A^-. The only way for our equilibrium to shift to the right is for more HA to ionize (dissolve). As a result, in very dilute solutions, even aspirin with an equilibrium constant of 1×10^{-4} can be extensively dissociated.

11.2.4 QUANTITATIVE ARGUMENTS: PERCENTAGE OF DISSOCIATION OF ASPIRIN

The problems in this section and the next will help you understand that a weak electrolyte (an electrolyte with a small *K* value) that does not ionize extensively at 1 M concentrations does ionize significantly when diluted.

Find the percentage of ionization of a 1 M solution of a weak acid, HA, whose equilibrium constant K *is* 1×10^{-4}.

The following calculations are a repetition of ideas detailed in the marriage analogy sections (see Chapter 10), so we'll just show the calculations without too much discussion:

$$HA \rightleftharpoons H^+ + A^-$$

$$K = \frac{[H^+][A^-]}{[HA]}$$

Let x equal the equilibrium hydrogen ion concentration.

$$10^{-4} = \frac{(x)(x)}{(1-x)} = \frac{(x)(x)}{(1)} = \frac{x^2}{1} = x^2$$

$$x^2 = 10^{-4}$$

$$x = 10^{-2} = [H^+]\text{equilibrium}$$

$$\text{percentage ionization} = \frac{[H^+]\text{equilibrium}}{[HA]\text{ initial}} \times 100\% = \frac{10^{-2}}{1} \times 100\%$$

$$= \frac{(10^{-2})(10^{2})\%}{1} = \frac{1\%}{1} = 1\%$$

All that we have shown here is what you intuitively knew; that is, because the *K* for aspirin is very small, very little of the aspirin dissociates.

11.2.5 QUALITATIVE ARGUMENTS: PERCENTAGE OF DISSOCIATION IN DILUTE SOLUTIONS, PART II

When you swallow two aspirin, there are complications. The first complication is that the aspirin is very dilute in your stomach. The human stomach has a volume of approximately 1 qt (liter). Aspirin has a molecular weight of 180 amu (180 g/mol aspirin), and the weight of an aspirin tablet (including the weight of the filler material) is approximately 0.35 g. Thus the concentration of aspirin in your stomach is very small. The following problem will complete our quantitative argument:

Find the percentage of ionization of a 1×10^{-4} *M solution of HA given that the equilibrium constant of HA is* 1×10^{-4}.

The percentage of ionization is 62%. Here's how we calculated that value:

$$HA \rightleftharpoons H^+ + A^-$$

$$K = \frac{[H^+][A^-]}{[HA]}$$

Let x equal the hydrogen ion concentration at equilibrium.

$$10^{-4} = \frac{(x)(x)}{(10^{-4} - x)}$$

We can't make the assumption that $(10^{-4} - x)$ is equal to 10^{-4} because x is fairly large relative to 10^{-4}. We will end up with a quadratic equation and use the quadratic formula to solve for x, as shown:

$$(10^{-4})(10^{-4} - x) = (x)(x) = x^2$$

$$10^{-8} - 10^{-4}x = x^2$$

$$x^2 + 10^{-4}x - 10^{-8} = 0$$

$$x = \frac{-b \pm \sqrt{b^2 - 4ac}}{2a} \quad \text{(quadratic formula)}$$

Substituting $a = 1$, $b = 10^{-4}$, $c = -10^{-8}$, we have

$$x = \frac{-10^{-4} \pm \sqrt{(10^{-4})^2 - (4)(1)(-10^{-8})}}{(2)(1)}$$

$$= \frac{-10^{-4} \pm \sqrt{10^{-8} + (4 \times 10^{-8})}}{2} = \frac{-10^{-4} \pm \sqrt{5 \times 10^{-8}}}{2}$$

$$= \frac{-10^{-4} \pm (2.24 \times 10^{-4})}{2}$$

Since x cannot be negative (you can't have a negative concentration),

$$x = \frac{-10^{-4} + (2.24 \times 10^{-4})}{2} = 0.62 \times 10^{-4} = 6.2 \times 10^{-5}$$

Since we let x be the hydrogen ion concentration at equilibrium,

$$[H^+] = x = 6.2 \times 10^{-5}$$

$$\text{percentage ionization} = \frac{[H^+]\text{ equilibrium}}{[HA]\text{ initial}} \times 100\%$$

$$= \frac{6.2 \times 10^{-5}}{1 \times 10^{-4}} \times 100\% = 62\%$$

Here's a tricky question. Keeping in mind all that we have done, what do you think happens when you take an aspirin? Does the aspirin sit in your stomach in an un-ionized form, or does the aspirin tend to dissociate (ionize) significantly?

The aspirin doesn't ionize significantly. You may think that because the aspirin is in a very dilute form in your stomach, it dissociates considerably as the previous discussions and calculations indicated. We'll see how those arguments and calculations can be used later. (They weren't a waste of time.) But first let's see why aspirin doesn't dissociate significantly in your stomach even though the aspirin is very dilute.

11.2.6 COMMON ION EFFECT WITH ASPIRIN

What is present in the digestive juices in your stomach that has an ion in common with aspirin (HA)?

That's right, hydrochloric acid. Your stomach is 0.1 M HCl and has a *p*H of 1.0, which is pretty acidic.

Would the presence of HCl in the stomach affect the aspirin equilibrium as shown in Figure 11.1?

Yes, it would. The high concentration of the hydrogen ion from the hydrochloric acid would cause the equilibrium in Figure 11.1 to shift to the left.

$$\xleftarrow{\text{adding } H^+ \text{ shifts equilibrium to the left}}$$
$$HA \rightleftharpoons H^+ + A^-$$

Le Chatelier's principle tells us that if we increase the hydrogen ion concentration, the equilibrium shifts to the left to reduce the hydrogen ion concentration. We can also demonstrate this with another sample problem that is a good review of the common ion effect.

Calculate the percentage of dissociation of a 1 M HA solution that is also 1 M HCl.

This is a very difficult problem for most people because it involves the common ion effect. The correct answer is 0.01%. Again, we are just going to outline the steps for the solution to this problem because this material is detailed in Section 10.5.1, "Marriage Analogy."

$$HCl \xrightarrow{100\%} H^+ + Cl^-$$

$$HA \rightleftharpoons H^+ + A^-$$

$$K = \frac{[H^+][A^-]}{[HA]}$$

Let x equal the anion (A^-) concentration at equilibrium.

$$10^{-4} = \frac{(x+1)(x)}{(1-x)}$$

The hydrogen ion concentration is $(1 + x)$ because hydrogen ion is supplied to this solution by two sources, HCl and HA. The HA molecules supply x mol of hydrogen ion per liter of solution, and the HCl, which fully dissociates, supplies 1 mol of hydrogen ion per liter of solution. The sum of the hydrogen ion concentrations from these two sources is $(1 + x)$. The value of x is very small with respect to 1.0, so we can say $(1 + x) = 1$ and $(1 - x) = 1$ (approximately).

$$10^{-4} = \frac{(x+1)(x)}{(1-x)} = \frac{(1)(x)}{(1)} = x = [A^-]_{\text{equilibrium}}$$

$$\text{percentage ionization} = \frac{[A^-]_{\text{equilibrium}}}{[HA]_{\text{initial}}} \times 100\% = \frac{10^{-4}}{1} \times 100\% = 0.01\%$$

Let's briefly review what we have shown. Aspirin is a very weak acid (electrolyte), which means that it doesn't dissociate (ionize) very much. We have shown, however, that aspirin will dissociate significantly in a dilute solution. We have also shown that if we add hydrogen ions to a dilute aspirin solution, the added hydrogen ion drives the dissociation reaction backward. Even a dilute solution of aspirin (HA) will not significantly dissociate if there is a high concentration of hydrogen ion present.

You should understand that when you take aspirin, the high concentration of hydrogen ion in your stomach keeps most of the aspirin in the molecular form (HA) as opposed to the ionized form (H^+ and A^-).

Can you figure out why aspirin is able to attack the stomach and cause bleeding?

In the un-ionized (relatively nonpolar*) form, aspirin can dissolve in the lipid layer (also nonpolar). Like dissolves like.† But why would this cause bleeding? For an acid to act like an acid, it must be significantly ionized. What would cause the aspirin to act like a strong acid (significantly dissociate) after the HA molecules pass through the lipid barrier?

Once the aspirin has passed through the protective membrane, it is no longer in an acidic medium because the *p*H of blood is 7.4, slightly alkaline and almost neutral. Thus, the aspirin molecules are free to ionize. Once the aspirin has passed through the lipid barrier, it is no longer in contact with the HCl inside the stomach. (Remember: It was the HCl that kept HA from dissociating.) The aspirin is a weak acid, but now it is present in a very dilute concentration (which causes it to ionize extensively). Although aspirin has a small equilibrium constant, a dilute solution of it will dissociate to a large extent. That's what we tried to show earlier with our qualitative and quantitative arguments.

In its ionized state, the aspirin is unable to pass back through the lipid barrier. Why?

That's right, you're getting the hang of this now. The aspirin is in a polar form and will not be dissolved by the nonpolar lipid barrier. The ionized aspirin attacks the stomach and causes bleeding. Hydrogen ion from the stomach's hydrochloric acid is also able to pass through the aspirin-destroyed lipid barrier, causing further damage. Figures 11.2, 11.3, and 11.4 illustrate the effect of aspirin on the stomach. Figure 11.2 shows normal stomach lining cells magnified 350 times. Figure 11.3 shows a portion of Figure 11.2 magnified 8000 times. Figure 11.4 shows what happens when aspirin gets past the "fat barrier."

FIGURE 11.2 Normal stomach lining cells (350 ×). (Photograph courtesy Jeanne M. Riddle, Henry Ford Hospital, Detroit.)

*Aspirin is slightly polar when un-ionized, but un-ionized aspirin is relatively nonpolar with respect to ionized aspirin.
†Absorption of drugs and food by the stomach is very complicated and cannot always be explained by "like dissolves like."

FIGURE 11.3 Section of Figure 11.2 (8000 ×). (Photograph courtesy Jeanne M. Riddle, Henry Ford Hospital, Detroit.)

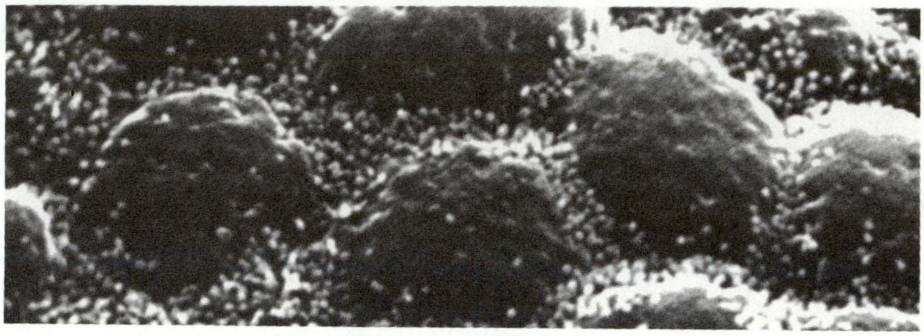

FIGURE 11.4 Damaged stomach cells (8000 ×). (Photograph courtesy Jeanne M. Riddle, Henry Ford Hospital, Detroit.)

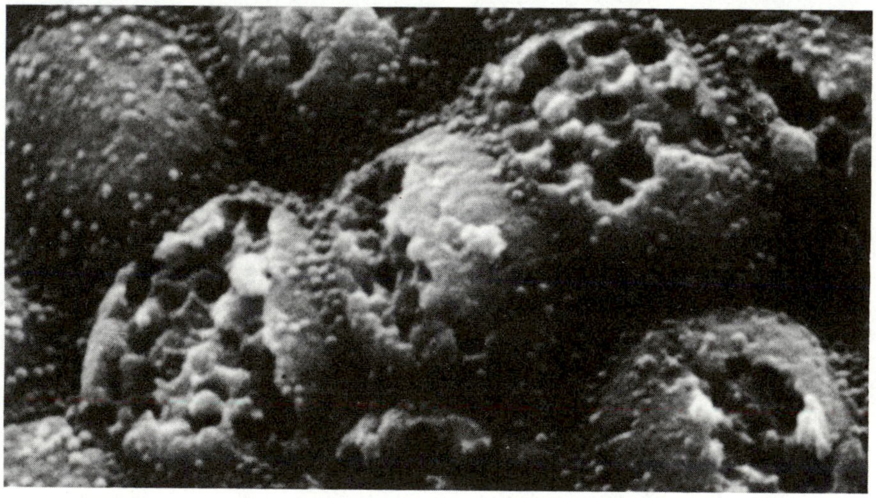

11.2.7 ASPIRIN BEFORE 1900

You might think, from the previous discussion, that aspirin is very dangerous and should be avoided. However, salicylic acid, the compound from which aspirin is synthesized (see Figure 11.5), has been used for many thousands of years, perhaps even millions of years. No, there wasn't a handy neighborhood pharmacy dispensing the drug, but primitive people did chew the leaves and bark of the willow tree to reduce their fevers. And, during the Middle Ages (A.D. 400–1400), the willow leaf and the willow roots were used to prepare potions to relieve pain. In 1898 Friedrich Bayer extracted and purified salicylic acid from the willow tree. In 1889 Felix Hoffmann first synthesized aspirin (see Figure 11.5), and by the early 1900's, acetylsalicylic acid (aspirin) was put in tablet form and sold as aspirin.

FIGURE 11.5 Molelcular structures of salicylic acid and aspirin

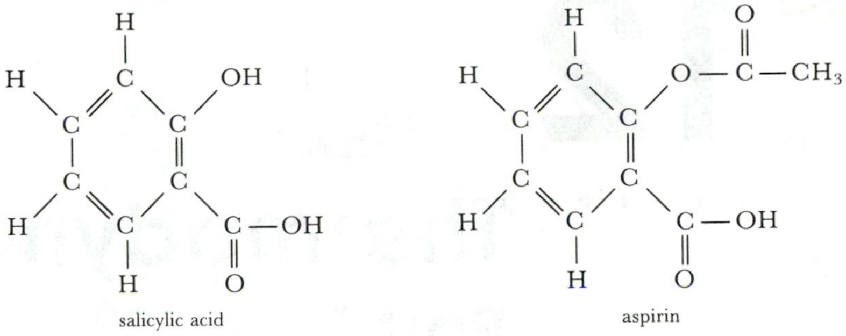

11.2.8 ASPIRIN TODAY

Before discussing some beneficial uses of aspirin and one common misuse, a word of caution: Some people are allergic or otherwise sensitive to aspirin (e.g., prone to excessive stomach bleeding) and should not use it. Also, the Food and Drug Administration (FDA) warns us that for those younger than 20, the use of aspirin may cause Reye's syndrome, a potentially fatal disease.

On the positive side, millions regularly and safely use aspirin, under their doctor's supervision, to reduce the chance of heart attack or colon cancer. Also, aspirin is relatively harmless when healthy people occasionally use it to relieve their aches and pains. However, the common use of aspirin to bring down fevers caused by the flu is not usually wise. A fever results when the body raises its temperature to combat an infection. A virus can only reproduce within a narrow temperature range. Using aspirin to lower a fever is counterproductive: it deprives the body of one of its important lines of defense. As a result, reducing a fever helps to prolong the life span of a viral infection. For children, this can be particularly dangerous. In fact, the American Academy of Pediatrics warned that the practice of reducing fevers in children is killing an average of one child per day in the United States. The Academy's advice is that, under normal circumstances, parents really don't have to jump on a fever with any drug.

There are abnormal circumstances in which we must reduce fevers. For example, since your heart rate increases eight to ten beats per minute for each degree of temperature increase above normal, heart patients with high fevers may suffer serious heart strain. Also, since fever increases oxygen demand, people with respiratory ailments also may be threatened by high fevers.

11.2.9 OTHER PAIN RELIEVERS

Some people buy non-aspirin pain relievers, such as Tylenol and Advil, to avoid the possible stomach damage associated with aspirin. However, as we just discussed, it is usually unwise to reduce fevers with any drug. Also, although these pain relievers may cause less stomach damage, the ones containing acetaminophen (Tylenol, Anacin-3), when taken at least once a week for a year or more, increase by 50 percent the likelihood of developing kidney diseases. Pain relievers containing ibuprofen (Advil, Nuprin) cause kidney failure in about 1 percent of the people who use them.

12

Thermodynamics Part 1

12.1 INTRODUCTION

Whenever any chemical reaction occurs, energy effects also occur (e.g., heat is given off, light is produced, electricity is generated). Energy effects accompany all chemical reactions. Thermochemistry (also called chemical thermodynamics) is the study of the energy effects that accompany chemical reactions. As innocent as this definition may sound, it is common to hear chemistry professors tell students not to expect to fully understand thermodynamics after only one year of study.

Solving thermodynamics problems at the freshman chemistry level can be relatively easy (e.g., relative to solving equilibrium problems). However, thermodynamic concepts are more difficult to grasp and comprehend than most of the material in this book because thermodynamics is based on mathematical models that are usually beyond the math background of freshmen students. For this reason, usually only the results of the mathematical models are presented to students. Students must memorize these results somewhat blindly (or, at least, not with full comprehension). Because this memorization is necessary, many professors question whether thermodynamics should be taught in first-year chemistry courses.

Other professors feel that thermodynamics offers valuable insights into chemistry, life, and the nature of our universe. Such insights are very interesting and educational.

To a chemist, one important function of thermodynamics is to predict which chemical reactions occur spontaneously (on their own). The subject of spontaneity can be extended to other areas, such as the beginning of the universe (was there a beginning?), evolution (could life begin on its own?), and stars (why do stars form?). Are these events spontaneous? Thermodynamic arguments have been used by religious leaders and in courts of law. St. Thomas Aquinas used thermodynamic-related concepts in one of his "proofs" of the existence of a creator. Recent court and congressional proceedings in California and Georgia have employed thermodynamics in their discussions of the merits of teaching evolution.

Thermodynamics is a natural extension of our equilibrium studies. The thermodynamic functions that we will discuss are different manifestations of the equilibrium constant K. We will restrict our discussion of thermodynamics to just three thermodynamic functions: enthalpy, entropy, and Gibbs free energy. We will limit thermodynamics problems to those that will expand your ability to interpret chemical equations. In these problems we will ask: Is a chemical reaction spontaneous, does a chemical reaction produce heat, and are the products of a reaction more disordered than the reactants?

12.2 ENTHALPY

Every substance (carbon, oxygen, carbon dioxide, and so forth) can be thought of as containing, in some way, a certain amount of heat or energy. How can substances (molecules and atoms) contain (hold) energy? If energy (e.g., heat) is added to a molecule, it can absorb (hold) the energy by vibrating, rotating, and translating (moving from one point to another point) more quickly.

There is a thermodynamic function called enthalpy that can be thought of as a measure of the amount of heat a substance contains under conditions of constant temperature and pressure. We will think of the word *enthalpy* and the phrase *heat content* as meaning the same thing. Every substance contains a certain amount of heat (enthalpy).

12.2.1 CUPS OF COFFEE ANALOGY

Assume that you have three coffee cups that are of different sizes. The first cup holds 250 mL of hot coffee, the second cup holds 50 mL of hot coffee, and the third cup is empty. The third cup can only hold a maximum of 200 mL of hot coffee. We want to pour the contents of the first two cups into the third cup.

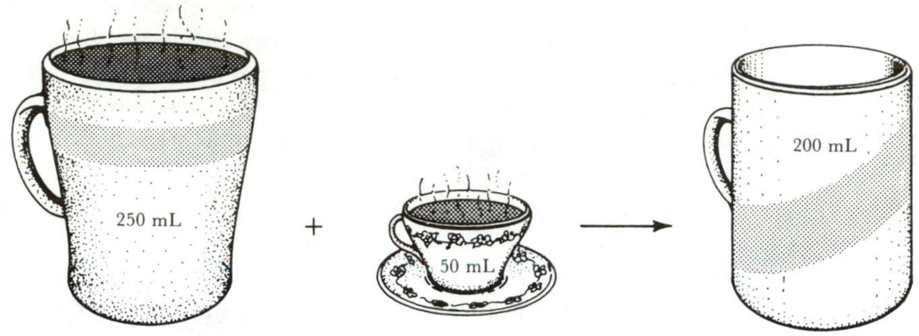

When the contents of the first two cups (a total volume of 300 mL of hot coffee) are poured into the third cup (which has a volume of 200 mL), how much coffee will overflow the third cup and spill out into the surroundings?

One hundred milliliters of hot coffee will overflow the sides of the third cup and spill onto the surrounding area (300 mL – 200 mL = 100 mL). This happens because the volume (for hot coffee) of the third cup is not as great as the total volume of coffee in the first two cups.

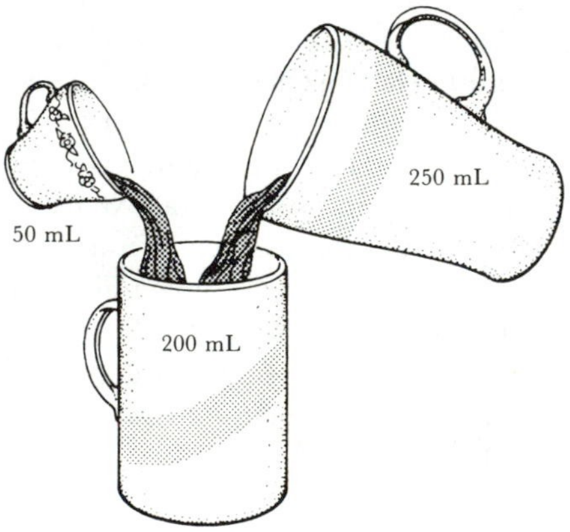

12.2.2 ENTHALPY CHANGE, HYPOTHETICAL REACTIONS

Now consider the hypothetical reaction A + B → C. The molecular weights of A, B, and C are 10 amu, 20 amu, and 30 amu, respectively. Each of the substances (A, B, and C) in this reaction can be thought of as being able to hold only a certain amount of energy or heat. This amount is called its enthalpy (heat content). The amount of heat that 1 mol (10 g) of A can contain is 250 kcal. We say that the molar heat content (the molar enthalpy) of A is 250 kcal per mole of A (250 kcal/mol A). If 1 mol of B (20 g) can hold 50 kcal, then we say that the molar enthalpy (heat content) of B is 50 kcal per mole of B (50 kcal/mol B). The heat content (enthalpy) of 1 mol (30 g) of C is 200 kcal. The total enthalpy (heat content) of A plus B is 300 kcal. When A and B react chemically to produce C, which has a heat content (enthalpy) of only 200 kcal/mol, some of the heat must spill over into the surroundings:

$$\overset{250\text{ kcal/mol}}{\text{A}} + \overset{50\text{ kcal/mol}}{\text{B}} \rightarrow \overset{200\text{ kcal/mol}}{\text{C}} + 100\text{ kcal}$$

The amount of heat that overflows the heat content of C is referred to as the heat of the reaction when pressure is kept constant during the reaction ($P_{\text{initial}} = P_{\text{final}}$).

How much heat is released when 1 mol of A reacts with 1 mol of B to produce 1 mol of C under conditions of constant pressure?

One hundred kilocalories of heat (300 kcal − 200 kcal = 100 kcal) is released into the atmosphere (surroundings) when this reaction occurs at constant pressure ($P_{\text{initial}} = P_{\text{final}}$).

When there is a difference between the sum of the enthalpies (heat contents) of the products and the sum of the enthalpies of the reactants, it is called the enthalpy change of the reaction. The enthalpy change for the preceding reaction is −100 kcal. The enthalpy change is negative when heat is produced by a chemical reaction because enthalpy changes are calculated by the formula

Δ means "change" Σ means "sum" or "total"

$$\Delta H = \Sigma H_{\text{products}} - \Sigma H_{\text{reactants}}$$

enthalpy change for a chemical reaction; total enthalpy (heat content) of all the products; total enthalpy (heat content) of all the reactants

EXAMPLE 1 One mole of A reacts with 1 mol of B to produce 1 mol of C.

$$\text{A} + \text{B} \rightarrow \text{C}$$

The enthalpies of A, B, and C are 250 kcal/mol, 50 kcal/mol, and 200 kcal/mol, respectively. Find the enthalpy change for this reaction. Is the reaction exothermic (does the reaction produce heat)?

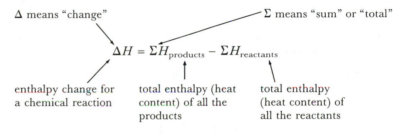

Solution

$$\text{A} + \text{B} \rightarrow \text{C}$$

$$\Delta H = \Sigma H_{\text{products}} - \Sigma H_{\text{reactants}}$$

$$= \left[(1\text{ mol C})\underbrace{\left(\frac{200\text{ kcal}}{\text{mol C}}\right)}_{\text{molar enthalpy of C}}\right] - \left[(1\text{ mol A})\underbrace{\left(\frac{250\text{ kcal}}{\text{mol A}}\right)}_{\text{molar enthalpy of A}} + (1\text{ mol B})\underbrace{\left(\frac{50\text{ kcal}}{\text{mol B}}\right)}_{\text{molar enthalpy of B}}\right]$$

$$= (200\text{ kcal}) - (250\text{ kcal} + 50\text{ kcal}) = 200\text{ kcal} - 300\text{ kcal} = -100\text{ kcal}$$

The enthalpy change for this reaction is −100 kcal. Heat is released into the surroundings when this reaction occurs because the heat content of the product formed is not as great as the combined heat contents of the two reactants. Whenever the enthalpy change for a chemical reaction is negative, the reaction is exothermic (heat is released by the reaction into the surroundings). Endothermic (heat-absorbing) reactions have positive enthalpy changes.

•
•

EXAMPLE 2 Determine the enthalpy change for the hypothetical reaction

$$2\text{E} + \text{G} \rightarrow 3\text{X}$$

The enthalpies (heat contents) for E, G, and X are 100 kcal/mol, 200 kcal/mol, and 300 kcal/mol, respectively. Is this reaction exothermic or endothermic?

Solution

$$\Delta H = \Sigma H_{\text{products}} - \Sigma H_{\text{reactants}}$$

$$= \left[(3\text{ mol X})\left(\frac{300\text{ kcal}}{\text{mol X}}\right)\right] - \left[(2\text{ mol E})\left(\frac{100\text{ kcal}}{\text{mol E}}\right) + (1\text{ mol G})\left(\frac{200\text{ kcal}}{\text{mol G}}\right)\right]$$

$$= 900\text{ kcal} - (200\text{ kcal} + 200\text{ kcal}) = 900\text{ kcal} - 400\text{ kcal} = 500\text{ kcal}$$

This reaction is endothermic (absorbs heat) because the enthalpy change is positive (+500 kcal). This endothermic reaction can be represented by

$$500\text{ kcal} + 2\text{E} + \text{G} \rightarrow 3\text{X}$$

or

$$2\text{E} + \text{G} \rightarrow 3\text{X} - 500\text{ kcal}$$

or

$$2\text{E} + \text{G} \rightarrow 3\text{X},\ \Delta H = +500\text{ kcal}$$

Some students understand this solution more clearly when the enthalpies of each substance are written over the chemical equation.

$$\begin{array}{ccccc} 200\text{ kcal} & & 200\text{ kcal} & & 900\text{ kcal} \\ 2\text{E} & + & \text{G} & \rightarrow & 3\text{X} \end{array}$$

$$(2\text{ mol E})\left(\frac{100\text{ kcal}}{\text{mol E}}\right) \qquad (3\text{ mol X})\left(\frac{300\text{ kcal}}{\text{mol X}}\right)$$

$$\Delta H = \Sigma H_{\text{products}} - \Sigma H_{\text{reactants}} = 900\text{ kcal} - 400\text{ kcal} = 500\text{ kcal}$$

•
•

EXAMPLE 3 Calculate the enthalpy change for the reaction

$$3\text{A} + 4\text{B} \rightarrow 2\text{C} + \text{D}$$

The heat contents of A, B, C, and D are 10 kJ/mol, 20 kJ/mol, 30 kJ/mol, and 40 kJ/mol, respectively. Is the reaction exothermic or endothermic?

Solution

$$\Delta H = \Sigma H_{\text{products}} - \Sigma H_{\text{reactants}}$$

$$= \left[(2\text{ mol C})\left(\frac{30\text{ kJ}}{\text{mol C}}\right) + (1\text{ mol D})\left(\frac{40\text{ kJ}}{\text{mol D}}\right)\right]$$

$$- \left[(3\text{ mol A})\left(\frac{10\text{ kJ}}{\text{mol A}}\right) + (4\text{ mol B})\left(\frac{20\text{ kJ}}{\text{mol B}}\right)\right]$$

$$= (60\text{ kJ} + 40\text{ kJ}) - (30\text{ kJ} + 80\text{ kJ}) = 100\text{ kJ} - 110\text{ kJ} = -10\text{ kJ}$$

This reaction is exothermic (produces heat) because the enthalpy change is negative. This exothermic reaction can be represented by

$$3A + 4B \rightarrow 2C + D + 10\ \text{kJ}$$

or

$$3A + 4B \rightarrow 2C + D,\ \Delta H = -10\ \text{kJ}$$

Writing the heat contents over the chemical equation, we have

$$\begin{array}{ccccccc} 30\ \text{kJ} & & 80\ \text{kJ} & & 60\ \text{kJ} & & 40\ \text{kJ} \\ 3A & + & 4B & \rightarrow & 2C & + & D \end{array}$$

$$\Delta H = \Sigma H_{\text{products}} - \Sigma H_{\text{reactants}} = 100\ \text{kJ} - 110\ \text{kJ} = -10\ \text{kJ}$$

EXAMPLE 4 The enthalpy change for the reaction

$$A + B \rightarrow E + 200\ \text{kcal}$$

is −200 kcal. The enthalpies of A and B are 250 kcal/mol and 500 kcal/mol, respectively. What is the molar enthalpy of E (i.e., what is the heat content of 1 mol of E)?

Solution

$$\Delta H = \Sigma H_{\text{products}} - \Sigma H_{\text{reactants}}$$

$$-200\ \text{kcal} = \Sigma H_{\text{products}} - \Sigma H_{\text{reactants}}$$

$$-200\ \text{kcal} = [(1\ \text{mol E})(H_E)]$$

(molar enthalpy of E)

$$- \left[(1\ \text{mol A})\left(\frac{250\ \text{kcal}}{\text{mol A}}\right) + (1\ \text{mol B})\left(\frac{500\ \text{kcal}}{\text{mol B}}\right)\right]$$

$$= (1\ \text{mol E})(H_E) - (250\ \text{kcal} + 500\ \text{kcal})$$

$$= (1\ \text{mol E})(H_E) - 750\ \text{kcal}$$

$$750\ \text{kcal} - 200\ \text{kcal} = (1\ \text{mol E})(H_E)$$

$$H_E = \frac{550\ \text{kcal}}{\text{mol E}} = 550\ \text{kcal/mol E}$$

EXAMPLE 5 Find the molar enthalpy (the heat content of 1 mol) of A. One mole of A reacts with 2 mol of B to produce 3 mol of C and 500 kcal of heat. The enthalpies of B and C are 200 kcal/mol and 400 kcal/mol, respectively.

Solution

$$A + 2B \rightarrow 3C + 500\ \text{kcal}$$

$$\Delta H = \Sigma H_{\text{products}} - \Sigma H_{\text{reactants}}$$

$$-500\ \text{kcal} = \Sigma H(\text{prod}) - \Sigma H(\text{react})$$

$$-500\ \text{kcal} = \left[(3\ \text{mol C})\left(\frac{400\ \text{kcal}}{\text{mol C}}\right)\right] - \left[(1\ \text{mol A})(H_A) + (2\ \text{mol B})\left(\frac{200\ \text{kcal}}{\text{mol B}}\right)\right]$$

(molar enthalpy of A)

$$= 1200\ \text{kcal} - [(1\ \text{mol A})(H_A) + 400\ \text{kcal}]$$

$$= 1200\ \text{kcal} - (1\ \text{mol A})(H_A) - 400\ \text{kcal}$$

$$= 800\ \text{kcal} - (1\ \text{mol A})(H_A)$$

$$-500 \text{ kcal} - 800 \text{ kcal} = -(1 \text{ mol A})(H_A)$$
$$-1300 \text{ kcal} = -(1 \text{ mol A})(H_A)$$
$$H_A = 1300 \text{ kcal/mol A}$$

EXAMPLE 6 One mole of A reacts with 1 mol of B to produce 2 mol of G and 400 kJ of heat. The molecular weights of A, B, and G are 10 amu, 20 amu, and 30 amu, respectively. How many kilojoules of heat are produced for every (per) 90 g of G that forms?

Solution

$$\text{A} + \text{B} \rightarrow 2\text{G} + 400 \text{ kJ}$$

This equation tells us that 400 kJ of heat is produced for every (per) 2 mol of G that forms.

$$(90 \text{ g G})\left(\frac{1.0 \text{ mol G}}{30 \text{ g G}}\right)\left(\frac{400 \text{ kJ heat produced}}{2.0 \text{ mol G forms}}\right) = 6.0 \times 10^2 \text{ kJ}$$

given — first conversion factor based on the molecular weight of G — second conversion factor obtained from the equation

For every 90 g of G that forms, about 600 kJ of heat is produced.

EXAMPLE 7 Calculate the number of kilocalories of heat produced for every gram of A that reacts if the enthalpies of A, B, and C are 100 kcal/mol, 200 kcal/mol, and 300 kcal/mol, respectively. The molecular weights of A, B, and C are 10 amu, 20 amu, and 30 amu, respectively. The reaction is represented by the chemical equation A + 2B → C.

Solution

$$\Delta H = \Sigma H_{\text{products}} - \Sigma H_{\text{reactants}}$$
$$= \left[(1 \text{ mol C})\left(\frac{300 \text{ kcal}}{\text{mol C}}\right)\right]$$
$$- \left[(1 \text{ mol A})\left(\frac{100 \text{ kcal}}{\text{mol A}}\right) + (2 \text{ mol B})\left(\frac{200 \text{ kcal}}{\text{mol B}}\right)\right]$$
$$= 300 \text{ kcal} - 500 \text{ kcal} = -200 \text{ kcal}$$

Thus,

$$\text{A} + 2\text{B} \rightarrow \text{C} + 200 \text{ kcal}$$

$$(1 \text{ g A reacts})\left(\frac{1.0 \text{ mol A}}{10 \text{ g A}}\right)\left(\frac{200 \text{ kcal evolved}}{1.0 \text{ mol A reacts}}\right) = 20 \text{ kcal}$$

Twenty kcal of heat is produced for every gram of A that reacts.

SELF-TEST

1. Two moles of A reacts with 1 mol of B to produce 1 mol of C. The enthalpies of A, B, and C are 300 kcal/mol, 50 kcal/mol, and 200 kcal/mol, respectively. Find the enthalpy change for this reaction. Is the reaction exothermic or endothermic?
2. Two moles of E reacts with 1 mol of G and 500 kcal of heat to produce 3 mol of X. The enthalpies of E and G are 100 kcal/mol and 200 kcal/mol, respectively. Find the molar enthalpy of X.
3. Calculate the number of kilojoules of heat produced for every gram of C that forms if the enthalpies of A, B, and C are 100 kJ/mol, 200 kJ/mol, and 300 kJ/mol, respectively. The mo-

lecular weights of A, B, and C are 10 amu, 20 amu, and 30 amu, respectively. The reaction is represented by the chemical equation $A + 2B \rightarrow C$.

ANSWERS **1.** −450 kcal; exothermic **2.** 300 kcal/mol X **3.** 6.67 kJ/g C

12.2.3 BODY TEMPERATURE ANALOGY

The average person in good health has a body temperature of 37°C (98.6°F). Metabolism efficiency decreases significantly at 40°C (104°F), and death can occur if the body temperature remains elevated by 3°C for a prolonged period of time.

Recall that the Celsius temperature scale was originally made by arbitrarily calling the freezing point of water 0°C and the boiling point of water 100°C. The space between these two points was marked with 100 equal divisions. When a temperature change equal to three of these divisions occurs ($\Delta T = 3°C$), we suspect a serious infection.

It would be just as valid to make another (imaginary) temperature scale by calling the freezing point of water 100°I (I stands for Imaginary) and the boiling point of water 200°I. The space between these two points can be marked with 100 equal divisions. Then the normal body temperature is 137°I, and illness is indicated by 140°I. It turns out that the actual temperature of a healthy person or a person who is seriously ill is not as important as the difference in temperatures between a healthy person and an ill person. A person is very ill when his or her body temperature increases by 3/100th (3%) of the temperature difference between the freezing and boiling points of water.

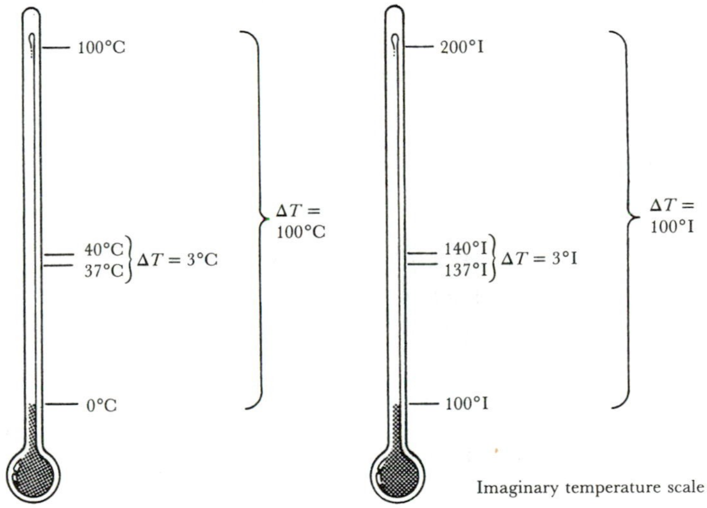

Celsius temperature scale

Imaginary temperature scale

And so it is with thermodynamic functions such as enthalpy (heat content). It would be very difficult, if not impossible, to determine the actual heat content of a substance (just as it is impossible to determine the actual temperature of boiling water). As it turns out, chemists aren't all that interested in actual heat contents; they are interested in differences between heat contents. We can make up heat contents of substances just as we can make up the boiling and freezing points of water. However, when making up heat contents, we must be careful that the difference in the heat content of products and reactants for a chemical reaction is consistent with the experimentally observed reaction heat.

Arbitrarily, all elements in their most stable form under conditions of 1 atm pressure and 25°C are said to have a standard molar enthalpy of formation (heat content) of exactly 0 kcal/mol. The symbol for this arbitrary standard molar enthalpy of formation (heat content) is

ΔH_f° — ° means "standard state conditions"

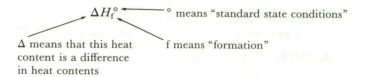

Δ means that this heat content is a difference in heat contents

f means "formation"

To show that the standard molar enthalpy of formation ΔH_f° of oxygen (gas) is equal to 0 kcal/mol O_2, we write

$$\Delta H^\circ_{f_{O_2(g)}} = 0 \text{ kcal/mol } O_2$$

Since we have defined a calorie as being exactly 4.184 J (see Section 3.4), the standard molar enthalpy of formation (heat content) for all elements in their most stable form is 0 kJ/mol. These values (in kilocalories per mole and kilojoules per mole) are shown in Table 12.1.

TABLE 12.1 Standard Molar Enthalpies of Formation, Standard Molar Gibbs Free Energies of Formation, and Absolute Standard Entropies (25°C, 1 atm)

	ΔH_f°		ΔG_f°		S°	
Substance	kcal/mol	kJ/mol	kcal/mol	kJ/mol	cal/mol K	J/mol K
C(graphite)	0	0	0	0	1.4	5.9
$CaCO_3(s)$	−288	−1204	−270	−1130	22	92
CaO(s)	−152	−636	−144	−602	10	42
$Cl_2(g)$	0	0	0	0	53	223
$CO_2(g)$	−94	−394	−94	−390	51	214
$H_2(g)$	0	0	0	0	31	131
HCl(g)	−22	−92	−23	−96	45	187
Hg(l)	0	0	0	0	18	75
HgO(s)	−22	−92	−14	−59	17	71
HI(g)	6.3	26	0.41	1.7	49	206
$H_2O(l)$	−68	−286	−57	−237	17	70
$H_2O(g)$	−58	−242	−55	−229	45	189
$H_2O_2(l)$	−45	−188	−29	−120	26	110
$I_2(s)$	0	0	0	0	28	117
$N_2(g)$	0	0	0	0	46	192
$N_2H_4(l)$	12	50	36	149	29	121
$O_2(g)$	0	0	0	0	49	205
Pb(s)	0	0	0	0	15	63
PbS(s)	−24	−100	−24	−99	22	92
S(s)	0	0	0	0	7.6	32
$SO_2(g)$	−71	−297	−72	−300	59	248

When assigning this arbitrary value of 0 kcal/mol (or 0 kJ/mol) to elements and when doing thermodynamics calculations, it is important always to specify

1. The elemental form (e.g., carbon can exist as both graphite and diamond at 1 atm and 25°C)
2. The state (phase) of the substance (e.g., H_2O can exist in both the liquid and gas state at 1 atm and 25°C)

Unless otherwise specified, always assume that the reactants and products in each problem in the thermodynamics chapters (Chapters 12 and 13) are at 25°C and 1 atm pressure. When a pure substance is at 25°C and 1 atm pressure, the substance is said to be in its standard state. The superscript ° in the symbol ΔH_f° indicates that the substance is in the standard state. The word *standard* in the name of the symbol (standard molar enthalpy of formation) refers to the standard state in which the substance exists.

From the arbitrary assignment of 0°C and 100°C for the freezing and boiling points of water, the relative temperatures of other objects (like human bodies) can be determined. Likewise, from the arbitrary assignment of 0 kcal/mol (or 0 kJ/mol) for the heat contents of elements in their most stable form under standard conditions, we can calculate the relative standard molar enthalpies of formation (heat contents) for compounds. The meaning of the delta (Δ) in the symbol ΔH_f° will be discussed in the solution to the next example.

12.2.4 ENTHALPY CHANGE, REAL REACTIONS

•
•

EXAMPLE 8 When 1 mol of carbon is burned with 1 mol of oxygen, 94 kcal of heat is produced. Find the standard molar enthalpy of formation (relative heat content) of carbon dioxide.

Solution

$$C(\text{graphite}) + O_2(g) \rightarrow CO_2(g) + 94\ \text{kcal}$$

$$\Delta H^\circ = \Sigma\Delta H^\circ_{f_{\text{products}}} - \Sigma\Delta H^\circ_{f_{\text{reactants}}}$$

- ΔH°: This is the standard enthalpy change for a reaction. ° means both products and reactants are at 25°C and 1 atm.
- Σ: total
- $\Sigma\Delta H^\circ_{f_{\text{products}}}$: total standard enthalpy of formation (relative heat content) of all the products
- $\Sigma\Delta H^\circ_{f_{\text{reactants}}}$: total standard enthalpy of formation (relative heat content) of all the reactants

$$-94\ \text{kcal} = [(1\ \text{mol}\ CO_2(g))(\Delta H^\circ_{f_{CO_2(g)}})]$$

$$- \left[(1\ \text{mol C(graphite)})\left(\frac{0\ \text{kcal}}{\text{mol C(graphite)}}\right) + (1\ \text{mol}\ O_2(g))\left(\frac{0\ \text{kcal}}{\text{mol}\ O_2(g)}\right)\right]$$

The standard molar enthalpy of formation for all elements in their most stable form is 0 kcal/mol.

$$= (1\ \text{mol}\ CO_2(g))\Delta H^\circ_{f_{CO_2(g)}} - [0\ \text{kcal} + 0\ \text{kcal}]$$

$$\Delta H^\circ_{f_{CO_2(g)}} = -94\ \text{kcal/mol}\ CO_2(g)$$

This value of the standard molar enthalpy of formation for $CO_2(g)$ (−94 kcal/mol $CO_2(g)$) is shown in Table 12.1. Remember: Heat contents determined by forming compounds from their elements in their standard state are called standard molar enthalpies of formation and given the symbol ΔH_f°. This then explains the presence in the symbol of the Δ, which means "change." Standard molar enthalpies of formation are actually enthalpy changes for the chemical reaction in which compounds are formed from their elements. For simplicity, sometimes we will refer to the standard molar enthalpy of formation of a compound as simply the relative heat content or the enthalpy of the compound.

•
•

The relative heat contents (standard molar enthalpies of formation) of many compounds are determined in the way illustrated in Example 8. A few of these relative heat contents are shown in the second and third columns of Table 12.1 (in units of both kilocalories per mole and kilojoules per mole).

EXAMPLE 9 When 1 mol of sulfur is burned with 1 mol of oxygen, the heat released by the reaction warms the surrounding 71 L of water from an initial temperature of 24.5°C to 25.5°C. Calculate the ΔH_f° for sulfur dioxide. The reaction can be represented by

$$S(s) + O_2(g) \rightarrow SO_2(g)$$

Solution

$$(71 \text{ L } H_2O)\left(\frac{1000 \text{ g } H_2O}{\text{L } H_2O}\right) = 71\,000 \text{ g } H_2O$$

$$(1.0°\text{C temp. increase in } 71\,000 \text{ g } H_2O)\left(\frac{1 \text{ cal absorbed}}{1°\text{C temp. increase in 1 g } H_2O}\right)\left(\frac{1 \text{ kcal}}{1000 \text{ cal}}\right)$$

$$= 71 \text{ kcal heat absorbed}$$

If the surrounding water absorbs 71 kcal of heat, the reaction produces 71 kcal.

$$S(s) + O_2(g) \rightarrow SO_2(g) + 71 \text{ kcal}, \ \Delta H^\circ = -71 \text{ kcal}$$

$$\Delta H^\circ = \Sigma\Delta H^\circ_{f_{\text{products}}} - \Sigma\Delta H^\circ_{f_{\text{reactants}}}$$

$$-71 \text{ kcal} = [(1 \text{ mol } SO_2(g))(\Delta H^\circ_{f_{SO_2(g)}})]$$

$$- \left[(1 \text{ mol S(s)})\left(\frac{0 \text{ kcal}}{\text{mol S(s)}}\right) + (1 \text{ mol } O_2(g))\left(\frac{0 \text{ kcal}}{\text{mol } O_2(g)}\right)\right]$$

$$\Delta H^\circ_{f_{SO_2(g)}} = -71 \text{ kcal/mol } SO_2(g)$$

EXAMPLE 10 Calculate the ΔH_f° for HgO(s) in kilojoules per mole of HgO. The reaction is represented by $2Hg(l) + O_2(g) \rightarrow 2HgO(s) + 184$ kJ.

Solution

$$\Delta H^\circ = \Sigma\Delta H^\circ_{f_{\text{products}}} - \Sigma\Delta H^\circ_{f_{\text{reactants}}}$$

$$-184 \text{ kJ} = [(2 \text{ mol HgO(s)})(\Delta H^\circ_{f_{HgO(s)}})]$$

$$- \left[(2 \text{ mol Hg(l)})\left(\frac{0 \text{ kJ}}{\text{mol Hg(l)}}\right) + (1 \text{ mol } O_2(g))\left(\frac{0 \text{ kJ}}{\text{mol } O_2(g)}\right)\right]$$

$$= (2 \text{ mol HgO(s)})(\Delta H^\circ_{f_{HgO(s)}}) - 0 \text{ kJ}$$

$$\Delta H^\circ_{f_{HgO(s)}} = -92 \text{ kJ/mol HgO(s)}$$

The standard molar enthalpy of formation for HgO(s) is −92 kJ/mol HgO(s).

EXAMPLE 11 Calculate the enthalpy change for the reaction

$$H_2(g) + I_2(s) \rightarrow 2HI(g)$$

using the following information: 12.6 g of I_2 reacts with 0.10 g of H_2 to produce 12.7 g of HI and 0.63 kcal of heat.

Solution First, determine the heat produced when 2 mol of HI forms.

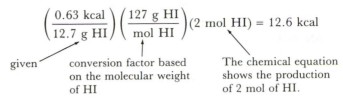

$$\left(\frac{0.63 \text{ kcal}}{12.7 \text{ g HI}}\right)\left(\frac{127 \text{ g HI}}{\text{mol HI}}\right)(2 \text{ mol HI}) = 12.6 \text{ kcal}$$

given — conversion factor based on the molecular weight of HI — The chemical equation shows the production of 2 mol of HI.

Therefore,

$$H_2(g) + I_2(s) \rightarrow 2HI(g) + 12.6 \text{ kcal}, \ \Delta H° = -12.6 \text{ kcal}$$

EXAMPLE 12 Use the information in the preceding problem to calculate the ΔH_f° for HI(g) in kilocalories per mole of HI(g).

Solution

$$H_2(g) + I_2(s) \rightarrow 2HI(g) + 12.6 \text{ kcal}$$

$$\Delta H° = \Sigma\Delta H^\circ_{f_{\text{products}}} - \Sigma\Delta H^\circ_{f_{\text{reactants}}}$$

$$-12.6 \text{ kcal} = [(2 \text{ mol HI(g)})(\Delta H^\circ_{f_{HI(g)}})] - \left[(1 \text{ mol } I_2(s))\left(\frac{0 \text{ kcal}}{\text{mol } I_2(s)}\right) + (1 \text{ mol } H_2(g))\left(\frac{0 \text{ kcal}}{\text{mol } H_2(g)}\right)\right]$$

$$= (2 \text{ mol HI(g)})(\Delta H^\circ_{f_{HI(g)}}) - 0 \text{ kcal}$$

$$\Delta H^\circ_{f_{HI(g)}} = -6.3 \text{ kcal/mol HI(g)}$$

EXAMPLE 13 Use Table 12.1 to calculate the standard enthalpy change for the reaction

$$CaO(s) + CO_2(g) \rightarrow CaCO_3(s)$$

Solution

$$\Delta H° = \Sigma\Delta H^\circ_{f_{\text{products}}} - \Sigma\Delta H^\circ_{f_{\text{reactants}}}$$

$$= \left[(1 \text{ mol } CaCO_3(s))\left(\frac{-288 \text{ kcal}}{\text{mol } CaCO_3(s)}\right)\right] - \left[(1 \text{ mol CaO(s)})\left(\frac{-152 \text{ kcal}}{\text{mol CaO(s)}}\right) + (1 \text{ mol } CO_2(g))\left(\frac{-94 \text{ kcal}}{\text{mol } CO_2(g)}\right)\right]$$

values obtained from Table 12.1

$$= (-288 \text{ kcal}) - (-246 \text{ kcal}) = 288 \text{ kcal} + 246 \text{ kcal} = -42 \text{ kcal}$$

SELF-TEST

1. When 1 mol of lead reacts with 1 mol of sulfur to produce 1 mol of lead sulfide (PbS), 24 kcal of heat is released. Calculate the molar enthalpy of lead sulfide.
2. Calculate the molar enthalpy of liquid water. The reaction that produces H_2O is represented by

$$2H_2(g) + O_2(g) \rightarrow 2H_2O(l) + 136 \text{ kcal}$$

3. Calculate the heat in kilocalories that is given off in the reaction

$$H_2 + Cl_2 \rightarrow 2HCl(g)$$

4. Find the enthalpy change in kilocalories for the reaction

$$N_2H_4(l) + 2H_2O_2(l) \rightarrow N_2(g) + 4H_2O(g)$$

ANSWERS **1.** −24 kcal **2.** −68 kcal **3.** −44 kcal **4.** −154 kcal

12.3 ENTROPY AS DISORDER

Entropy is one of the thermodynamic functions that is more difficult to comprehend. To simplify this concept, we will use the words *entropy* and *disorder* interchangeably. (It is not always correct to interchange these words, but we'll throw caution to the wind.) In a later section we will see how the words *entropy*, *disorder*, and *probability* can be used interchangeably. However, for now, think of all substances as possessing (containing) a certain amount of disorder. The disorder of a substance is called its entropy.

Consider the equilibrium between liquid water and water vapor:

$$H_2O(l) \rightleftharpoons H_2O(g)$$

Which of the two phases (states) of water do you think is more disordered?

That's right. Most people think of a gas as having more disorder than a liquid because a liquid is confined to the shape of the bottom of its container (relatively ordered) while a gas can occupy the entire space of its container (relatively disordered).

Unlike standard molar enthalpies of formation, absolute (not relative) standard entropy values can be determined. Just as with standard molar enthalpies of formation, there are tables that show the absolute standard entropies of many substances (see the sixth and seventh columns of Table 12.1). These tables are used to calculate standard absolute entropy changes for chemical reactions in much the same way that standard enthalpy changes are calculated.

Use Table 12.1 to find the absolute standard entropy (abbreviated $S°$) of 1 mol of water vapor (water in the gas phase).

Good. The absolute standard entropy of 1 mol of water vapor is 45 cal/mol K (or 189 J/mol K).

If the absolute standard entropy (disorder) of water vapor is 45 cal/mol K, and water vapor is more disordered than liquid water, would you expect the absolute standard entropy of 1 mol of liquid water to be greater than or less than 45 cal/mol K? (Don't cheat by looking at Table 12.1.)

Right again! The absolute standard entropy for 1 mol of liquid water would have to be less than 45 cal/mol K because the absolute standard entropy (disorder) of liquid water is less than the disorder of water vapor.

What is the absolute standard entropy of liquid water?

Table 12.1 gives the absolute standard entropy of liquid water as 17 cal/mol H_2O K. Now, for the reaction

$$H_2O(l) \rightarrow H_2O(g)$$

we can talk about an absolute standard entropy change in much the same way that we spoke about standard enthalpy changes. We define the absolute standard entropy change of a chemical reaction as

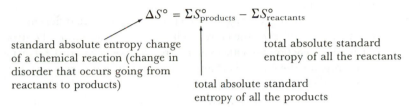

$$\Delta S^\circ = \Sigma S^\circ_{\text{products}} - \Sigma S^\circ_{\text{reactants}}$$

ΔS°: standard absolute entropy change of a chemical reaction (change in disorder that occurs going from reactants to products)

$\Sigma S^\circ_{\text{products}}$: total absolute standard entropy of all the products

$\Sigma S^\circ_{\text{reactants}}$: total absolute standard entropy of all the reactants

12.3.1 ENTROPY CHANGE CALCULATIONS

EXAMPLE 14 Using the standard absolute entropy values found in Table 12.1, calculate the standard absolute entropy change for the reaction

$$H_2O(l) \rightarrow H_2O(g)$$

Solution

$$\Delta S^\circ = \Sigma S^\circ_{\text{products}} - \Sigma S^\circ_{\text{reactants}}$$

$$= (1 \text{ mol } H_2O(g))\left(\frac{45 \text{ cal}}{\text{mol } H_2O(g) \text{ K}}\right) - (1 \text{ mol } H_2O(l))\left(\frac{17 \text{ cal}}{\text{mol } H_2O(l) \text{ K}}\right)$$

$$= +28 \text{ cal/K}$$

Look at the sign of the entropy change (it's positive, +28 cal/K). Whenever a chemical reaction occurs with a positive entropy change, the products must be in a state of greater disorder than the reactants. (In this case, water vapor is in a state of greater disorder than liquid water.) You should try to understand that if the total entropy (disorder) of the products is greater than the total entropy of the reactants, the entropy change (ΔS°) for that reaction will be positive.

EXAMPLE 15 One mol of A reacts with 1 mol of B to produce 1 mol of C. The absolute standard entropies of A, B, and C are 10 J/mol K, 20 J/mol K, and 40 J/mol K, respectively. Find the standard absolute entropy change for this reaction. Are the products of this reaction more disordered than the reactants?

Solution

$$A + B \rightarrow C$$

$$\Delta S^\circ = \Sigma S^\circ_{\text{products}} - \Sigma S^\circ_{\text{reactants}}$$

$$= \left[(1 \text{ mol C})\left(\frac{40 \text{ J}}{\text{mol C K}}\right)\right]$$

$$- \left[(1 \text{ mol A})\left(\frac{10 \text{ J}}{\text{mol A K}}\right) + (1 \text{ mol B})\left(\frac{20 \text{ J}}{\text{mol B K}}\right)\right]$$

$$= \frac{40 \text{ J}}{\text{K}} - \left(\frac{10 \text{ J}}{\text{K}} + \frac{20 \text{ J}}{\text{K}}\right) = \frac{40 \text{ J}}{\text{K}} - \frac{30 \text{ J}}{\text{K}}$$

$$= \frac{10 \text{ J}}{\text{K}} = 10 \text{ J/K}$$

Since the entropy change is positive (+10 J/K), the products are more disordered than the reactants.

It is common to refer to a chemical reaction as a system. We say that the disorder for this system has increased because the entropy change ($\Delta S°$) is positive. This system has become more disordered. If the entropy change for a system is positive, the entropy (disorder) of the system has increased.

Notice that the units for the entropy change are different from the units for the enthalpy change. The units for the entropy change (absolute standard entropy change) that accompanies a chemical reaction are either calories per degree kelvin (cal/K) or joules per degree kelvin (J/K). The units for the enthalpy change (standard enthalpy change) that accompanies a chemical reaction are either kilocalories or kilojoules. Also notice that the absolute standard entropies of elements are not numerically equal to zero. (Look back at Table 12.1 if you failed to notice this.) Unlike enthalpy, we can determine the absolute (actual) value of an element's entropy under standard conditions. The absolute standard entropies of elements do not have arbitrary values.

•
•

EXAMPLE 16 Determine the absolute standard entropy change for the reaction

$$2E + G \rightarrow 3X$$

The absolute standard entropies for E, G, and X are 10 J/mol K, 20 J/mol K, and 30 J/mol K, respectively. When the reaction occurs, does the disorder of the system increase or decrease?

Solution

$$\begin{aligned}\Delta S° &= \Sigma S°_{\text{products}} - \Sigma S°_{\text{reactants}}\\ &= \left[(3 \text{ mol X})\left(\frac{30 \text{ J}}{\text{mol X K}}\right)\right]\\ &\quad - \left[(2 \text{ mol E})\left(\frac{10 \text{ J}}{\text{mol E K}}\right) + (1 \text{ mol G})\left(\frac{20 \text{ J}}{\text{mol G K}}\right)\right]\\ &= \frac{90 \text{ J}}{\text{K}} - \left[\frac{20 \text{ J}}{\text{K}} + \frac{20 \text{ J}}{\text{K}}\right] = \frac{90 \text{ J}}{\text{K}} - \frac{40 \text{ J}}{\text{K}}\\ &= \frac{50 \text{ J}}{\text{K}} = 50 \text{ J/K}\end{aligned}$$

The disorder of this system increases because the disorder (entropy) of the products is greater than the disorder of the reactants. The disorder of any system increases when the entropy change for the system is positive. The entropy change for the system in this example is +50 J/K.

•
•

EXAMPLE 17 The absolute standard entropy change for the reaction

$$A + B \rightarrow E$$

is −20 cal/K. The absolute standard entropies of A and B are 25 cal/mol K and 50 cal/mol K, respectively. What is the absolute standard entropy of E in calories per mole per degree kelvin? Is E more disordered than the total disorder of A plus B?

Solution

$$\begin{aligned}\Delta S° &= \Sigma S°_{\text{products}} - \Sigma S°_{\text{reactants}}\\ \frac{-20 \text{ cal}}{\text{K}} &= \Sigma S°_{\text{products}} - \Sigma S°_{\text{reactants}}\\ &= [(1 \text{ mol E})(S°_{\text{E}})]\\ &\quad - \left[(1 \text{ mol A})\left(\frac{25 \text{ cal}}{\text{mol A K}}\right) + (1 \text{ mol B})\left(\frac{50 \text{ cal}}{\text{mol B K}}\right)\right]\end{aligned}$$

$$S^\circ_E = \frac{55\text{ cal}}{\text{mol E K}} = 55\text{ cal/mol E K}$$

The product (E) is less disordered than the total disorder of all of the reactants (A plus B). The entropy change for the reaction is negative ($\Delta S° = -20$ cal/K), which means that the disorder of this system decreases as the reaction proceeds from reactants to products.

EXAMPLE 18 Calculate the entropy change for every gram of A that reacts if the absolute standard entropies of A, B, and C are 10 cal/mol K, 20 cal/mol K, and 30 cal/mol K, respectively. The molecular weights of A, B, and C are 5 amu, 10 amu, and 15 amu, respectively. The reaction is represented by

$$A + 2B \rightarrow C$$

Solution

$$\begin{aligned}\Delta S^\circ &= \Sigma S^\circ_{\text{products}} - \Sigma S^\circ_{\text{reactants}}\\ &= \left[(1\text{ mol C})\left(\frac{30\text{ cal}}{\text{mol C K}}\right)\right]\\ &\quad - \left[(1\text{ mol A})\left(\frac{10\text{ cal}}{\text{mol A K}}\right) + (2\text{ mol B})\left(\frac{20\text{ cal}}{\text{mol B K}}\right)\right]\\ &= \frac{30\text{ cal}}{\text{K}} - \frac{50\text{ cal}}{\text{K}} = \frac{-20\text{ cal}}{\text{K}}\end{aligned}$$

$$(1.0\text{ g A})\left(\frac{1.0\text{ mol A}}{5.0\text{ g A}}\right)\left(\frac{-20\text{ cal}}{\text{mol A K}}\right) = \frac{-4.0\text{ cal}}{\text{K}} = -4.0\text{ cal/K}$$

The disorder change that occurs whenever 1 g of A reacts is -4.0 cal/g A K.

EXAMPLE 19 Find the entropy change for the reaction that occurs when 1 mol of graphite is burned with 1 mol of oxygen to produce 1 mol of carbon dioxide. Discuss the disorder change of this system.

Solution

$$C(\text{graphite}) + O_2(g) \rightarrow CO_2(g)$$

$$\Delta S^\circ = \Sigma S^\circ_{\text{products}} - \Sigma S^\circ_{\text{reactants}}$$

The $S°$ values for C(graphite), $O_2(g)$, and $CO_2(g)$ are found in Table 12.1.

$$\begin{aligned}\Delta S^\circ &= \left[(1\text{ mol }CO_2(g))\left(\frac{214\text{ J}}{\text{mol }CO_2(g)\text{ K}}\right)\right]\\ &\quad - \left[(1\text{ mol C(gr)})\left(\frac{6\text{ J}}{\text{mol C(gr) K}}\right) + (1\text{ mol }O_2(g))\left(\frac{205\text{ J}}{\text{mol }O_2(g)\text{ K}}\right)\right]\\ &= \frac{214\text{ J}}{\text{K}} - \frac{211\text{ J}}{\text{K}} = \frac{3\text{ J}}{\text{K}} = 3\text{ J/K}\end{aligned}$$

The entropy change for this reaction is positive. This means that the disorder of this system increases as the reaction occurs.

EXAMPLE 20 Find the absolute standard entropy change for the reaction

$$2Hg(l) + O_2(g) \rightarrow 2HgO(s)$$

Does the disorder of this system increase? Can you determine if the disorder of this system increases before you calculate the entropy change for this system?

Solution You might guess that, because the products of this reaction are in the solid phase (little disorder) while some of the reactants are in the liquid phase (more disorder) and other reactants are in the gas phase (even more disorder), the reaction produces a decrease in disorder. We would expect that the entropy change for this reaction is negative.

$$\Delta S° = \Sigma S°_{\text{products}} - \Sigma S°_{\text{reactants}}$$

The $S°$ values for Hg(l), O_2(g), and HgO(s) are found in Table 12.1.

$$\Delta S° = \left[(2 \text{ mol HgO(s)})\left(\frac{17 \text{ cal}}{\text{mol HgO(s) K}}\right)\right]$$

$$- \left[(2 \text{ mol Hg(l)})\left(\frac{18 \text{ cal}}{\text{mol Hg(l) K}}\right) + (1 \text{ mol } O_2\text{(g)})\left(\frac{49 \text{ cal}}{\text{mol } O_2\text{(g) K}}\right)\right]$$

$$= \frac{34 \text{ cal}}{\text{K}} - \left(\frac{36 \text{ cal}}{\text{K}} + \frac{49 \text{ cal}}{\text{K}}\right) = \frac{34 \text{ cal}}{\text{K}} - \frac{85 \text{ cal}}{\text{K}}$$

$$= \frac{-51 \text{ cal}}{\text{K}} = -51 \text{ cal/K}$$

The disorder of this system decreases because the entropy change for this system is negative.

EXAMPLE 21 Find the entropy change per gram of HgO(s) that forms according to the reaction given in Example 20.

Solution

$$\left(\frac{-51 \text{ cal/K}}{2 \text{ mol HgO}}\right)\left(\frac{1 \text{ mol HgO}}{217 \text{ g HgO}}\right) = \frac{-0.12 \text{ cal/K}}{\text{g HgO}}$$

$$= \frac{-0.12 \text{ cal}}{\text{g HgO K}}$$

SELF-TEST

1. Find the absolute standard entropy change in cal/K for the reaction that occurs when 1 mol of S(s) is burned with 1 mol of O_2(g) producing 1 mol of SO_2(g) according to the reaction

$$S(s) + O_2(g) \rightarrow SO_2(g)$$

Discuss the disorder change accompanying this reaction.

2. Find the absolute standard entropy change in cal/K for the reaction

$$H_2(g) + I_2(s) \rightarrow 2HI(g)$$

Discuss the disorder change accompanying this reaction.

3. Find the absolute standard entropy change in cal/K per gram of $CaCO_3$(s) that forms in the reaction

$$CaO(s) + CO_2(g) \rightarrow CaCO_3(s)$$

Discuss the disorder change accompanying this reaction.

ANSWERS **1.** 2 cal/K; the system disorder increases **2.** 39 cal/K; the system disorder increases **3.** −0.39 cal/g $CaCO_3$ K; the system disorder decreases

12.3.2 THREE LAWS OF THERMODYNAMICS

There are three laws of thermodynamics. The first law of thermodynamics simply says that energy is always conserved. This means you cannot create or destroy energy.

The second law of thermodynamics, the one with which we are now most concerned, says that every time a spontaneous event occurs (that is, occurs all by itself without outside help), the entropy (disorder) of the entire universe increases. Whenever something happens spontaneously (e.g., a chemical reaction, the birth of a star, evolution), the universe must become more disordered. The entropy change (ΔS) of the universe is positive whenever an event occurs spontaneously.

The third law of thermodynamics tells us that at absolute zero (−273.15°C, 0 K), the entropy (disorder) of many substances becomes exactly zero. At absolute zero, many substances exist in a state of perfect order. At −273.15°C, many substances have no disorder at all. The one way in which many substances can exist at −273.15°C is perfectly ordered.

12.3.3 SPONTANEITY

The second law of thermodynamics interests chemists because it indicates a way to determine whether a chemical reaction can occur spontaneously (all by itself). According to the second law, if we can calculate the entropy change of the universe when a reaction occurs, and if the entropy change of the universe is positive, then the chemical reaction is a spontaneous reaction.

There is a difference between the entropy change of the universe when a chemical reaction occurs and the entropy change for the chemical reaction. In thermodynamics we distinguish between a system, which is whatever is being studied (e.g., a chemical reaction), and the surroundings, which is everything else in the universe besides the system (see Figure 12.1). The universe is then defined as the sum of the system plus the surroundings.

FIGURE 12.1 System, surroundings, and universe

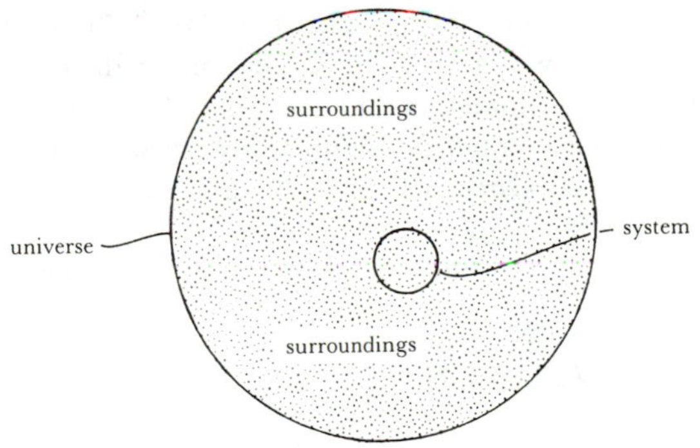

When we calculate values for the entropy change of a reaction, that is,

$$\Delta S^\circ_{\text{reaction}} = \Sigma S^\circ_{\text{products}} - \Sigma S^\circ_{\text{reactants}}$$

we are calculating the entropy change for the system, not the universe.

$$\Delta S^\circ_{\text{reaction}} = \Delta S_{\text{system}}$$

To calculate the entropy change of the universe, we would have to calculate the entropy change of the surroundings (which is beyond the scope of this book) and then add the entropy change of the reaction (system) to the entropy change of the surroundings.

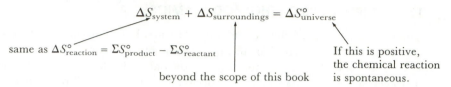

$$\Delta S_{\text{system}} + \Delta S_{\text{surroundings}} = \Delta S^{\circ}_{\text{universe}}$$

same as $\Delta S^{\circ}_{\text{reaction}} = \Sigma S^{\circ}_{\text{product}} - \Sigma S^{\circ}_{\text{reactant}}$

beyond the scope of this book

If this is positive, the chemical reaction is spontaneous.

Fortunately, there is an easier way to determine if a chemical reaction is spontaneous without having to worry about the surroundings. We'll discuss this other method in Section 13.2, "Gibbs Free Energy."

12.3.4 CREATION OF THE UNIVERSE, PART I

Philosophers and cosmologists have debated for centuries whether the universe had a beginning or whether it simply always existed. Scientists have used the second law of thermodynamics to show that the universe had to have had a beginning.

The second law of thermodynamics states that in all spontaneous events, the entropy of the universe increases. Never does the entropy of the universe decrease. As we spontaneously move forward in time, the entropy (disorder) of the universe is constantly increasing. One indication of this is that the universe is expanding with time. With the passage of time, the stars and galaxies are distributed over greater distances, and their disorder increases.

past — today — future →

order — disorder

It is logical, therefore, to expect that if we could travel through time into the past, we would see the universe moving more and more toward an ordered system. It is also logical, however, that there must be a limit to how much something can be ordered (organized). We would eventually reach a point where the universe is as ordered as possible. This would be the point of maximum order for the universe, and we could travel no further into the past. This implies that the universe had a beginning.

St. Thomas Aquinas (A.D. 1225–1274), an Italian theologian, used similar arguments to prove there was a moment of creation for the universe. Today it is fashionable for scientists to refer to that moment as the big bang. According to the big bang theory, about 15 to 20 billion years ago a singularity (an infinitely dense, dimensionless point) exploded. All that exists today and all that existed before the birth of the universe was once within that singularity. We'll discuss this theory in greater detail when we study Section 15.11, "Creation of the Universe, Part II: A Review" in the nuclear chemistry chapter.

12.3.5 EVOLUTION

Evolution means different things to different people. One popular idea associated with evolution is that nonliving molecules spontaneously changed into living one-celled life-forms that spontaneously changed into more complicated life-forms that spontaneously changed into monkeys and finally into humans. To many biologists, however, evolution is nothing more than an organism's ability to adapt to its surroundings as it passes from one generation to another. This adaptation ability is the result of a large number of genes (which determine certain traits) contained in an organism's chromosomes (DNA). Offspring with undesirable traits (relative to their environment) have a higher fatality rate and pass on those undesirable traits to successive generations with decreasing frequency. Offspring with desirable traits (traits that help the organism cope with the environment) have a higher survival rate, and they pass on their desirable traits with increasing frequency to successive generations. You may remember that Le Chatelier's principle tells us that even inanimate matter (e.g., chemical equilibrium systems) has an ability to react to changes in the surrounding conditions.

Evolution is a very emotional and controversial issue for many people. A large number of people with deep religious convictions do not want to see their children "brainwashed" with anti-religious propaganda, while many scientists do not want to see state agencies, religious and political groups, and nonscientists dictating the contents of science textbooks.

Beginning in the early 1970's, some state legislatures and boards of education (e.g., California and Louisiana) required that all science textbooks present divine creation theory (also called creation science) as a valid alternative theory for evolution. The American Chemical Society and the American Association for the Advancement of Science joined others in legally challenging this requirement because they believed that creation science is not, in fact, science. It wasn't until 1987 that the U.S. Supreme Court struck down state laws requiring this equal-time approach.

In trying to convince people that evolution was invalid, creation scientists argued that the evolutionary idea of complex forms of life spontaneously arising from simpler forms of life violated the second law of thermodynamics (see Figure 12.2).

FIGURE 12.2 Human evolution?

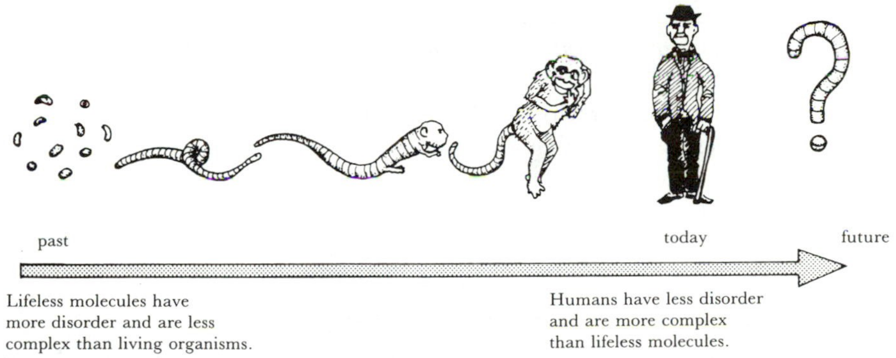
past
today
future
Lifeless molecules have
more disorder and are less
complex than living organisms.
Humans have less disorder
and are more complex
than lifeless molecules.

First of all, it is interesting that not all people agree that humans are more organized (ordered) than one-celled animals. This is one reason why scientists use words like *entropy* (*entropy* has a precise mathematical meaning) instead of words like *disorder* (*disorder* can have different meanings to different people).

Let's assume, for the point of argument, that a human is objectively more organized and less disordered than a single cell.

Can a single cell spontaneously (on its own) evolve to form a human? More importantly, would such a spontaneous change violate the second law of thermodynamics?

You began as a single cell (one of your mother's fertilized eggs) and spontaneously developed into your present handsome state.

Why doesn't this violate the second law of thermodynamics?

The second law of thermodynamics says that the disorder of the universe must increase when something occurs spontaneously. The second law does not say that the disorder of the system (reaction) must increase. As humans progress from a single cell at the moment of conception to their present state, many carbohydrates (solid and liquid food molecules) are converted into more disordered gaseous waste products (e.g., carbon dioxide and water vapor). This conversion disorders the surroundings extensively. The total disorder (you and your surroundings) increases as your life continues. In 1977 the Nobel prize for chemistry was given to Ilya Prigogine, who came up with a thermodynamic explanation of how life could have come into being spontaneously in apparent defiance of the second law of thermodynamics.

12.4 ENTROPY AS PROBABILITY

In this section we will examine the important relationship between entropy, disorder, and probability.

12.4.1 CRAPS SHOOTING ANALOGY

How many different ways can you roll a two (snake eyes) with a pair of dice?

There is only one way that you can roll a two. Each die (singular form of dice) must show a one (single spot) in order for you to roll a two.

How many different ways can you roll a (lucky) seven with a pair of dice?

There are six different ways to roll a seven with a pair of dice. One possible way to roll a seven is for the first die to show a one (one spot) and the second die to show a six.

← This is the only way to roll snake eyes.

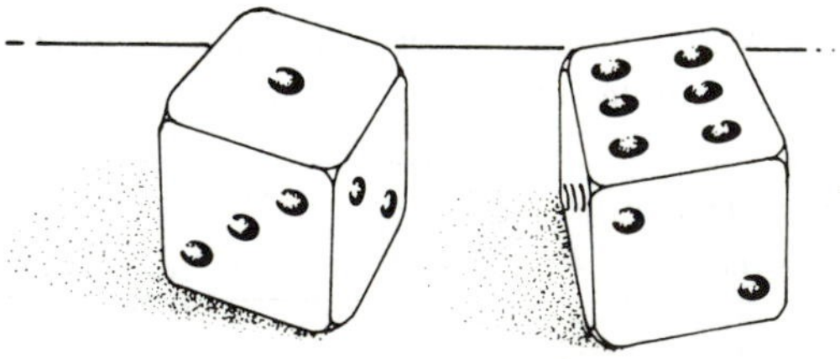

← This is one of six different ways to roll a seven.

You can also roll a seven if a two comes up on the first die and a five comes up on the second die.

← This is another way to roll a seven.

We can summarize the six different ways of rolling a seven and the only way of rolling a two with the following table.

Die 1	Die 2	
1	6	There are six different ways to roll a seven.
2	5	
3	4	
4	3	
5	2	
6	1	
1	1	There is only one way to roll a two.

Because there are more ways to roll a seven than there are to roll a two, we say that it is more probable that a seven will be rolled.

And so it is with other systems. There are more ways in which systems (e.g., your bedroom) can be in a state of disorder than a state of order. It is more probable that systems will be found in a state of disorder than in a state of order.

12.4.2 WATER VAPOR VS. LIQUID WATER, PART I

Consider a hypothetical and unrealistic situation in which we have two containers, one with two molecules of liquid water and the other with two molecules of water vapor (gas) (see Figure 12.3). There are more ways for two water molecules to be arranged as a gas than as a liquid because the two liquid molecules must be confined to the bottom of the container and they must touch one another. There are many more ways to arrange two gas water molecules because gas molecules do not have to touch one another and they are not confined to any one part of the container.

FIGURE 12.3 Water vapor vs. liquid water

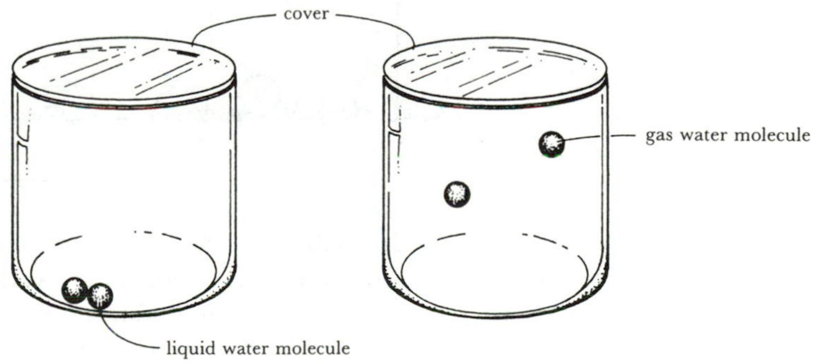

Compare these two beakers with the rolling of a two and a seven. There are more ways for two water molecules to exist as a gas (there are more ways for a pair of dice to exist as a seven) than there are ways for two water molecules to exist as a liquid (than there are ways for a pair of dice to exist as a two). It is more probable that we will find water existing as a gas (it is more probable to roll a seven). We are more likely to have water vapor than liquid water because there are more ways to arrange water molecules as a gas than there are ways to arrange water molecules as a liquid.

You may wonder why liquid water is more abundant on earth than water vapor. Good observation! Unfortunately, you will have to wait until we discuss Gibbs free energy in the next chapter before we can fully explain this apparent violation of the laws of probability. For now, however, here's a partial explanation: The earth is quite an insignificant part of the universe. The vast majority of water (over 99%) in the universe does exist as a gas. Here is an analogy: The laws of probability say that if you flip a coin you will get heads with the same frequency as you will get tails, and yet it is possible to get ten tails in a row because the laws of probability are more accurate over long periods of time than they are over short periods of time.

You should begin to see the relationship between entropy (disorder) and probability. Water vapor is more disordered than liquid water, and water vapor is more likely to exist than liquid water because there are many more ways to arrange (disorder) water vapor than there are ways to arrange liquid water.

12.4.3 ENERGY LEVELS

Before we can discuss entropy any further, you must have some feeling for the concept of energy levels. We say that an object suspended 10 ft from the ground is at a higher energy level (state) than an object sitting on the ground (see Figure 12.4). Objects in higher energy states are less stable than objects in lower energy states. Object A is less stable (it can fall to the ground) than object B (it will tend to just sit there). We say that the potential energy of A is greater than the potential energy of B.

FIGURE 12.4 Potential energy levels

If you should ever be in doubt as to which of two energy states is at a higher level (less stable), just remember that it always requires energy (e.g., work) to place an object in a higher energy state. In Figure 12.4, somebody had to lift object A to its position 10 ft off the ground. It is important for you to remember that if an object goes from a higher energy level to a lower energy level, energy is always given off (released to the surroundings).

When object A falls to the ground, what energy will be given off?

When object A hits the ground, a loud noise (sound energy) will be heard. There will also be a small amount of heat (from friction) released as the object falls and strikes the ground.

EXAMPLE 22 Two pairs of magnets are placed on a table. In one pair (state 1), the north pole of one magnet is touching the south pole of the other magnet. In the other pair (state 2), the north pole of one magnet is facing but separated from the south pole of the other magnet.

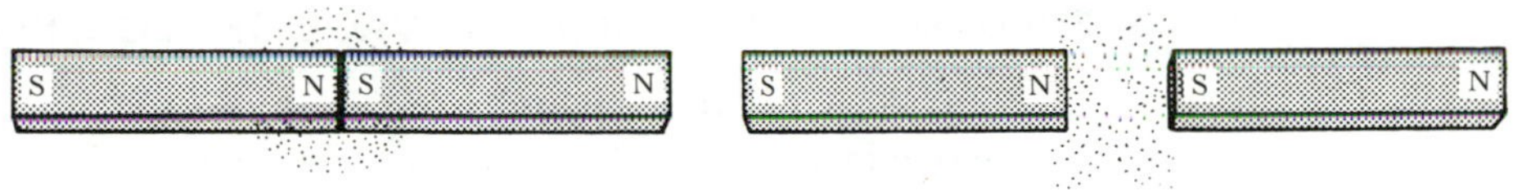

state 1 state 2

Which state is in a higher energy level? Which state is less stable?

Solution Because unlike magnetic poles (north and south) of a magnet attract each other, state 1 is at a lower energy level and is more stable than state 2. It requires work to separate the north pole of one magnet from the south pole of another magnet. The potential energy of state 2 is higher than the potential energy of state 1.

EXAMPLE 23 Two pairs of positively charged particles are placed together.

state 1 state 2

In one pair (state 1), the two positive charges are touching each other. In the other pair (state 2), the positive charges are separated from each other.

Which state is at a higher energy level? Which state is less stable?

Solution It takes work to push two separated positive charges together because positive (like) charges repel each other. State 2 is at the lower energy level and more stable than state 1. The positive particles in state 1 have a greater potential energy than the positive particles in state 2.

Many years ago people thought that spontaneous events happened when systems tried to move to lower energy levels. People thought that spontaneous events occurred to minimize (decrease) the energy of a system. This explains why objects fall to the ground, for example, and why north and south magnetic poles move together, and why similarly charged (e.g., two positively charged) particles move away from each other. Heat (energy) must be added to liquid water to change it into a gas (water vapor). Water vapor is in a higher energy state (has a higher potential energy) than liquid water. When water vapor condenses, energy (heat) is released and the water "falls" to a lower energy level.

Although it is true that many spontaneous events can be explained by the tendency of objects to decrease (minimize) their energy levels, many spontaneous events occur in which the system (which is made up of objects) goes from a lower to a higher energy level. For example, many spontaneous chemical reactions are exothermic; that is, they lose heat energy to their surroundings as the chemicals go from a higher energy state to a lower energy state. However, not all spontaneous chemical reactions are exothermic. Some spontaneous chemical reactions are endothermic; that is, they require heat.

It turns out that objects have two tendencies: (1) they try to lower their energy levels (e.g., fall to a lower energy level), and (2) they try to increase their disorder (maximize their entropy). Much of what is observed in the universe can be explained by keeping these two tendencies in mind. The relative importance of these two tendencies, how they assist each other and how they can be in opposition, is discussed in greater detail in the next chapter.

12.4.4 FLUID MOTION IN A PARTIALLY EVACUATED BOX, PART I

Consider a hypothetical box with a partition separating neutral (uncharged) gas molecules from a pure vacuum (see Figure 12.5). When the partition is removed, gas molecules expand into the right-hand side of the box, completely and evenly filling the entire box. This phenomenon cannot be explained as the system trying to minimize its energy because the energy of states 1 and 2 are the same (if their temperatures are the same).

FIGURE 12.5 Gas expansion

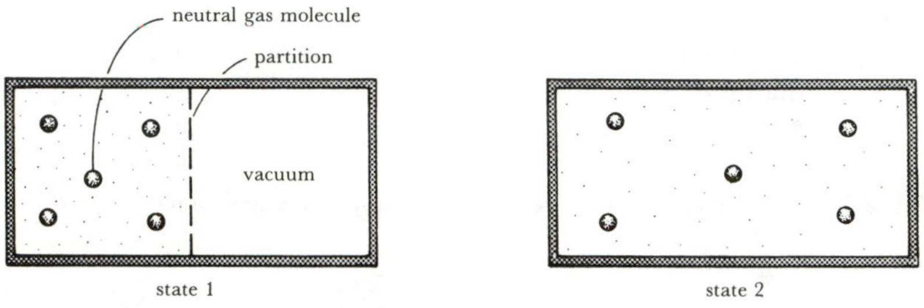

Why do the gas molecules expand to fill the container? Why can't the gas spontaneously move from state 2 to state 1 (spontaneously forming a perfect vacuum in the right-hand side of the box)?

Some events occur spontaneously as objects attempt to increase their disorder. An expanding gas increases its disorder (and the disorder of the universe as well). In Figure 12.5, state 2 is more likely to occur than state 1 because there are more ways to arrange (disorder) gas molecules using the entire box than there are ways to arrange gas molecules confined to one side of the box.

You should try to comprehend this last answer in two different ways, relative to (1) the second law of thermodynamics and (2) the relationship between disorder and probability. Can you imagine the number of people living in fear of spontaneous suffocation before the second law of thermodynamics was passed?

PROBLEM SET 12

The problems in Problem Set 12 parallel the examples in Chapter 12. For example, if you should have trouble working Problem 5, go back to Example 5 in this chapter to get help. The correct answers to these problems are given at the end of this book.

1. One mole of A reacts with 1 mol of B to produce 1 mol of C. The enthalpies of A, B, and C are 200 kcal/mol, 100 kcal/mol, and 50 kcal/mol, respectively. Find the enthalpy change for this reaction. Is the reaction exothermic or endothermic?
2. Determine the enthalpy change for the hypothetical reaction

$$X + 2W \rightarrow 4Y$$

The enthalpies for X, W, and Y are 70 kcal/mol, 80 kcal/mol, and 90 kcal/mol, respectively. Is this reaction exothermic or endothermic?

3. Calculate the enthalpy change for the reaction

$$4A + 2B \rightarrow C + 2D$$

The heat contents of A, B, C, and D are 5 kJ/mol, 10 kJ/mol, 15 kJ/mol, and 20 kJ/mol, respectively. Is the reaction exothermic or endothermic?

4. The enthalpy change for the reaction

$$A + B \rightarrow E + 400 \text{ kcal}$$

is −400 kcal. What is the molar enthalpy of E? The enthalpies of A and B are 250 kcal/mol and 500 kcal/mol, respectively.

5. Find the molar enthalpy of A. One mole of A reacts with 2 mol of B to produce 3 mol of C and 800 kcal of heat. The enthalpies of B and C are 200 kcal/mol and 400 kcal/mol, respectively.
6. One mole of A reacts with 1 mol of B to produce 2 mol of G and 500 kJ of heat. The molecular weights of A, B, and G are 20 amu, 30 amu, and 40 amu, respectively. How many kilojoules of heat are produced for every 90 g of G that forms?
7. Calculate the number of kilocalories of heat produced for every gram of A that reacts if the enthalpies of A, B, and C are 10 kcal/mol, 20 kcal/mol, and 30 kcal/mol, respectively. The molecular weights of A, B, and C are 10 amu, 20 amu, and 30 amu, respectively. The reaction is represented by

$$A + 2B \rightarrow C$$

8. When 2 mol of calcium is allowed to react with 1 mol of oxygen, 2 mol of calcium oxide forms. Find the standard molar enthalpy of formation of calcium oxide. The reaction is represented by

$$2Ca(s) + O_2(g) \rightarrow 2CaO(s) + 304 \text{ kcal}$$

9. When 1 mol of lead reacts with 1 mol of sulfur to produce lead sulfide, the heat released by the reaction warms the surrounding 24 L of water from an initial temperature of 24.5°C to 25.5°C. Calculate the ΔH_f° for PbS. The reaction is represented by

$$Pb(s) + S(s) \rightarrow PbS(s)$$

10. When 2 mol of hydrogen reacts with 1 mol of oxygen, 2 mol of water forms. Find the standard molar enthalpy of formation of water. The reaction is represented by

$$2H_2(g) + O_2(g) \rightarrow 2H_2O(l) + 572 \text{ kJ}$$

11. Calculate the enthalpy change for the reaction

$$H_2(g) + I_2(s) \rightarrow 2HI(g)$$

using the following information: 10.16 g of I_2 reacts with 0.0800 g of H_2 to produce 10.16 g of HI and 0.504 kcal of heat.
12. Use the information given in Problem 11 to calculate the ΔH_f° for HI(g) in kilocalories per mole of HI(g).
13. Use Table 12.1 to calculate the standard enthalpy change in kilojoules for the reaction

$$CaO(s) + CO_2(g) \rightarrow CaCO_3(s)$$

14. Using the standard absolute entropy values found in Table 12.1, calculate the standard absolute entropy change for the reaction

$$H_2O(g) \rightarrow H_2O(l)$$

in calories per degree kelvin.
15. One mole of C decomposes into 1 mol of A and 1 mole of B. The absolute standard entropies of A, B, and C are 10 J/mol K, 20 J/mol K, and 40 J/mol K. Find the standard absolute entropy change for this reaction. Are the products of this reaction more disordered than the reactants?
16. Determine the absolute standard entropy change for the reaction

$$3X \rightarrow 2E + G$$

The absolute standard entropies for E, G, and X are 10 J/mol K, 20 J/mol K, and 30 J/mol K, respectively. When the reaction occurs, does the disorder of the system increase or decrease?
17. The absolute standard entropy change for the reaction

$$A + B \rightarrow E$$

is −30 cal/K. The absolute standard entropies of A and B are 37.5 cal/mol K and 75 cal/mol K, respectively. What is the absolute standard entropy of E in calories per mole per degree kelvin? Is E more disordered than the total disorder of A plus B?
18. Calculate the entropy change for every gram of A that reacts if the absolute standard entropies of A, B, and C are 8.0 cal/mol K, 16 cal/mol K, and 24 cal/mol K, respectively. The molecular weights of A, B, and C are 4.0 amu, 8.0 amu, and 12 amu, respectively. The reaction is represented by

$$A + 2B \rightarrow C$$

19. Find the entropy change for the reaction that occurs when 1 mol of sulfur is burned with 1 mol of oxygen to produce 1 mol of sulfur dioxide. Discuss the disorder change of this system. Use Table 12.1 to obtain entropy values for S(s), $O_2(g)$, and $SO_2(g)$ in kilojoules.
20. Find the absolute standard entropy change for the reaction

$$2HgO(s) \rightarrow 2Hg(l) + O_2(g)$$

in calories per degree kelvin. Does the disorder of this system increase? Can you determine if the disorder of this system increases before you calculate the entropy change for this reaction?
21. Find the entropy change per gram of O_2 that forms according to the reaction given in Problem 20.
22. Two pairs of magnets are placed on a table, as shown. Which state is more stable?

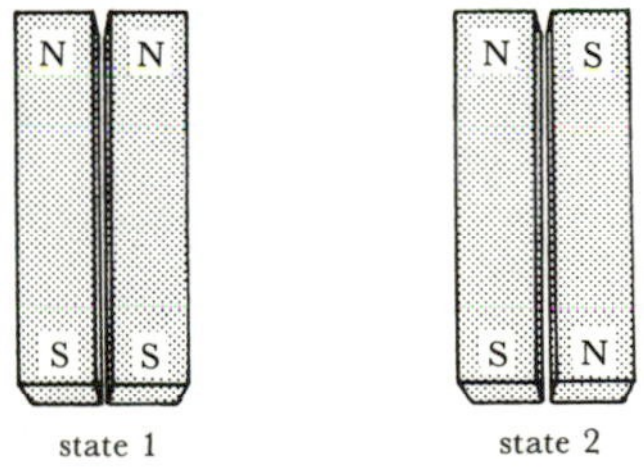

23. Two pairs of negatively charged particles are placed together as shown. Which state is more stable?

13

Thermodynamics Part II

13.1 INTRODUCTION

You can determine if a reaction is exothermic by comparing the sum of the enthalpies of the products with the sum of the enthalpies of the reactants. If the total product enthalpy (heat content) is less than the total reactant enthalpy, the reaction is exothermic. Unfortunately, we cannot determine if a reaction is spontaneous by comparing the entropy of the products with the entropy of the reactants. Entropy can be used to determine the spontaneity of a reaction, but this involves calculating both the entropy change of the reaction and the entropy change of the surroundings. These two calculations are added together to determine the entropy change of the universe. The entropy change of the universe (not of the reaction) is related to reaction spontaneity. Fortunately, there is a thermodynamic function, the Gibbs free energy, that eliminates the need to perform calculations on the surroundings to determine reaction spontaneity.

13.2 GIBBS FREE ENERGY

The standard molar Gibbs free energy of formation, abbreviated ΔG_f°, is a thermodynamic function that is used to determine the spontaneity of a chemical reaction under conditions of constant temperature and pressure. Just as physical objects can have different potential energies (see Section 12.4.3), chemicals can also exist at different levels of chemical potential. If the reactants exist at a higher chemical potential than the products, useful work can be done by the energy released when the reactants at their higher chemical potential "drop" to the lower chemical potential of the products. These ideas are discussed in more detail in the next section, Section 13.2.1.

13.2.1 POTENTIAL ENERGY VS. GIBBS FREE ENERGY

You can think of the Gibbs free energy of a substance as its chemical potential in much the same way that you can think of an object's distance from the ground as its physical potential energy. Consider a 5-lb object raised 10 ft from the floor, as shown in Figure 13.1. The 5.0-lb object will fall spontaneously to the ground because the final state (fallen object) has a lower potential energy than the initial state (raised object). The potential energy change, ΔPE, is always negative for any physical process than can occur spontaneously.

$$\Delta PE = PE_{products} - PE_{reactants}$$

- ΔPE: change in potential energy going from the 10-ft height to the floor
- $PE_{products}$: PE of the final state
- $PE_{reactants}$: PE of the initial state

$$\Delta PE = 0 \text{ ft lb} - 50 \text{ ft lb}$$

$$= -50 \text{ ft lb}$$

And so it is with the Gibbs free energy of formation. Whenever the total Gibbs free energy (chemical potential) of the reactants is greater than the total Gibbs free energy of the products, the change in Gibbs free energy, ΔG, will be negative. Whenever the Gibbs free energy change is negative, the reaction can occur spontaneously.

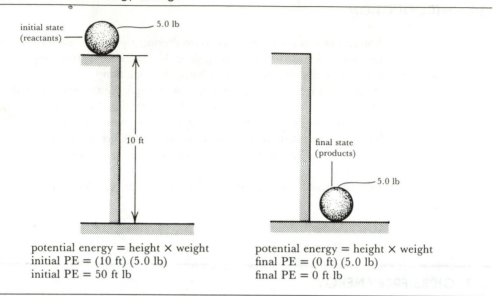

FIGURE 13.1 Potential energy change

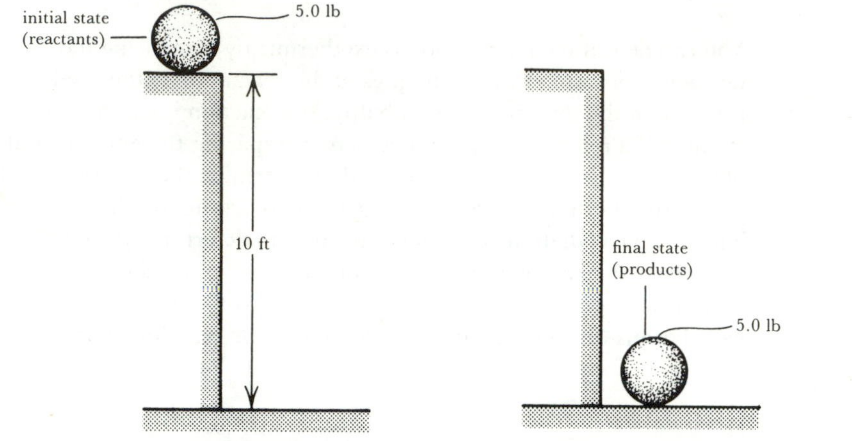

$$\Delta G = \Sigma G_{\text{products}} - \Sigma G_{\text{reactants}}$$

- $\Sigma G_{\text{products}}$: total chemical potential of all the products
- $\Sigma G_{\text{reactants}}$: total chemical potential of all the reactants
- ΔG: chemical potential change for a chemical reaction
- If ΔG is negative, the reactants are at a higher chemical potential than the products and the reaction can occur all by itself.

13.2.2 GIBBS FREE ENERGY CHANGE CALCULATIONS, HYPOTHETICAL REACTIONS

•
•

EXAMPLE 1 One mole of A reacts with 1 mol of B to produce 1 mol of C. The Gibbs free energies of A, B, and C are 250 kJ/mol, 50 kJ/mol, and 200 kJ/mol, respectively. Find the Gibbs free energy change for this reaction. Is the reaction spontaneous?

Solution

$$A + B \rightarrow C$$

$$\begin{aligned}\Delta G &= \Sigma G_{\text{products}} - \Sigma G_{\text{reactants}} \\ &= \left[(1\text{ mol C})\left(\frac{200\text{ kJ}}{\text{mol C}}\right)\right] - \left[(1\text{ mol A})\left(\frac{250\text{ kJ}}{\text{mol A}}\right) + (1\text{ mol B})\left(\frac{50\text{ kJ}}{\text{mol B}}\right)\right] \\ &= (200\text{ kJ}) - (250\text{ kJ} + 50\text{ kJ}) = 200\text{ kJ} - 300\text{ kJ} = -100\text{ kJ}\end{aligned}$$

The reaction is spontaneous because the Gibbs free energy change is negative. The products are at a lower chemical potential than the reactants.

•
•

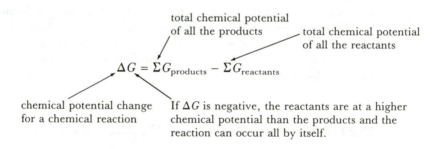

EXAMPLE 2 Determine the Gibbs free energy change for the hypothetical reaction

$$2E + G \rightarrow 3X$$

The Gibbs free energies (chemical potentials) of E, G, and X are 100 kcal/mol, 200 kcal/mol, and 300 kcal/mol, respectively. Does this reaction occur spontaneously?

Solution

$$\Delta G = \Sigma G_{\text{products}} - \Sigma G_{\text{reactants}}$$

$$= \left[(3 \text{ mol X})\left(\frac{300 \text{ kcal}}{\text{mol X}}\right)\right] - \left[(2 \text{ mol E})\left(\frac{100 \text{ kcal}}{\text{mol E}}\right) + (1 \text{ mol G})\left(\frac{200 \text{ kcal}}{\text{mol G}}\right)\right]$$

$$= 900 \text{ kcal} - (200 \text{ kcal} + 200 \text{ kcal}) = 900 \text{ kcal} - 400 \text{ kcal} = 500 \text{ kcal}$$

This reaction is not spontaneous. The reverse reaction

$$3X \rightarrow 2E + G$$

is spontaneous because it has a Gibbs free energy change of −500 kcal. See the next example for details.

•
•

EXAMPLE 3 Calculate the Gibbs free energy change for the reverse reaction shown in Example 2.

Solution

$$3X \rightarrow 2E + G$$

$$\Delta G = \Sigma G_{\text{products}} - \Sigma G_{\text{reactants}}$$

$$= \left[(2 \text{ mol E})\left(\frac{100 \text{ kcal}}{\text{mol E}}\right) + (1 \text{ mol G})\left(\frac{200 \text{ kcal}}{\text{mol G}}\right)\right] - \left[(3 \text{ mol X})\left(\frac{300 \text{ kcal}}{\text{mol X}}\right)\right]$$

$$= (200 \text{ kcal} + 200 \text{ kcal}) - 900 \text{ kcal} = 400 \text{ kcal} - 900 \text{ kcal} = -500 \text{ kcal}$$

The reverse reaction

$$3X \rightarrow 2E + G$$

is spontaneous because it has a negative Gibbs free energy change. Its reactants are at a higher chemical potential than its products.

•
•

EXAMPLE 4 The Gibbs free energy change for the hypothetical reaction

$$A + B \rightarrow E$$

is −200 kJ. The Gibbs free energies of A and B are 250 kJ/mol and 500 kJ/mol, respectively. What is the molar Gibbs free energy of E?

Solution

$$\Delta G = \Sigma G_{\text{products}} - \Sigma G_{\text{reactants}}$$

$$-200 \text{ kJ} = \Sigma G_{\text{products}} - \Sigma G_{\text{reactants}}$$

$$= [(1 \text{ mol E})(G_{\text{E}})]$$

(arrow to G_{E}: molar Gibbs free energy of E)

$$- \left[(1 \text{ mol A})\left(\frac{250 \text{ kJ}}{\text{mol A}}\right) + (1 \text{ mol B})\left(\frac{500 \text{ kJ}}{\text{mol B}}\right)\right]$$

$$= (1 \text{ mol E})(G_{\text{E}}) - (250 \text{ kcal} + 500 \text{ kcal})$$

$$= (1 \text{ mol E})(G_{\text{E}}) - 750 \text{ kJ}$$

$$750 \text{ kJ} - 200 \text{ kJ} = (1 \text{ mol E})(G_E)$$
$$G_E = 550 \text{ kJ/mol E}$$

13.2.3 GIBBS FREE ENERGY CHANGE CALCULATIONS, REAL REACTIONS

Recall that we cannot know the absolute enthalpy (heat content) of substances, and so we arbitrarily define the standard molar enthalpy of formation, abbreviated ΔH_f°, for any element in its most stable form under conditions of 1 atm pressure and 25°C as being exactly 0 kcal/mol. Likewise, the standard molar Gibbs free energy of formation, abbreviated ΔG_f°, for any element in its most stable form under conditions of 1 atm pressure and 25°C is also arbitrarily set equal to exactly 0 kcal/mol (or 0 kJ/mol). For simplicity, sometimes we will refer to the standard molar Gibbs free energy of formation as simply the relative chemical potential, or Gibbs free energy, of the compound. Just as we define the standard enthalpy change of a reaction in terms of the standard molar enthalpies of formation of the products and of the reactants, we can define the standard Gibbs free energy change for a chemical reaction (abbreviated ΔG°) as shown in the next example.

EXAMPLE 5 One mole of carbon (graphite) burns with 1 mol of oxygen. Find the standard Gibbs free energy change in kilocalories for this reaction.

$$C(\text{graphite}) + O_2(g) \rightarrow CO_2(g)$$

Solution Values of standard molar Gibbs free energies of formation are shown in Table 12.1.

$$\Delta G^\circ = \Sigma \Delta G^\circ_{f_{\text{products}}} - \Sigma \Delta G^\circ_{f_{\text{reactants}}}$$

ΔG°: standard Gibbs free energy change

$\Sigma \Delta G^\circ_{f_{\text{products}}}$: total standard Gibbs free energies of formation (relative chemical potentials) of all the products

$\Sigma \Delta G^\circ_{f_{\text{reactants}}}$: total standard Gibbs free energies of formation (relative chemical potentials) of all the reactants

$$= \left[(1 \text{ mol } CO_2)\left(\frac{-94 \text{ kcal}}{\text{mol } CO_2}\right)\right]$$
$$- \left[(1 \text{ mol C})\left(\frac{0 \text{ kcal}}{\text{mol C}}\right) + (1 \text{ mol } O_2)\left(\frac{0 \text{ kcal}}{\text{mol } O_2}\right)\right]$$
$$= -94 \text{ kcal} - (0 \text{ kcal} + 0 \text{ kcal}) = -94 \text{ kcal} + 0 \text{ kcal} = -94 \text{ kcal}$$

This reaction is spontaneous because the Gibbs free energy change is negative.

EXAMPLE 6 Find the standard molar Gibbs free energy of formation of HgO(s) in kilojoules per mole. The standard Gibbs free energy change for the reaction

$$2Hg(l) + O_2(g) \rightarrow HgO(s)$$

is −118 kJ.

Solution

$$\Delta G^\circ = \Sigma \Delta G^\circ_{f_{\text{products}}} - \Sigma \Delta G^\circ_{f_{\text{reactants}}}$$

$$-118 \text{ kJ} = \Sigma\Delta G^\circ_{f_{\text{products}}} - \Sigma\Delta G^\circ_{f_{\text{reactants}}}$$

$$= [(2 \text{ mol HgO(s)})(\Delta G^\circ_{f_{\text{HgO(s)}}})]$$

$$- \left[(2 \text{ mol Hg(l)})\left(\frac{0 \text{ kJ}}{\text{mol Hg(l)}}\right) + (1 \text{ mol } O_2(g))\left(\frac{0 \text{ kJ}}{\text{mol } O_2(g)}\right)\right]$$

$$= [(2 \text{ mol HgO(s)})(\Delta G^\circ_{f_{\text{HgO(s)}}})] - 0 \text{ kJ}$$

$$\Delta G^\circ_{f_{\text{HgO(s)}}} = \frac{-118 \text{ kJ}}{2 \text{ mol HgO(s)}} = \frac{-59 \text{ kJ}}{\text{mol HgO(s)}} = -59 \text{ kJ/mol HgO(s)}$$

13.2.4 RELATING $\Delta G°$, $\Delta H°$, $\Delta S°$, AND T

The standard Gibbs free energy change ($\Delta G°$) is related to the two other thermodynamic function changes ($\Delta H°$ and $\Delta S°$) and to absolute temperature (T in K) as follows:

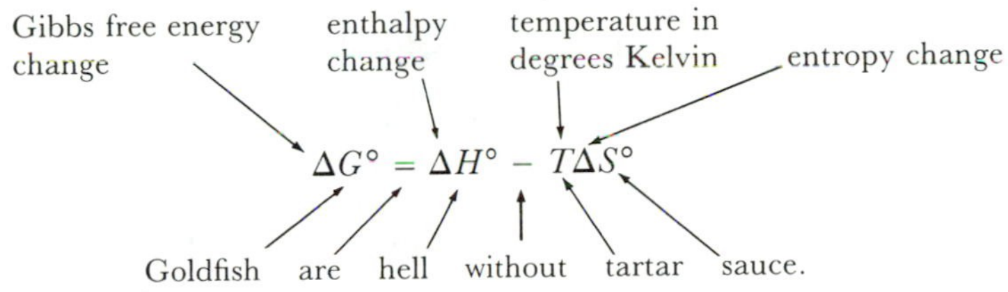

This relation should be easy to memorize because of the mnemonic (memory trick) below the equation. This equation gives us a second method for determining the standard Gibbs free energy change.

EXAMPLE 7 The standard enthalpy change for the reaction

$$C(\text{graphite}) + O_2(g) \rightarrow CO_2(g)$$

is −94 kcal at 25°C and 1 atm pressure. The standard absolute entropy change for this reaction is 1 cal/K. Calculate the standard Gibbs free energy change for this reaction without the use of Table 12.1.

Solution

$$\Delta G° = \Delta H° - T\Delta S°$$

$$= (-94 \text{ kcal}) - (298 \text{ K})\left(\frac{1 \text{ cal}}{\text{K}}\right)$$

$$= -94 \text{ kcal} - 298 \text{ cal} = -94 \text{ kcal} - 0.298 \text{ kcal} = -94 \text{ kcal}$$

This reaction is spontaneous because the Gibbs free energy change is negative. (Carbon does react with oxygen spontaneously to produce carbon dioxide.)

EXAMPLE 8 The standard enthalpy change for the reaction

$$S(s) + O_2(g) \rightarrow SO_2(g)$$

is −71 kcal at 25°C. The $\Delta S°$ value for this reaction is 2 cal/K. Calculate the standard Gibbs free energy change for this reaction without using Table 12.1.

Solution

$$\Delta G^\circ = \Delta H^\circ - T\Delta S^\circ$$

$$= (-71 \text{ kcal}) - (298 \text{ K})\left(\frac{2 \text{ cal}}{\text{K}}\right)$$

$$= -71 \text{ kcal} - 596 \text{ cal} = -71 \text{ kcal} - 0.596 \text{ kcal} = -72 \text{ kcal}$$

Sulfur spontaneously (on its own) reacts with oxygen to form sulfur dioxide.

EXAMPLE 9 Consider the hypothetical reaction A → B, which occurs at 25°C and 1 atm pressure. The ΔH_f° values for A and B are 10 kcal/mol and 20 kcal/mol, respectively. The S° values for A and B are 5 cal/mol K and 15 cal/mol K, respectively. Calculate the standard Gibbs free energy change for this reaction. Is this reaction spontaneous?

Solution There are three steps to the solution of this problem. First, calculate the enthalpy change. Second, calculate the entropy change. Third, using the calculated enthalpy and entropy changes, calculate the Gibbs free energy change.

Step 1 Calculate the enthalpy change:

$$\Delta H^\circ = \Sigma\Delta H^\circ_{f_{\text{products}}} - \Sigma\Delta H^\circ_{f_{\text{reactants}}}$$

$$= \left[(1 \text{ mol B})\left(\frac{20 \text{ kcal}}{\text{mol B}}\right)\right] - \left[(1 \text{ mol A})\left(\frac{10 \text{ kcal}}{\text{mol A}}\right)\right]$$

$$= 20 \text{ kcal} - 10 \text{ kcal} = 10 \text{ kcal (endothermic)}$$

Step 2 Calculate the entropy change:

$$\Delta S^\circ = \Sigma S^\circ_{\text{products}} - \Sigma S^\circ_{\text{reactants}}$$

$$= \left[(1 \text{ mol B})\left(\frac{15 \text{ cal}}{\text{mol B K}}\right)\right] - \left[(1 \text{ mol A})\left(\frac{5 \text{ cal}}{\text{mol A K}}\right)\right]$$

$$= \frac{15 \text{ cal}}{\text{K}} - \frac{5 \text{ cal}}{\text{K}} = \frac{10 \text{ cal}}{\text{K}} \text{ (disorder increases)}$$

Step 3 Calculate the Gibbs free energy change:

$$\Delta G^\circ = \Delta H^\circ - T\Delta S^\circ$$

$$= 10 \text{ kcal} - (298 \text{ K})\left(\frac{10 \text{ cal}}{\text{K}}\right)$$

$$= 10 \text{ kcal} - 2980 \text{ cal} = 10 \text{ kcal} - 2.980 \text{ kcal}$$

$$= 7 \text{ kcal (not spontaneous)}$$

The reaction is not spontaneous because the Gibbs free energy change is positive. However, the reverse reaction (B → A) is spontaneous because its Gibbs free energy change is negative (−7 kcal). The reverse reaction is exothermic, with an enthalpy change of −10 kcal.

SELF-TEST

1. The ΔG_f° values for A and B are 15 kcal/mol and 30 kcal/mol, respectively. Find the standard Gibbs free energy change for the hypothetical reaction 3A → 2B. Is this reaction spontaneous?

2. The $\Delta G°$ value for the hypothetical reaction X + Y → Z is −100 kcal. The $\Delta G_f°$ values for X and Y are 40 kcal/mol and 30 kcal/mol, respectively. Find the $\Delta G_f°$ value of Z in kilocalories per mol of Z.
3. The $\Delta S°$ value for the hypothetical reaction M + 2N → 4R is 20 cal/K at 25°C and 1 atm pressure. The standard enthalpy change for this reaction is 300 kcal. Find the $\Delta G°$ value for this reaction. Is the reaction spontaneous?

ANSWERS **1.** 15 kcal; not spontaneous **2.** −30 kcal/mol Z **3.** 294 kcal; not spontaneous

13.2.5 TWO UNIVERSAL DRIVING FORCES

The Gibbs free energy change is made up of two terms: (1) an energy term (the enthalpy change) and (2) a disorder term (entropy change times temperature).

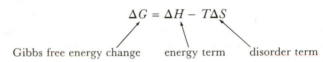

$$\Delta G = \Delta H - T\Delta S$$

Gibbs free energy change (ΔG) — energy term (ΔH) — disorder term ($T\Delta S$)

These two terms represent the two driving forces in nature: All systems (events) try to (1) decrease their energy (have a negative enthalpy change) and (2) increase their disorder (increase their entropy, have a positive entropy change, exist in a more probable state). For example, objects fall to the ground to decrease their energy, and gases expand to become more disordered.

Both of these tendencies (energy minimization and disorder maximization) are reflected by the Gibbs free energy change. The enthalpy change (ΔH term) reflects the energy change of a system (reaction), and the entropy change (ΔS term) reflects the disorder change of a system.

These two forces (tendencies), energy minimization and disorder maximization, can work together to drive a system to a common goal, or they can oppose each other. In the next two sections we will study each of these possibilities.

13.2.6 JIGSAW PUZZLE ANALOGY

Let's look at a situation where energy minimization and disorder maximization work together. If you drop a jigsaw puzzle (either assembled or not), two things happen: (1) The puzzle falls to the floor and then (2) scatters across the floor as hundreds of individual pieces. That is, the puzzle minimizes its energy (goes from a higher energy level to a lower energy level) and then becomes more disordered.

Why is it more probable that a jigsaw puzzle will be found in a scattered (not assembled) state than in an assembled state?

There are many more ways in which a jigsaw puzzle can exist scattered (not assembled) than there are ways for it to exist intact. Therefore, it is more probable that a jigsaw puzzle will be found as many individual pieces, scattered on the floor.

13.2.7 WATER VAPOR VS. LIQUID WATER, PART II

Sometimes energy minimization and disorder maximization work in opposition to each other. The equilibrium between water vapor and liquid water is such an example.

$$H_2O(l) \rightleftharpoons H_2O(g)$$

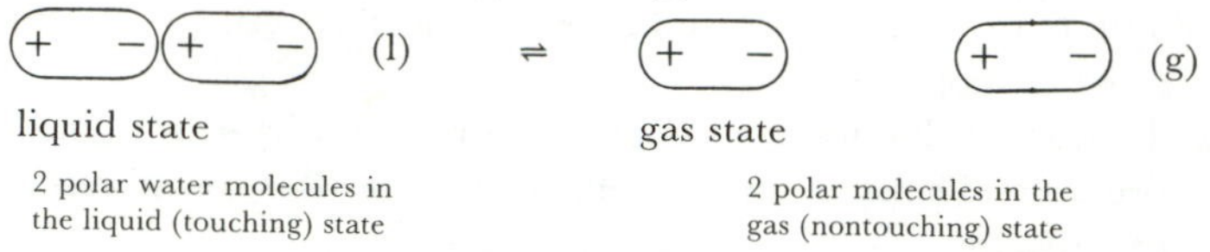

In which state (liquid or gas) is water more disordered?

Right. Water is more disordered in the gas state and less disordered in the liquid state.

Which state exists at a lower energy level?

The liquid state is at a lower energy level than the gas state because the positive pole of one water molecule attracts the negative pole of the other molecule, just as the south pole of one magnet attracts the north pole of another magnet. Pulling apart two polar water molecules (or two magnets) requires energy. The separated water molecules (or magnets) are in a higher energy state. Water vapor molecules (separated water molecules) are in a higher energy state than liquid water molecules.

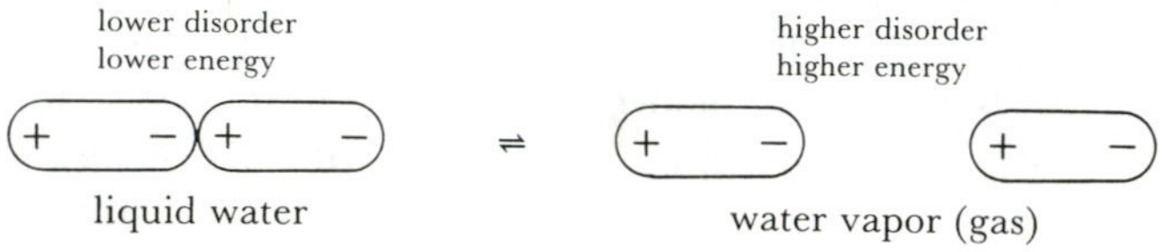

If the tendency to maximize disorder is followed, then water should exist as a gas. If the tendency to minimize energy is followed, then water should exist as a liquid. Which of these two tendencies dominates? As we will see in the following section, the temperature of the system determines which of these two tendencies becomes the dominant tendency.

13.2.8 IMPORTANT ROLE OF TEMPERATURE

When the two forces, energy minimization and disorder maximization, are in opposition, it is the system's temperature that determines the dominant force. Notice the temperature in the Gibbs free energy change equation:

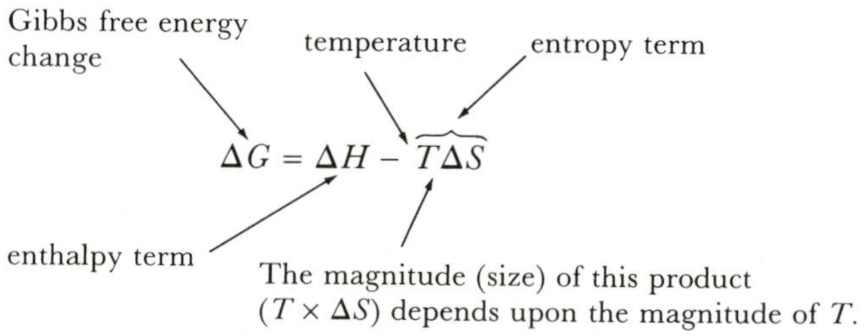

$$\Delta G = \Delta H - \overbrace{T\Delta S}$$

At very high temperatures, the product of temperature times entropy change is larger than the enthalpy change. We say that at high temperatures the entropy term dominates the enthalpy term. At very high temperatures, the tendency to maximize disorder dominates the tendency to minimize energy. Therefore, water is found as a gas (disorder maximized) at high temperatures.

At very low temperatures, the product of temperature times entropy change is small relative to the enthalpy change, the enthalpy term dominates the entropy term, and the tendency to minimize energy dominates the tendency to maximize disorder. Therefore, water is found as a liquid or solid (ice) at lower temperatures.

13.3 APPLICATIONS

13.3.1 CONVECTION CURRENTS

Suppose we have a large container (a classroom) with a match burning at one end (see Figure 13.2). Warm air from the match rises and cooler air at the ceiling falls to the floor. This movement of air is referred to as convection currents.

FIGURE 13.2 Convection currents produced by a burning match

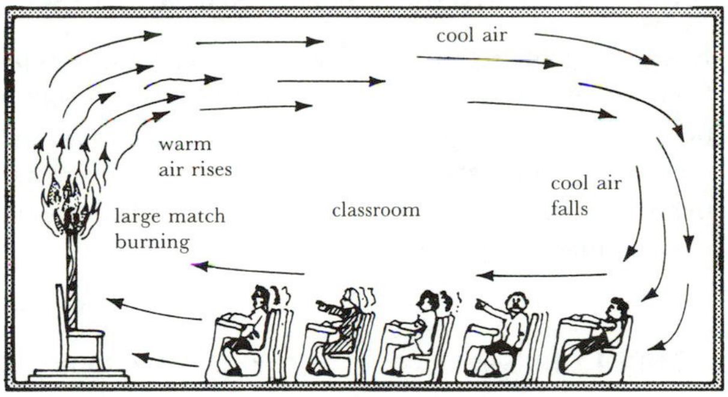

Why does this happen? Why doesn't the tendency to maximize disorder cause warm air around the match to spread out evenly across the entire room?

To answer this question, ask yourself another question.

Why does a pencil, released from the ceiling, fall to the floor (see Figure 13.3)?

FIGURE 13.3 Pencils and cold air fall to lower energy levels.

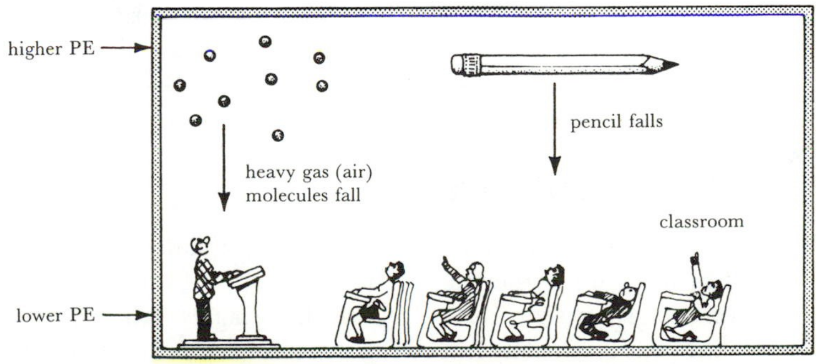

A pencil falls because it is heavier than the surrounding air. The pencil falls to lower its potential energy. Likewise, cool air is heavier than warm air. Cool air falls to minimize its potential energy. At room temperature, energy minimization dominates disorder maximization for these two objects.

At a very high temperature (e.g., one million degrees), the pencil will not fall! Under these conditions, the tendency to minimize energy (falling) does not dominate. At very high temperatures, the tendency to maximize disorder (entropy) dominates the tendency to minimize energy. Thus, the pencil is converted into a gas, and this gas evenly distributes itself (maximizes its disorder) around the room.

At very low temperatures (e.g., −273°C), air molecules in the room turn into a solid crystalline structure and fall to the floor (minimize energy). At very low temperatures, the tendency to minimize energy (fall) dominates the tendency to maximize disorder (spread out). The solid air molecules lower their potential energy by falling from higher energy levels (between the floor and ceiling) to lower energy levels at the floor.

13.3.2 FLUID MOTION IN A PARTIALLY EVACUATED BOX, PART II

Reconsider the hypothetical box with a partition separating neutral (uncharged) gas molecules from a pure vacuum, shown in Figure 13.4. At room temperature (295 K), a relatively high temperature for a gas, air molecules fill the entire box (state 2) when the partition is removed from the original box (state 1). However, at very low temperatures (such as −273°C), the molecules behind the partition do not spread out to fill the room. The molecules fall to the floor (state 3) to minimize their energy.

FIGURE 13.4 Gas behavior at high and low temperatures

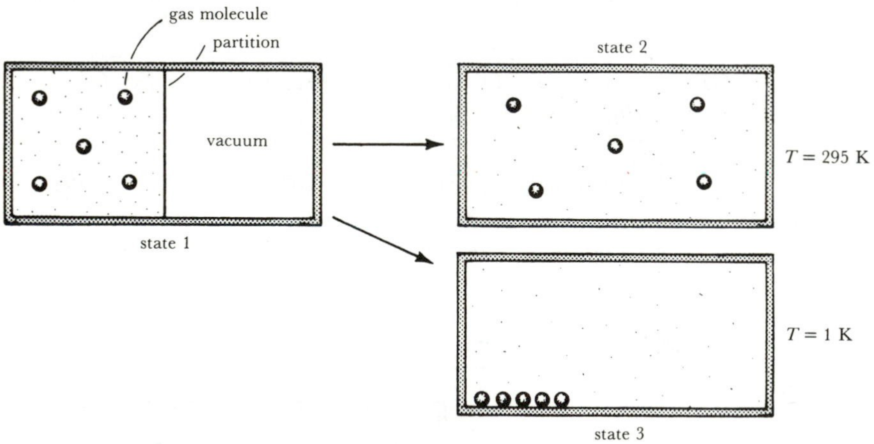

13.3.3 APOLLO I DISASTER

Let's see how these ideas can be used to explain a mystery associated with the Apollo I tragedy. In 1967, three American astronauts were killed in a spacecraft fire while the craft (Apollo I) was on the ground (on earth). As was the practice of our space program, the atmosphere inside the craft was essentially pure oxygen. A fire accidentally started inside the craft, quickly developed out of control, and killed the three American astronauts inside.

In an attempt to learn how to deal with such emergencies, the United States government flew unmanned (robot-controlled) missions in which matches were repeatedly struck by remote control in a weightless environment and a 100% oxygen atmosphere. A fire could not be set in these early attempts. In the 1990's, astronauts continued to experiment with confined fires aboard space shuttles to further study the spread of flames in space. Why couldn't investigators start a fire in their first attempts, and why are they now successful?

Under weightless conditions, free from the earth's gravity, the tendency to minimize energy (by having objects fall) doesn't exist because there is no potential energy difference between the floor and the ceiling. Only the tendency to maximize disorder is at work. A pencil (or astronaut) floats in outer space (all objects are essentially weightless). Objects cannot change their potential energy by moving from one place to another. There is no way for objects to lower their energy levels (see Figure 13.5).

FIGURE 13.5 Gravity is not present in outer space to make pencils and cold air fall.

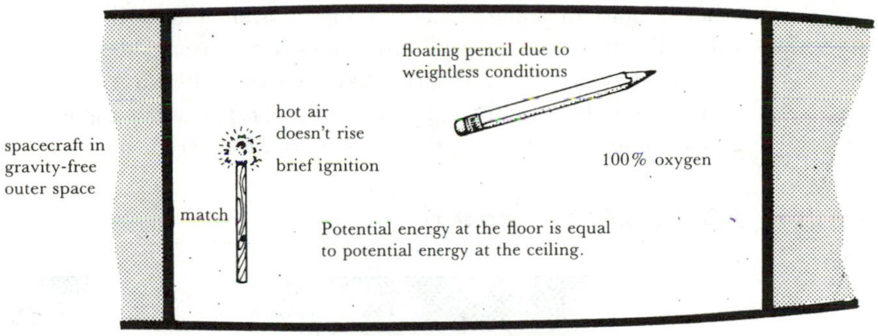

On earth, gravity aids combustion. Warm air rises (because it is lighter than the surrounding air) while cooler air (with fresh oxygen) falls, because it is heavier than the surrounding air. Convection currents (rising hot air being replaced with falling cooler air) keep a match burning by supplying the match with oxygen, as shown in Figure 13.6.

Cold air, pencils, and astronauts do not fall in outer space. Combustion products (e.g., carbon dioxide) hover around a match when it begins to burn. This layer of combustion products extinguishes the match in much the same way that a blanket or a carbon dioxide fire ex-

FIGURE 13.6 Convection currents on earth

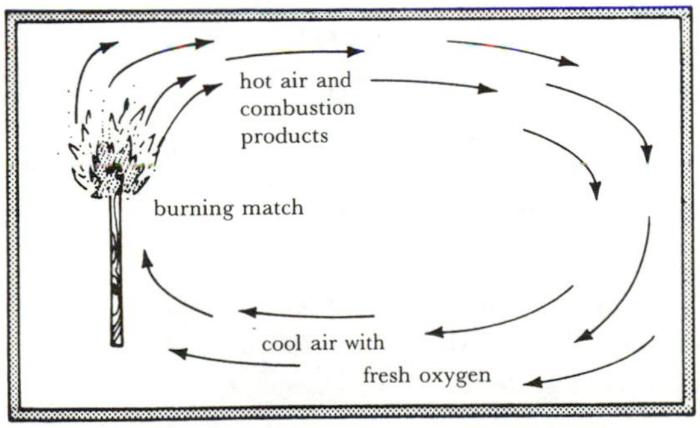

tinguisher on earth does. The combustion products do not rise away from the match because they are not lighter than the surrounding air (because nothing has weight, nothing can be heavier or lighter). The layer of surrounding combustion gases deprives the match of needed oxygen.

Without ventilating fans, you would be killed by your own exhaled breath under weightless conditions. On earth your warm exhaled breath rises away from your face and is replaced by cooler fresh air containing oxygen. If you were to stay in one place for too long under the weightless conditions of a spacecraft, you would fall down dead. (Actually, you would sort of float away dead.)

13.3.4 STARS, PART I

Interstellar space (the space between stars) is essentially 100% hydrogen (about 1 hydrogen atom or molecule per cubic centimeter of space). The temperature of outer space is about 3 K (−270°C, or −455°F).

In Figure 13.7, state 1 shows hydrogen under normal conditions of outer space, and state 2 shows hydrogen packed more closely together. Hydrogen molecules in outer space attract each other because of gravity. There is a gravitational attraction between any two objects in the universe. Because gravitational attractions are so weak, they usually go unnoticed unless one or more of the objects is very large (like the earth). Because of the gravitational attraction between hydrogen molecules in outer space, state 1 represents a higher potential energy state than state 2 (see Figure 13.8). Also, state 1 is more disordered than state 2.

FIGURE 13.7 Birth of a star

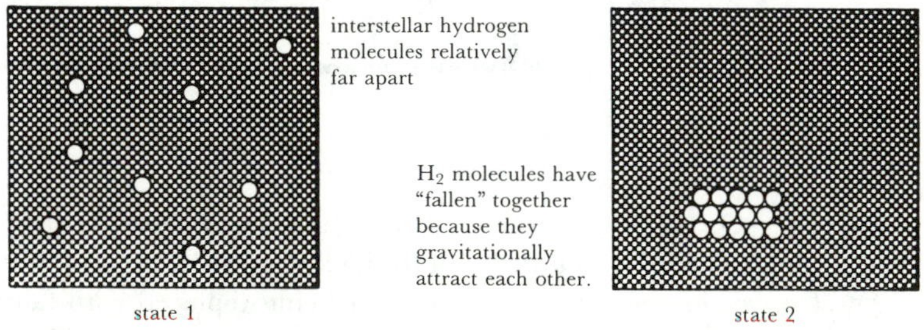

FIGURE 13.8 Disorder and potential energy in interstellar hydrogen molecules

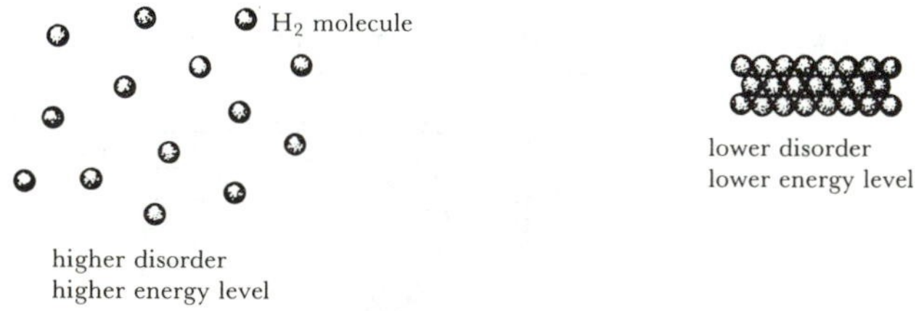

The two forces (energy minimization and disorder maximization) are working in opposition here. The tendency to maximize disorder favors state 1, and the tendency to minimize energy favors state 2.

Which of these two tendencies dominates in interstellar space?

Because the temperature of outer space is so low (−270°C), the tendency to minimize energy dominates the tendency to maximize disorder. Large volumes of hydrogen gas spontaneously condense (fall together) in outer space because of gravitational attractions between the molecules. The final condensed mass of hydrogen has a volume 100 000 times smaller than the original disordered mass of hydrogen.

Whenever objects lower their energy levels, energy is released. The heat released when interstellar hydrogen condenses raises the temperature of the hydrogen to about 20 million degrees Celsius. Fusion occurs at this temperature. The fusion process joins (fuses) hydrogen atoms together and converts the hydrogen to helium. Mass is always lost (mass is converted into energy) in fusion reactions. Our sun converts 5 million tons of mass per second into energy this way. It sends 5 million tons of mass (in the form of energy) into the universe at the speed of light. This increases the disorder of the universe. The tendency to disorder is very strong at high temperatures.

13.4 EQUILIBRIUM CONSTANTS

13.4.1 RELATING K, $\Delta G°$, $\Delta H°$, $\Delta S°$, AND T

Each of the thermodynamic functions is a different manifestation (form) of the equilibrium constant K. The relationship between K and the Gibbs free energy change for a reaction is shown by

temperature in degrees Kelvin

$$\Delta G° = -2.303\ RT \log K$$

gas law constant (1.987 cal/mol K)

equilibrium constant

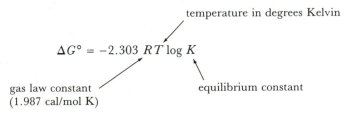

At a given temperature, the Gibbs free energy change is equal to the product of four constants: −2.303, R (the gas law constant), T, and the log of the equilibrium constant K. The Gibbs free energy is just a different manifestation of the equilibrium constant. You can think of the Gibbs free energy change as the equilibrium constant in disguise

•
•

EXAMPLE 10 The equilibrium constant for the hypothetical reaction A → B at 40°C is equal to 1.00×10^{-5}. Find the change in the Gibbs free energy for this reaction. Is the reaction spontaneous?

Solution

$$40°\text{C} = 313\ \text{K}$$

$$\Delta G° = -2.303\ RT \log K$$

$$= -2.303 \left(\frac{1.987\ \text{cal}}{\text{mol K}}\right)(313\ \text{K}) \log (10^{-5})$$

$$\Delta G° = -2.303 \left(\frac{1.987\ \text{cal}}{\text{mol K}}\right)(313\ \text{K})(-5.00) = +7162\ \text{cal/mol}$$

This is interpreted as calories per mole of equation, that is, for the equation A → B.

$$= +7.2\ \text{kcal (for the equation A} \rightarrow \text{B)}$$

The reaction is not spontaneous because the Gibbs free energy change is positive.

•
•

EXAMPLE 11 The pK for the hypothetical reaction $G \rightarrow H$ at 25°C is 50.00. Find the standard Gibbs free energy change for this reaction. Is this reaction spontaneous?

Solution

$$pK = 50.00$$
$$K = 1.0 \times 10^{-50}$$
$$\log K = -50.00$$
$$25°\text{C} = 298 \text{ K}$$
$$\Delta G° = -2.303\ RT \log K$$
$$= -(2.303)\left(\frac{1.987 \text{ cal}}{\text{mol K}}\right)(298 \text{ K})(-50.00) = 68\,183 \text{ cal/mol}$$
$$= +68 \text{ kcal (for the equation } G \rightarrow H)$$

The reaction is not spontaneous because the Gibbs free energy change is positive.

•
•

EXAMPLE 12 The standard Gibbs free energy change for the hypothetical reaction $R \rightarrow X$ is 7.2 kcal at 40°C. What is the equilibrium constant for this reaction?

Solution

$$40°\text{C} = 313 \text{ K}$$
$$\Delta G° = -2.303\ RT \log K$$
$$-2.303\ RT \log K = \Delta G°$$
$$\log K = \frac{\Delta G°}{-2.303\ RT} = \frac{7200 \text{ cal/mol}}{-2.303\left(\frac{1.987 \text{ cal}}{\text{mol K}}\right)(313 \text{ K})} = -5.0$$
$$K = \text{antilog}(-5.0) = 10^{-5.0} = 1.0 \times 10^{-5} = \frac{[\text{X}]}{[\text{R}]}$$

•
•

EXAMPLE 13 Find the equilibrium constant for the reaction

$$S(s) + O_2(g) \rightarrow SO_2(g)$$

at 25°C and 1 atm pressure.

Solution Use Table 12.1 to obtain the standard molar Gibbs free energy of formation for $SO_2(g)$. (At 25°C and 1 atm pressure, the ΔG_f° values of S(s) and $O_2(g)$ are 0 kcal/mol.)

$$\Delta G° = \Sigma \Delta G°_{f_{\text{products}}} - \Sigma \Delta G°_{f_{\text{reactants}}}$$
$$= \left[(1 \text{ mol } SO_2(g))\left(\frac{-72 \text{ kcal}}{\text{mol } SO_2(g)}\right)\right]$$
$$- \left[(1 \text{ mol S(s)})\left(\frac{0 \text{ kcal}}{\text{mol S(s)}}\right) + (1 \text{ mol } O_2(g))\left(\frac{0 \text{ kcal}}{\text{mol } O_2(g)}\right)\right]$$
$$= -72 \text{ kcal} - (0 \text{ kcal} + 0 \text{ kcal}) = -72 \text{ kcal} = -72\,000 \text{ cal}$$
$$\log K = \frac{-\Delta G°}{2.303\ RT} = \frac{-(-72\,000 \text{ cal/mol})}{(2.303)\left(\frac{1.987 \text{ cal}}{\text{mol K}}\right)(298 \text{ K})} = 52.80$$

$$K = \text{antilog}\ (52.80) = 10^{52.80} = 10^{52} \times 10^{0.80} = 10^{52} \times \text{antilog}\ (0.80) = 6.3 \times 10^{52}$$

EXAMPLE 14 Find the equilibrium constant for the reaction

$$C(\text{graphite}) + O_2(g) \rightarrow CO_2(g)$$

at 25°C and 1 atm pressure.

Solution Use Table 12.1 to obtain the ΔG_f° value of $CO_2(g)$.

$$\Delta G^\circ = \Sigma \Delta G^\circ_{f_{\text{products}}} - \Sigma \Delta G^\circ_{f_{\text{reactants}}}$$
$$= (-94 \text{ kcal}) - (0 \text{ kcal}) = -94 \text{ kcal}$$

$$\log K = \frac{-\Delta G^\circ}{2.303\ RT} = \frac{-(-94\,000 \text{ cal/mol})}{2.303\left(\dfrac{1.987 \text{ cal}}{\text{mol K}}\right)(298 \text{ K})} = 68.93$$

$$K = \text{antilog}\ 68.93 = 10^{68.93} = 8.5 \times 10^{68}$$

SELF-TEST

1. The equilibrium constant K for the hypothetical reaction G → A at 10°C is equal to 1.00×10^{-20}. Find the Gibbs free energy change for this reaction.
2. The pK for the hypothetical reaction D → X at 60°C is equal to 70.00. Find the Gibbs free energy change for this reaction.
3. The Gibbs free energy change for the hypothetical reaction A + B → C at 85°C is equal to 106.5 kcal. Find the pK for this reaction.

ANSWERS **1.** 25.9 kcal **2.** 106.7 kcal **3.** 65.0

Enthalpy and entropy are also related to the equilibrium constant. Since

$$\Delta G^\circ = -2.303\ RT \log K$$

and

$$\Delta G^\circ = \Delta H^\circ - T\Delta S^\circ$$

then

$$\Delta H^\circ - T\Delta S^\circ = -2.303\ RT \log K$$

13.4.2 CALCULATING VERY LARGE AND VERY SMALL *K*'s

In the last section, we saw that the equilibrium constant for the reaction in which carbon burns in oxygen to form carbon dioxide is very large (8.5×10^{68}). We defined K to be very large whenever it is greater than 10. In this section we want to discuss the difficulty of determining very large or very small equilibrium constants using laboratory measurements.

Consider the hypothetical reaction

$$A \rightarrow B$$

whose equilibrium constant is equal to 10^{12}. That is,

$$K = \frac{[B]}{[A]} = 1 \times 10^{12}$$

If A and B are chemicals dissolved in a 1-L solution, and if the concentration of B equals 1 M (1 mol of B per liter of solution), what is the concentration of A?

The concentration of A must be 10^{-12} M.

$$K = \frac{[B]}{[A]}$$

$$10^{12} = \frac{(1)}{[A]}$$

$$[A] = \frac{1}{10^{12}} = 10^{-12} \text{ mol A/L soln}$$

It would be very difficult (if not impossible) to determine the concentration of A (one trillionth of a mole of A per liter of solution) in order to calculate this equilibrium constant. In situations such as this (when concentrations of reactants or products are very low and difficult to determine), the Gibbs free energy change is used to calculate the equilibrium constant, using the equation

$$\Delta G° = -2.303\ RT \log K$$

Such calculations were illustrated in the preceding section.

13.4.3 STOMACH ACID, PART II*

EXAMPLE 15 Your stomach makes about 3 qt of 0.1 M hydrochloric acid every day. Your body prepares the acid by concentrating the hydrogen ion in your blood. The *p*H of your blood is about 7.4. The *p*H of your stomach acid is about 1. How much work is done by your body when it concentrates the hydrogen ion concentration from 3.98×10^{-8} mol of H^+ per liter (if *p*H = 7.4, $[H^+] = 3.98 \times 10^{-8}$) to 0.1 mol of H^+ per liter (if *p*H = 1, $[H^+] = 0.1$)?

Solution

$$H_i^+ \rightleftharpoons H_f^+$$

initial $[H^+]$ in blood (→ H_i^+) final $[H^+]$ in stomach (→ H_f^+)

$$K = \frac{[H_f^+]}{[H_i^+]} = \frac{(0.100)}{(3.98 \times 10^{-8})} = 2.51 \times 10^6$$

$$\Delta G° = -2.303\ RT \log K$$

98.6°F = 37°C = 310 K

$$= -2.303\left(\frac{1.987 \text{ cal}}{\text{K mol } H^+}\right)(310 \text{ K}) \log (2.51 \times 10^6)$$

$$= \frac{-10\,000 \text{ cal}}{\text{mol } H^+ \text{ concentrated}}$$

$$(3.00 \text{ L soln})\left(\frac{0.100 \text{ mol } H^+}{1.00 \text{ L soln}}\right) = 0.3 \text{ mol } H^+$$

Therefore, 3 L of 0.1 M H^+ contains 0.3 mol of H^+. Your body concentrates 0.3 mol of H^+ per day.

$$\left(\frac{0.3 \text{ mol } H^+}{\text{day}}\right)\left(\frac{10\,000 \text{ cal}}{\text{mol } H^+ \text{ conc}}\right) = \frac{3000 \text{ cal}}{\text{day}} = 3000 \text{ cal/day}$$

*Adapted from J. A. Campbell, "Eco-Chem Question Number 244," *Journal of Chemical Education* 53 (June 1976):371.

Every day your body uses (burns) about 3000 calories (food) just to prepare the 3 qt of stomach acid needed to help digest the food you eat. If you're an average American, your diet provides you with 2000 to 3000 kcal/day (between 2 and 3 million calories per day). Recall from Chapter 3 that the food *Calorie* is actually a kilocalorie.

13.5 REVIEW PROBLEM

The following problem reviews many thermodynamic calculations and ideas. Hopefully, in addition to being a review, it will convince you how easy (mechanically) thermodynamic problems can be because of the similar approaches used to calculate $\Delta H°$, $\Delta S°$, and $\Delta G°$.

EXAMPLES 16–22 Consider the hypothetical reaction

$$A(s) + B(s) \rightarrow 2C(g)$$

The enthalpies, entropies, and Gibbs free energies of A, B, and C are given in the accompanying table:

Compound	$\Delta H_f°$ (kcal/mol)	$S°$ (kcal/mol K)	$\Delta G_f°$ (kcal/mol)
A(s)	−10	0.010	3.0
B(s)	−20	0.020	22.0
C(g)	−30	0.030	−7.0

The solutions that follow address these questions related to the reaction:

16. What is the standard enthalpy change?
17. What is the absolute standard entropy change?
18. Calculate $\Delta G°$ for this reaction in two different ways.
19. What is the value for the equilibrium constant?
20. Discuss the heat changes.
21. Is this reaction spontaneous?
22. Discuss the disorder changes for the reaction, the universe, and the surroundings.

Solution 16 The enthalpy change is

$$\Delta H° = \Sigma \Delta H°_{f_{\text{products}}} - \Sigma \Delta H°_{f_{\text{reactants}}}$$

$$= \left[(2 \text{ mol C(g)})\left(\frac{-30 \text{ kcal}}{\text{mol C(g)}}\right)\right]$$

$$- \left[(1 \text{ mol A(s)})\left(\frac{-10 \text{ kcal}}{\text{mol A(s)}}\right) + (1 \text{ mol B(s)})\left(\frac{-20 \text{ kcal}}{\text{mol B(s)}}\right)\right]$$

$$= (-60 \text{ kcal}) - [(-10 \text{ kcal}) + (-20 \text{ kcal})]$$

$$= -60 \text{ kcal} - (-30 \text{ kcal}) = -60 \text{ kcal} + 30 \text{ kcal} = -30 \text{ kcal}$$

Solution 17 The entropy change is

$$\Delta S^\circ = \Sigma S^\circ_{\text{products}} - \Sigma S^\circ_{\text{reactants}}$$

$$= \left[(2\text{ mol C(g)})\left(\frac{0.030\text{ kcal}}{\text{mol C(g) K}}\right)\right]$$

$$- \left[(1\text{ mol A(s)})\left(\frac{0.010\text{ kcal}}{\text{mol A(s) K}}\right) + (1\text{ mol B(s)})\left(\frac{0.020\text{ kcal}}{\text{mol B(s) K}}\right)\right]$$

$$\Delta S^\circ = \frac{0.060\text{ kcal}}{\text{K}} - \left(\frac{0.010\text{ kcal}}{\text{K}} + \frac{0.020\text{ kcal}}{\text{K}}\right)$$

$$= \frac{0.060\text{ kcal}}{\text{K}} - \frac{0.030\text{ kcal}}{\text{K}} = 0.030\text{ kcal/K}$$

•
•

Solution 18 The Gibbs free energy change is

$$\Delta G^\circ = \Sigma\Delta G^\circ_{f_{\text{products}}} - \Sigma\Delta G^\circ_{f_{\text{reactants}}}$$

$$= \left[(2\text{ mol C(g)})\left(\frac{-7.0\text{ kcal}}{\text{mol C(g)}}\right)\right]$$

$$- \left[(1\text{ mol A(s)})\left(\frac{3.0\text{ kcal}}{\text{mol A(s)}}\right) + (1\text{ mol B(s)})\left(\frac{22.0\text{ kcal}}{\text{mol B(s)}}\right)\right]$$

$$= -14.0\text{ kcal} - (3.0\text{ kcal} + 22.0\text{ kcal}) = -14.0\text{ kcal} - 25.0\text{ kcal} = -39.0\text{ kcal}$$

An alternative method for finding the Gibbs free energy change is

$$\Delta G^\circ = \Delta H^\circ - T\Delta S^\circ$$

$$= (-30\text{ kcal}) - (298\text{ K})\left(\frac{0.030\text{ kcal}}{\text{K}}\right)$$

(−30 kcal: from Solution 16; 298 K: 25°C assumed, 25°C = 298 K; 0.030 kcal/K: from Solution 17)

$$= -30\text{ kcal} - 9\text{ kcal} = -39\text{ kcal}$$

Notice that both methods of calculating the Gibbs free energy change produce the same answer. You should expect this if both methods are valid.

•
•

Solution 19 The equilibrium constant is

$$\Delta G^\circ = -2.303\ RT\log K$$

$$\log K = \frac{-\Delta G^\circ}{2.303\ RT} = \frac{-(-39\,000\text{ cal/mol})}{2.303\left(\dfrac{1.987\text{ cal}}{\text{mol K}}\right)(298\text{ K})} = 28.60$$

(−39 000 cal/mol: from Solution 18)

$$K = \text{antilog } 28.60 = 10^{28.60} = 10^{28} \times 10^{0.60}$$

$$10^{0.60} = \text{antilog } 0.60 = 4.0$$

$$K = 4.0 \times 10^{28}$$

•
•

Solution 20 The enthalpy change is negative (−30 kcal). The heat content of the product is not as large as the sum of the heat contents of the two reactants. Some heat spills over into the surroundings. The reaction is exothermic. The reverse reaction would be endothermic, with a standard enthalpy change of +30 kcal.

•
•

Solution 21 The Gibbs free energy change is negative (−39 kcal). The product has a lower chemical potential than the sum of the chemical potentials of the two reactants. The reaction occurs spontaneously in the forward direction as written. The reverse reaction would not be spontaneous. The standard Gibbs free energy change for the reverse reaction is +39 kcal.

•
•

Solution 22 The entropy change for this reaction is positive (+0.030 kcal/K). The disorder of the products is greater than the disorder of the reactants. When this reaction occurs, the disorder of the system increases. This is reasonable because both of the reactants are solids (low disorder) and the product is a gas (high disorder). The entropy change for this reaction is also referred to as the entropy change of the system and the local entropy change.

$$\Delta S_{\text{reaction}} = \Delta S_{\text{system}} = \Delta S_{\text{local}}$$

The total entropy change refers to the entropy change of the universe. The total entropy change is equal to the entropy change of the system plus the entropy change of the surroundings.

$$\Delta S_{\text{universe}} = \Delta S_{\text{total}} = \Delta S_{\text{system}} + \Delta S_{\text{surroundings}}$$

The entropy change of the universe must be positive because this reaction is spontaneous. (The reaction is spontaneous because the Gibbs free energy change for the reaction is negative.) The entropy change of the universe is always positive when a spontaneous reaction occurs. From this information, we cannot say whether the entropy change of the surroundings is positive or negative.

•
•

PROBLEM SET 13

The problems in Problem Set 13 parallel the examples in Chapter 13. For example, if you should have trouble working Problem 5, refer to Example 5 in this chapter to get help. The correct answers to the problems are given at the end of this book.

1. One mole of A reacts with 1 mol of B to produce 1 mol of C. The Gibbs free energies of A, B, and C are 300 kJ/mol, 400 kJ/mol, and 500 kJ/mol, respectively. Find the Gibbs free energy change for this reaction. Is the reaction spontaneous?
2. Determine the Gibbs free energy change for the hypothetical reaction

   $$2E + G \rightarrow 3X$$

   The Gibbs free energies of E, G, and X are 200 kcal/mol, 300 kcal/mol, and 500 kcal/mol, respectively. Does this reaction occur spontaneously?
3. Calculate the Gibbs free energy change for the reverse reaction shown in the preceding problem. Is this reverse reaction spontaneous?
4. The Gibbs free energy change for the hypothetical reaction

   $$A + B \rightarrow E$$

   is −300 kJ. The Gibbs free energies of A and B are 375 kJ/mol and 750 kJ/mol, respectively. What is the molar Gibbs free energy of E?
5. Find the standard Gibbs free energy change in kilocalories for the reaction

   $$CO_2(g) \rightarrow C(\text{graphite}) + O_2(g)$$

   Use Table 12.1 to obtain the values of the standard molar Gibbs free energies of formation that you may need. Is this reaction spontaneous?

6. Find the standard molar Gibbs free energy of formation of $SO_2(g)$ in kilojoules per mole. The standard Gibbs free energy change for the reaction

$$S(s) + O_2(g) \rightarrow SO_2(g)$$

is −300 kJ.

7. The standard enthalpy change for the reaction

$$CO_2(g) \rightarrow C(\text{graphite}) + O_2(g)$$

is 94 kcal at 25°C and 1 atm pressure. The standard absolute entropy change for this reaction is −1 cal/K. Calculate the standard Gibbs free energy change for this reaction without using Table 12.1. Is the reaction spontaneous?

8. The standard enthalpy change for the reaction

$$SO_2(g) \rightarrow S(s) + O_2(g)$$

is +71 kcal at 25°C. The $\Delta S°$ value for this reaction is −2 cal/K. Calculate the standard Gibbs free energy change for this reaction without using Table 12.1. Is this reaction spontaneous?

9. Consider the hypothetical reaction $A \rightarrow B$, which occurs at 25°C and 1 atm pressure. The $\Delta H_f°$ values for A and B are 8.0 kcal/mol and 16 kcal/mol, respectively. The $S°$ values for A and B are 4 cal/mol K and 12 cal/mol K, respectively. Calculate the standard Gibbs free energy change for this reaction. Is the reaction spontaneous?

10. The equilibrium constant for the hypothetical reaction $A \rightarrow B$ at 40°C is equal to 1.50×10^{-5}. Find the change in the Gibbs free energy for this reaction. Is the reaction spontaneous?

11. The pK for the hypothetical reaction $G \rightarrow H$ at 25°C is equal to 40.00. Find the standard Gibbs free energy change for this reaction. Is this reaction spontaneous?

12. The standard Gibbs free energy change for the hypothetical reaction $R \rightarrow X$ is 10.8 kcal at 40°C. What is the equilibrium constant for this reaction?

13. Find the equilibrium constant for the reaction

$$SO_2(g) \rightarrow S(s) + O_2(g)$$

at 25°C and 1 atm pressure. Use Table 12.1.

14. Find the equilibrium constant for the reaction

$$CO_2(g) \rightarrow C(\text{graphite}) + O_2(g)$$

at 25°C and 1 atm pressure. Use Table 12.1.

15. TR-5 is an alien from the planet D2L6 in Galaxy WAYAWAY. TR-5's stomach makes 4 qt of 0.01 M nitric acid every day by concentrating the hydrogen ion present in his borco. The pH of the borco solution is 8.5, and that of his stomach is 2. How much work is done by TR-5's body when it concentrates hydrogen ion to make nitric acid? TR-5's body temperature is 50°C.

16. Consider the hypothetical reaction

$$A(s) + B(s) \rightarrow 2C(g)$$

The enthalpies, entropies, and Gibbs free energies of A, B, and C are given in the following table.

Compound	$\Delta H_f°$ **(kcal/mol)**	$S°$ **(kcal/mol K)**	$\Delta G_f°$ **(kcal/mol)**
A(s)	−15	0.015	4.5
B(s)	−30	0.030	33.0
C(g)	−45	0.045	−10.5

When this reaction occurs, what is the standard enthalpy change?

Refer to the reaction and the table described in Problem 16 when working Problems 17–22.

17. What is the absolute standard entropy change for the reaction?
18. Calculate $\Delta G°$ for the reaction in two different ways.
19. What is the value for the equilibrium constant for the reaction?
20. Discuss the heat changes that occur in the reaction.
21. Is the reaction spontaneous?
22. Discuss the disorder changes for the reaction relative to the reaction, the universe, and the surroundings.

14

Electrochemistry

14.1 INTRODUCTION

Most of the topics we have been studying deepen our understanding of chemical reactions. In the preceding two chapters (about thermodynamics) we learned that energy effects accompany chemical reactions, and some of these effects can be used to predict the spontaneity of a chemical reaction. In this chapter we will study the electrical effects (e.g., the movements of electrons) that also accompany chemical reactions. At the end of this journey we will discover how our old friend, the equilibrium constant, is related to these electrical effects.

Many chemical reactions occur when electrons move from one substance (compound or element) to another substance. For example, consider the reaction between zinc (Zn) and copper nitrate [$Cu(NO_3)_2$]. When a zinc atom in zinc metal makes contact with a copper ion (Cu^{2+}) in copper nitrate, the zinc atom loses two of its electrons to the copper ion. (Zinc loses two electrons and Cu^{2+} gains those two electrons.) The products of this reaction (the results of this electron transfer) are zinc nitrate [$Zn(NO_3)_2$] and copper metal (Cu atoms). This reaction can be represented by either of the following chemical equations:

$$\underbrace{Zn \quad + \quad Cu}_{2\,e^-\ \text{transferred}}(NO_3)_2 \rightarrow Zn(NO_3)_2 + Cu$$

or

$$\underbrace{Zn \quad + \quad Cu}_{2\,e^-\ \text{transferred}}^{2+} \rightarrow Zn^{2+} + Cu$$

If Zn and $Cu(NO_3)_2$ are separated by several inches and connected with a wire (in a way that will be shown), the reaction (electron transfer) can still occur if electrons travel through the wire from zinc to copper nitrate. This does happen. We call the flow of electrons through a wire an electric current. If a light bulb were correctly connected to the connecting wire, the light bulb would begin to glow. An electric bell correctly wired to this circuit would ring.

A chemical reaction can cause an electric current to flow through a wire. It is also possible for an electric current to cause a chemical reaction to occur. We will study both of these situations. A chemical reaction causing an electric current to flow is called a voltaic cell. (An ordinary flashlight battery is an example of a voltaic cell.) An electric current causing a chemical reaction to occur is called an electrolytic cell. (Gold plating of jewelry is an example of an electrolytic cell. An electric current causes gold ions in solution to accept electrons and become metallic atoms of gold.)

14.2 FUNDAMENTAL DEFINITIONS

Before we discuss electrolytic and voltaic cells in more detail, let's go over some basic electrochemical vocabulary. A coulomb is an amount of charge. One electron has a charge of -1.6022×10^{-19} coulombs (-1.6022×10^{-19} C). It is usually more convenient to work with a large number of electrons, such as 1 mol (6.022×10^{23}) of electrons.

•
•

EXAMPLE 1 *What is the charge of 1 mol of electrons?*

Solution The charge of 1 mol of electrons is -9.649×10^4 C.

$$\underset{\text{given}}{(6.022 \times 10^{23}\ e^-)}\underset{\text{conversion factor}}{\left(\frac{-1.6022 \times 10^{-19}\ C}{e^-}\right)} = -9.649 \times 10^4\ C$$

The charge of 1 mol of electrons is usually written without the negative sign and rounded off to 96 500 C. We will follow this practice in the remaining problems of this chapter.

EXAMPLE 2 How many electrons would be required to build up a charge of 5000 C?

Solution 3.120×10^{22} electrons have a charge of 5000 C.

$$(5000\ C)\left(\frac{1\ \text{mol}\ e^-}{96\,500\ C}\right)\left(\frac{6.022 \times 10^{23}\ e^-}{\text{mol}\ e^-}\right) = 3.120 \times 10^{22}\ e^-$$

A current of 1 ampere (1 A) flowing through a wire for 1 sec will cause 1 C of charge to pass by any point on the wire (see Figure 14.1).

FIGURE 14.1 Coulomb, ampere, and second relationship

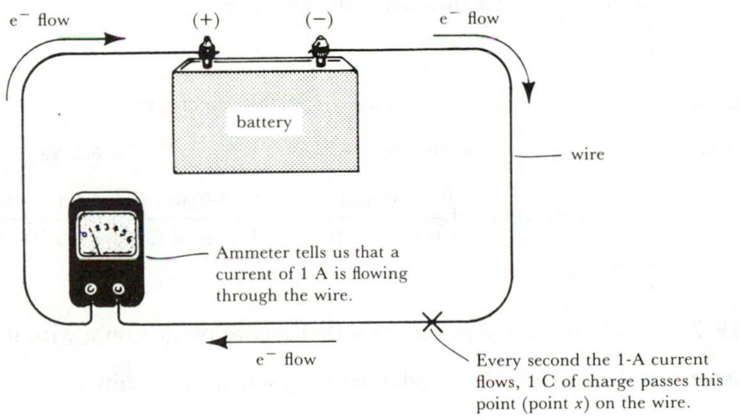

We say that 1 C of charge equals 1 A sec.

$$1\ \text{coulomb} = (1\ \text{ampere})(1\ \text{second})$$

$$1\ C = 1\ A\ \text{sec}$$

EXAMPLE 3 How many electrons pass point x in 1 sec if a 1-A current is flowing through the wire?

Solution Every second, 6×10^{18} electrons pass by point x.

$$(1\ A)(1\ \text{sec})\left(\frac{1\ C}{A\ \text{sec}}\right)\left(\frac{6.022 \times 10^{23}\ e^-}{96\,500\ C}\right) = 6 \times 10^{18}\ e^-$$

There are 6×10^{18} electrons in 1 C of charge; that is, 1 C charge = $6 \times 10^{18}\ e^-$ = 1 A sec.

EXAMPLE 4 How many electrons flow by point x in 5 sec if the current is increased to 3 A?

Solution In 5 sec, at a current of 3 A, 9×10^{19} electrons pass point x.

$$(3\text{ A})(5\text{ sec})\left(\frac{1\text{ C}}{\text{A sec}}\right)\left(\frac{6.022 \times 10^{23}\text{ e}^-}{96\,500\text{ C}}\right) = 9 \times 10^{19}\text{ e}^-$$

EXAMPLE 5 A current of 10 A is allowed to go through a wire for 1 hr. How many electrons pass point x during this time?

Solution In 1 hr, at a current of 10 A, 2×10^{23} electrons pass point x on the wire. This is about one-third of a mole of electrons.

$$(10\text{ A})(1\text{ hr})\left(\frac{60\text{ min}}{\text{hr}}\right)\left(\frac{60\text{ sec}}{\text{min}}\right)\left(\frac{1\text{ C}}{\text{A sec}}\right)\left(\frac{6.022 \times 10^{23}\text{ e}^-}{96\,500\text{ C}}\right) = 2 \times 10^{23}\text{ e}^-$$

$$(2 \times 10^{23}\text{ e}^-)\left(\frac{1\text{ mol e}^-}{6 \times 10^{23}\text{ e}^-}\right) = 0.3\text{ mol e}^-$$

The charge of 1 mol of electrons, 96 500 C, is used so frequently that it is given a special name, the faraday. We say that 1 faraday is the charge of 1 mol of electrons: 1 faraday = 96 500 C, and 1 faraday = 96 500 A sec.

EXAMPLE 6 Find the charge in faradays of 20 000 electrons.

Solution The charge of 20 000 electrons is 3.32×10^{-20} faradays.

$$(20\,000\text{ e}^-)\left(\frac{1\text{ mol e}^-}{6.022 \times 10^{23}\text{ e}^-}\right)\left(\frac{96\,500\text{ C}}{\text{mol e}^-}\right)\left(\frac{1\text{ faraday}}{96\,500\text{ C}}\right) = 3.32 \times 10^{-20}\text{ faradays}$$

EXAMPLE 7 Find the charge in faradays that passes point x on a wire if a current of 10 A flows for 45 min.

Solution A charge of 0.28 faraday passes point x in 45 min.

$$(10\text{ A})(45\text{ min})\left(\frac{60\text{ sec}}{\text{min}}\right)\left(\frac{1\text{ faraday}}{96\,500\text{ A sec}}\right) = 0.28\text{ faraday}$$

EXAMPLE 8 How long will it take for 1 mol of electrons to pass by point x if a current of 1 A is flowing through the wire?

Solution It will take 96 500 sec for 1 mol of electrons to pass point x if the current is 1 A.

$$\left(\frac{1\text{ mol e}^-}{1\text{ A}}\right)\left(\frac{96\,500\text{ C}}{\text{mol e}^-}\right)\left(\frac{1\text{ A sec}}{\text{C}}\right) = 96\,500\text{ sec}$$

Check this answer to see if it is reasonable:

$$(1\text{ A})(96\,500\text{ sec}) = 96\,500\text{ A sec} = 96\,500\text{ C} = 1\text{ faraday} = \text{charge of 1 mol e}^-$$

EXAMPLE 9 How many amperes are required to make 1 mol of electrons pass point x in 1 sec?

Solution In theory, a current of 96 500 A running for 1 sec will send 1 mol of electrons past point *x* in 1 sec.

$$\left(\frac{1 \text{ mol e}^-}{1 \text{ sec}}\right)\left(\frac{96\,500 \text{ C}}{\text{mol e}^-}\right)\left(\frac{1 \text{ A sec}}{\text{C}}\right) = 96\,500 \text{ A}$$

Check this answer to see if it is reasonable:

$$(96\,500 \text{ A})(1 \text{ sec}) = 96\,500 \text{ A sec} = 96\,500 \text{ C} = 1 \text{ faraday} = \text{charge of 1 mol e}^-$$

EXAMPLE 10 How long will it take 1 C of electrons to pass point *x* with a current of 1 A flowing through the wire?

Solution It will take 1 sec for a charge of 1 C of electrons to pass point *x* if the current is 1 A.

$$\left(\frac{1 \text{ C}}{1 \text{ A}}\right)\left(\frac{1 \text{ A sec}}{\text{C}}\right) = 1 \text{ sec}$$

SELF-TEST

1. Calculate the charge of 0.6 mol of electrons if the charge of one electron is 1.60×10^{-19} C.
2. Find the number of electrons required to build up a charge of 1×10^3 C.
3. How many electrons pass by point *x* on a wire in 30 min with a 5-A current flowing through the wire?
4. Find the charge in faradays that passes by point *x* on a wire if a 5-A current flows for 10 min.

ANSWERS **1.** 6×10^4 C **2.** 6×10^{21} electrons **3.** 6×10^{22} electrons **4.** 3×10^{-2} faraday

The chemical equation

$$Cu^{2+} + 2e^- \rightarrow Cu$$

can be interpreted as saying. "One mol of copper ions accepts 2 mol of electrons and produces 1 mol of copper atoms (metallic copper)." We know that 1 mol of electrons has a charge of 96 500 C (96 500 C = 1 faraday = 96 500 A sec), so we can reinterpret the preceding equation in any of the ways shown in Table 14.1.

In any column of Table 14.1, all five lines are equivalent (e.g., in Column 3, 1 mol Cu = 64 g metallic copper = 6×10^{23} Cu atoms). We can interpret line 3 in Table 14.1 as (reading across from left to right) "One mole of copper ion reacts with 2 faradays of charge (2 moles of electrons) to produce 64 grams of metallic copper." If we divide each number in line 5 by 2, we get "Thirty-two grams of copper ions plus 96 500 ampere seconds of electricity produces 32 grams of copper."

TABLE 14.1 Four Possible Interpretations of the Equation $Cu^{2+} + 2e^- \rightarrow Cu$

	Column 1		Column 2		Column 3
Line 1	Cu^{2+}	+	$2e^-$	$\longrightarrow$	Cu
Line 2	64 g Cu^{2+}	+	2 mol e^-	yields	1 mol Cu
Line 3	1 mol Cu^{2+}	+	2 faradays	yields	64 g Cu
Line 4	6×10^{23} Cu^{2+} ions	+	$2 \times 96\,500$ C	yields	6×10^{23} Cu atoms
Line 5	64 g Cu^{2+}	+	$2 \times 96\,500$ A sec	yields	64 g Cu

14.3 ELECTROLYTIC CELLS

In an electrolytic cell, the flow of an electric current causes a chemical reaction to occur. For example, if somehow (we'll see how in a moment) electrons were made to flow through a solution of copper ions (e.g., a copper sulfate solution), the copper ions would react with the electrons to form copper metal. This reaction is represented by the chemical equation

$$Cu^{2+} + 2e^- \rightarrow Cu$$

Figure 14.2 shows an electrolytic cell in which this reaction occurs.

FIGURE 14.2 Electrolytic cell

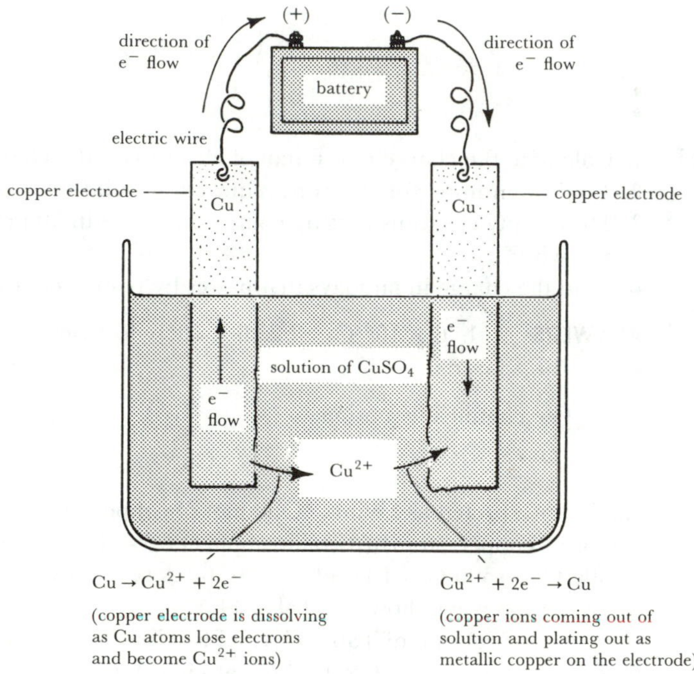

Figure 14.2 shows two metal bars of copper inserted into a solution of copper ions (copper sulfate solution). The copper bars are called electrodes in this electrolytic cell. The right-hand copper electrode gains weight as metallic copper forms (plates out) there, adhering to the surface of this electrode.

•
•

EXAMPLE 11 In the electrolytic cell shown in Figure 14.2, a current of 5 A is passed through the cell for 2 hr. How many grams of copper will plate out on the right-hand electrode? (How much weight will the right-hand electrode gain?)

Solution After 2 hr, about 10 g (1×10^1 g) of copper metal will plate out on the right-hand electrode. The electrode on the right-hand side will gain about 10 g of copper in 2 hr with a 5-A current flowing through the electrolytic cell.

$$(5\text{ A})(2\text{ hr})\left(\frac{60\text{ min}}{\text{hr}}\right)\left(\frac{60\text{ sec}}{\text{min}}\right)\left(\frac{1\text{ C}}{\text{A sec}}\right)\left(\frac{1\text{ faraday}}{96\,500\text{ C}}\right)\left(\frac{1\text{ mol e}^-}{\text{faraday}}\right)\left(\frac{64\text{ g Cu}}{2\text{ mol e}^-}\right) = 1 \times 10^1\text{ g Cu}$$

The following are two shorter and equally correct approaches:

$$(5\ \text{A})(2\ \text{hr})\left(\frac{3600\ \text{sec}}{\text{hr}}\right)\left(\frac{1\ \text{mol e}^-}{96\,500\ \text{A sec}}\right)\left(\frac{64\ \text{g Cu}}{2\ \text{mol e}^-}\right) = 1 \times 10^1\ \text{g Cu}$$

1 mol e^- = 1 faraday = 96 500 C = 96 500 A sec — see line 2 of Table 14.1

or

$$(5\ \text{A})(2\ \text{hr})\left(\frac{3600\ \text{sec}}{\text{hr}}\right)\left(\frac{32\ \text{g Cu}}{96\,500\ \text{A sec}}\right) = 1 \times 10^1\ \text{g Cu}$$

from dividing line 5 in Table 14.1 by 2

EXAMPLE 12 An electrolytic cell, similar to the one shown in Figure 14.2, has zinc electrodes immersed in a solution of zinc nitrate (Zn^{2+} ions). A current of 3 A is passed through the cell for 10 min. How many grams of zinc plate out? The reaction in which Zn^{2+} ions plate out is represented by the chemical equation

$$Zn^{2+}(aq) + 2e^- \rightarrow Zn$$

Solution

$$(3\ \text{A})(10\ \text{min})\left(\frac{60\ \text{sec}}{\text{min}}\right)\left(\frac{1\ \text{C}}{\text{A sec}}\right)\left(\frac{1\ \text{faraday}}{96\,500\ \text{C}}\right)\left(\frac{1\ \text{mol e}^-}{\text{faraday}}\right)\left(\frac{65\ \text{g Zn}}{2\ \text{mol e}^-}\right) = 1\ \text{g Zn}$$

or

$$(3\ \text{A})(10\ \text{min})\left(\frac{60\ \text{sec}}{\text{min}}\right)\left(\frac{1\ \text{mol e}^-}{96\,500\ \text{A sec}}\right)\left(\frac{65\ \text{g Zn}}{2\ \text{mol e}^-}\right) = 1\ \text{g Zn}$$

EXAMPLE 13 Aluminum is prepared in electrolytic cells by plating out metallic aluminum from solutions of Al^{3+} ions. How long would it take to plate out 10 lb of aluminum with a current of 5 A? The reaction in which aluminum ions plate out is represented by the chemical equation

$$Al^{3+}(aq) + 3e^- \rightarrow Al$$

Solution It would take about 100 days to plate out 10 lb of aluminum with a 5-A current.

$$(10\ \text{lb Al})\left(\frac{454\ \text{g}}{\text{lb}}\right)\left(\frac{3\ \text{mol e}^-}{27\ \text{g Al}}\right)\left(\frac{96\,500\ \text{A sec}}{\text{mol e}^-}\right)\left(\frac{1}{5\ \text{A}}\right) = 1 \times 10^7\ \text{sec}$$

$$(1 \times 10^7\ \text{sec})\left(\frac{1\ \text{min}}{60\ \text{sec}}\right)\left(\frac{1\ \text{hr}}{60\ \text{min}}\right)\left(\frac{1\ \text{day}}{24\ \text{hr}}\right) = 1 \times 10^2\ \text{days}$$

EXAMPLE 14 Hydrogen gas forms at one electrode when a 5.0-A current is passed for 5.0 min through a hydrochloric acid solution using two inert platinum electrodes (platinum doesn't dissolve in HCl). What volume of hydrogen forms at this electrode if the gas is collected at 25°C and 750 mm Hg? The reaction in which hydrogen gas forms in this electrolytic cell is represented by the chemical equation

$$2H^+(aq) + 2e^- \rightarrow H_2(g)$$

Solution About 200 mL of hydrogen gas forms under these conditions.

$$(5.0\ \text{A})(5.0\ \text{min})\left(\frac{60\ \text{sec}}{\text{min}}\right)\left(\frac{1\ \text{mol e}^-}{96\,500\ \text{A sec}}\right)\left(\frac{2\ \text{g H}_2}{2\ \text{mol e}^-}\right) = 0.016\ \text{g H}_2$$

$$(0.016 \text{ g H}_2)\left(\frac{22.4 \text{ L H}_2 \text{ at STP}}{2.0 \text{ g H}_2}\right)\left(\frac{1000 \text{ mL}}{\text{L}}\right) = 1.8 \times 10^2 \text{ mL H}_2 \text{ at STP}$$

$$(1.8 \times 10^2 \text{ mL H}_2 \text{ at STP})\left(\frac{298 \text{ K}}{273 \text{ K}}\right)\left(\frac{760 \text{ mm Hg}}{750 \text{ mm Hg}}\right) = 2.0 \times 10^2 \text{ mL H}_2$$

EXAMPLE 15 Chlorine gas forms at the other electrode in the electrolytic cell described in Example 14. What is the volume of chlorine that forms after 7.0 min? The reaction in which chlorine gas forms from a solution of chloride ions is represented by the chemical equation

$$2Cl^-(aq) \rightarrow Cl_2 + 2e^-$$

Solution

$$(5.0 \text{ A})(7.0 \text{ min})\left(\frac{60 \text{ sec}}{\text{min}}\right)\left(\frac{1 \text{ mol e}^-}{96\,500 \text{ A sec}}\right)\left(\frac{71 \text{ g Cl}_2}{2 \text{ mol e}^-}\right) = 0.77 \text{ g Cl}_2$$

$$(0.77 \text{ g Cl}_2)\left(\frac{22.4 \text{ L Cl}_2 \text{ at STP}}{71 \text{ g Cl}_2}\right)\left(\frac{1000 \text{ mL}}{\text{L}}\right) = 2.4 \times 10^2 \text{ mL Cl}_2 \text{ at STP}$$

$$(2.4 \times 10^2 \text{ mL Cl}_2 \text{ at STP})\left(\frac{298 \text{ K}}{273 \text{ K}}\right)\left(\frac{760 \text{ mm Hg}}{750 \text{ mm Hg}}\right) = 2.7 \times 10^2 \text{ mL Cl}_2$$

SELF-TEST

1. A 10-A current is passed through an electrolytic cell for 2 hr. How many grams of copper will plate out according to the equation

$$Cu^{2+}(aq) + 2e^- \rightarrow Cu$$

2. How many days would it take to plate out 5.0 lb of aluminum from a solution of Al^{3+} using a 10-A current?
3. What volume of hydrogen gas forms in an electrolytic cell using a 5.0-A current for 20 min? The reduction of hydrogen ion in this cell reaction is represented by

$$2H^+(aq) + 2e^- \rightarrow H_2$$

ANSWERS **1.** 2×10^1 g Cu **2.** 28 days **3.** 7.0×10^2 mL H_2

14.3.1 ELECTROCHEMICAL MACHINING

The discovery of three-million-year-old stone tools in the Hadar of the Afar region of northeastern Ethiopia indicates that humans have been machining for millions of years. Machining is the grinding, chipping, or scooping of a softer object (called a workpiece) by a harder object (called a tool) for the purpose of shaping the softer object. Sharpening a knife blade with a file is a simple example of machining. The knife is the workpiece and the file is the tool. Usually the tool is harder than the workpiece.

There have always been problems (disadvantages) with conventional machining:

1. Machining very hard objects can be a very slow process that requires many hours or days to complete.
2. Conventional machining leaves relatively rough surfaces (each scrape of a file removes billions upon billions of metal atoms at a time).
3. The number of ways that a workpiece can be shaped is limited with conventional machining. For example, you cannot drill a square blind hole into very hard material. (A square blind hole is a square-shaped hole that only partially penetrates a workpiece.)

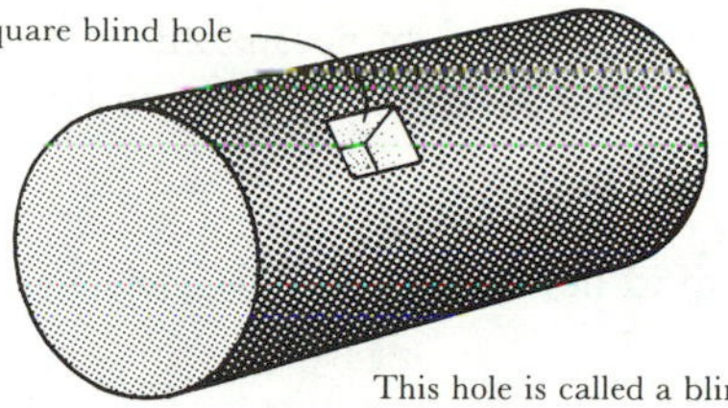

4. Conventional machining cannot be used to shape thin foils. Contact between the tool and the thin foil workpiece would cause the foil to bend or become distorted.

Recently, a new machining method has been developed. It eliminates most of the problems of conventional machining. The new method is called electrochemical machining. In its simplest form, the setup for electrochemical machining is almost the same as the setup for an ordinary electrolytic cell (see Figure 14.3). A dissolving electrode will assume a shape that is determined by the shape of the other electrode when the two electrodes are placed very close together, about 0.005 cm to 0.1 cm apart (i.e., a few thousandths of an inch apart), and if the current is very high (50 A to 500 A per square centimeter).

FIGURE 14.3 Electrochemical machining setup

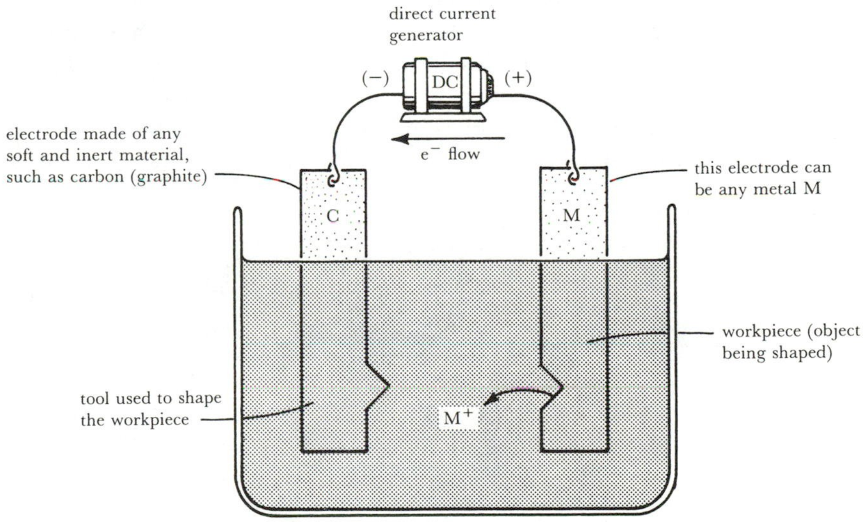

Electrochemical machining eliminates many of the disadvantages of conventional machining. The advantages of electrochemical machining are as follows:

1. The workpiece has a very smooth machined surface (it is dissolved away atom by atom).
2. Thin foils can be machined without distortion.

Why doesn't electrochemical machining distort thin foil workpieces?

Thin foils can be machined without foil distortion because there is no physical contact between the foil and the tool.

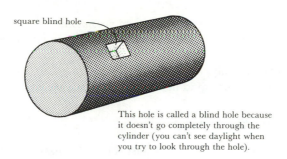

3. The hardness of the workpiece doesn't affect the rate at which it is machined (dissolved). Electrochemical machining only requires a few minutes, even with very hard metals or alloys.
4. Theoretically, electrochemical machining can produce an unlimited number of shapes, even square blind holes.

14.3.2 RAISE THE *TITANIC**

The *Titanic*, built in 1912 as an "unsinkable" luxury ocean liner, sank on its maiden voyage after colliding with an iceberg. More than 1500 people died.

Today, the *Titanic* lies 2 miles beneath the ocean's surface some 100 miles south of the Great Banks of Newfoundland, Canada.

What is the pressure at this depth?

The pressure at 2 mi beneath the ocean's surface is about 300 atm.

$$(2\ \text{mi})\left(\frac{5280\ \text{ft}}{\text{mi}}\right)\left(\frac{1\ \text{atm}}{33\ \text{ft}}\right) = 3 \times 10^2\ \text{atm}$$

There is considerable interest in raising the *Titanic*. Although there is no major scientific justification for raising the ship, there will be some commercial value because people are interested in seeing things of historical significance. However, the pressure at this depth is about 300 atm, which is too extreme for underwater divers. Nevertheless, some scientists want the challenge of the advanced technological problems that such a feat presents. These scientists want to develop the technology to locate more important objects in the deep ocean. Scientists also want to expand the technology for deep-sea photography.

In the past, sunken ships have been raised by attaching inflatable buoys (big balloons) to them and filling the buoys with air. Compressors located on surface ships supply the air needed

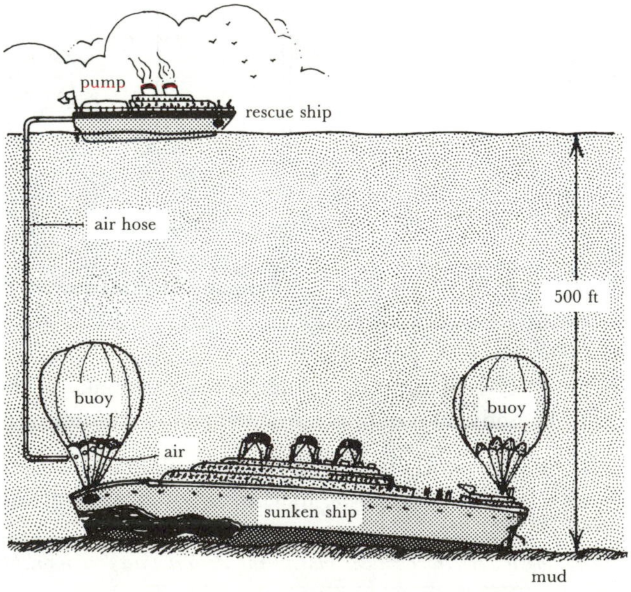

*Adapted from Robert C. Plumb, "Raising the *Titanic* by Electrolysis," *Journal of Chemical Education* 50 (Jan. 1973):61.

to inflate the buoys. Unfortunately, commercial pumps are not available to pump air at pressures of 300 to 350 atmospheres over distances of two to three miles.

If you can't use a pump to fill the buoys with gas, how else could you inflate the buoys at this depth?

By substituting a direct current (DC) generator for the pump and a wire for the hose, the hydrogen ions present in the ocean (from dissociation of ocean water) will react according to the chemical equation

$$2H^+ + 2e^- \rightarrow H_2$$

to produce hydrogen gas at one electrode. This gas can be used to inflate the buoys.

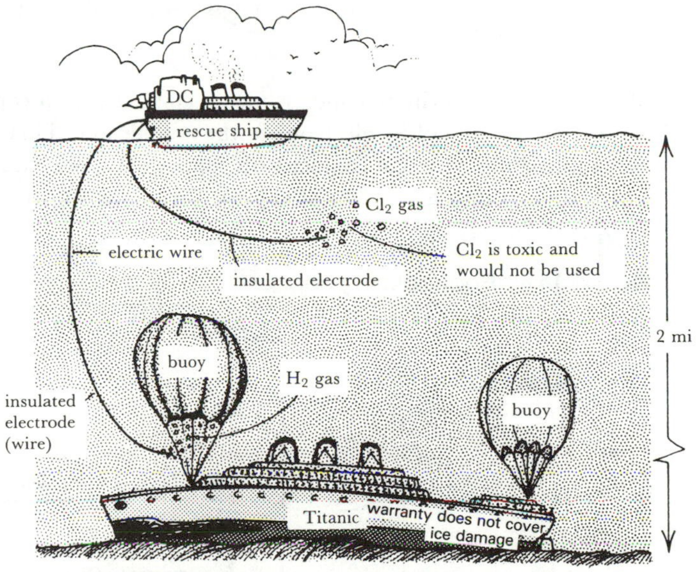

Scientists have calculated that it would take a 20-Mw (20-million-watt) generator operating continuously more than three years to produce enough hydrogen to raise the *Titanic*.

14.3.3 DETERMINATION OF AVOGADRO'S NUMBER

Many methods have been used to calculate the value of Avogadro's number. One of these methods makes use of ideas from electrochemistry.

The charge of 1 mol of electrons is very accurately known to be 96 485 C. This value is obtained by carefully measuring the current and the time needed to plate out 107.87 g of silver (1 mol of silver by the reaction $Ag^+ + e^- \rightarrow Ag$). The charge of one electron is also very accurately known (1.6022×10^{-19} C). This value is obtained by performing the Millikan oil drop experiment, which many chemistry and physics textbooks describe in detail.

Use dimensional analysis (the manipulation of units to obtain what you want) and the information in this section to calculate the number of electrons in 1 mol of electrons (Avogadro's number).

$$\left(\frac{96\,485 \text{ C}}{\text{mol of e}^-}\right)\left(\frac{1 \text{ e}^-}{1.6022 \times 10^{-19} \text{ C}}\right) = \frac{6.0220 \times 10^{23} \text{ e}^-}{\text{mol of e}^-}$$

14.4 VOLTAIC CELLS

In the last section we studied the situation in which chemical reactions are caused by electric currents. Such a situation is called an electrolytic cell. In this section we will consider the reverse situation in which electric currents are caused by chemical reactions. In such a situation, called a voltaic cell, a chemical reaction causes an electric current to flow.

Metallic zinc will react with copper nitrate as represented by the chemical equation

$$\underbrace{Zn + Cu}_{2\,e^-}(NO_3)_2 \rightarrow Zn(NO_3)_2 + Cu^{2+}$$

or

$$\underbrace{Zn + Cu}_{2\,e^-}{}^{2+}(aq) \rightarrow Zn^{2+}(aq) + Cu$$

When a zinc atom (in zinc metal) makes contact with a copper ion (in copper nitrate), the zinc atom can lose two of its electrons to the copper ion. This reaction can occur when zinc atoms come into direct physical contact with copper ions. For example, this reaction will occur if a zinc bar is dipped into a solution of copper nitrate, as shown in Figure 14.4.

FIGURE 14.4 Electroplating copper

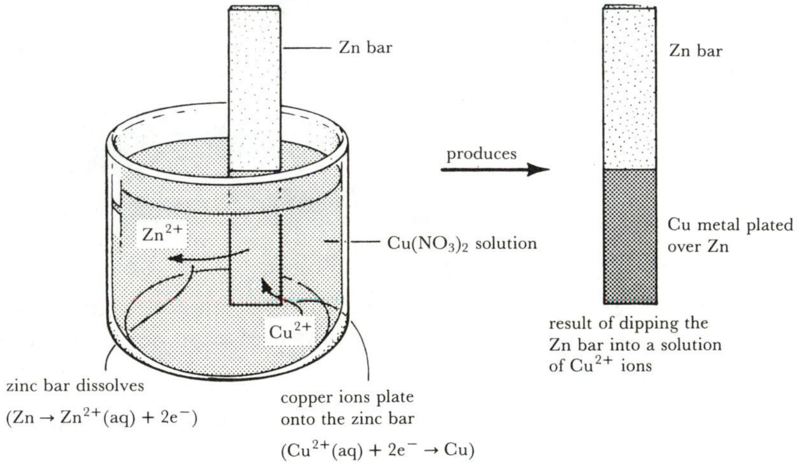

When a zinc bar is dipped into a solution of copper ions, the direct physical contact between zinc atoms and copper ions results in electrons being transferred from Zn to Cu^{2+}. When a zinc atom (Zn) loses two electrons, it becomes a zinc ion (Zn^{2+}). The zinc ion dissolves into the copper nitrate solution. When a copper ion (Cu^{2+}) picks up (gains) two electrons from zinc, the copper ion turns into metallic copper (comes out of solution) and plates out onto the zinc bar.

When the zinc bar is pulled out of the copper nitrate solution, the lower part of the bar (which was immersed) is coated with metallic copper (copper plating). If the zinc electrode and the copper solution were separated by a few inches and connected with a wire (see Figure 14.5), electrons would travel from the zinc, through the connecting wire, to the copper ions in the copper nitrate solution. We say that the chemical reaction causes electrons to flow through the connecting wire; that is, the chemical reaction causes an electric current to flow through the wire.

FIGURE 14.5 Voltaic cell

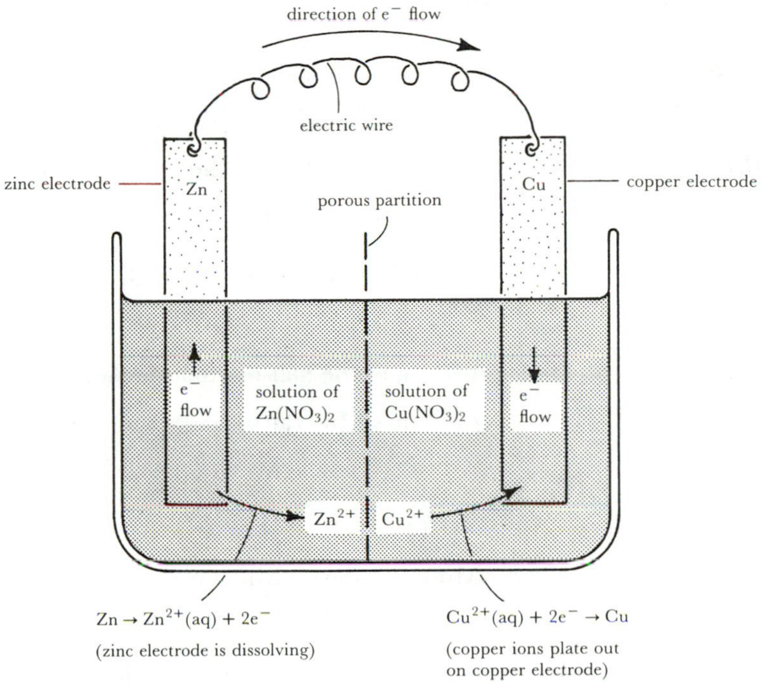

The voltaic cell shown in Figure 14.5 contains a porous partition (which could be just an ordinary piece of paper) that prevents the mechanical mixing of the two solutions but allows the passage of ions under the influence (force) of an electric current. In the voltaic cell shown in Figure 14.5, the copper electrode gains weight as copper plates out from the copper nitrate solution. Copper ions gain electrons to become copper atoms. A reaction in which electrons are gained is called a reduction reaction. We say that copper ions are reduced (undergo reduction) to copper atoms. The zinc electrode in Figure 14.5 loses weight as zinc dissolves into the zinc nitrate solution. The zinc atoms making up the zinc electrode lose electrons to become zinc ions. A reaction in which electrons are lost is called an oxidation reaction. We say that zinc atoms are oxidized (undergo oxidation) to zinc ions. You can remember this if you know that LEO the lion goes GER, where LEO means Losing Electrons is Oxidation, and GER means Gaining Electrons is Reduction.

The electrode at which oxidation occurs is called the anode. The zinc electrode is the anode. (An easy way to remember this is to know that the words "oxidation always occurs only at an anode" all begin with vowels.) The electrode at which reduction occurs is called the cathode. (The words *cathode* and *reduction* both begin with consonants.)

The two reactions shown under the beakers

$$Cu^{2+}(aq) + 2e^- \rightarrow Cu \quad \text{and} \quad Zn \rightarrow Zn^{2+}(aq) + 2e^-$$

are called half-reactions. A half-reaction shows either the oxidation "half" or the reduction "half" of a chemical reaction. The "whole" chemical reaction is called a redox (short for reduction-oxidation) reaction.

The chemical reaction

$$Zn + Cu^{2+}(aq) \rightarrow Zn^{2+}(aq) + Cu$$

is called a redox reaction. Redox reactions can be created by adding two half-reactions (one oxidation and one reduction) together.

EXAMPLE 16 Write the redox reaction whose half-reactions are

$$Al \rightarrow Al^{3+}(aq) + 3e^- \quad \text{and} \quad Cr^{3+}(aq) + 3e^- \rightarrow Cr$$

Solution

$Al \rightarrow Al^{3+}(aq) + 3e^-$ ← oxidation half-reaction

+

$Cr^{3+}(aq) + 3e^- \rightarrow Cr$ ← reduction half-reaction

=

$Al + Cr^{3+}(aq) \rightarrow Al^{3+}(aq) + Cr$ ← redox reaction

EXAMPLE 17 Write the redox reaction whose half-reactions are

$$Mn \rightarrow Mn^{2+}(aq) + 2e^- \quad \text{and} \quad 2H^+(aq) + 2e^- \rightarrow H_2$$

Solution

$Mn \rightarrow Mn^{2+}(aq) + 2e^-$ ← oxidation half-reaction

+

$2H^+(aq) + 2e^- \rightarrow H_2$ ← reduction half-reaction

=

$Mn + 2H^+(aq) \rightarrow Mn^{2+}(aq) + H_2$ ← redox reaction

Table 14.2 lists a few reduction half-reactions. A standard electrode potential (voltage) is given for each reduction half-reaction.

TABLE 14.2 Reduction Half-Reactions

Element	Reduction Reaction	Standard Electrode Potential $E°$ in Volts
Magnesium	$Mg^{2+}(aq) + 2e^- \rightarrow Mg$	−2.37
Aluminum	$Al^{3+}(aq) + 3e^- \rightarrow Al$	−1.66
Manganese	$Mn^{2+}(aq) + 2e^- \rightarrow Mn$	−1.18
Zinc	$Zn^{2+}(aq) + 2e^- \rightarrow Zn$	−0.76
Chromium	$Cr^{3+}(aq) + 3e^- \rightarrow Cr$	−0.74
Iron	$Fe^{2+}(aq) + 2e^- \rightarrow Fe$	−0.44
Cobalt	$Co^{2+}(aq) + 2e^- \rightarrow Co$	−0.28
Nickel	$Ni^{2+}(aq) + 2e^- \rightarrow Ni$	−0.25
Tin	$Sn^{2+}(aq) + 2e^- \rightarrow Sn$	−0.14
Lead	$Pb^{2+}(aq) + 2e^- \rightarrow Pb$	−0.13
Hydrogen	$2H^+(aq) + 2e^- \rightarrow H_2$	0.00
Copper	$Cu^{2+}(aq) + 2e^- \rightarrow Cu$	+0.34
Oxygen	$O_2 + 4H^+(aq) + 4e^- \rightarrow 2H_2O$	+1.23
Chlorine	$Cl_2 + 2e^- \rightarrow 2Cl^-(aq)$	+1.36
Gold	$Au^+(aq) + e^- \rightarrow Au$	+1.68

EXAMPLE 18 What is the standard electrode potential for the reduction of $Cu^{2+}(aq)$ according to the reaction

$$Cu^{2+}(aq) + 2e^- \rightarrow Cu$$

Solution Table 14.2 shows the standard electrode potential for this reduction half-reaction as +0.34 volt (+0.34 V).

•
•

EXAMPLE 19 What is the standard electrode potential of the half-reaction for the oxidation of copper to $Cu^{2+}(aq)$ according to the reaction

$$Cu \rightarrow Cu^{2+}(aq) + 2e^-$$

Solution This half-reaction is an oxidation half-reaction (electrons are lost). Table 14.2 shows reduction half-reactions. However, you can determine the standard electrode potential of an oxidation half-reaction by changing the sign of the appropriate reduction half-reaction as follows: Since

$$Cu^{2+}(aq) + 2e^- \rightarrow Cu \qquad E° = +0.34\ V$$

then

$$Cu \rightarrow Cu^{2+}(aq) + 2e^- \qquad E° = -0.34\ V$$

•
•

EXAMPLE 20 What is the standard electrode potential for the oxidation of iron to the +2 state according to the reaction

$$Fe \rightarrow Fe^{2+}(aq) + 2e^-$$

Solution This half-reaction is an oxidation half-reaction (electrons are lost). Since

$$Fe^{2+}(aq) + 2e^- \rightarrow Fe \qquad E° = -0.44\ V$$

then

$$Fe \rightarrow Fe^{2+}(aq) + 2e^- \qquad E° = +0.44\ V$$

•
•

EXAMPLE 21 What is the standard electrode potential for the half-reaction

$$Co^{2+}(aq) + 2e^- \rightarrow Co$$

Solution Table 14.2 shows that the reduction standard electrode potential for $Co^{2+}(aq)$ equals −0.28 V.

•
•

Whenever you can add together two half-reactions (one oxidation and one reduction) such that the sum of their voltages is positive, the resulting redox reaction will occur spontaneously.

•
•

EXAMPLE 22 A voltaic cell is made with metallic electrodes of Zn and Cu immersed in solutions of 1 M Zn^{2+} and 1 M Cu^{2+}, respectively, at 760 mm Hg and 25°C. What spontaneous redox reaction occurs? What is the voltage of this voltaic cell?

Solution Look up the reduction half-reactions of Zn and Cu in Table 14.2. They are

$$Zn^{2+}(aq) + 2e^- \rightarrow Zn \qquad E° = -0.76\ V$$

$$Cu^{2+}(aq) + 2e^- \rightarrow Cu \qquad E° = +0.34\ V$$

There are four possible ways to combine these two half-reactions. Only the fourth way is correct.

First Possible Combination:

$Zn^{2+}(aq) + 2e^- \rightarrow Zn$	$E° = -0.76$ V
$Cu^{2+}(aq) + 2e^- \rightarrow Cu$	$E° = +0.34$ V
$Zn^{2+}(aq) + Cu^{2+}(aq) + 4e^- \rightarrow Zn + Cu$	$E° = -0.42$ V

This reaction cannot occur spontaneously. It is not a redox reaction because both half-reactions are reduction half-reactions. One half-reaction must be a reduction half-reaction, and the other must be an oxidation half-reaction.

If something gains electrons (reduction) in a redox reaction, something else has to lose electrons (oxidation). Electrons lost by the oxidation half-reaction are the same electrons gained by the reduction half-reaction.

Second Possible Combination:

$Zn \rightarrow Zn^{2+}(aq) + 2e^-$	$E° = +0.76$ V
$Cu \rightarrow Cu^{2+}(aq) + 2e^-$	$E° = -0.34$ V
$Zn + Cu \rightarrow Zn^{2+}(aq) + Cu^{2+}(aq) + 4e^-$	$E° = +0.42$ V

Although this "whole" reaction has a positive voltage (+0.42 V), this reaction will not occur spontaneously because it is not a redox reaction. Both Zn and Cu cannot be oxidized.

Whenever an oxidation occurs (electrons are lost) in a redox reaction, a reduction must also occur to pick up (gain) the lost electrons. When a redox reaction is broken down into two half-reactions, one of the half-reactions must be an oxidation half-reaction, and the other half-reaction must be a reduction half-reaction.

Third Possible Combination:

$Zn^{2+}(aq)\ 2e^- \rightarrow Zn$	$E° = -0.76$ V
$Cu \rightarrow Cu^{2+}(aq) + 2e^-$	$E° = -0.34$ V
$Zn^{2+}(aq) + Cu \rightarrow Zn + Cu^{2+}(aq)$	$E° = -1.10$ V

This redox reaction is not spontaneous. Although this reaction does show something being oxidized and something being reduced, the voltage for this redox reaction is negative. A redox reaction needs a positive voltage to be spontaneous.

Fourth (and Correct) Possible Combination:

$Zn \rightarrow Zn^{2+}(aq) + 2e^-$	$E° = +0.76$ V
$Cu^{2+}(aq) + 2e^- \rightarrow Cu$	$E° = +0.34$ V
$Zn + Cu^{2+}(aq) \rightarrow Zn^{2+}(aq) + Cu$	$E° = +1.10$ V

This redox reaction occurs spontaneously. This redox reaction is the sum of an oxidation half-reaction and a reduction half-reaction. The voltage of the redox reaction is positive.

The overall (whole) reaction of this voltaic cell is

$$Zn + Cu^{2+}(aq) \rightarrow Zn^{2+}(aq) + Cu$$

The voltage of this voltaic cell is 1.10 V. The diagram for this voltaic cell is shown in Figure 14.5.

•
•

EXAMPLE 23 A voltaic cell is made with metallic electrodes of iron and nickel immersed in solutions of 1 M Fe^{2+} and 1 M Ni^{2+}, respectively, at 760 mm Hg and 25°C. What spontaneous redox reaction occurs? What is the voltage of this cell?

Solution Look up the reduction half-reactions for Fe and Ni in Table 14.2.

$$Fe^{2+}(aq) + 2e^- \rightarrow Fe \qquad E° = -0.44\ V$$

$$Ni^{2+}(aq) + 2e^- \rightarrow Ni \qquad E° = -0.25\ V$$

The only correct way to combine these half-reactions so that their sum is a redox reaction with a positive voltage is

$$\begin{array}{rl} Fe \rightarrow Fe^{2+}(aq) + 2e^- & E° = +0.44\ V \\ Ni^{2+}(aq) + 2e^- \rightarrow Ni & E° = -0.25\ V \\ \hline Fe + Ni^{2+}(aq) \rightarrow Fe^{2+}(aq) + Ni & E° = +0.19\ V \end{array}$$

•
•

The diagram for the voltaic cell discussed in Example 23 is shown here:

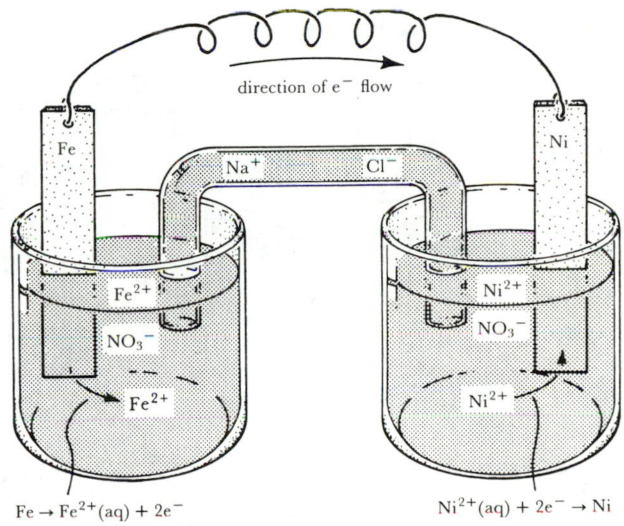

An electric current with a voltage of 0.19 V flows through the wire connecting the nickel and iron electrodes. While the electric current flows, the iron electrode dissolves (loses weight) as Fe is oxidized to $Fe^{2+}(aq)$. The nickel electrode gains weight as $Ni^{2+}(aq)$ ions in the solution of $Ni(NO_3)_2$ are reduced to metallic nickel.

Which electrode is the anode?

Right. The iron electrode is the anode. Oxidation always occurs at an anode.

•
•

SELF-TEST

1. A voltaic cell is made with metallic electrodes of lead and tin immersed in solutions of 1 M Pb^{2+} and 1 M Sn^{2+}, respectively, at 760 mm Hg and 25°C. What spontaneous chemical reaction occurs?
2. What voltage is produced by the voltaic cell in the preceding problem?
3. A voltaic cell is made with metal electrodes of manganese and nickel immersed in solutions of 1 M Mn^{2+} and 1 M Ni^{2+}, respectively, at 25°C. What spontaneous chemical reaction occurs?
4. What voltage is produced by the voltaic cell in the preceding problem?

ANSWERS **1.** $Sn + Pb^{2+}(aq) \rightarrow Sn^{2+}(aq) + Pb$ **2.** 0.01 V
3. $Mn + Ni^{2+}(aq) \rightarrow Mn^{2+}(aq) + Ni$ **4.** 0.93 V

Gases and liquids can also be used as "electrodes" in voltaic cells. A chlorine electrode is shown in Figure 14.6. Platinum is used as part of the chlorine electrode because it is inert (doesn't react). The platinum bar is used to make electrical contact with the chlorine gas (Cl_2) and the chloride (Cl^-) ions in the solution.

FIGURE 14.6 Chlorine electrode in a voltaic cell

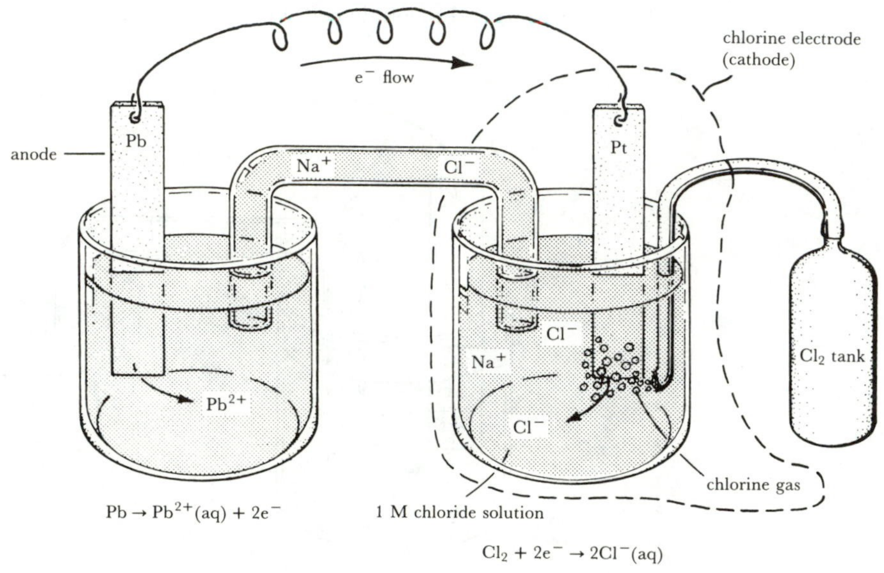

EXAMPLE 24 A voltaic cell is prepared by connecting a bar of lead, immersed in a solution of 1 M Pb^{2+} at 25°C, to a chlorine electrode. The chlorine electrode consists of a platinum bar immersed in a solution of 1 M chloride ion at 25°C through which chlorine gas is bubbled at 1 atm pressure. What spontaneous chemical reaction occurs? What is the voltage of this cell? (The voltaic cell described in this example is the cell shown in Figure 14.6.)

Solution From Table 14.2,

$$\begin{array}{rl} Pb \rightarrow Pb^{2+}(aq) + 2e^- & E° = +0.13\ V \\ Cl_2 + 2e^- \rightarrow 2Cl^-(aq) & E° = +1.36\ V \\ \hline Pb + Cl_2 \rightarrow Pb^{2+}(aq) + 2Cl^-(aq) & E° = +1.49\ V \end{array}$$

EXAMPLE 25 A voltaic cell is made by connecting a bar of zinc, immersed in a solution of 1 M Zn^{2+} at 25°C, to an oxygen electrode. The oxygen electrode consists of a bar of platinum immersed in water at 25°C through which oxygen gas is bubbled at 1 atm pressure. What spontaneous reaction occurs? What is the voltage of this cell?

Solution

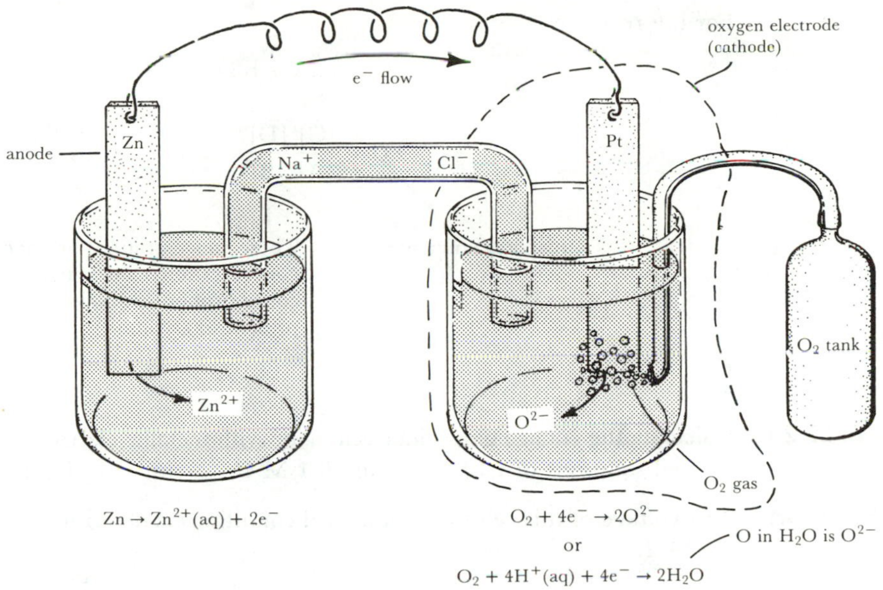

From Table 14.2,

$$
\begin{array}{rl}
Zn \rightarrow Zn^{2+}(aq) + 2e^- & E° = +0.76\ V \\
\tfrac{1}{2}O_2 + 2H^+(aq) + 2e^- \rightarrow H_2O & E° = +1.23\ V \\
\hline
Zn + \tfrac{1}{2}O_2 + 2H^+(aq) \rightarrow Zn^{2+}(aq) + H_2O & E° = +1.99\ V
\end{array}
$$

SELF-TEST

1. A voltaic cell is made by connecting a copper electrode, immersed in a solution of 1 M Cu^{2+} at 760 mm Hg and 25°C, to a chlorine electrode. The chlorine electrode consists of a platinum bar immersed in a solution of 1 M chloride ion at 25°C through which chlorine gas is bubbled at 1 atm pressure. What spontaneous chemical reaction occurs?

2. What voltage is produced by the voltaic cell described in the preceding problem?

ANSWERS **1.** $Cu + Cl_2 \rightarrow Cu^{2+}(aq) + 2Cl^-$ **2.** +1.02 V

14.4.1 NERNST EQUATION

When we add half-reactions in Table 14.2 to determine the voltage of a voltaic cell, we assume that the cell temperature is 25°C, that the atmospheric pressure is 760 mm Hg, and that all solutes are at a concentration of 1 M. If the solute concentrations are not 1 M, the Nernst equation is used to calculate the cell voltage.

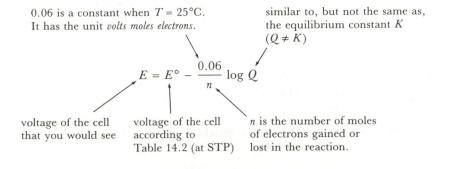

0.06 is a constant when $T = 25°C$. It has the unit *volts moles electrons*.

similar to, but not the same as, the equilibrium constant K ($Q \neq K$)

$$E = E° - \frac{0.06}{n} \log Q$$

E: voltage of the cell that you would see

$E°$: voltage of the cell according to Table 14.2 (at STP)

n is the number of moles of electrons gained or lost in the reaction.

For the reaction

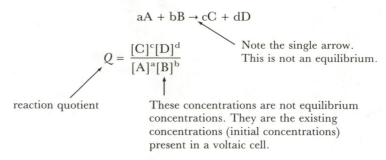

$$aA + bB \rightarrow cC + dD$$

Note the single arrow. This is not an equilibrium.

$$Q = \frac{[C]^c[D]^d}{[A]^a[B]^b}$$

reaction quotient

These concentrations are not equilibrium concentrations. They are the existing concentrations (initial concentrations) present in a voltaic cell.

EXAMPLE 26 Calculate the voltage of a voltaic cell made with metallic electrodes of magnesium and cobalt immersed solutions of 1 M Mg^{2+} and 0.1 M Co^{2+}, respectively, at 25°C and 760 mm Hg.

Solution The voltage of this cell under standard conditions is determined first using information in Table 14.2.

n = 2

$Mg \rightarrow Mg^{2+}(aq) + 2e^-$	$E° = +2.37$ V
$Co^{2+}(aq) + 2e^- \rightarrow Co$	$E° = -0.28$ V
$Mg + Co^{2+}(aq) \rightarrow Mg^{2+}(aq) + Co$	$E° = +2.09$ V

voltage of cell if $[Co^{2+}(aq)]$ and $[Mg^{2+}(aq)]$ = 1 M

$$Q = \frac{[Mg^{2+}][Co]}{[Mg][Co^{2+}]} = \frac{(1)(1)}{(1)(0.1)} = 10$$

Recall that molarity values are approximate values for unitless activity values (see Section 9.2.2). By definition, the activity of any pure solid, liquid, or gas at 25°C and 760 mm Hg is equal to 1 (1.00).

$$E = E° - \frac{0.06}{n}\log Q$$

$$= 2.09\text{ V} - \frac{0.06\text{ V mol e}^-}{2\text{ mol e}^-}\log 10$$

$$= 2.09\text{ V} - 0.03\text{ V }(1) = 2.09\text{ V} - 0.03\text{ V} = 2.06\text{ V}$$

EXAMPLE 27 Calculate the voltage of a voltaic cell made with metallic electrodes of aluminum and chromium immersed in solutions of 0.001 M Al^{3+} and 0.01 M Cr^{3+}, respectively, at 25°C and 760 mm Hg.

Solution First determine the standard voltage of this cell, $E°$, by assuming that all solutes are at concentrations of 1 M.

n = 3

$$
\begin{array}{lr}
\mathrm{Al \rightarrow Al^{3+}(aq) + 3e^-} & E^\circ = +1.66\ \mathrm{V} \\
\mathrm{Cr^{3+}(aq) + 3e^- \rightarrow Cr} & E^\circ = -0.74\ \mathrm{V} \\
\hline
\mathrm{Al + Cr^{3+}(aq) \rightarrow Al^{3+}(aq) + Cr} & E^\circ = +0.92\ \mathrm{V}
\end{array}
$$

$$Q = \frac{[\mathrm{Al^{3+}}][\mathrm{Cr}]}{[\mathrm{Al}][\mathrm{Cr^{3+}}]} = \frac{(10^{-3})(1)}{(1)(10^{-2})} = 10^{-1}$$

These are pure solids. Their activities are equal to 1.

$$E = E^\circ - \frac{0.06}{n}\log Q$$

$$= 0.92\ \mathrm{V} - \frac{0.06\ \mathrm{V\ mol\ e^-}}{3\ \mathrm{mol\ e^-}}\log 10^{-1}$$

$$= 0.92\ \mathrm{V} - 0.02\ \mathrm{V}\,(-1) = 0.92\ \mathrm{V} + 0.02\ \mathrm{V} = 0.92\ \mathrm{V}$$

EXAMPLE 28 Calculate the voltage of a voltaic cell made with metallic electrodes of iron and platinum immersed in solutions of 0.1 M Fe^{2+} and 0.1 M H^+, respectively, at 25°C and 760 mm Hg. Hydrogen gas is bubbled through the hydrogen ion solution over the platinum electrode at a pressure of 760 mm Hg.

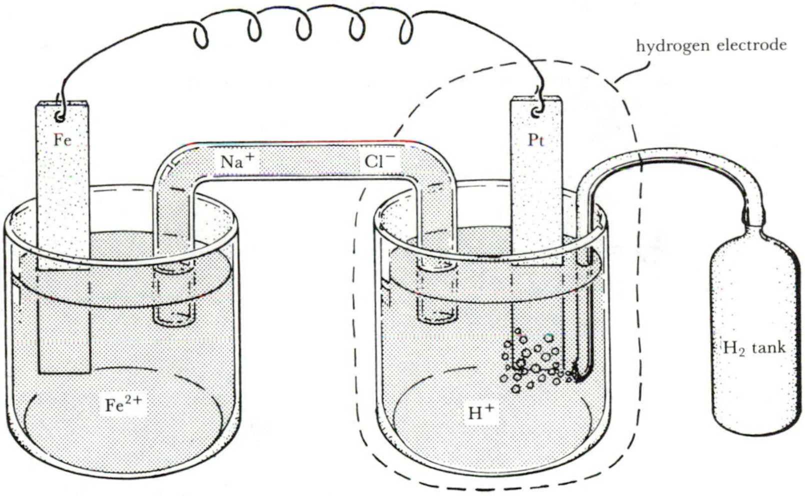

Solution The standard voltage for this cell is calculated using information in Table 14.2.

$$
\begin{array}{lr}
\mathrm{Fe \rightarrow Fe^{2+}(aq) + 2e^-} & E^\circ = +0.44\ \mathrm{V} \\
\mathrm{2H^+(aq) + 2e^- \rightarrow H_2} & E^\circ = \ \ 0.00\ \mathrm{V} \\
\hline
\mathrm{Fe + 2H^+(aq) \rightarrow Fe^{2+}(aq) + H_2} & E^\circ = +0.44\ \mathrm{V}
\end{array}
$$

pure gas at 1 atm

$$Q = \frac{[\mathrm{Fe^{2+}}][\mathrm{H_2}]}{[\mathrm{Fe}][\mathrm{H^+}]^2} = \frac{(0.1)(1)}{(1)(0.1)^2} = 10$$

$$E = E° - \frac{0.06}{n} \log Q$$

$$= 0.44 \text{ V} - \frac{0.06 \text{ V mol e}^-}{2 \text{ mol e}^-} \log 10$$

$$= 0.44 \text{ V} - 0.03 \text{ V } (1) = 0.44 \text{ V} - 0.03 \text{ V} = 0.41 \text{ V}$$

The hydrogen electrode was originally used to determine all of the half-reaction standard electrode potentials shown in Table 14.2. The hydrogen electrode is given an arbitrary standard potential of 0.00 V and then attached to another electrode. The standard potential of the attached electrode is determined by measuring the voltage of the voltaic cell.

EXAMPLE 29 A voltaic cell is made with metallic electrodes of iron and platinum immersed in solutions of 1.0 M Fe^{2+} and 1.0 M H^+, respectively, at 25°C and 760 mm Hg. Hydrogen gas is bubbled through the hydrogen ion solution over the platinum electrode at a pressure of 760 mm Hg. The cell voltage is measured and found to be +0.44 V. The iron electrode dissolves while the cell operates. Calculate the standard electrode potential for the reduction of iron according to the reaction

$$Fe^{2+}(aq) + 2e^- \rightarrow Fe$$

Solution Since the iron electrode dissolves while the cell operates, the iron electrode must be undergoing oxidation, and the cell redox reaction must be

$$Fe + 2H^+(aq) \rightarrow Fe^{2+}(aq) + H_2 \qquad E° = +0.44 \text{ V}$$

The two half-reactions for this redox reaction are

$$Fe \rightarrow Fe^{2+}(aq) + 2e^- \qquad E° = +0.44 \text{ V}$$

$$2H^+(aq) + 2e^- \rightarrow H_2 \qquad E° = \ \ 0.00 \text{ V} \leftarrow \text{arbitrarily assigned}$$

The standard reduction potential for iron is

$$Fe^{2+}(aq) + 2e^- \rightarrow Fe \qquad E° = -0.44 \text{ V}$$

SELF-TEST

1. Calculate the voltage of a voltaic cell made with manganese and tin electrodes immersed in solutions of 1.0 M Mn^{2+} and 0.01 M Sn^{2+}, respectively, at 25°C and 760 mm Hg.
2. Calculate the voltage of a voltaic cell made with zinc and lead electrodes immersed in solutions of 0.01 M Zn^{2+} and 0.001 M Pb^{2+}, respectively, at 25°C and 760 mm Hg.
3. Calculate the voltage of a voltaic cell made with iron and copper electrodes immersed in solutions of 0.1 M Fe^{2+} and 0.01 M Cu^{2+}, respectively, at 25°C and 760 mm Hg.

ANSWERS **1.** 0.98 V **2.** 0.60 V **3.** 0.75 V

14.4.2 ARTIFICIAL HEART PACEMAKERS: TURNING THE HUMAN BODY INTO A BATTERY

Your heart has its own natural pacemaker that sends nerve impulses (pulses of electric current) throughout the heart at a rate of approximately 72 times per minute. These electric pulses cause your heart muscles to contract (beat), which pumps blood through your body. The fibers that carry the nerve impulses are very delicate and can be damaged by diseases, old age, drugs, heart attacks, and surgery. When these heart fibers are damaged, the heart may run too slowly, stop temporarily, or stop altogether.

To correct damaged fibers, artificial heart pacemakers (shown in Figure 14.7) are surgically inserted in the human body. A pacemaker (pacer) is a battery-driven device that sends an electric current (pulse) to the heart about 72 times per minute. Over 300 000 Americans are now wearing artificial pacemakers, with an additional 30 000 pacemakers installed each year (that's about 100 every day).

FIGURE 14.7 Artificial heart pacemaker

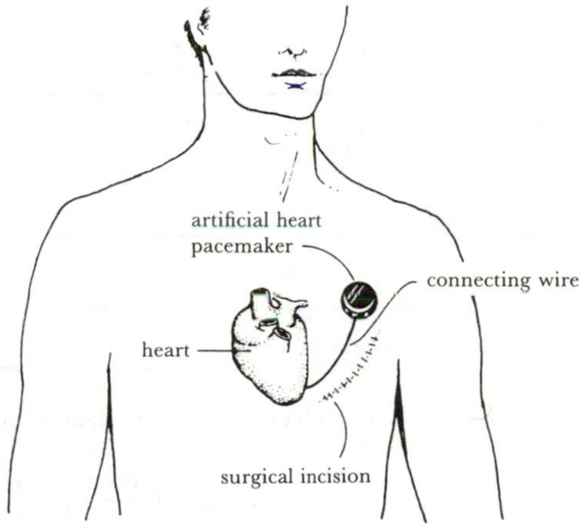

Yearly operations used to be necessary to surgically replace a pacemaker's batteries. Today, pacemaker wearers use improved batteries that last much longer, but even these must be replaced eventually.

It would be desirable to develop a permanent battery to run a pacemaker. Some scientists have begun working on ways of converting the human body itself into a battery (voltaic cell) to power an artificial pacemaker. There would be numerous side benefits as well. For example, just think of all those cold winter mornings when you have trouble starting your car!

Several methods for using the human body as a voltaic cell have been suggested. One of these is to insert platinum and zinc electrodes into the human body, as diagrammed in Figure 14.8. However, the pacemaker and the electrodes would be worn internally. This "body battery" could easily generate the small amount of current (5×10^{-5} A) that is required by most pacemakers. There is some concern, however, that the dissolving zinc electrode may have toxic effects.

•
•

EXAMPLE 30 Calculate the number of grams of zinc that will dissolve if the battery shown in Figure 14.8 runs for one year.

Solution

$$(5 \times 10^{-5}\ \text{A})(1\ \text{yr})\left(\frac{365\ \text{day}}{\text{yr}}\right)\left(\frac{24\ \text{hr}}{\text{day}}\right)\left(\frac{3600\ \text{sec}}{\text{hr}}\right)\left(\frac{33\ \text{g Zn}}{96\,500\ \text{A sec}}\right) = 0.5\ \text{g Zn}$$

The human body normally has about 4 g of Zn and can excrete additional amounts at a rate of more than 0.5 g of Zn per year.

•
•

FIGURE 14.8 Arrangement using the human body as a voltaic cell

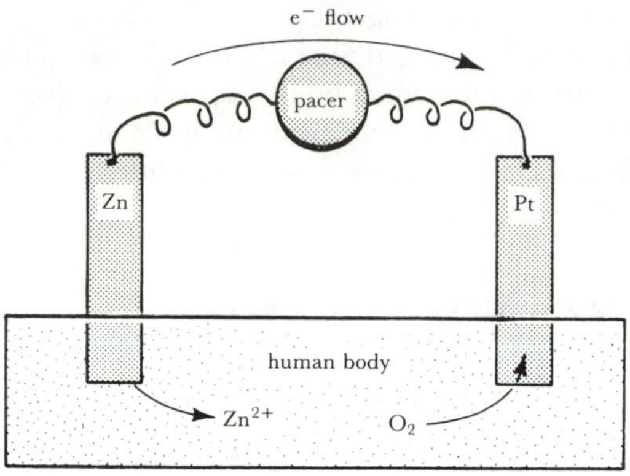

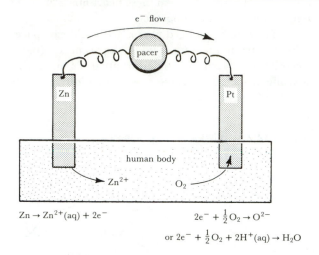

$Zn \rightarrow Zn^{2+}(aq) + 2e^-$

$2e^- + \frac{1}{2}O_2 \rightarrow O^{2-}$

or $2e^- + \frac{1}{2}O_2 + 2H^+(aq) \rightarrow H_2O$

How long would a 1-oz Zn electrode last?

A 1-oz (28-g) zinc electrode might last about 60 years.

$$(1 \text{ oz Zn})\left(\frac{28 \text{ g}}{\text{oz}}\right)\left(\frac{1 \text{ yr}}{0.5 \text{ g}}\right) = 6 \times 10^1 \text{ yr}$$

If you were afraid that you might live more than 60 additional years, you could always install a larger zinc electrode proportional in weight to your optimism. This "body battery" we have discussed has been tested on animals for periods of time in excess of four months without noticeable problems.

14.4.3 DENTAL FILLINGS AND SHOCKING CANDY BAR WRAPPERS*

Most cavities are filled with a substance known as dental amalgam. Dental amalgam contains mercury, silver, tin, copper, and zinc. Some people experience sharp pain when they accidentally chew part of an aluminum foil candy bar wrapper. This pain is caused by an electric shock produced by a voltaic cell made up of the aluminum and dental amalgam electrodes (see Figure 14.9).

14.4.4 GOLD-CAPPED TOOTH ERROR†

Some dentists, ignorant of chemistry, have inserted a gold cap over a tooth adjacent to a dental filling. The filling becomes a voltaic cell anode and sends electrons to the gold cap cathode (see Figure 14.10). As the dental filling dissolves, the victim (professional ignorance is a crime) notices a constant unpleasant metallic (tin) taste.

*Adapted from Richard S. Treptow, "Dental Fillings Discomforts Illustrate Electrochemical Potential of Metals," *Journal of Chemical Education* 55 (Mar. 1978):189.

†*Ibid.*

FIGURE 14.9 Voltaic cell made of aluminum and dental amalgam electrodes

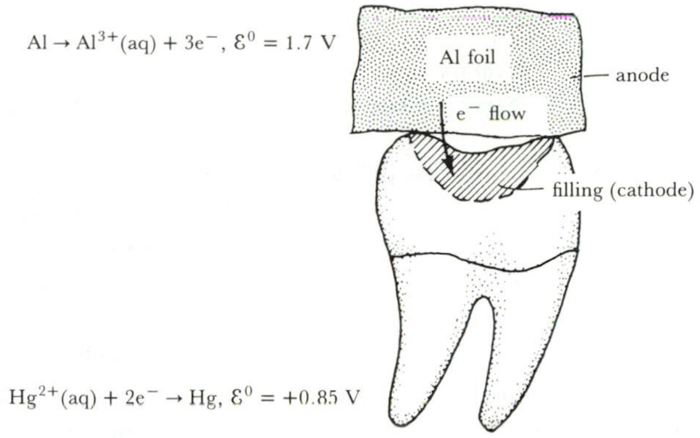

FIGURE 14.10 Voltaic cell made of filling and gold cap electrodes

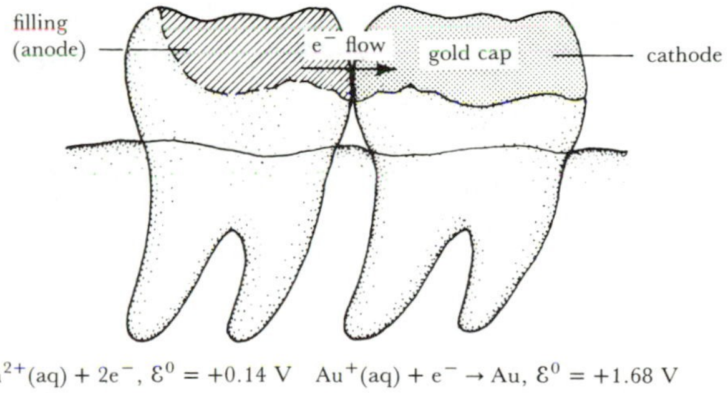

14.4.5 RACE CAR BATTERIES

As you use your car battery, it discharges and ultimately results in a decreased voltage. The voltage of your car battery is important to keep your car running. The relationship between battery voltage, its state of charge, and the sulfuric acid concentration within the battery is shown here:

electrodes

$$PbO_2(s) + Pb(s) + 2H^+(aq) + 2HSO_4^-(aq) \rightarrow 2PbSO_4(s) + 2H_2O(l)$$

2 e^- transferred

This is the chemical reaction inside your car battery that produces electricity.

Set the activities of pure solids and liquids equal to 1.

$$Q = \frac{[PbSO_4]^2[H_2O]^2}{[PbO_2][Pb][H^+]^2[HSO_4^-]^2} = \frac{(1)^2(1)^2}{(1))(1)[H^+]^2[HSO_4^-]^2} = \frac{1}{[H^+]^2[HSO_4^-]^2}$$

Activities of pure solids are set equal to 1.

$$E = E^\circ - \frac{0.06}{2}\log\frac{1}{[H^+]^2[HSO_4^-]^2}$$

voltage delivered by your car battery (E)

Concentrations of H^+ and HSO_4^- affect the value of E.

The concentration of sulfuric acid decreases as the chemical reaction in the battery proceeds in the forward direction. This is why auto mechanics frequently determine battery charge with a hydrometer, which is a device that measures the specific gravity (the density relative to that of water) of liquids. The specific gravity of the sulfuric acid solution in a fully charged battery is greater than the specific gravity of the sulfuric acid solution in a discharged battery.

The average driver is unaware of minor voltage fluctuations produced by his or her car battery as the battery wears down. However, racing cars run more efficiently if their electrical parts always receive an electric current at a constant voltage. This increased efficiency can make the difference between winning and losing a race. Therefore, it is desirable for a racing car to have a battery whose voltage remains constant as the battery's chemical reaction, shown above, goes in the forward direction.

How could you construct a voltaic cell (battery) whose voltage does not depend upon the concentrations of the products and the reactants (the battery's state of charge)?

If all the products and reactants of the battery were pure solids or liquids, their concentrations would remain constant, and the battery voltage (E) would always equal the standard voltage (E°) at 25°C.

$$E = E^\circ - \frac{0.06}{n}\log 1$$

$$= E^\circ - \frac{0.06}{n}(0) = E^\circ - 0 = E^\circ$$

$\log 1 = 0$

If all reactants and products are liquids and solids, their concentrations remain constant, and their activities equal 1. This makes Q have a value of 1.

Some race cars use a silver-zinc battery.

electrodes

$$Ag_2O(s) + Zn(s) + H_2O(l) \rightarrow 2Ag(s) + Zn(OH)_2(s)$$

$2\,e^-$ transferred

All products and reactants in this reaction are solids and liquids. The concentrations of solids and liquids remain constant as the reaction progresses in the forward direction, thereby maintaining a constant voltage in the battery.

$$Q = \frac{[Ag]^2[Zn(OH)_2]}{[Ag_2O][Zn][H_2O]} = \frac{(1)^2(1)}{(1)(1)(1)} = 1$$

$$E = E^\circ - \frac{0.06}{n}\log Q = E^\circ - \frac{0.06}{n}\log 1$$

$$= E^\circ - \frac{0.06}{n}(0) = E^\circ - 0 = E^\circ$$

14.5 EQUILIBRIUM CONSTANTS, SPONTANEITY, AND CELL VOLTAGE

You may have wondered why you spent so much time on thermodynamics to determine spontaneity when all you have to do to predict spontaneity is to look at the voltage of a reaction. The voltage of a reaction is nothing more than the Gibbs free energy in disguise (a different manifestation of the Gibbs free energy).

$$\Delta G^\circ = -nFE^\circ$$

cell voltage

Gibbs free energy change

number of moles of electrons gained or lost

value of the faraday

The only difference between the Gibbs free energy change of a reaction and the reaction voltage (standard electrode potential) is the factor $-nF$.

We have already shown that the Gibbs free energy change is another manifestation of the equilibrium constant K.

$$\Delta G^\circ = -2.303RT \log K$$

Thus, voltage is just another manifestation of the equilibrium constant.

$$\Delta G^\circ = -nFE^\circ$$

$$= -2.303RT \log K$$

$$-2.303RT \log K = -nFE^\circ$$

$$\log K = \frac{nFE^\circ}{2.303RT}$$

$$K = \text{antilog}\ \frac{nFE^\circ}{2.303RT}$$

PROBLEM SET 14

The problems in Problem Set 14 parallel the examples in Chapter 14. For example, if you should have trouble working Problem 5, go back to Example 5 in this chapter to get help. The correct answers to these problems are given at the end of this book.

1. What is the charge of 3.0×10^{23} electrons?
2. How many electrons would be required to build up a charge of 7500 C?
3. How many electrons pass a fixed point on a wire in 1 sec if a 2-A current is flowing through the wire?
4. How many electrons flow by a fixed point on a wire in 6 sec if the current is 6 A?
5. A current of 15 A is allowed to go through a wire for 1 hr. How many electrons pass a fixed point on the wire during this time?
6. Find the charge in faradays of 16 000 electrons.
7. Find the charge in faradays that passes a fixed point on a wire if a current of 15 A flows for 45 min.
8. How long will it take for 1 mol of electrons to pass a fixed point on a wire if the current flowing through the wire is 2.5 A?
9. How many amperes are required to make 0.5 mol of electrons pass point x in 4 sec?
10. How long would it take 2 C of electrons to pass a fixed point on a wire with a current of 0.5 A flowing through the wire?
11. In the electrolytic cell shown in Figure 14.2, a current of 4 A is passed through a cell for 1.6 hr. How many grams of copper will plate out on the right-hand electrode?
12. An electrolytic cell, similar to the one shown in Figure 14.2, has zinc electrodes immersed in a solution of

zinc nitrate. A current of 2.4 A is passed through the cell for 8.0 min. How many grams of zinc plate out? The reaction in which zinc ions plate out is represented by

$$Zn^{2+}(aq) + 2e^- \rightarrow Zn$$

13. Aluminum is prepared in electrolytic cells by plating out metallic aluminum from solutions of Al^{3+} ions. How long would it take to plate out 15 lb of aluminum with a current of 7.5 A?
14. Hydrogen gas forms at one electrode when a 4.0-A current is passed for 5.0 min through a hydrochloric acid solution using two inert platinum electrodes. What volume of hydrogen forms at this electrode if the gas is collected at 25°C and 750 mm Hg? The reaction in which hydrogen gas forms in this electrolytic cell is represented by

$$2H^+(aq) + 2e^- \rightarrow H_2$$

15. Chlorine gas forms at the other electrode in the electrolytic cell discussed in Problem 14. What is the volume of chlorine formed after 5.6 min? The reaction in which chlorine gas forms from a solution of chloride ions is represented by the chemical equation

$$2Cl^-(aq) \rightarrow Cl_2 + 2e^-$$

16. Write the redox reaction whose half-reactions are

$$Cu \rightarrow Cu^{2+}(aq) + 2e^- \quad \text{and} \quad Cl_2 + 2e^- \rightarrow 2Cl^-(aq)$$

17. Write the redox reaction whose half-reactions are

$$Zn \rightarrow Zn^{2+}(aq) + 2e^- \quad \text{and} \quad 2H^+(aq) + 2e^- \rightarrow H_2$$

18. What is the standard electrode potential for the reduction of Zn^{2+} to Zn?
19. What is the standard electrode potential of the half-reaction for the oxidation of zinc to Zn^{2+}?
20. What is the standard electrode potential for the oxidation of aluminum to the +3 state?
21. What is the standard electrode potential for the half-reaction

$$Mg^{2+}(aq) + 2e^- \rightarrow Mg$$

22. A voltaic cell is made with metallic electrodes of Mg and Zn immersed in solutions of 1 M Mg^{2+} and 1 M Zn^{2+}, respectively, at 25°C and 760 mm Hg. What spontaneous redox reaction occurs? What is the voltage of this voltaic cell?
23. A voltaic cell is made with metallic electrodes of iron and cobalt immersed in solutions of 1 M Fe^{2+} and 1 M Co^{2+}, respectively, at 25°C and 760 mm Hg. What spontaneous redox reaction occurs? What is the cell voltage?
24. A voltaic cell is prepared by connecting a bar of iron, immersed in a solution of 1 M Fe^{2+} at 25°C, to a chlorine electrode. The chlorine electrode consists of a platinum bar immersed in a solution of 1 M chloride ion at 25°C through which chlorine gas is bubbled at 1 atm pressure. What spontaneous chemical reaction occurs? What is the cell voltage?
25. A voltaic cell is made by connecting a bar of copper, immersed in a solution of 1 M Cu^{2+} at 25°C, to an oxygen electrode. The oxygen electrode consists of a bar of platinum immersed in water at 25°C through which oxygen gas is bubbled at 1 atm pressure. What spontaneous reaction occurs? What is the cell voltage?
26. Calculate the voltage of a voltaic cell made with metallic electrodes of magnesium and cobalt immersed in solutions of 1 M Mg^{2+} and 0.01 M Co^{2+}, respectively, at 25°C and 760 mm Hg.
27. Calculate the voltage of a voltaic cell made with metallic electrodes of aluminum and chromium immersed in solutions of 0.1 M Al^{3+} and 0.001 M Cr^{3+}, respectively, at 25°C and 760 mm Hg.
28. Calculate the voltage of a voltaic cell made with metallic electrodes of iron and platinum immersed in solutions of 0.01 M Fe^{2+} and 0.01 M H^+, respectively, at 25°C and 760 mm Hg. Hydrogen gas is bubbled through the hydrogen ion solution over the platinum electrode at a pressure of 760 mm Hg.
29. A voltaic cell is made with metallic electrodes of chromium and platinum immersed in solutions of 1.0 M Cr^{3+} and 1.0 M H^+, respectively, at 25°C and 760 mm Hg. Hydrogen gas is bubbled through the hydrogen ion solution over the platinum electrode at a pressure of 760 mm Hg. The cell voltage is measured and found to be +0.74 V. The chromium electrode dissolves while the cell operates. Calculate the standard electrode potential for the reduction of chromium (Cr^{3+}) to Cr.
30. A heart pacer can run off of a "body battery" if platinum and zinc electrodes are inserted into the human body as diagrammed in Figure 14.8. If the pacer uses a current of 3×10^{-5} A, how much zinc will dissolve in one year of pacer use?

15

Nuclear Chemistry

15.1 INTRODUCTION

In the last chapter we saw how a piece of metallic zinc becomes coated with copper metal when it is inserted into a copper sulfate solution. In the early days of alchemy, a similar demonstration (iron nails dipped into a copper sulfate solution) was used to "prove" that it is possible to change one element into another element (it appeared that the iron nails were changed into copper nails). This "proof" helped to maintain the enthusiasm of the alchemists, who were trying to change base metals, such as lead, into gold.

Today, we *can* change elements into other elements. These changes are called nuclear reactions because they involve changes within the nuclei (the plural form of nucleus) of the reacting atoms. Regular (traditional) chemical reactions usually involve changes in the number of outer electrons in the atomic orbitals of the reacting species (elements, atoms, molecules, or ions). When a zinc bar is dipped into a copper sulfate solution, outer electrons from zinc atoms are transferred to vacant outer atomic orbitals of copper ions. Most chemical reactions can be explained in terms of the sharing or transferring of electrons between two or more substances (elements, atoms, molecules, or ions). Nuclear chemistry concerns reactions in which the number of nucleons (neutrons and protons) within the nucleus of an atom is changed.

15.2 BASIC DEFINITIONS AND IDEAS

The four corners surrounding the symbol for a chemical element are reserved for specific information. The upper right-hand corner is reserved for the charge of the element. If a charge is not shown, a zero charge (no charge) is understood. The three symbols

$$O^{2-} \quad O^{1-} \quad O$$

show oxygen with three different charges (negative two, negative one, and zero).

The number of atoms present is indicated by a number in the lower right-hand corner. If no number is shown, a 1 is assumed. The three symbols

$$O_3 \quad O_2 \quad O$$

show the ozone molecule (O_3) made up of 3 oxygen atoms, the oxygen molecule (O_2) made up of 2 oxygen atoms, and a single oxygen atom (O_1 is understood).

Nuclear information is shown in the left-hand corners. The number of protons present in the nucleus of an atom is written in the lower left-hand corner.

•
•

EXAMPLE 1 How many protons are present in the nuclei of the following atoms?

$${}_{6}C \quad {}_{7}N \quad {}_{8}O$$

Solution Carbon has 6 protons present in its nucleus. Nitrogen has 7 protons, and oxygen has 8 protons.

•
•

The number of protons present in the nucleus of an atom is called the atomic number of the atom. The atomic number is given the symbol Z.

•
•

EXAMPLE 2 What is the atomic number of carbon? Find Z for nitrogen and oxygen.

Solution The atomic number of carbon is 6. The atomic numbers of nitrogen and oxygen are 7 and 8 ($Z = 7$ and $Z = 8$), respectively.

•
•

Protons are positively charged particles. The magnitude (strength) of a proton's charge is equal to but opposite in sign to the charge of an electron. We say that a proton has a charge of +1 and that an electron has a charge of −1. Elements (as neutral atoms) always have the same number of electrons in their atomic orbitals as they have protons in their nuclei. An element can be identified by the number of protons inside its nucleus. For example, any atom or ion that has 8 protons inside its nucleus is called oxygen. If 1 proton is somehow removed (we'll see how later) from the nucleus of an oxygen atom, 7 protons are left behind, and the atom is now called nitrogen.

•
•

EXAMPLE 3 Which element has an atomic number of 14 ($Z = 14$)? How many electrons are in the atomic orbitals of an atom of this element?

Solution From a periodic table, we learn that the only element that can have 14 protons is silicon. A silicon atom has 14 electrons in its atomic orbitals, the same number as the number of protons in its nucleus.

•
•

EXAMPLE 4 If 2 protons were added to the nucleus of a silicon atom, what new element would result?

Solution The new atom would have 16 protons ($14 + 2 = 16$) in its nucleus. Any atom with 16 protons in its nucleus—atomic number (Z) = 16—is called a sulfur atom. An atom of sulfur also has 16 electrons in its atomic orbitals.

•
•

The number of nucleons (protons plus neutrons) present in the nucleus of an atom is called the mass number of the atom and is given the symbol A. The mass number is written in the upper left-hand corner of a chemical symbol.

•
•

EXAMPLE 5 Find the mass number (A) for each of the following atoms:

$$^{16}O \qquad ^{17}O \qquad ^{18}O$$

Solution These three oxygen atoms contain 16, 17, and 18 nucleons (protons plus neutrons), respectively. Their mass numbers are 16, 17, and 18 ($A = 16$, $A = 17$, and $A = 18$), respectively. Each of these oxygen atoms has an atomic number of 8 ($Z = 8$); that is, each has 8 protons in its nucleus.

•
•

When you focus your attention on the nuclear properties of an atom, the atom is sometimes referred to as a nuclide. When two or more atoms of the same element (i.e., nuclides with equal Z values) have different mass numbers (i.e., different A values), the atoms are called isotopes. The nuclides $^{16}_{8}O$, $^{17}_{8}O$, and $^{18}_{8}O$ are called isotopes of oxygen.

We will use the symbol N to represent the number of neutrons inside a nucleus. The number of neutrons in $^{16}_{8}O$ is 8 ($N = 8$). The number of neutrons in a nucleus is calculated by subtracting the atomic number from the mass number ($N = A - Z$).

•
•

EXAMPLE 6 Find the values of A, Z, and N for $^{17}_{8}O$ and $^{18}_{8}O$.

Solution The nucleus of $^{17}_{8}O$ (called oxygen-17 or O-17) contains 8 protons and 17 nucleons (neutrons plus protons). The number of neutrons present in the nucleus of an oxygen-17 atom is equal to $17 - 8 = 9$. For oxygen-17, $A = 17$, $Z = 8$, and $N = 9$. For oxygen-18, $A = 18$, $Z = 8$, and $N = 10$.

Recall that 1 amu (atomic mass unit) is exactly equal to one-twelfth of the mass of a $^{12}_{6}C$ atom. The atomic weight of almost every isotope is very close to a whole number (within 0.01 amu). The atomic weights of oxygen-16, oxygen-17, and oxygen-18 are equal to 15.995 amu, 16.999 amu, and 17.999 amu, respectively. In the future we will frequently use the same whole number to represent both the nuclidic mass (in atomic mass units) and the mass number (number of nucleons). For example, we will say that the nuclidic masses of oxygen-16, oxygen-17, and oxygen-18 are equal to 16 amu, 17 amu, and 18 amu, respectively.

EXAMPLE 7 Find the nuclidic mass, mass number, atomic number, number of neutrons, number of protons, and the number of electrons in one atom of the isotope $^{59}_{27}Co$.

Solution The number written in the upper left-hand corner is the mass number (A). The mass number of cobalt-59 is 59 (the number of protons and neutrons). The nuclidic mass for this isotope is 59 amu. The number written in the lower left-hand corner is the atomic number (number of protons) in the nucleus. The atomic number (Z) for this isotope is 27. There are 27 protons in the nucleus of a cobalt-59 atom. This is a neutral atom (no charge is written in the upper right-hand corner), so there are also 27 electrons in the atomic orbitals of this atom. The number of neutrons is equal to the difference between the mass number (A) and the atomic number (Z).

$$N = A - Z = 59 - 27 = 32$$

This isotope has 32 neutrons in its nucleus.

EXAMPLE 8 Find the values of A, Z, and N for $^{31}_{15}P$. How many electrons are in one atom of P-31, and what is the nuclidic mass of this isotope?

Solution The value of A (the mass number) is written in the upper left-hand corner. The nucleus of P-31 contains 31 nucleons ($A = 31$). The value of Z is written in the lower left-hand corner. There are 15 protons in the nucleus ($Z = 15$). The value of N (number of neutrons) is equal to $A - Z$.

$$N = A - Z = 31 - 15 = 16$$

There are 16 neutrons in the nucleus. The number of protons in the nucleus equals the number of electrons outside the nucleus in a neutral atom. There are 15 electrons going around the nucleus of an atom of P-31. The nuclidic mass of this isotope is 31 amu. It is $\frac{31}{12}$ times the mass of one atom of carbon-12.

The density of all known nuclei is relatively constant and is equal to 250 million tons per cubic centimeter. (The density of a helium nucleus is about the same as the density of a lead nucleus.) If the diameter of an atom could be enlarged to 1 mi (about the diameter of a town with 5000 people), the nucleus would have a diameter of only two-thirds of an inch.

The positive protons within the nucleus exert a repulsive electrostatic force on each other (two protons repel each other), and yet the protons within a nucleus do not fly apart because of a more powerful nuclear force that binds the nucleons together. As an analogy, you can think of the neutrons as acting like glue, helping to keep the protons together. If you were to pour a bucketful of glue over an envelope and then try to attach a postage stamp, you would have trouble getting the stamp to stick (it would float away). Too little glue will also result in a stamp that

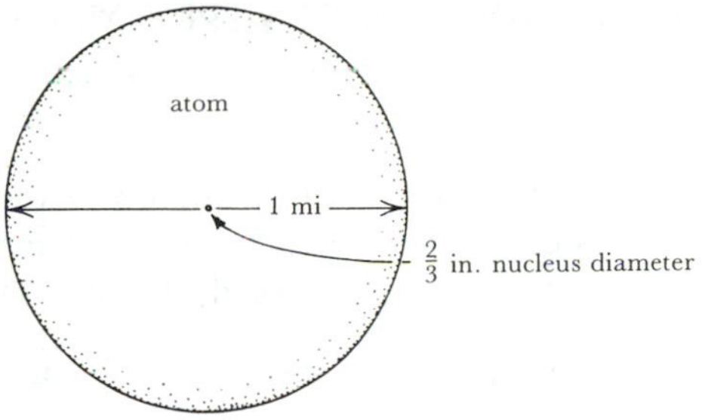

doesn't properly stick to the envelope. And so it appears to be with neutrons. Too many neutrons or too few neutrons seem to produce unstable nuclei that fly apart. We observe that lighter elements (such as oxygen, nitrogen, and carbon) have about an equal number of protons and neutrons (there's about one neutron for every proton present). Heavier elements (such as gold, mercury, and lead) seem to have nuclei with about 60% neutrons and 40% protons. For the majority of stable elements it appears that the number of protons present within their nucleus is about equal to or less than the number of neutrons (there are one or more neutrons for every proton).

We observe that nuclei with too little or too much "glue" (less than 50% neutrons or greater than 65% neutrons) do not "stick together" very well, and these nuclei are said to be unstable. Unstable nuclei can become stable by giving off radiation (details will follow). An unstable nucleus is said to be radioactive.

No nucleus can be stable with more than 83 protons, no matter how many neutrons (which act as "glue") are present. What is the last stable element on the periodic table?

Bismuth ($^{209}_{83}Bi$) is the last stable element.

How would you explain the appearance of elements such as thorium and uranium, which have 90 and 92 protons, respectively, in their nucleus?

These elements can exist, but they are not stable. They are radioactive and decay into more stable elements with the passage of time. There are several ways in which unstable radioactive elements can decay into more stable elements. We'll look into some of these methods later in this chapter.

15.3 ISOTOPES

If two atoms have the same atomic number (number of protons) but different mass numbers (number of nucleons), the atoms are called isotopes. Most elements exist in nature as a mixture of two or more isotopes. Although we generally think of hydrogen as having one proton in its nucleus, any naturally occurring sample of hydrogen will also contain a small number of hydrogen atoms that have one or two neutrons inside their nuclei. These three isotopes of hydrogen are $^{1}_{1}H$, $^{2}_{1}H$, and $^{3}_{1}H$.

You have probably noticed that the atomic weights of most elements shown on the periodic table are not whole numbers. Some elements have atomic weights that are not even close to a whole number. For example, the atomic weight of chlorine is 35.5 amu. One reason for non-whole number atomic weights is the occurrence of isotopes. The examples in the next two sections will shed more light on this idea.

15.3.1 IF THE HUMAN RACE WERE 90% FEMALE

•
•

EXAMPLE 9 The members of the human race come in two varieties, male and female. Assume that the human race is exactly 50% male and 50% female. Also, assume that all males weigh exactly 200 lb, and all females weigh exactly 100 lb. What is the average weight of a human being?

Solution If you had two people, one male and one female (a 50/50 mixture), their total weight would be 300 lb, and their average weight would be 150 lb.

$$\begin{array}{rl} 200\text{ lb} & \leftarrow \text{weight of one man} \\ +100\text{ lb} & \leftarrow \text{weight of one woman} \\ \hline 300\text{ lb} & \leftarrow \text{total weight} \end{array}$$

$$\frac{300\text{ lb}}{2} = 150\text{ lb} \leftarrow \text{average weight}$$

For reasons that will become clear in later examples, let's solve this same problem in a different way.

$$\left(\frac{50\text{ males}}{100\text{ people}}\right)\left(\frac{200\text{ lb}}{\text{male}}\right) + \left(\frac{50\text{ females}}{100\text{ people}}\right)\left(\frac{100\text{ lb}}{\text{female}}\right) = \frac{100\text{ lb}}{\text{person}} + \frac{50\text{ lb}}{\text{person}} = \frac{150\text{ lb}}{\text{person}}$$

50% male = 50 males per 100 people (pointing to $\frac{50\text{ males}}{100\text{ people}}$); average weight of a human being (pointing to $\frac{150\text{ lb}}{\text{person}}$)

•
•

EXAMPLE 10 Now let's assume that the human race is exactly 90% female and 10% male. All males weigh exactly 200 lb, and all females weigh exactly 100 lb. What is the average weight of a person?

Solution

$$\left(\frac{90\text{ females}}{100\text{ people}}\right)\left(\frac{100\text{ lb}}{\text{female}}\right) + \left(\frac{10\text{ males}}{100\text{ people}}\right)\left(\frac{200\text{ lb}}{\text{male}}\right) = \frac{90\text{ lb}}{\text{person}} + \frac{20\text{ lb}}{\text{person}} = \frac{110\text{ lb}}{\text{person}}$$

average weight of a human being (pointing to $\frac{110\text{ lb}}{\text{person}}$)

•
•

EXAMPLE 11 Assume that a random sample of exactly 1000 humans were weighed. Assume that this sample contains a normal distribution of males and females (exactly 90% female, 10% male). What is the total weight of this sample of 1000 humans?

Solution This sample of 1000 people weighs 110 000 lb.

$$(1000\text{ people})\left(\frac{110\text{ lb}}{\text{person}}\right) = 110\,000\text{ lb}$$

from Example 10 (pointing to $\frac{110\text{ lb}}{\text{person}}$)

or

$$(1000\text{ people})\left(\frac{90\text{ females}}{100\text{ people}}\right) = 900\text{ females}$$

$$(1000\text{ people})\left(\frac{10\text{ males}}{100\text{ people}}\right) = 100\text{ males}$$

$$(900 \text{ females})\left(\frac{100 \text{ lb}}{\text{female}}\right) + (100 \text{ males})\left(\frac{200 \text{ lb}}{\text{male}}\right) = 90\,000 \text{ lb} + 20\,000 \text{ lb}$$
$$= 110\,000 \text{ lb}$$

•
•

15.3.2 ATOMIC WEIGHT DETERMINATIONS

The atomic weight of an element shown on a periodic table is the average weight of all the isotopes of that element based upon their percentage of abundance (just as the average weight of a human depends upon the percentage of abundance of the two types of humans, male and female).

•
•

EXAMPLE 12 The element boron exists primarily in two isotopic forms, boron-10 ($^{10}_{5}B$) and boron-11 ($^{11}_{5}B$). About 20% of all boron atoms are boron-10. What is the average weight of a boron atom?

Solution

$$\left(\frac{20\ ^{10}_{5}\text{B atoms}}{100 \text{ B atoms}}\right)\left(\frac{10.0 \text{ amu}}{^{10}_{5}\text{B atom}}\right) + \left(\frac{80\ ^{11}_{5}\text{B atoms}}{100 \text{ B atoms}}\right)\left(\frac{11.0 \text{ amu}}{^{11}_{5}\text{B atom}}\right)$$
$$= 2.0 \text{ amu/B atom} + 8.8 \text{ amu/B atom} = 10.8 \text{ amu/B atom}$$

(10.8 amu/B atom ← average weight of a B atom)

The atomic weight of boron shown on a periodic table is 10.8 amu. The atomic weight reported on the periodic table is the average weight.

•
•

EXAMPLE 13 About 7.4% of all lithium exists as the $^{6}_{3}Li$ isotope. The rest exists as $^{7}_{3}Li$. What is the atomic weight of lithium?

Solution

$$\left(\frac{7.4\ ^{6}_{3}\text{Li atoms}}{100 \text{ Li atoms}}\right)\left(\frac{6.0 \text{ amu}}{^{6}_{3}\text{Li atom}}\right) + \left(\frac{92.6\ ^{7}_{3}\text{Li atoms}}{100 \text{ Li atoms}}\right)\left(\frac{7.0 \text{ amu}}{^{7}_{3}\text{Li atom}}\right)$$
$$= 0.44 \text{ amu/Li atom} + 6.5 \text{ amu/Li atom} = 6.9 \text{ amu/Li atom}$$

(6.9 amu/Li atom ← average weight of a lithium atom)

The atomic weight of lithium shown on a periodic table is 6.9 amu.

•
•

EXAMPLE 14 The element neon exists in three naturally occurring isotopic forms: neon-20 (90.92%), neon-21 (0.257%), and neon-22 (8.82%). What atomic weight is reported on a periodic table for neon?

Solution

$$\left(\frac{90.92\ ^{20}\text{Ne atoms}}{100 \text{ Ne atoms}}\right)\left(\frac{20.0 \text{ amu}}{^{20}\text{Ne atom}}\right) + \left(\frac{0.257\ ^{21}\text{Ne atoms}}{100 \text{ neon atoms}}\right)\left(\frac{21.0 \text{ amu}}{^{21}\text{Ne atom}}\right)$$
$$+ \left(\frac{8.82\ ^{22}\text{Ne atoms}}{100 \text{ Ne atoms}}\right)\left(\frac{22.0 \text{ amu}}{^{22}\text{Ne atom}}\right) = \frac{18.2 \text{ amu}}{\text{Ne atom}} + \frac{0.054 \text{ amu}}{\text{Ne atom}} + \frac{1.94 \text{ amu}}{\text{Ne atom}}$$
$$= 20.2 \text{ amu/Ne atom}$$

The atomic weight of neon shown on a periodic table is 20.2 amu, the average atomic weight of a neon atom.

•
•

SELF-TEST

1. Find the atomic weight, the mass number, the atomic number, the number of electrons, and the number of neutrons in an atom of ${}^{9}_{5}B$.
2. The hypothetical element Q exists in two isotopic forms: 20% exists as ${}^{55}Q$, and the rest as ${}^{57}Q$. What mass would be reported for Q on a periodic table of hypothetical elements?

ANSWERS 1. atomic weight = 9 amu; $A = 9$; $Z = 5$; 5 electrons; 4 neutrons
2. The atomic weight of Q is 56.6 amu.

15.4 RADIOACTIVITY

Radioactivity is the spontaneous (which means it occurs on its own) decay (falling apart, changing into something else) of an unstable nucleus. A nucleus can be unstable because of a poor neutron-to-proton ratio (too much glue or not enough glue), or because the nucleus contains more than 83 protons.

15.4.1 TYPES OF DECAY

When a nucleus is unstable (radioactive) it can decay (change into another element) in several different ways. We will discuss the following three methods of decay in detail: (1) alpha decay, (2) beta decay, and (3) K-capture. We will also discuss a fourth radioactive process that occurs when an unstable (radioactive) nucleus gives off gamma radiation.

An alpha particle is a helium nucleus (i.e., it is made up of 2 protons and 2 neutrons and carries a +2 charge). An alpha particle is represented by the Greek letter α (alpha), or ${}^{4}_{2}He^{2+}$, or, simply, ${}^{4}_{2}He$.

The boron isotope ${}^{12}_{5}B$ is unstable. Can you explain why?

The nuclei of lighter elements contain about 50% neutrons and 50% protons. Boron-12 has too many neutrons (too much glue) in its nucleus.

•
•

EXAMPLE 15 The boron isotope ${}^{12}_{5}B$ is unstable and decays by emitting an alpha particle from its nucleus. What element does boron-12 turn into after losing an alpha particle?

Solution

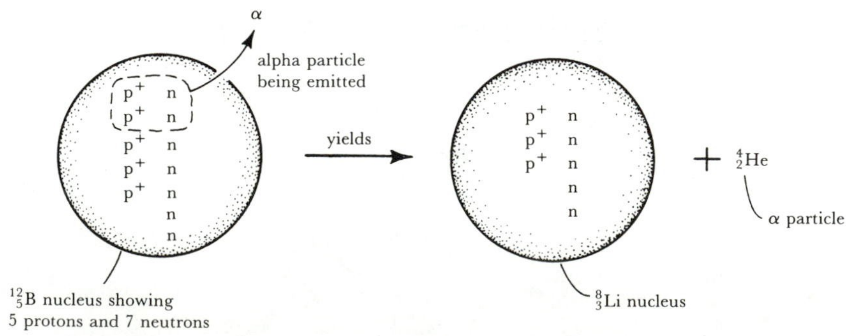

or

$$^{12}_{5}B \rightarrow {}^{8}_{3}Li + {}^{4}_{2}He$$

When a boron-12 nucleus loses 2 protons and 2 neutrons (in the form of an alpha particle, or helium nucleus), the number of protons within the boron-12 nucleus is decreased from 5 to 3, and the number of neutrons is decreased from 7 to 5. The mass number of boron-12 is decreased by 4 (from 12 to 8). Whenever a nucleus of an element gains or loses protons, a new element is formed. Boron-12 loses 2 protons and is left with 3 protons. Atoms with 3 protons in their nuclei are called lithium atoms. It turns out that lithium-8 is also unstable (has too many neutrons) and decays by another process (beta decay) that we will discuss later.

•
•

The isotope of polonium $^{210}_{84}Po$ is unstable. Can you explain why?

Polonium-210 is unstable because it contains more than 83 protons in its nucleus. All nuclei with more than 83 protons are unstable and decay.

•
•

EXAMPLE 16 Polonium-210 decays by giving off an alpha particle. What element forms when Po-210 decays?

Solution

$$^{210}_{84}Po \rightarrow {}^{206}_{82}Pb + {}^{4}_{2}He$$

When the nucleus of polonium-210 loses 2 protons (an alpha particle is made up of 2 protons and 2 neutrons), a new element, lead, forms. The total mass number of the polonium-210 nucleus decreases by 4 (from 210 to 206), which is the mass of an alpha particle.

•
•

A beta particle is an electron that is emitted from an unstable nucleus of an atom. How can an electron come from a nucleus? A neutron in an unstable nucleus can decay into a proton and an electron.

$$^{1}_{0}n \rightarrow {}^{1}_{1}H + {}^{0}_{-1}e$$

The process by which a neutron decays into a proton and an electron is not fully understood. When a neutron within a nucleus decays into a proton and an electron, the electron is emitted from the nucleus and is called a beta particle, or beta radiation. The process is called beta decay. A beta particle is represented by the Greek letter β (beta).

The carbon isotope $^{14}_{6}C$ is unstable. Why might you have suspected that it is unstable?

Carbon-14 contains more than 50% neutrons. Stable lightweight elements tend to have about equal numbers of neutrons and protons. Heavier elements (elements with larger atomic weights) tend to have more neutrons than protons.

•
•

EXAMPLE 17 The carbon isotope $^{14}_{6}C$ is unstable and decays by emitting a beta particle. What element does the carbon-14 nuclide turn into after releasing a beta particle?

Solution

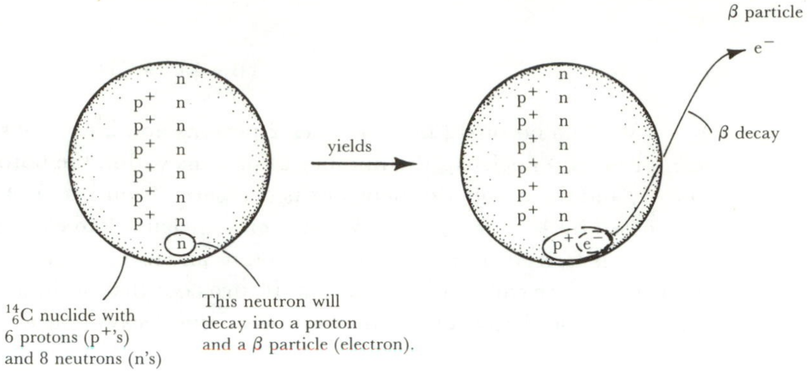

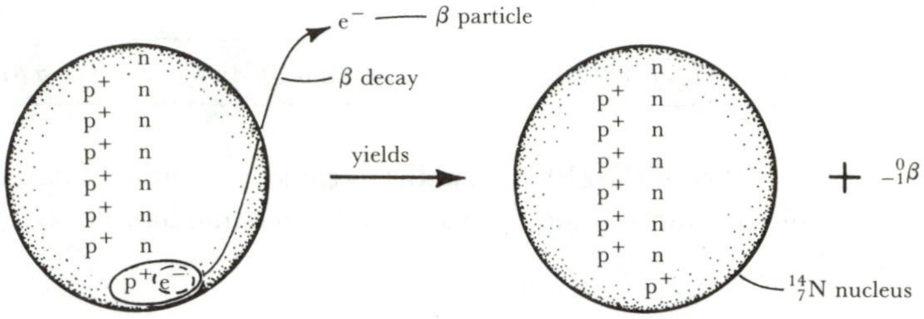

or

$$^{14}_{6}C \rightarrow ^{14}_{7}N + ^{0}_{-1}\beta$$

When one of the 8 neutrons of carbon-14 decays into a proton and an electron (beta particle), the carbon-14 nuclide gains 1 proton (goes from 6 protons to 7 protons). Notice that the mass number, 14, does not change. The mass number of the neutron that decays and the resulting proton are both 1. The mass number of the nucleus is the sum of the protons and neutrons. In beta decay, the sum of the protons and neutrons does not change.

•
•

Helium-6 is unstable. Why might you suspect this?

The nuclei of the lighter elements tend to have about the same number of protons and neutrons. Helium-6 ($^{6}_{2}He$) has many more neutrons than protons (it has too much glue).

•
•

EXAMPLE 18 Helium-6 is unstable and decays by giving off a beta particle. What does He-6 decay into?

Solution

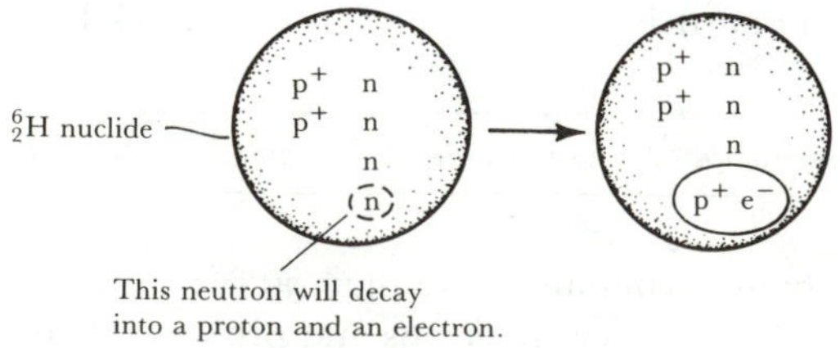

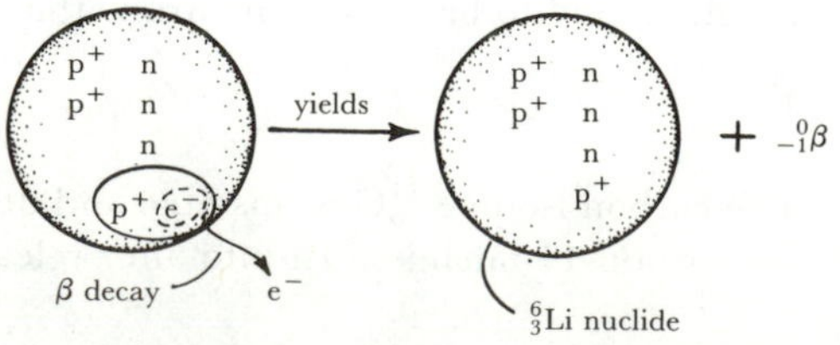

or

$$^{6}_{2}He \rightarrow ^{6}_{3}Li + ^{0}_{-1}\beta$$

•
•

You can think of K-capture as the reverse of beta decay. Recall that the main energy shells (levels) around a nucleus are called the K shell, the L shell, the M shell, and so forth. The K shell is the main energy shell (level) closest to the nucleus. Electrons in the K shell are in 1*s* orbitals, which are spherical in shape. Electrons in a 1*s* orbital are more likely to be found somewhere within this spherical shape, the volume of which includes the nucleus of the atom. An electron from a K shell can exist within the nucleus of an atom. If the nucleus of an atom has too many protons, a K shell electron can combine with an extra proton and convert the proton into a neutron (increasing the glue supply).

Beryllium-7 is unstable. Why might you suspect this?

Beryllium-7 has 3 neutrons and 4 protons. We observe that stable elements appear to have equal numbers of neutrons and protons or more neutrons than protons. Beryllium-7 has more protons than neutrons (it has too little glue).

•
•

EXAMPLE 19 Beryllium-7 is unstable and undergoes K-capture. What element forms as a result of the K-capture?

Solution

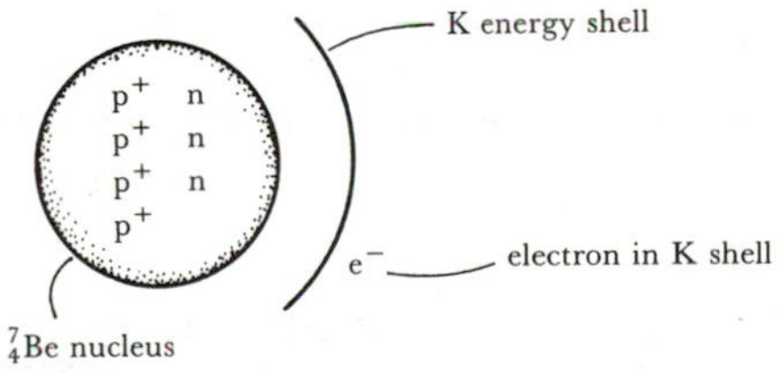

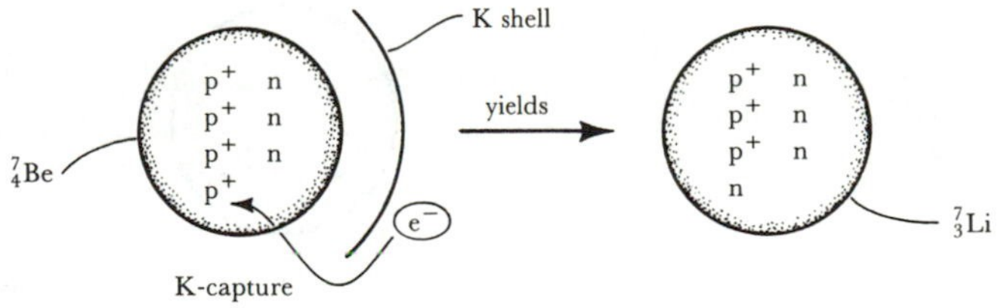

or

$$^{0}_{-1}e^{-} + ^{7}_{4}Be \xrightarrow{\text{K-capture}} ^{7}_{3}Li$$

In K-capture, a proton is changed into a neutron. The total number of protons inside the Be-7 nucleus is decreased by 1 (changes from 4 protons to 3 protons) while the total number of neutrons is increased by 1 (changes from 3 neutrons to 4 neutrons). The mass number remains constant.

•
•

EXAMPLE 20 Mercury-197 ($^{197}_{80}Hg$) is unstable and undergoes K-capture. What element forms?

Solution

$$^{0}_{-1}e^{-} + ^{197}_{80}Hg \xrightarrow{\text{K-capture}} ^{197}_{79}Au$$

One of the 80 protons of mercury is changed into a neutron by one of its own K shell electrons. The number of protons decreases by 1 (changes from 80 protons to 79 protons). Any nucleus with 79 protons is called gold. This reaction is one in which a base metal (mercury) is changed into gold, the ancient dream of the alchemist. However, producing gold this way is more expensive than digging it out of the ground.

•
•

Nucleons (neutrons and protons) within a nucleus can exist in different energy levels, just as electrons in atomic orbitals around the nucleus do. Figure 15.1 shows a nucleus with 1 neutron and 1 proton, each in the ground (lowest energy) state.

FIGURE 15.1 Nucleus with nucleons in the ground state

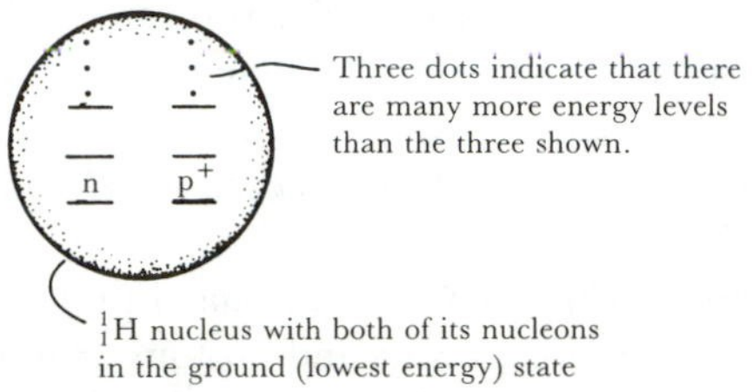
Three dots indicate that there are many more energy levels than the three shown.
n
p+
$^{1}_{1}$H nucleus with both of its nucleons in the ground (lowest energy) state

Some nuclear processes (for example, fission) produce nuclei in which nucleons are in excited (higher energy) states, such as the nucleon shown in Figure 15.2. When the excited neutron (the neutron in the second energy level) "falls" to a lower energy level (the first level), the nucleus becomes more stable. Whenever we go from a less stable (higher energy) situation to a more stable (lower energy) situation, energy is always released. When nucleons fall to lower energy levels, the energy released is called gamma radiation (see Figure 15.3). Gamma radiation is represented by the Greek symbol γ (gamma).

FIGURE 15.2 Nucleus with a nucleon in an excited state

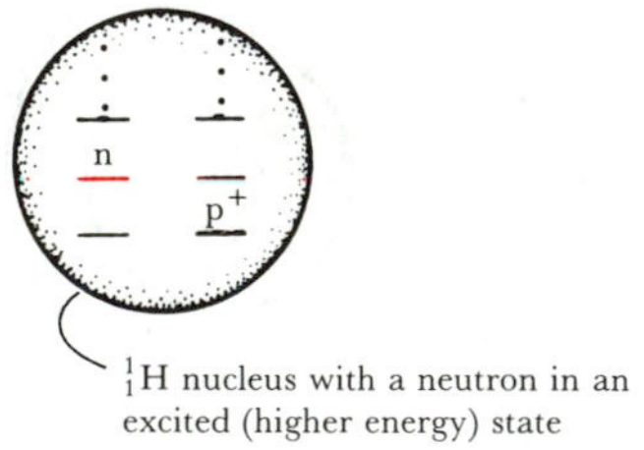
n
p+
$^{1}_{1}$H nucleus with a neutron in an excited (higher energy) state

FIGURE 15.3 Release of gamma radiation

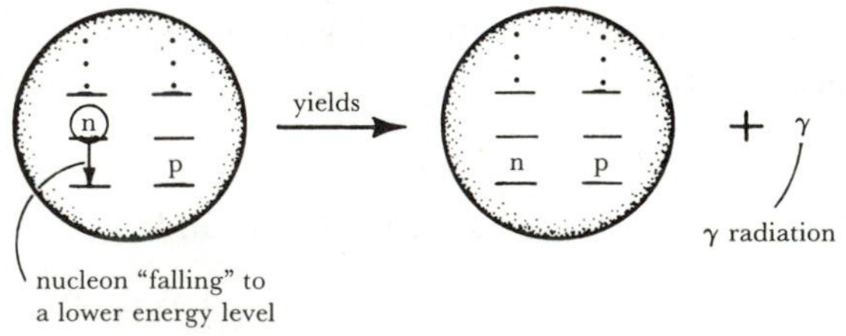

•
•

EXAMPLE 21 $^{1}_{1}H$ decays by emitting a gamma ray. Write an equation showing this decay.

Solution

$$^{1}_{1}H \rightarrow {}^{1}_{1}H + \gamma$$

Notice that when gamma radiation is given off by an unstable nucleus, the number of neutrons and protons doesn't change. The mass number remains constant. The atomic number remains constant. The element doesn't change into a different element.

•
•

EXAMPLE 22 Uranium-236 decays by releasing gamma radiation. Write an equation showing this decay.

Solution

$$^{236}_{92}U \rightarrow {}^{236}_{92}U + \gamma$$

•
•

What is gamma radiation? To understand what gamma radiation is, you must know that ordinary (visible) light can be represented as a wave (see Figure 15.4), as can radio waves (sent out by a radio station). The distance between any two crests (bumps) of a wave is called the wavelength of the wave. The only difference between visible (ordinary) light and radio waves is their wavelengths. Radio waves have a much longer wavelength than visible light.

FIGURE 15.4 Wavelength

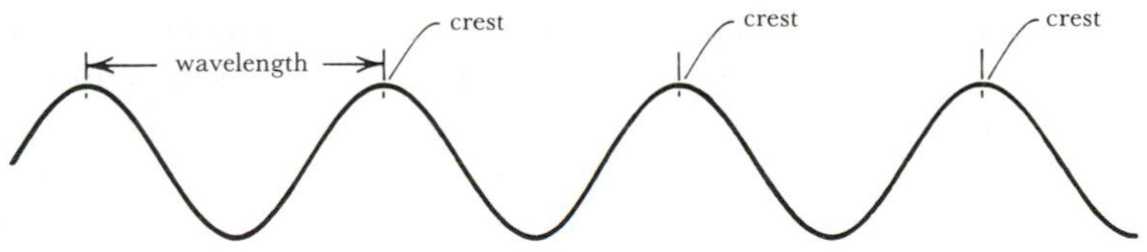

Both radio waves and visible light are called electromagnetic radiation because they have electrical (charged) and magnetic components that peak (crest) and dip with a wavelike motion. Radio waves, TV, radar waves, microwaves, heat (infrared radiation, also called IR), visible light, ultraviolet radiation (UV), X rays, gamma rays, and cosmic rays are all examples of electromagnetic radiation. They differ from each other because each has its own particular range of wavelengths, as shown in the following diagram:

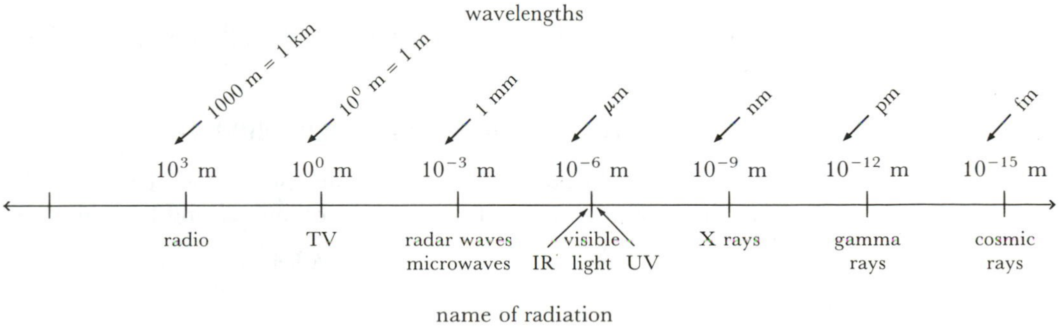

You can think of gamma rays as being similar to X rays but with a shorter wavelength. The shorter the wavelength, the more energy the wave has. Gamma rays have more energy than X rays.

•
•

EXAMPLE 23 When the hypothetical radioactive element D undergoes radioactive decay, its atomic number and the number of neutrons in its nucleus decrease. What type of decay would account for these changes?

Solution When alpha decay occurs, both the number of protons (atomic number) and the number of neutrons decrease because an alpha particle is made up of 2 protons and 2 neutrons. When a nucleus emits an alpha particle, both the atomic number and the mass number decrease. Beta decay, gamma decay, and K-capture decay do not cause both Z and A to decrease.

•
•

EXAMPLE 24 When the hypothetical radioactive element E undergoes radioactive decay, the mass number remains constant and the atomic number increases. What type of decay would account for the change?

Solution When beta decay occurs, a neutron changes into a proton. This change (from n to p) increases the atomic number (Z increases) but does not change the mass number (A remains constant). Alpha decay, gamma decay, and K-capture decay do not cause N to decrease by an amount equal to the increase in Z.

•
•

EXAMPLE 25 When the hypothetical radioactive element G undergoes radioactive decay, both the mass number and the atomic number remain constant. What type of decay would account for this?

Solution When gamma decay occurs, the mass number and the atomic number of a nucleus do not change. A gamma ray is emitted when a nucleon falls from a less stable to a more stable energy level within the nucleus. Alpha decay, beta decay, and K-capture decay cause changes in Z and/or A.

•
•

SELF-TEST

1. When the hypothetical radioactive element J undergoes radioactive decay, the mass number and the number of neutrons remain constant. What type of decay accounts for these observations?
2. When the hypothetical radioactive element L undergoes radioactive decay, the number of neutrons increases and A remains constant. What type of nuclear decay accounts for the change?
3. Write an equation showing the alpha decay of $^{146}_{62}Sm$. Use the periodic table to determine the products of this decay.
4. Write an equation showing the beta decay of $^{143}_{58}Ce$.

ANSWERS **1.** gamma **2.** K-capture **3.** $^{146}_{62}Sm \rightarrow ^{142}_{60}Nd + ^{4}_{2}He$
4. $^{143}_{58}Ce \rightarrow ^{143}_{59}Pr + ^{0}_{-1}\beta$

15.4.2 DETERMINATION OF AVOGADRO'S NUMBER

Radium-226 is radioactive and decays by giving off alpha particles (He^{2+} ions). The helium ions readily pick up electrons from their surroundings and form atoms of gaseous helium.

A helium nucleus is a helium ion (He^{2+}).
↓

$$^{226}_{88}Ra \rightarrow ^{222}_{86}Rn + ^{4}_{2}He$$

$$He^{2+} + 2e^{-} \rightarrow He$$

↑ helium ion ↑ helium atom

A sample of radium-226 can float on a column of mercury inside an inverted test tube, as shown in Figure 15.5. Helium gas is collected inside the inverted test tube. The volume of the helium gas is easily measured if the inverted test tube is graduated (has volume marks). Gas laws are used to determine the He gas volume at standard temperature and pressure conditions (0°C and 1 atm pressure). A Geiger counter is used to measure the number of helium ions given off every second.

FIGURE 15.5 Apparatus used to determine Avogadro's number

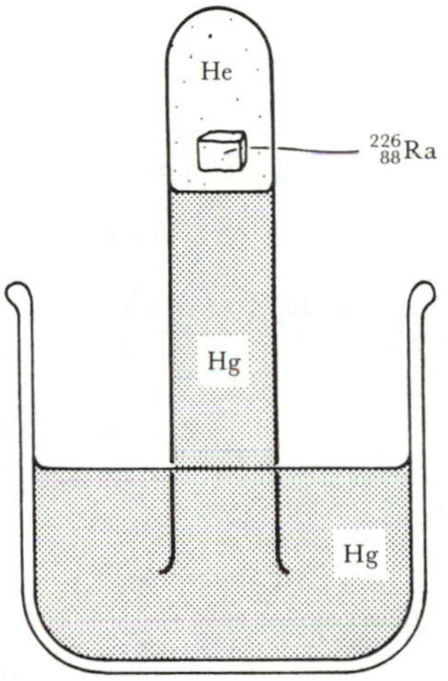

EXAMPLE 26 It has been found that a sample of radium-226 gives off 9.27×10^{12} helium ions per second, and that the volume of helium gas collected at STP over a period of one year is 10.88 mL. Calculate the value of Avogadro's number.

Solution Start by finding the number of moles of helium produced in one second.

$$\left(\frac{10.88\text{ mL He at STP}}{\text{yr}}\right)\left(\frac{1\text{ yr}}{365\text{ day}}\right)\left(\frac{1\text{ day}}{24\text{ hr}}\right)\left(\frac{1\text{ hr}}{3600\text{ sec}}\right)\left(\frac{1\text{ L}}{1000\text{ mL}}\right)\left(\frac{1\text{ mol}}{22.4\text{ L at STP}}\right)$$

$$= \frac{1.54 \times 10^{-11}\text{ mol He}}{\text{sec}}$$

Then, using dimensional analysis, calculate the number of He atoms per mole of helium.

$$\left(\frac{9.27 \times 10^{12}\text{ He atoms}}{\text{sec}}\right)\left(\frac{1\text{ sec}}{1.54 \times 10^{-11}\text{ mol He}}\right) = \frac{6.03 \times 10^{23}\text{ He atoms}}{\text{mol He}}$$

15.5 BINDING ENERGY

Most chemical phenomena can be explained by using a model of matter that says (1) all matter is made up of atoms and molecules, (2) molecules are made up of atoms, and (3) atoms are made up of protons, neutrons, and electrons. The masses of the three fundamental particles of atoms are

Particle	Mass in Atomic Mass Units
proton	1.007 28
neutron	1.008 67
electron	0.000 549

where 1 atomic mass unit (amu) is defined as one-twelfth of the mass of one carbon-12 atom ($^{12}_{6}C$).

An oxygen-16 atom ($^{16}_{8}O$) is made up of 8 protons, 8 neutrons, and 8 electrons. Yet, an oxygen-16 atom weighs less than the total weight of its parts (8e + 8p + 8n), as shown in the following example.

EXAMPLE 27 An oxygen-16 atom has a mass of 15.994 91 amu. Compare its mass with the total mass of the protons, neutrons, and electrons from which oxygen-16 is made.

Solution First find the total mass of 8 electrons, 8 protons, and 8 neutrons.

$$(8e)\left(\frac{0.000\,549 \text{ amu}}{e}\right) + (8p)\left(\frac{1.007\,28 \text{ amu}}{p}\right) + (8n)\left(\frac{1.008\,67 \text{ amu}}{n}\right)$$

$$= \underbrace{0.004\,392 \text{ amu}}_{\text{mass of 8 electrons}} + \underbrace{8.058\,24 \text{ amu}}_{\text{mass of 8 protons}} + \underbrace{8.069\,36 \text{ amu}}_{\text{mass of 8 neutrons}}$$

$$= 16.131\,99 \text{ amu} \longleftarrow \text{total mass of } 8e + 8p + 8n$$

$$\begin{array}{rl} 16.131\,99 \text{ amu} & \longleftarrow \text{mass of } 8e + 8p + 8n \\ -15.994\,91 \text{ amu} & \longleftarrow \text{mass of an } ^{16}_{8}O \text{ atom} \\ \hline 0.137\,08 \text{ amu} & \longleftarrow \text{mass difference} \end{array}$$

It appears that when 8 electrons, 8 protons, and 8 neutrons are combined to make one atom of oxygen-16, some mass (0.137 08 amu) is lost. What happened to the missing 0.137 08 amu of mass?

The law of conservation of mass and energy tells us that mass and energy cannot be created or destroyed, but they can be interconverted (i.e., mass can be converted into energy, and energy can be converted into mass). The relationship between mass and energy was shown in a formula presented by Albert Einstein in 1905. That formula is

$$E = mc^2$$

where E stands for energy, m for mass, and c for the velocity of light. According to this formula, 1 amu of mass has an energy equivalent of 932 million electron volts (932 MeV). What is an electron volt? Well, if a single electron were placed between two electrodes, one positively charged and the other negatively charged, and if a 1-V potential existed between these electrodes, the electron would "fall" to the lower energy state next to the positively charged electrode because unlike charges attract. This is diagrammed in Figure 15.6. Whenever a system moves from a higher energy state to a lower energy state (from a less stable condition to a more stable condition), energy is given off. In this case, when the electron "falls" to a lower energy state, 3.82×10^{-20} cal of energy is released. This amount of energy is called 1 electron volt (1 eV). The unit *electron volt* is used to eliminate the need (and the mess) for scientific notation. (That's why you measure a pencil in inches rather than in miles: it's easier to write a length as 6 in. rather than as 9.47×10^{-5} mi.)

FIGURE 15.6 Electron movement between oppositely charged plates

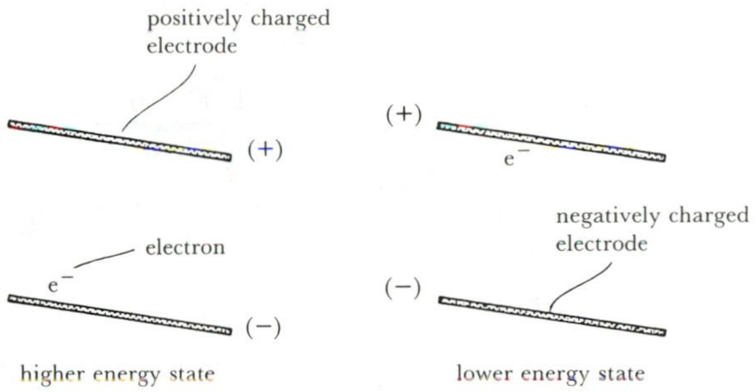

Let's get back to our mystery of the missing 0.137 08 amu of mass when an oxygen-16 atom is formed from 8e + 8p + 8n. We can calculate the energy equivalent of this missing mass using the relationship 1 amu = 932 MeV.

$$\underbrace{(0.137\,08 \text{ amu})}_{\text{missing mass}}\left(\frac{932 \text{ MeV}}{\text{amu}}\right) = 128 \text{ MeV}$$

When 8 protons, 8 electrons, and 8 neutrons combine to form an atom of oxygen-16, 128 MeV of energy is released.

$$8e + 8p + 8n \rightarrow {}^{16}_{8}O + 128 \text{ MeV}$$

This reaction is exothermic. The energy released in this reaction is supplied by converting matter into energy. As a result of this conversion, the product weighs less than the reactants.

It is possible to decompose oxygen-16 into its basic parts by supplying an atom of oxygen-16 with 128 MeV of energy.

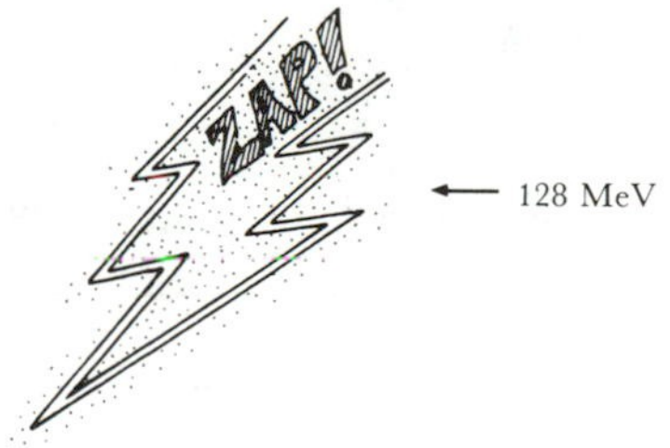

$${}^{16}_{8}O \rightarrow 8n + 8p + 8e$$

In this endothermic reaction, the products weigh more than the reactant. The energy supplied to the atom of oxygen-16 (128 MeV) is converted into mass by this reaction. We call the 128 MeV the binding energy of the oxygen-16 nucleus. It is a measure of the stability of the nucleons (how tightly they are bonded together) within the oxygen-16 nucleus. The oxygen-16 nucleus must be more stable than the overall stability of 8 individual electrons, 8 individual protons, and 8 individual neutrons because energy is released when these individual parts combine to form an atom of oxygen-16. Energy is always released when a system goes from a less stable condition to a more stable condition.

•
•

EXAMPLE 28 The mass of helium-4 is 4.002 60 amu. Find the binding energy of ${}^{4}_{2}He$.

Solution First, find the difference in mass between a helium-4 atom and the combined mass of the 2 protons, 2 neutrons, and 2 electrons from which a helium-4 atom is made.

$$(2e)\left(\frac{0.000\,549 \text{ amu}}{e}\right) + (2p)\left(\frac{1.007\,28 \text{ amu}}{p}\right) + (2n)\left(\frac{1.008\,67 \text{ amu}}{n}\right) = 4.033\,00 \text{ amu}$$

$$\begin{array}{rl} 4.033\,00 \text{ amu} & \leftarrow \text{mass of } 2e + 2p + 2n \\ -4.002\,60 \text{ amu} & \leftarrow \text{mass of a } {}^{4}_{2}\text{He atom} \\ \hline 0.030\,40 \text{ amu} & \leftarrow \text{mass difference} \end{array}$$

Convert the mass difference to binding energy in millions of electron volts using the relationship 1 amu = 932 MeV.

$$\underbrace{(0.030\,40 \text{ amu})}_{\text{mass difference}}\left(\frac{932 \text{ MeV}}{\text{amu}}\right) = 28.3 \text{ MeV}$$

EXAMPLE 29 The mass of lithium-7 is 7.016 01 amu. Find the binding energy of ${}^{7}_{3}Li$.

Solution First, find the difference in mass between a lithium-7 atom and the combined mass of the 3 protons, 3 electrons, and 4 neutrons from which a lithium-7 atom is made.

$$(3p)\left(\frac{1.007\,28 \text{ amu}}{p}\right) + (3e)\left(\frac{0.000\,549 \text{ amu}}{e}\right) + (4n)\left(\frac{1.008\,67 \text{ amu}}{n}\right) = 7.058\,17 \text{ amu}$$

$$\begin{array}{rl} 7.058\,17 \text{ amu} & \leftarrow 3p + 3e + 4n \\ -7.016\,01 \text{ amu} & \leftarrow \text{Li-7} \\ \hline 0.042\,16 \text{ amu} & \leftarrow \text{mass difference} \end{array}$$

$$\underbrace{(0.042\,16 \text{ amu})}_{\text{mass difference}}\left(\frac{932 \text{ MeV}}{\text{amu}}\right) = 39.3 \text{ MeV}$$

The binding energy of helium-4 is 28.3 MeV. The binding energy of lithium-7 is 39.3 MeV.

Which atom is more stable?

You probably (and incorrectly) said that lithium-7 is more stable because more energy is released when it forms than when helium-4 forms. In the next section we will see what is wrong with this type of reasoning.

15.5.1 BINDING ENERGY PER NUCLEON: A PARTY GAME WITH RUBBER BANDS

Dr. I. M. Dum, Professor of Psychology at Loweyeque University, has suggested a controversial form of group therapy that has also gained popularity as a party game. The participants are divided into two groups. Each group is tied together by a different number of rubber bands.

Let's assume that group A contains 10 people around which have been wrapped 50 rubber bands, and group B contains 30 people around which have been wrapped 60 rubber bands (see Figure 15.7). It may appear to you that group B is more tightly bonded together than group

FIGURE 15.7 Party game with rubber bands

A because group B is held in place by more rubber bands. However, it is not the total number of rubber bands holding a group together that determines how tightly a group is bonded together. The number of rubber bands per person is a more important factor. The number of rubber bands per person determines how tightly a group is bonded together.

$$\frac{\text{50 rubber bands}}{\text{10 people}} = \frac{\text{5 rubber bands}}{\text{person}} \qquad \frac{\text{60 rubber bands}}{\text{30 people}} = \frac{\text{2 rubber bands}}{\text{person}}$$

group A　　　　group B

In group A, there are 5 rubber bands per person holding the group together. In group B, there are only 2 rubber bands per person holding the group together. The people in group A are more tightly bonded together than the people in group B.

And so it is with nuclei. The binding energy per nucleon is more important than the total binding energy when comparing the stability of two nuclei.

Consider two hypothetical nuclei, A and B (corresponding to groups A and B at the party). Nucleus A is made up of 5 neutrons and 5 protons and has a binding energy (BE) of 50 MeV. Nucleus B is made up of 15 protons and 15 neutrons and has a binding energy of 60 MeV.

Which nucleus is more stable?

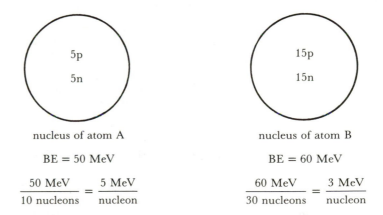

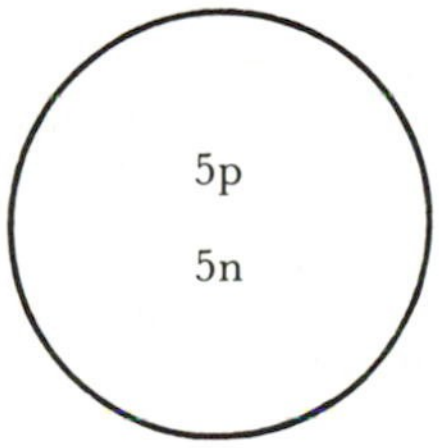

nucleus of atom A

BE = 50 MeV

$$\frac{\text{50 MeV}}{\text{10 nucleons}} = \frac{\text{5 MeV}}{\text{nucleon}}$$

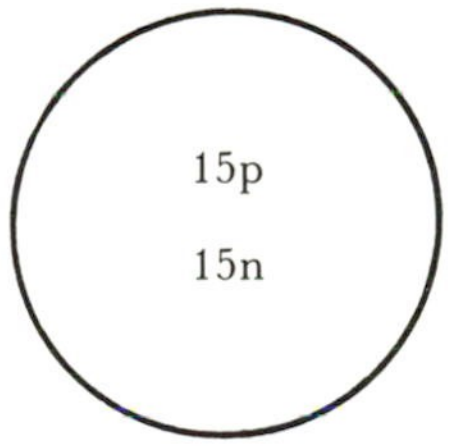

nucleus of atom B

BE = 60 MeV

$$\frac{\text{60 MeV}}{\text{30 nucleons}} = \frac{\text{3 MeV}}{\text{nucleon}}$$

The binding energy per nucleon for atom A is larger than the binding energy per nucleon in atom B. Nucleus A is more stable than nucleus B.

Let's go back to the question we left unanswered in the last section. Which is more stable, a helium-4 nucleus with a binding energy of 28.3 MeV or a lithium-7 nucleus with a binding energy of 39.3 MeV?

•
•

EXAMPLE 30 Determine which nucleus is more stable, He-4 or Li-7.

Solution Calculate the binding energy per nucleon of each atom. For $^{4}_{2}He$,

$$\text{BE/nucleon} = \frac{28.3\ \text{MeV}}{4\ \text{nucleons}} = \frac{7.08\ \text{MeV}}{\text{nucleon}}$$

For $^{7}_{3}Li$,

$$\text{BE/nucleon} = \frac{39.3\ \text{MeV}}{7\ \text{nucleons}} = \frac{5.61\ \text{MeV}}{\text{nucleon}}$$

A helium-4 nucleus is more stable than a lithium-7 nucleus because helium-4 has a higher binding energy per nucleon than lithium-7.

•
•

EXAMPLE 31 Calculate the binding energy per nucleon of uranium-238. The mass of a uranium-238 atom is 238.031 28 amu.

Solution First, calculate the mass of the 92 protons, 92 electrons, and 146 neutrons that make up an atom of uranium-238.

$$(92\text{p})\left(\frac{1.007\,28\ \text{amu}}{\text{p}}\right) + (92\text{e})\left(\frac{0.000\,549\ \text{amu}}{\text{e}}\right) + (146\text{n})\left(\frac{1.008\,67\ \text{amu}}{\text{n}}\right) = 239.986\,09\ \text{amu}$$

Then, calculate the mass difference.

$$\begin{array}{rl} 239.986\,09\ \text{amu} & \leftarrow 92\text{p} + 92\text{e} + 146\text{n} \\ -238.031\,28\ \text{amu} & \leftarrow {}^{238}_{92}\text{U} \\ \hline 1.954\,81\ \text{amu} & \leftarrow \text{mass difference} \end{array}$$

Finally, calculate the binding energy per nucleon.

$$(1.954\,81\ \text{amu})\left(\frac{932\ \text{MeV}}{\text{amu}}\right) = 1822\ \text{MeV} \leftarrow \text{binding energy}$$

$$\frac{1822\ \text{MeV}}{238\ \text{nucleons}} = \frac{7.65\ \text{MeV}}{\text{nucleon}} \leftarrow \text{binding energy per nucleon}$$

•
•

EXAMPLE 32 Calculate the binding energy per nucleon of an iron-56 nucleus. The mass of one iron-56 atom is 55.934 93 amu.

Solution 26p + 26e + 30n = 56.463 65 amu
mass difference = 0.528 72 amu
binding energy = 493 MeV
binding energy per nucleon = 8.80 MeV/nucleon

•
•

EXAMPLE 33 Which is more stable, a nucleus of uranium-238 with a binding energy of 1822 MeV or a nucleus of iron 56 with a binding energy of 493 MeV?

Solution The nucleus of iron-56 is more stable than the nucleus of uranium-238. The binding energy per nucleon of a nucleus of iron-56 is 8.80 MeV/nucleon. This is greater than the binding energy per nucleon of a nucleus of uranium-238, which is only 7.65 MeV/nucleon.

•
•

We can plot the binding energies in units of million electron volts per nucleon for all the known atoms, as shown in Figure 15.8. According to this plot, the most stable element in the known universe is iron. The least stable element shown is the isotope of hydrogen called deuterium ($^{2}_{1}H$).

FIGURE 15.8 Binding energy per nucleon vs. mass number

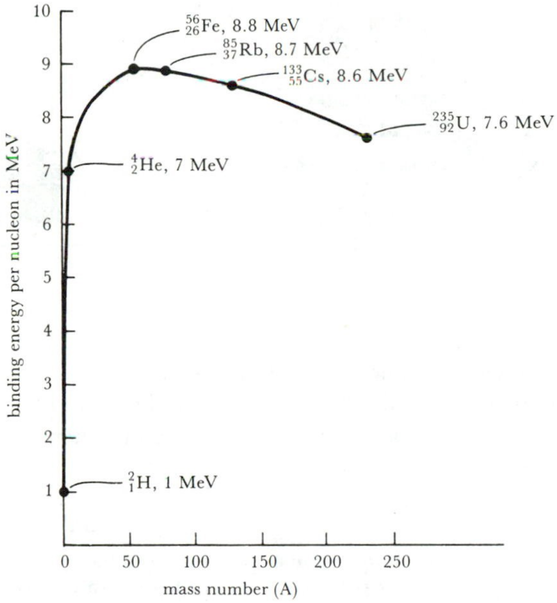

Try to estimate the binding energy per nucleon for a hydrogen-1 nucleus.

The nucleus of hydrogen-1 ($^{1}_{1}H$) contains only one nucleon (a single proton), so it is meaningless to discuss the energy that binds this single proton to itself.

•
•

SELF-TEST

1. Calculate the binding energy per nucleon of hydrogen-2. One atom of $^{2}_{1}H$ has a mass of 2.014 10 amu.
2. Calculate the binding energy per nucleon of beryllium-9. One atom of $^{9}_{4}Be$ has a mass of 9.012 19 amu.
3. Which nucleus is more stable, hydrogen-2 or beryllium-9?

ANSWERS **1.** 1.12 MeV/nucleon **2.** 6.47 MeV/nucleon **3.** Be-9

15.6 BASIC EQUATIONS

There are many interesting calculations that can be performed with only an elementary knowledge of nuclear chemistry. We will be performing some of the calculations that reflect the current excitement of those involved with nuclear chemistry.

Geiger counters are instruments used to detect the radiation given off by radioactive material. Radiation entering the detector of a Geiger counter ionizes gas molecules within the detector. A Geiger counter is designed to let a current flow every time its detector gas molecules are ionized. The current flow makes the Geiger counter click or flash lights, which indicates the presence of radioactive material. Some Geiger counters total (count) the number of detector ionizations and display the total counts in digital form. The total number of counts recorded per minute is related to the total number of radioactive atoms that have decayed per minute.

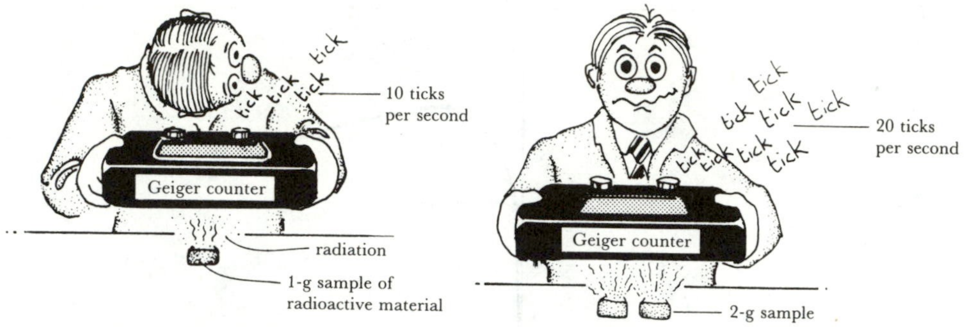

Let's put a radioactive sample weighing 1 g under a Geiger counter detector, as shown. The Geiger counter ticks 10 times per second, indicating the decay of 10 radioactive atoms each second. If we double the amount of radioactive sample, the number of ticks per second is also doubled. This observation can be stated mathematically as

$$\text{rate of decay} = kN \leftarrow \text{first basic equation}$$

- rate of decay: e.g., number of atoms that decay per second
- k: proportionality constant, called the rate constant
- N: number of radioactive atoms present in the sample

This formula is the first of three basic equations we will use to solve all of the nuclear chemistry problems in this chapter.

•
•

EXAMPLE 34 A gold sample contains 8.00×10^{30} atoms of Au-191 and decays at a rate of 1.60×10^{30} atoms/hr. Find the rate constant for gold-191.

Solution Solve the first basic equation (rate of decay = kN) for k.

$$k = \frac{\text{rate of decay}}{N} = \frac{\dfrac{1.60 \times 10^{30} \text{ Au-191 atoms}}{\text{hr}}}{8.00 \times 10^{30} \text{ Au-191 atoms}} = \frac{0.200}{\text{hr}}$$

The rate constant for gold-191 is 0.200/hr.

•
•

EXAMPLE 35 A 2.0-g sample of tin-120 decays at a rate of 4.3×10^{20} tin-120 atoms/min. Find the rate constant for Sn-120.

Solution Find the number of Sn-120 atoms present in the 2.0-g sample of Sn-120.

$$(2.0 \text{ g Sn-120})\left(\frac{6.0 \times 10^{23} \text{ Sn-120 atoms}}{120 \text{ g Sn-120}}\right) = 1.0 \times 10^{22} \text{ Sn-120 atoms}$$

Then, use the first basic equation (rate of decay = kN) to determine k.

$$k = \frac{\text{rate of decay}}{N} = \frac{\dfrac{4.3 \times 10^{20} \text{ Sn-120 atoms}}{\text{min}}}{1.0 \times 10^{22} \text{ Sn-120 atoms}} = \frac{0.043}{\text{min}}$$

The rate constant for tin-120 is 0.043/min. Notice that the units for this rate constant are not the same as the units for gold-191 in the previous example.

EXAMPLE 36 Nickel-59 has a rate constant of 9×10^{-6}/yr. Find the rate of decay of a 10.0-g sample of Ni-59.

Solution First, find the number of Ni-59 atoms present in 10.0 g of Ni-59.

$$(10.0 \text{ g Ni-59})\left(\frac{6.02 \times 10^{23} \text{ Ni-59 atoms}}{59.0 \text{ g Ni-59}}\right) = 1.02 \times 10^{23} \text{ Ni-59 atoms}$$

Then, use the first basic equation (rate of decay = kN) to calculate the rate of decay.

$$\text{rate of decay} = \left(\frac{9 \times 10^{-6}}{\text{yr}}\right)(1.02 \times 10^{23} \text{ Ni-59 atoms})$$

$$= \frac{9 \times 10^{17} \text{ Ni-59 atoms}}{\text{yr}}$$

EXAMPLE 37 The rate constant for aluminum-25 is 0.095/sec. Find the mass of a sample of aluminum-25 if it has a decay rate of 8.00×10^3 atoms/sec.

Solution Use the first basic equation (rate of decay = kN) to find the number of aluminum-25 atoms that are present.

$$N = \frac{\text{rate of decay}}{k} = \frac{\dfrac{8.00 \times 10^3 \text{ Al-25 atoms}}{\text{sec}}}{\dfrac{0.095}{\text{sec}}} = 8.4 \times 10^4 \text{ Al-25 atoms}$$

Next, find the weight of 8.4×10^4 aluminum-25 atoms.

$$(8.4 \times 10^4 \text{ Al-25 atoms})\left(\frac{25.0 \text{ g Al-25}}{6.02 \times 10^{23} \text{ Al-25 atoms}}\right) = 3.5 \times 10^{-18} \text{ g Al-25}$$

The sample of aluminum-25 that has a decay rate of 8.00×10^3 atoms/sec must weigh 3.5×10^{-18} g.

SELF-TEST

1. Find the rate constant for the hypothetical isotope M-150 if 5.6×10^8 atoms of M-150 decay at a rate of 54 atoms/hr.

2. Find the mass of an M-150 sample whose rate of decay is 208 atoms/hr.

ANSWERS **1.** 9.6×10^{-8}/hr **2.** 5.5×10^{-13} g M-150

It is a relatively simple two-step process to go from the first basic equation

$$\text{rate of decay} = kN$$

to the second basic equation

initial number of radioactive atoms

rate constant

time

$$\log \frac{N_0}{N} = \frac{kt}{2.303} \quad \leftarrow \text{second basic equation}$$

final number of radioactive atoms

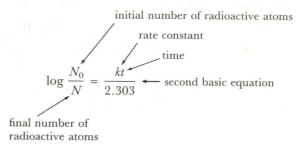

(after you've taken one or two calculus courses). In this second basic equation, N_0 stands for the initial number of radioactive atoms, N is the number of radioactive atoms left after time t has passed, t is the time that it takes to go from N_0 radioactive atoms to N atoms, and k is the rate constant. N and N_0 can also be the initial and final counts per second (from a Geiger counter), initial and final masses (e.g., in grams), or any other initial and final properties of the radioactive substance that are directly proportional to the initial and final numbers of radioactive atoms.

•
•

EXAMPLE 38 A sample of nitrogen-13 contains 980 atoms of N-13. How long must you wait for the sample to drop down to 330 atoms of N-13 if the rate constant for N-13 is 0.0693/min?

Solution Solve the second basic formula [log $(N_0/N) = kt/2.303$] for t.

$$t = \frac{2.303}{k} \log \frac{N_0}{N} = \frac{2.303}{\frac{0.0693}{\text{min}}} \log \frac{980 \text{ N-13 atoms}}{330 \text{ N-13 atoms}}$$

$$= (33.2 \text{ min})(\log 2.970) = (33.2 \text{ min})(0.473) = 15.7 \text{ min}$$

It takes 15.7 min for all but 330 of the atoms in a sample with 980 N-13 atoms to decay.

•
•

EXAMPLE 39 A sample of nitrogen-13 weighs 98.0 g. How long must you wait for the sample to weigh 33.0 g? The rate constant for N-13 is 0.0693/min.

Solution Solve the second basic formula [log $(N_0/N) = kt/2.303$] for t.

$$t = \frac{2.303}{k} \log \frac{N_0}{N} = \frac{2.303}{\frac{0.0693}{\text{min}}} \log \frac{98.0 \text{ g Ni-13}}{33.0 \text{ g Ni-13}}$$

$$= (33.2 \text{ min})(\log 2.970) = (33.2 \text{ min})(0.473) = 15.7 \text{ min}$$

It takes 15.7 min for all but 33.0 g of the mass of a 98.0-g sample of Ni-13 to decay.

•
•

EXAMPLE 40 A sample of oxygen-19 weighs 4.0 g. How much will this sample weigh after 15.0 sec? The rate constant for oxygen-19 is 0.0236/sec.

Solution Substitute the given information into the second basic equation [log $(N_0/N) = kt/2.303$].

$$\log \frac{4.0 \text{ g O-19}}{N} = \frac{\left(\frac{0.0236}{\text{sec}}\right)(15 \text{ sec})}{2.303} = 0.154$$

Take the antilog of both sides

$$\text{antilog}\left(\log \frac{4.0 \text{ g O-19}}{N}\right) = \text{antilog } 0.154$$

$$\frac{4.0 \text{ g O-19}}{N} = 1.43$$

$$N = \frac{4.0 \text{ g O-19}}{1.43} = 2.8 \text{ g O-19}$$

After 15.0 sec, the 4.0-g sample of O-19 will weigh 2.8 g.

EXAMPLE 41 A sample of oxygen-19 has an activity of 8000 cps. What will the activity of this sample be after 15.0 sec? The rate constant for O-19 is 0.0236/sec.

Solution Substitute the given information into the second basic equation [$\log (N_0/N) = kt/2.303$].

$$\log \frac{8000 \text{ counts/sec}}{N} = \frac{\left(\frac{0.0236}{\text{sec}}\right)(15.0 \text{ sec})}{2.303} = 0.154$$

Take the antilog of each side, giving

$$\frac{8000 \text{ cps}}{N} = 1.43$$

$$N = \frac{8000 \text{ cps}}{1.43} = 5.59 \times 10^3 \text{ cps}$$

After 15.0 sec, the activity of the sample will fall from 8000 cps to about 5590 cps.

SELF-TEST

1. A sample of the hypothetical element R-100 has a mass of 20.0 g. R-100 has a rate constant of 0.0500/day. Find the mass of this sample after 10 days.

2. How many days do you have to wait for the 20.0-g sample of R-100 to decrease to 19.0 g?

ANSWERS **1.** 12.1 g R-100 **2.** 1.03 day

15.7 HALF-LIFE

Let's assume that at 8:00 A.M. you weighed a sample of a hypothetical radioactive element, R, and found its weight to be 100 g. At 10:00 A.M. you weighed the sample again and found only 50.0 g of R present.

100 g of R	50 g of R
8:00 A.M.	10:00 A.M.

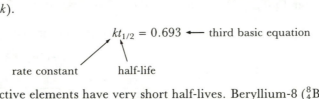

We say that the half-life (abbreviated $t_{1/2}$) of the radioactive element R is 2 hr ($t_{1/2} = 2$ hr). The half-life of a radioactive element is the time it takes for one-half of that element to decay.

How many grams of R will be present at 12:00 noon?

At 12:00 noon there should be 25.0 g of R left, half the amount that was present 2 hr earlier.

How much R will be left at 2:00 P.M.?

That's right. You've got the idea now. At 2:00 P.M. there will be only 12.5 g of R remaining, half of that which was present at 12:00 noon.

There is an important relationship between the half-life ($t_{1/2}$) of a radioactive substance and the rate constant (k).

$$kt_{1/2} = 0.693 \longleftarrow \text{third basic equation}$$

rate constant (k) half-life ($t_{1/2}$)

Some radioactive elements have very short half-lives. Beryllium-8 ($^{8}_{4}Be$) has a half-life of 1×10^{-16} sec. Lithium-5 ($^{5}_{3}Li$) has a half-life of 1×10^{21} sec. Some radioactive elements have very long half-lives. Uranium-238 ($^{238}_{92}U$) has a half-life of about 5×10^{9} yr (5 billion years). Vanadium-50 ($^{50}_{23}V$) has a half-life of 6×10^{15} yr.

•
•

EXAMPLE 42 Gold-191 has a half-life of 3.4 hr. Find the rate constant for gold-191.

Solution Solve the third basic equation ($kt_{1/2} = 0.693$) for k.

$$kt_{1/2} = 0.693$$

$$k = \frac{0.693}{t_{1/2}} = \frac{0.693}{3.4\ \text{hr}} = \frac{0.20}{\text{hr}} = 0.20/\text{hr}$$

The rate constant for gold-191 is 0.20/hr.

•
•

EXAMPLE 43 Tin-120 has a half-life of 16 min. Find the rate constant for tin-120.

Solution Solve the third basic equation ($kt_{1/2} = 0.693$) for k.

$$kt_{1/2} = 0.693$$

$$k = \frac{0.693}{t_{1/2}} = \frac{0.693}{16\ \text{min}} = \frac{0.043}{\text{min}} = 0.043/\text{min}$$

•
•

EXAMPLE 44 The rate constant for nickel-59 is 9×10^{-6}/yr. Find the half-life of nickel-59.

Solution Solve the third basic equation ($kt_{1/2} = 0.693$) for $t_{1/2}$.

$$kt_{1/2} = 0.693$$

$$t_{1/2} = \frac{0.693}{k} = \frac{0.693}{\dfrac{9 \times 10^{-6}}{\text{yr}}} = (0.693)\left(\frac{\text{yr}}{9 \times 10^{-6}}\right) = 8 \times 10^4 \text{ yr}$$

EXAMPLE 45 A sample of oxygen-19 weighs 4.0 g. How much will this sample weigh after 15.0 sec? The half-life of oxygen-19 is 29.4 sec.

Solution Use the third basic equation ($kt_{1/2} = 0.693$) to find the rate constant of oxygen-19.

$$k = \frac{0.693}{t_{1/2}} = \frac{0.693}{29.4 \text{ sec}} = \frac{0.0236}{\text{sec}}$$

Then, use the second basic equation [$\log (N_0/N) = kt/2.303$] to find the weight of the sample after 15.0 sec.

$$\log \frac{N_0}{N} = \frac{kt}{2.303}$$

$$\log \frac{4.0 \text{ g O-19}}{N} = \frac{\left(\dfrac{0.0236}{\text{sec}}\right)(15.0 \text{ sec})}{2.303}$$

$$= 0.154$$

Take the antilog of both sides.

$$\text{antilog}\left(\log \frac{4.0 \text{ g O-19}}{N}\right) = \text{antilog } 0.154$$

$$\frac{4.0 \text{ g O-19}}{N} = 1.43$$

$$N = \frac{4.0 \text{ g O-19}}{1.43}$$

$$= 2.8 \text{ g O-19}$$

After 15.0 sec, only 2.8 g of oxygen-19 will remain.

EXAMPLE 46 A sample of oxygen-19 contains 8000 atoms of O-19. How many atoms of O-19 will be present after 15.0 sec? The half-life of O-19 is 29.4 sec.

Solution Use the third basic equation ($kt_{1/2} = 0.693$) to find the rate constant for O-19.

$$k = \frac{0.693}{t_{1/2}} = \frac{0.693}{29.4 \text{ sec}} = \frac{0.0236}{\text{sec}}$$

Then, use the second basic equation [$\log (N_0/N) = kt/2.303$] to find the number of atoms of O-19 (N) that remain after 15.0 sec.

$$\log \frac{N_0}{N} = \frac{kt}{2.303}$$

$$\log \frac{8000 \text{ O-19 atoms}}{N} = \frac{\left(\dfrac{0.0236}{\text{sec}}\right)(15.0 \text{ sec})}{2.303} = 0.154$$

Take the antilog of both sides of the equation.

$$\frac{8000 \text{ O-19 atoms}}{N} = 1.43$$

$$N = \frac{8000 \text{ O-19 atoms}}{1.43} = 5.59 \times 10^3 \text{ O-19 atoms}$$

About 5590 atoms of oxygen-19 remain after 15.0 sec.

•
•

SELF-TEST **1.** Calculate the half-life of the hypothetical isotope Z-20 if the rate constant for this isotope is 0.020/min.

2. Find the time it would take for a 30.0-g sample of Z-20 to be reduced to 10.0 g of Z-20.

ANSWERS **1.** 35 min **2.** 55 min

15.7.1 DATING THE EARTH AND THE UNIVERSE

The half-life concept has been useful in obtaining the ages of many objects, including the earth and the universe.

For example, if you found a rock with half of its uranium-238 missing (decayed), how old do you think it would be? The rate constant for uranium-238 is 1.386 × 10^{-10}/yr.

Insert the value of the rate constant of U-238 in the third basic equation ($kt_{1/2} = 0.693$) to find the half-life of uranium-238. The half-life of uranium-238 is 5 billion years. Since half of the uranium-238 is missing in our rock sample, the rock must be 5 billion years old. Simple, isn't it?

You may wonder how you could tell that half of the rock's uranium is missing. Well, uranium decays into thorium-90. Thorium-90 is also radioactive, and it decays into protactinium-91. Protactinum-91 is radioactive. All in all, a rock that may have begun as 100% pure uranium would end up being made up of 15 different atoms (uranium-238 plus 14 of its decay products). The age of a rock can be determined by measuring the relative amounts of uranium and one or more of the other decay products. In this way, uranium mines found on earth have been determined to be about 5 billion years old.

Similar techniques are used to determine the age of the universe. The radioactive isotope $^{187}_{75}Re$, with a half-life of 40 billion years, has been used to date the universe. Experiments with rhenium-187 have shown that the universe is about 20 billion years old.

Of course, just because the oldest rock that man has ever found is 20 billion years old doesn't mean that the universe can't be older. There may be older rocks that we haven't found. However, the age of the universe has been determined by using several different methods, all of which appear to be in agreement. For example, the age of the universe has been determined by measuring its size and its rate of expansion. These measurements indicate that the universe is at least 16 billion years old. There are uncertainties in all methods, and the universe is usually accepted as being somewhere between 15 and 20 billion years old.

15.7.2 COBALT-60 AND CANCER: CALCULATING FUTURE SUPPLIES

There are 600 000 new cancer cases every year in the United States. Half of these victims are treated with radiation. Such treatment frequently involves the unstable radioactive isotope ^{60}Co (cobalt-60) and is referred to as cobalt treatment. Cobalt-60 gives off radiation of just the right energy that does maximum damage to cancer cells while causing minimum damage to normal cells. Cobalt-60 has a half-life of 5.27 years. A hospital could treat a million people a day or nobody at all over a period of 5.27 years and still lose half of all its Co-60. It is important for hospitals to be able to calculate the amount of Co-60 that will be on hand so that supplies can be replenished.

EXAMPLE 47 A hospital has 500 g of Co-60. How much will they have 1.00 yr from today? The half-life of Co-60 is 5.27 yr.

Solution Use the third basic equation ($kt_{1/2} = 0.693$) to determine the rate constant for Co-60.

$$kt_{1/2} = 0.693$$

$$k = \frac{0.693}{t_{1/2}} = \frac{0.693}{5.27\ \text{yr}} = \frac{0.131}{\text{yr}}$$

Then, use the second basic formula [$\log (N_0/N) = kt/2.303$] to determine the amount of Co-60 that will remain after 1 yr.

$$\log \frac{N_0}{N} = \frac{kt}{2.303}$$

$$\log \frac{500\ \text{g Co-60}}{N} = \frac{\left(\frac{0.131}{\text{yr}}\right)(1.00\ \text{yr})}{2.303} = 0.0571$$

Take the antilog of both sides of this equation.

$$\frac{500\ \text{g Co-60}}{N} = 1.141$$

$$N = \frac{500\ \text{g Co-60}}{1.141} = 438\ \text{g Co-60}$$

EXAMPLE 48 A hospital has a 500-g supply of Co-60. Additional Co-60 must be purchased whenever the supply of Co-60 is reduced to 200 g. How many years will pass before the hospital's supply of Co-60 is reduced to 200 g?

Solution Use the second basic equation [$\log (N_0/N) = kt/2.303$] to determine the time (t).

$$t = \frac{2.303}{k} \log \frac{N_0}{N} = \frac{2.303}{\frac{0.131}{\text{yr}}} \log \frac{500\ \text{g Co-60}}{200\ \text{g Co-60}} = \frac{2.303}{\frac{0.131}{\text{yr}}} \log 2.5$$

$$= \frac{2.303}{\frac{0.131}{\text{yr}}}(0.398) = 7.00\ \text{yr}$$

15.8 RADIOCARBON DATING

The earth is constantly receiving cosmic radiation from outer space. The cosmic radiation from outer space consists of a heterogeneous array of particles (just about every type of atomic particle that you can think of, from protons and neutrons to heavier atoms, such as lead and uranium).

As these particles, traveling over a million miles per hour, strike atoms and molecules in the earth's atmosphere, nuclear reactions occur that produce other particles and electromagnetic radiation, which has a shorter wavelength than gamma radiation.

Our atmosphere is made up of about 80% nitrogen. When cosmic radiation neutrons collide with atmospheric nitrogen, some of the nitrogen is converted into radioactive carbon-14.

$${}^{1}_{0}n + {}^{14}_{7}N \rightarrow {}^{14}_{6}C + {}^{1}_{1}H$$

Carbon-14 and ordinary carbon (carbon-12) are essentially chemically equivalent, both reacting with oxygen in the air to produce carbon dioxide.

$${}^{14}C + O_2 \rightarrow {}^{14}CO_2$$

$${}^{12}C + O_2 \rightarrow {}^{12}CO_2$$

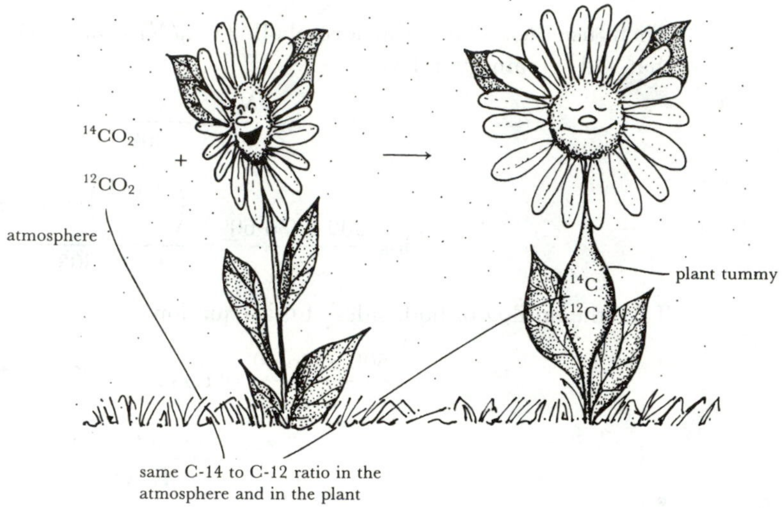

The ratio of $^{14}CO_2$ to $^{12}CO_2$ in the atmosphere appears to remain relatively constant; that is, carbon-14 is being produced by cosmic rays just as rapidly as carbon-14 decays into something else. Plants use carbon dioxide from the atmosphere in photosynthesis. Plant tissue is made up of hydrocarbon derivatives (compounds that are primarily made up of hydrogen and carbon). Plant tissue contains both carbon-12 and radioactive carbon-14. As long as a plant is alive, its tissue contains the same ratio of C-14 to C-12 as that found in the atmosphere. When a plant dies, photosynthesis stops, and the carbon-14 within the plant tissue decreases with time.

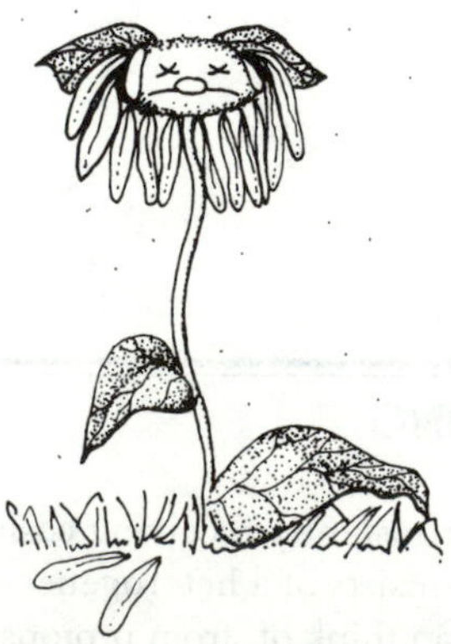

The half-life of carbon-14 is about 5730 yr. If you found a dead plant (in the form of wood, ashes, cloth, corn cobs, charcoal, peat, or grain) that only had half of the carbon-14 that a living plant contains, how old would the dead plant be?

Right. It would be about 5730 yr old. Living plants have a radioactivity of about 15.3 carbon-14 atoms decaying each minute for every gram of carbon present. A plant sample, converted to pure carbon, would cause a Geiger counter to tick 15.3 times per minute per gram of carbon present. This is abbreviated 15.3 cpm/gC (15.3 counts per minute per gram of carbon).

How many counts per minute would be produced by a carbon sample prepared from a wooden object that is about 6000 yr old?

A 6000-yr-old object (close to the half-life of carbon-14, which is 5730 yr) would have half of its original carbon-14 left. Its activity would be about one-half of 15.3 cpm/gC because rate of decay depends upon the number of radioactive atoms present (our first basic equation). The 6000-yr-old wooden object would have an activity of about 7.6 cpm/gC.

Radiocarbon dating is also used to date beeswax, antlers, shells (from ocean animals as well as from land animals), and bones.

•
•

EXAMPLE 49 A sample of wood was found to have an activity of 7.00 cpm/gC. If fresh wood has an activity of 15.3 cpm/gC, and the half-life of carbon-14 is 5730 yr, how old is the wooden sample?

Solution Use the third basic equation ($kt_{1/2} = 0.693$) to find the rate constant for carbon-14.

$$k = \frac{0.693}{5730 \text{ yr}} = \frac{1.21 \times 10^{-4}}{\text{yr}}$$

Then, use the second basic equation [$\log (N_0/N) = kt/2.303$] to determine the age (t) of the wooden object.

$$t = \frac{2.303}{k} \log \frac{N_0}{N} = \frac{2.303}{\dfrac{1.21 \times 10^{-4}}{\text{yr}}} \log \frac{15.3 \text{ cpm/gC}}{7.00 \text{ cpm/gC}}$$

$$= (1.90 \times 10^4 \text{ yr})(\log 2.19) = (1.90 \times 10^4 \text{ yr})(0.340) = 6.47 \times 10^3 \text{ yr}$$

The wooden object is about 6470 yr old.

•
•

EXAMPLE 50 Campfire ashes were found to have an activity of 1.02 cpm/gC. How long ago was the campfire lit?

Solution

$$t = \frac{2.303}{k} \log \frac{N_0}{N} = \frac{2.303}{\dfrac{1.21 \times 10^{-4}}{\text{yr}}} \log \frac{15.3 \text{ cpm/gC}}{1.02 \text{ cpm/gC}}$$

$$= (1.90 \times 10^4 \text{ yr})(\log 15.0) = (1.90 \times 10^4 \text{ yr})(1.176) = 2.23 \times 10^4 \text{ yr}$$

The ashes are about 22 300 yr old.

•
•

EXAMPLE 51 How many years must pass for a piece of cloth to lose 80% of its original activity?

Solution For this problem, $N_0 = 100\%$ activity and $N = 20\%$ activity.

$$t = \frac{2.303}{k} \log \frac{100\% \text{ activity}}{20\% \text{ activity}} = \frac{2.303}{\frac{1.21 \times 10^{-4}}{\text{yr}}} \log 5$$

$$= (1.90 \times 10^4 \text{ yr})(0.699) = 1.33 \times 10^4 \text{ yr}$$

About 13 300 yr must pass for a cloth sample to lose 80% of its activity.

•
•

EXAMPLE 52 Corn cobs were found that only had two-thirds of their original activity. How old are the corn cobs?

Solution For this problem, $N_0 = \frac{3}{3}$ and $N = \frac{2}{3}$.

$$t = \frac{2.303}{\frac{1.21 \times 10^{-4}}{\text{yr}}} \log \frac{\frac{3}{3} \text{ activity}}{\frac{2}{3}\text{activity}} = (1.90 \times 10^4 \text{ yr})(\log 1.5)$$

$$= (1.90 \times 10^4 \text{ yr})(0.176) = 3.35 \times 10^3 \text{ yr}$$

The corn cobs are about 3350 yr old.

•
•

SELF-TEST

1. A bone was found to have a carbon-14 activity of 12.9 cpm/gC. Find the age of this bone if the half-life of C-14 is 5730 yr.
2. A layer of peat is found to have an activity of 15.08 cpm/gC. Find the age of this layer of peat.

ANSWERS **1.** 1.42×10^3 yr **2.** 121 yr

15.8.1 NOAH'S ARK, PART II

Some people believe that the remains of Noah's ark are buried in a glacier near the top of the 17 000-foot Mt. Ararat in eastern Turkey (see Chapter 2, "Noah's Ark, Part I"). The object at the top of Mt. Ararat has been known to exist for hundreds of years. While traveling in this area in the 1300's, Marco Polo mentioned seeing what he called in his diary "the Ark of Noah." On July 2, 1840, there was a powerful earthquake in the Mt. Ararat area. Workmen went into the mountain to build defenses to protect the area from avalanches. When the men returned they told of seeing the prow of an enormous boat protruding from a glacier. The Turkish government sent a special team to confirm the report. The team found a structure that was fairly well preserved. In fact, they were able to enter three of the structure's compartments. Only one section of the structure was seriously damaged.

French explorer Fernand Navarra, along with his 11-year-old son, Rafael, climbed Mt. Ararat in 1955 and removed a 5-ft piece of wood from a 150-ft hand-tooled beam. (A 150-ft beam is half the length of a football field.) Samples of this wood were sent for analysis to the Madrid Institute of Forestry in Spain, the Centre Technique de Bois in Paris, the University of California at Los Angeles (UCLA Isotope Laboratory), the Geochron Laboratories in Cambridge, Massachusetts, and the University of Pennsylvania. The last three of these institutions analyzed the wood using the technique called carbon dating. This technique was developed by W. F. Libby in 1949 and earned him the Nobel Prize.

•
•

EXAMPLE 53 In 1978, scientists at the University of Pennsylvania determined the activity of the wooden sample taken from Mt. Ararat to be 13.19 cpm/gC. It is known that wood taken from a living tree has a radioactivity of 15.3 cpm/gC. The half-life of carbon-14 is 5730 yr. Determine the age of "Noah's ark."

Solution Use the third basic equation ($kt_{1/2} = 0.693$) to find k.

$$k = \frac{0.693}{t_{1/2}} = \frac{0.693}{5730\ \text{yr}} = 1.21 \times 10^{-4}/\text{yr}$$

Use the second basic equation—$\log (N_0/N) = kt/2.303$—to find t.

$$t = \frac{2.303}{k} \log \frac{N_0}{N} = \frac{2.303}{\frac{1.21 \times 10^{-4}}{\text{yr}}} \log \frac{15.3\ \text{cpm/gC}}{13.19\ \text{cpm/gC}}$$

$$= (1.90 \times 10^4\ \text{yr})(\log 1.1600) = (1.90 \times 10^4\ \text{yr})(0.06446) = 1228\ \text{yr}$$

The latest value reported by the University of Pennsylvania for the wood's age is 1230 ± 60 yr. Biblical scholars say that the age of the ark is about 4400 yr. The Madrid Institute of Forestry in Spain and the Centre Technique de Bois in Paris estimate the age of the wood at 5000 yr and 4500 yr, respectively. These higher estimates of the wood's age were determined, in part, by the degree of lignite formation (lignite is a form of coal intermediate between peat and bituminous, or soft coal, in which the texture of the original wood is distinct), the gain in density of the wood, the cell modification, and the degree of fossilization.

There is quite a controversy over the carbon dating results. Those who believe the object is the actual remains of Noah's ark have supplied numerous theories to explain why the normally reliable carbon dating method produces such a young age. Others who accept the carbon dating findings (as gospel truth) suggest that perhaps the object is the remains of a monument to the flood story.

15.8.2 SHROUD OF TURIN: IMAGE OF CHRIST?

In 1357 a linen cloth about 14 ft long and 3 ft wide was exhibited in Lirey, France. This cloth shows an image of a crucified man and was proclaimed to be the burial shroud of Jesus Christ. The Bishop of Troyes, who had jurisdiction in Lirey in 1357, denounced the cloth as a forgery. The matter was never settled, and to this day the shroud continues to attract interest.

Scientists have not yet been able to explain how a forger could create certain characteristics of the shroud. The image on the cloth appears to be a photographic negative whose positive has three-dimensional qualities that are never found in paintings or photographs. The wounds on the pictured man show the nails running through the wrists, something that a medieval forger would not have thought to picture. It had always been assumed that the nails were placed through the hands. The color appears only on the surface of the shroud; it has not penetrated into the cloth as it would if it were a painting or a stain. The picture appears to have been made by scorch marks.

There have been many suggestions to explain how the image may have been forged, including one that suggests that a stone or metal life-sized statue of Christ was first heated in a bread oven, then laid on its side and wrapped with a dampened cloth. This would produce the scorch marks, the negative image, and the three-dimensional qualities observed.

Scientists from around the world have traveled to the Royal Chapel of the Cathedral of St. John the Baptist in Turin, Italy, where the shroud has been kept for the last 400 years. For years church authorities would not give their permission for a carbon dating analysis of the shroud because such an analysis would destroy a portion of the shroud (a small part of the shroud would have to be converted to carbon). However, permission was finally granted because carbon dating methods have become so refined that extremely small amounts of sample destruction are now required. In 1988, scientists at Oxford University detected a carbon-14 activity of 14.16 cpm/gC. Calculate the age of the shroud by exactly following the steps shown in Example 53, but substitute 14.16 cpm/gC in place of 13.19 cpm/gC. Your calculation should show the shroud's age as being 640 years old.

EXAMPLE 54 Assume that the Shroud of Turin is 2000 years old. What activity should scientists measure?

Solution Solve the second basic equation—$\log(N_0/N) = kt/2.303$—for N.

$$\log \frac{N_0}{N} = \frac{kt}{2.303}$$

$$\log \frac{15.3 \text{ cpm/gC}}{N} = \frac{\left(\dfrac{1.21 \times 10^{-4}}{\text{yr}}\right)(2000 \text{ yr})}{2.303} = 0.105$$

Take the antilog of both sides of this equation.

$$\frac{15.3 \text{ cpm/gC}}{N} = 1.273$$

$$N = \frac{15.3 \text{ cpm/gC}}{1.273} = 12.0 \text{ cpm/gC}$$

If the shroud is 2000 years old, scientists should find its present activity to be about 12.0 cpm/gC.

15.8.3 CALIFORNIA EARTHQUAKES: PREDICTING THE NEXT BIG ONE

Historians have recorded only two major earthquakes in California, and both of those occurred when the state had a small population. Now that California is the most populated state in the Union, another major earthquake could be devastating both economically and in terms of the loss of human life. Were the California earthquakes quirks of nature or part of a regular, repeating series of earthquakes? If the latter is true, it could lead to the prediction of the next major California quake.

Scientists excavated a 13-foot trench in a marsh about 60 miles northeast of Los Angeles. There they found places where the sedimentary layers had shifted and separated at several locations (see Figure 15.9). Scientists have assumed that each strata break of several feet repre-

FIGURE 15.9 Strata. (a) Normal strata in sand, silt, and peat deposited at the bottom of the marsh. (b) Breaks in the strata represent sudden shifts of 3 to 6 ft between the land masses on opposite sides of the fault.

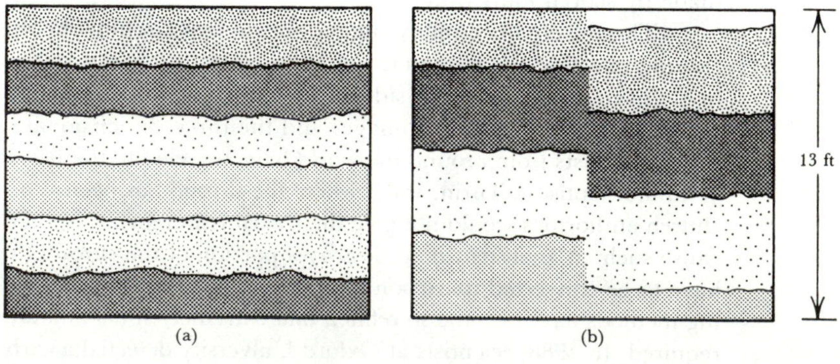

sents the occurrence of a major earthquake. By carbon-dating the remains of once-living organisms at each of these breaks, scientists have determined the approximate time when each earthquake took place.

EXAMPLE 55 In 1993, the activities of the organic matter in these separated sedimentary layers had activities in counts per minute per gram of carbon that ranged from 12.89 to 15.06 cpm/gC tested. Calculate the date of the oldest known major earthquake in California.

Solution The oldest layer is the layer with the least activity. The activity of the oldest layer is 12.89 cpm/gC.

$$t = \frac{2.303}{k} \log \frac{N_0}{N} = \frac{2.303}{\dfrac{1.21 \times 10^{-4}}{\text{yr}}} \log \frac{15.3}{12.89} = 1417 \text{ yr}$$

The age of the oldest layer is 1417 yr. The earthquake occurred back in A.D. 576.

We now know from data such as this that over the last 1400 years, major California earthquakes have taken place about every 150 years. Since the last large earthquake in California took place in 1857 (the 1990 earthquake you may remember was relatively tame), some think that the next major one is due around the year 2010. However, the degree of strata separation suggests to others that the 1857 quake was much smaller than previous major earthquakes. This would make the 1720 earthquake the last major quake. If this is true, the next big quake has been overdue for the last 130 years. Since the largest previous interval between quakes was 275 years, Californians may be due for another one just about any day now.

15.9 NEUTRON ACTIVATION ANALYSIS

Carbon dating would not be possible were it not for a natural form of neutron activation analysis. The term *neutron activation analysis* means just about what it says: neutrons are used to activate atoms (make them radioactive) so that those atoms can be detected (analyzed) by the radiation that they emit as they decay into more stable atoms.

In nature, cosmic ray neutrons change our atmospheric nitrogen into carbon-14. Scientists use neutrons to determine which chemical elements are present in unknown samples and how much of each element is present (they can even detect amounts as little as 1×10^{-20} g). When bombarded with neutrons, elements in a treated sample absorb some of the neutrons and turn into radioactive isotopes. Each radioactive isotope emits its own characteristic radiation, and by making careful measurements to determine the type of radiation being emitted, scientists can determine which isotopes are present. Since the intensity of radiation depends only on the amount of radioactive material present, it is also possible to determine how much of each radioactive element is present in the unknown sample. If the unknown sample is human hair, much information about the person can be deduced.

Your hair is a living diary because it's where your body deposits all chemicals that you have ever taken in. If you drank soda from an aluminum can last week, your body deposited in your hair some of the aluminum ions that were dissolved in the soda. Each day your hair grows about 1/2 mm, and each day's growth contains trace amounts of whatever you have consumed. Figure 15.10 illustrates this phenomenon.

FIGURE 15.10 Human hair, separated into segments of one day's growth

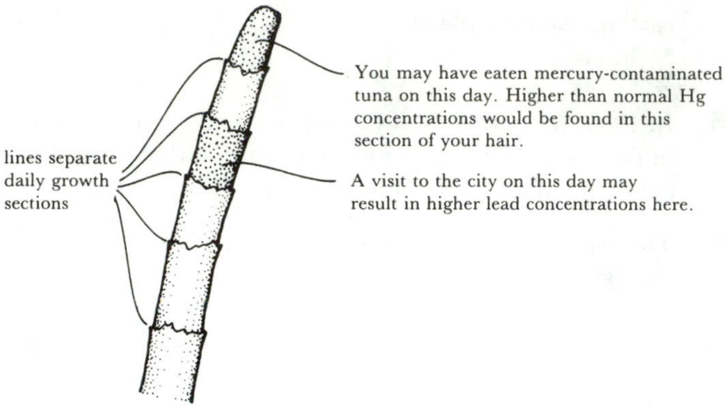
You may have eaten mercury-contaminated tuna on this day. Higher than normal Hg concentrations would be found in this section of your hair.
lines separate daily growth sections
A visit to the city on this day may result in higher lead concentrations here.
human hair

15.9.1 WAS ZACHARY TAYLOR THE FIRST ASSASSINATED U.S. PRESIDENT?

In 1991, scholars suggested that Zachary Taylor, the twelfth president of the United States, had been assassinated. It was suggested that because Taylor opposed slavery in new states seeking admission to the Union, his enemies may have added arsenic to Taylor's food. If true, the history books claiming Abraham Lincoln as the first assassinated president would have to be rewritten. After exhuming Taylor's body, scientists bombarded samples of his hair with neutrons. Any arsenic in his hair would have undergone the following change:

$$^{75}\text{As} + 5\text{n} \rightarrow {}^{80}\text{As}$$

The stable arsenic-75 is converted to the radioactive (unstable) arsenic-80, which has a half-life of 15.0 sec.

Why do you think arsenic-80 is unstable?

Right! It has too much glue (too many neutrons).

As it turned out, the neutron activation analysis of Taylor's hair showed only normal trace amounts of arsenic that were far too small to have killed him.

•
•

EXAMPLE 56 A hair sample shows a radioactivity of 5000 disintegrations per second (5000 atoms of arsenic-80 are decaying each second). How many grams of arsenic-80 are present in the hair sample?

Solution First, use the third basic equation ($kt_{1/2} = 0.693$) to determine the rate constant for arsenic-80.

$$k = \frac{0.693}{t_{1/2}} = \frac{0.693}{15.0\text{ sec}} = \frac{4.62 \times 10^{-2}}{\text{sec}}$$

Then, solve the first basic equation (rate of decay = kN) for the number of radioactive arsenic-80 atoms present in the hair sample.

$$N = \frac{\text{rate of decay}}{k} = \frac{\dfrac{5000\text{ As-80 atoms}}{\text{sec}}}{\dfrac{4.62 \times 10^{-2}}{\text{sec}}} = 1.08 \times 10^{5}\text{ As-80 atoms}$$

Finally, convert the number of arsenic-80 atoms present in the hair to grams of arsenic-80 present in the hair.

$$(1.08 \times 10^5 \text{ As-80 atoms})\left(\frac{80.0 \text{ g As-80}}{6.02 \times 10^{23} \text{ As-80 atoms}}\right) = 1.44 \times 10^{-17} \text{ g As-80}$$

The ability to detect such a small mass is quite impressive. Think about the balances that you use in your chemistry laboratory. Those balances probably cost around five or six thousand dollars and can weigh objects with a mass no less than 0.0001 g. Compare that with the mass of arsenic found in the previous calculation (0.000 000 000 000 000 014 4 g). Neutron activation analysis is used for detecting the presence of, or "weighing," minute amounts of elements.

15.9.2 HAIR ANALYSIS: FROM LEARNING DISABILITIES IN CHILDREN TO ADULT DRUG ABUSE

Hair analysis may become a frequently used and valuable tool in early disease detection programs. Hair normally contains relatively fixed amounts of several trace elements, such as calcium, magnesium, potassium, sodium, cadmium, cobalt, copper, iron, lead, manganese, zinc, chromium, lithium, and mercury. Abnormally high or low concentrations of these and other elements in the hair can be an indication of many abnormalities or diseases, such as air and water pollution, cystic fibrosis, diet deficiencies, diabetes, Down's Syndrome, schizophrenia, and learning disabilities. Crime labs can also detect certain drugs in the hair, such as barbiturates, amphetamines, morphine, and heroin. It is also possible to determine the approximate time when the drug was taken or injected.

Hair analysis has many advantages over the more widely used blood and urine analyses. Collecting hair is less painful and less embarrassing. Hair can be stored for longer periods without deterioration. (It is this storage advantage that made the investigation of Taylor's "assassination" possible.) Also, substances are accumulated in hair with a concentration of at least ten times that found in blood or urine.

15.9.3 DINOSAUR EXTINCTION

Let's now consider how neutrons were used to deduce the demise of the dinosaurs. You may have read the theory that dinosaurs became extinct after a large comet struck the earth 65 million years ago. The comet broke into two pieces that landed in Iowa and the Yucatan Peninsula. Dust from the impacts circled the globe and blocked out the sun for several years. During that time, photosynthesis stopped, land and marine plants died, and most animal life vanished. Eventually the dust, rich in the element iridium, settled evenly on the earth's surface.

Now let's look at how this theory developed. Scientists used neutrons to bombard sedimentary rock layers that had deposited 65 million years ago, about the time of the dinosaur extinction. The neutrons reacted with iridium in the 65-million-year-old rock layers as follows:

$$^{193}\text{Ir} + \text{n} \rightarrow {}^{194}\text{Ir}$$

Iridium-193 is not radioactive, but iridium-194 is. Iridium-194 is so radioactive that scientists can easily detect billionths of a gram per cm^3 of soil. The scientists found that the concentration of iridium in the sedimentary rock increased by 2300% 65 million years ago. Since most iridium present in the earth's crust comes from extraterrestrial sources, the scientists developed the giant comet theory.

15.10 NEUTROGRAPHY: PARTICLES VS. WAVES

The following discussion is designed to give you more insight into electromagnetic radiation (such as X rays and gamma rays) and particle radiation (such as pions and neutrons).

Neutrography is the technique of taking a picture of an object using a beam of neutrons. A particle beam of neutrons can be used to take a picture of an object in much the same way that pictures are taken with X rays. Neutron pictures are taken by placing an object to be photographed between a neutron source and a sensitive photographic film. Regions of higher neutron absorption in the object cause less blackening of the film than do regions of lower neutron absorption. The result is a neutron picture, which is referred to as a neutrograph.

Consider the elements hydrogen, carbon, and lead. Hydrogen and carbon are of particular interest because these are the elements of which we are primarily composed. Skin, protein, carbohydrates, enzymes, amino acids, and many other components of our bodies are hydrocarbon derivatives. (Hydrocarbons are compounds made up of just hydrogen and carbon.) Plastics, such as polyethylene, are also hydrocarbons or hydrocarbon derivatives.

How many electron shells (energy levels) are found in H, C, and Pb atoms?

Hydrogen has one shell of electrons around it, carbon has two, and lead has six. You can determine these numbers by looking at a periodic table and observing the period to which each of these elements belongs. Each period contains a number of electron shells equal to the period number. For example, the first period elements (H and He) have just one electron shell, the K shell, around their central nuclei. All of the second period elements contain two shells, the K shell and the L shell, around their nuclei. The energy shells of H, C, and Pb are illustrated in Figure 15.11.

FIGURE 15.11 Energy shells of hydrogen, carbon, and lead

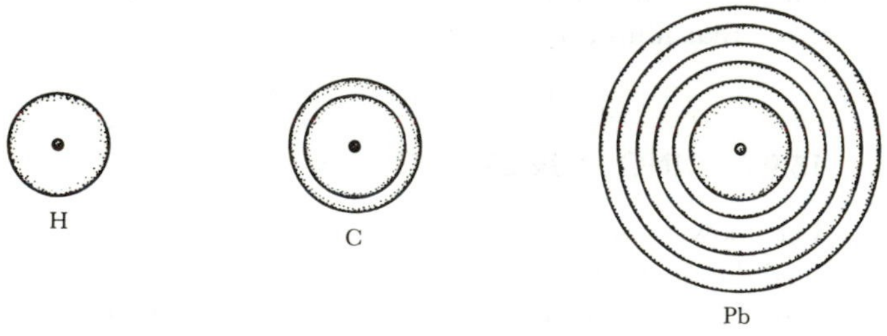

If neutrons were fired through a plastic shield (a hydrocarbon made up of hydrogen and carbon atoms) and a lead shield (a bunch of lead atoms), which shield would the neutrons penetrate more easily?

You probably thought that the neutrons would get through hydrogen and carbon atoms more easily than they would get through lead atoms, but this is not correct. Consider Figure 15.12, which shows a box of hydrogen atoms and a box of lead atoms.

Which box contains the higher nuclei concentration? (The nuclei concentration is defined as the number of nuclei per unit volume.)

The hydrogen box contains more nuclei per unit volume than the box containing lead.

FIGURE 15.12 Nuclei concentration in H and Pb compared

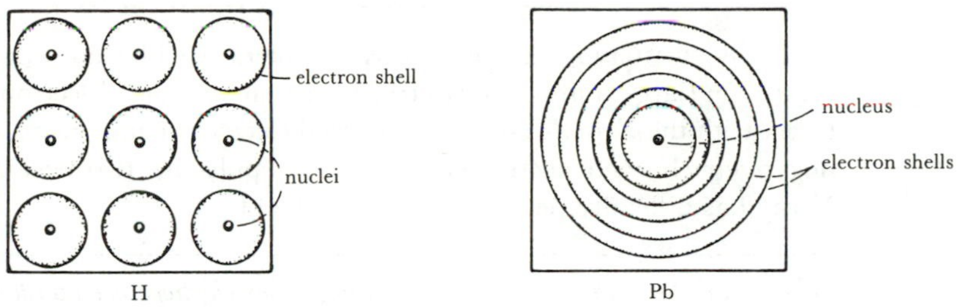

Do you see why neutrons would be more easily stopped by the hydrogen box than by the lead box?

A neutron is a particle with a mass approximately 2000 times that of an electron. In a direct collision between a neutron and an electron, the neutron would hardly feel the crash (like a 2000-lb car hitting a 1-lb rock). The neutron is neutral, so it does not electrically interact with charged particles, such as electrons and protons. Neutrons would zip through the lead box easily because the lead box is primarily filled with electrons (charged particles with a very small mass relative to the neutron). As shown in Figure 15.12, the lead box has one nucleus, and the only way that a neutron would be stopped from penetrating this box is if the neutron collided with that nucleus. Remember that the nucleus of an atom occupies very little space. If atoms, such as H or Pb, could be enlarged so that their diameters were 1 mile, their nuclei would have diameters of only two-thirds of an inch. Therefore, neutrons can penetrate the lead box very easily.

Why would neutrons have trouble getting through the hydrogen box? (Actually, the box in Figure 15.12 is a hydrocarbon box with only the hydrogen atoms shown for simplicity.)

We know that electrons cannot stop neutrons because electrons are too light and their negative charges don't interact with neutral neutrons. Only a head-on collision with a nucleus would stop a neutron. In the hydrogen box there are many more nuclei per unit volume relative to the lead box. There is a greater chance that the neutron would collide with a nucleus in the H box and be stopped.

Why does the lead box stop X rays?

An X ray can be thought of as a wave with positive and negative properties (see Figure 15.13). The electrical part of the X ray fluctuates (changes) regularly between positive (+) and negative (−) values. You can think of an X ray as a wave that sometimes acts as if it were positively charged and sometimes acts as if it were negatively charged.

FIGURE 15.13 Electrical nature of an X ray

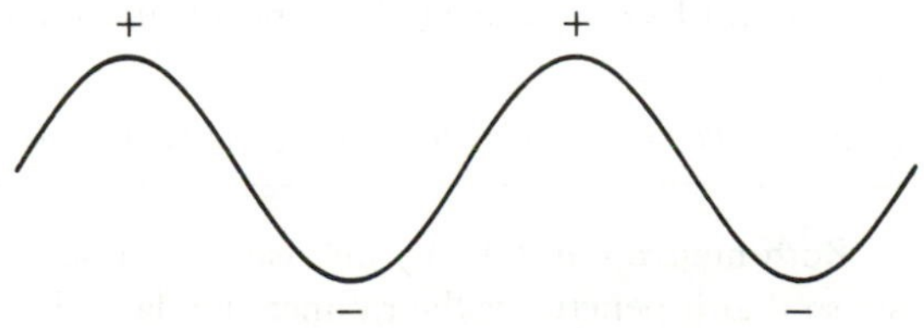

Now can you explain why lead stops X rays?

The fluctuating charge of an X ray interacts with lead because lead has a high electron concentration (large number of electrons per unit volume). Electrons are negative, and they will interact with the X ray's positively charged electrical component (attraction) and with the X ray's negatively charged electrical component (repulsion). It is because of these interactions that the X ray has difficulty penetrating a lead shield.

Why would the X ray get through the hydrogen box (hydrocarbon) with relative ease?

As you can see in Figure 15.12, X rays can get through a hydrogen sample more easily than they can get through a lead sample because a hydrogen sample contains fewer electrons per atom. (There are fewer electrons with which the X ray can interact in a hydrogen sample than in a lead sample.)

In summary, X rays are stopped by heavier elements (such as lead) that have higher electron concentrations. X rays can more easily pass through lighter elements (such as carbon) that have lower electron concentrations. On the other hand, neutrons can more easily penetrate heavier elements (such as lead) that have low nuclei concentrations. But neutrons are stopped by lighter elements (such as carbon) that have higher nuclei concentrations. It is important for you to understand this summary before you read any further.

15.10.1 MEDICAL APPLICATIONS

When physicians want to check for a possible broken bone, they use X rays. Bone contains calcium.

Can you explain why X rays are used?

Calcium atoms contain twenty electrons in four main energy levels (four electron shells). Skin is a hydrocarbon derivative made up (primarily) of hydrogen and carbon, which have one and two shells of electrons, respectively. X rays can penetrate elements of low electron concentration (such as hydrogen and carbon, and thus skin), but X rays are stopped (absorbed) by elements with higher electron concentrations (such as calcium). Skin is less visible to X rays than are bones.

When a bone is bombarded with a stream of X rays, usually denser parts of the bone absorb (stop) more X rays. After passing through the bone, the X rays strike a sensitive film. The film shows the denser parts of the bone as darker areas, the less dense parts as lighter areas.

Suppose you thought you had a tumor (a thick hydrocarbon derivative) growing inside one of your bones. Why would neutrography be used to look at the tumor inside your bone?

Neutrons would be stopped by the tumor inside your bone because the tumor is primarily hydrogen and carbon, both of which have relatively high nuclei concentrations. The bone material calcium has a lower nuclei concentration and would be invisible to the neutron beam.

Why doesn't the skin interfere with seeing the tumor if both the tumor and skin are hydrocarbon derivatives?

Both neutron and X-ray intensities can be varied. The intensity of the neutron beam can be set so that it penetrates the thinner skin layer but is stopped by the thicker tumor in the bone.

15.10.2 INDUSTRIAL APPLICATIONS

Neutrography is able to detect plastic objects (plastics are hydrocarbons or hydrocarbon derivatives) inside metal pipes, water drops inside metal structures (water is made of hydrogen and oxygen), and illegal drugs and explosives (both drugs and explosives are made from hydrocarbons and hydrocarbon derivatives) in sealed steel drums or containers.

Figure 15.14 shows three illustrations of a bullet. On the left is an illustration of an X-radiograph (an X-ray picture). In the middle is an illustration of a neutron radiograph (a picture taken with a beam of neutrons). On the right, the X-radiograph and neutron radiograph are shown superimposed. Notice how the X ray shows the outline of the bullet. The exterior part of a bullet (casing, jacket) is made of copper or steel. Also notice how the neutron picture shows the grains of gunpowder inside the bullet (gunpowder contains carbon).

FIGURE 15.14 Bullet. (a) Illustration of an X-radiograph. (b) Illustration of a neutron radiograph. (c) Superimposing the X-radiograph and the neutron radiograph.

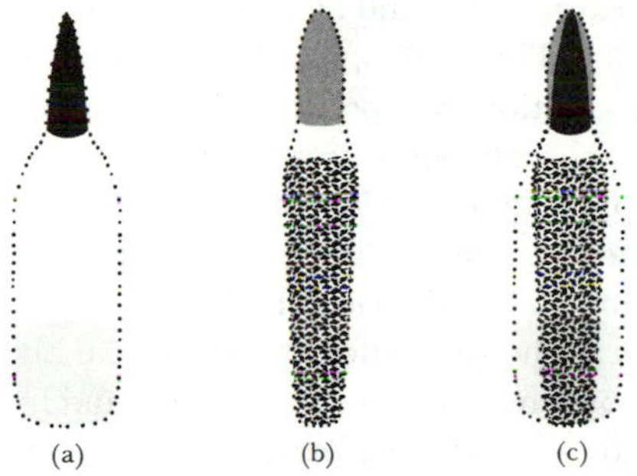

Notice the heads (top part) of each bullet picture. The head in the middle appears to be larger than the head on the left. Can you guess why?

A bullet contains a lead slug in its head. Lead has a greater electron concentration than iron (steel is made from iron). Lead has 6 shells of electrons containing 82 electrons while iron has 4 shells of electrons that contain only 26 electrons. The X rays are stopped by the inner lead slug, which is visible in the X-ray picture. The neutrons are stopped by the copper casing, which is visible in the neutron picture.

Notice that the X rays don't penetrate the bottom of the shell (darker area). This indicates that the bottom of the shell is thicker than the rest of the shell. This explanation is supported by looking at the bottom portion of the shell in the neutron picture. The column of gunpowder narrows at the bottom of the shell.

Although both X rays and neutron pictures are useful, they become more useful when they are used together. They complement each other, as illustrated in Figure 15.14(c).

15.10.3 NEUTRON BOMB

The United States has considered developing and perfecting a neutron bomb.

The neutron bomb kills people but doesn't destroy buildings and cities. Can you explain why?

A high-intensity neutron beam would interfere with the higher nuclei density of your body tissue (made of hydrocarbon derivatives). The neutrons would slip through buildings and other structural components of a city, which are made of iron and concrete.

15.11 CREATION OF THE UNIVERSE, PART II: A REVIEW

There are many different explanations of how the universe began. These explanations range from divine creation to the view that matter and energy always were and always will be. One of the more popular scientific theories today has enough loose ends and uncertainties to encompass many opposing opinions concerning the moment of creation. That theory is called the Singularity—Big Bang—Black Hole theory (hereafter referred to as the S3BH theory).

We discuss S3BH theory here because it serves as a review of many of the concepts that we have studied. The S3BH theory is not being presented as objective truth (see Chapter 12, "Creation of the Universe, Part I").

15.11.1 BIG BANG

Prior to the birth of our universe there existed a singularity. The singularity was infinitely dense. All that exists today and all that existed before the birth of the universe was compressed into that singularity. Time and space did not exist until after the singularity exploded. Some believe that if divine creation ever occurred it was to make the singularity, because the universe as we know it today spontaneously arose from the singularity.

Approximately 15 to 20 billion years ago, the singularity exploded, a moment that scientists refer to as the big bang. It was the moment of creation, the birth of the universe, the beginning of time and space. Today scientists are continuously making measurements to test the accuracy of the predictions made by the big bang theory. They have found, for example, that the proportions of hydrogen, helium, and lithium isotopes that we now measure in the universe are exactly what the big bang theory predicts they should be. Also, in the early 1990's, NASA's Cosmic Background Explorer (CORE) looked back with its optical instrumentation to within a year of the big bang. The results of CORE's experiments also support the big bang theory.

Temperatures were very high during the early moments of the universe, and we know little about the chemistry that occurred. Some suggest that the initial explosion sent out approximately 1×10^{80} neutrons that, being unstable in free space (outside a nuclear environment), decayed into protons and electrons.

$$n_x \rightarrow xn \longleftarrow \text{big bang}$$

$$n \rightarrow p^+ + e^- \longleftarrow \text{decay of neutrons into protons and electrons shortly after the big bang}$$

The electrostatic (electric charge) attraction between oppositely charged protons and electrons caused them to form hydrogen atoms. (Remember that a hydrogen atom has a single proton for its nucleus and a single electron in its first main energy level.)

$$p^+ + e^- \rightarrow H \longleftarrow \text{formation of hydrogen}$$

Shortly after the beginning, hydrogen atoms filled the universe. Even today, interstellar space is essentially 100% hydrogen (atoms and molecules). The entire universe (including the stars) is about 80% hydrogen and 20% helium.

15.11.2 STARS, PART II

We have already shown (see Section 13.3.4, "Stars, Part I") how stars form from gravitationally condensing particles of hydrogen. (In 1977, visual evidence of hydrogen clouds collapsing to form a star was reported for the first time.) We have also briefly mentioned the process of nuclear fusion in stars, which converts hydrogen into helium.

Nuclear fusion is not restricted to the process in which hydrogen nuclei fuse together to form helium. As a star's supply of fuel (hydrogen) runs out, the star begins burning (fusing) helium nuclei, which produces carbon.

$$3\,^{4}_{2}He \rightarrow \,^{12}_{6}C \leftarrow \text{fusion of 3 helium nuclei to form carbon}$$

Carbon nuclei can fuse together under the right conditions and form magnesium nuclei.

$$2\,^{12}_{6}C \rightarrow \,^{24}_{12}Mg \leftarrow \text{fusion of 2 carbon nuclei to form magnesium}$$

The fusion process can continue in some stars, forming all of the known elements through iron. Why can't elements with higher mass numbers than iron be formed by fusion?

You may recall that iron is the most stable element, that is, it has the highest binding energy per nucleon. Fusion is a spontaneous process of forming a more stable situation from a less stable situation.

While a star is burning (while nuclear fusion occurs), the star is being constantly subjected to two large and opposing forces: nuclear fusion and gravity. Nuclear explosions (fusion reactions) are outward forces that tend to rip the star apart. Opposing this outward force is the inward force of gravity that tends to keep all of the stellar material together.

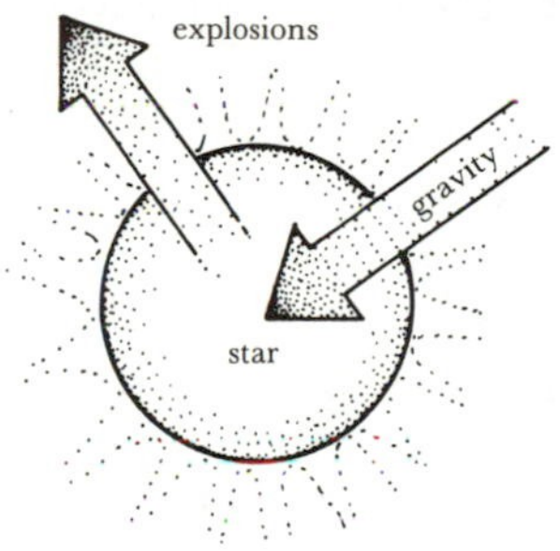

Eventually, all stars exhaust their fuel supply (nuclei for fusion), and gravity dominates. When this happens, the star collapses almost instantaneously with such a strong force that electrons are driven into stellar nuclei. This forms a large mass of neutrons called a neutron star, or pulsar.

$$p^{+} + e^{-} \rightarrow n \leftarrow \text{star collapses, driving electrons into protons within the star's nuclei}$$

The density of a neutron star is about 250 000 000 tons/cc, the same as the density of all nuclei.

15.11.3 BLACK HOLES

The more a star collapses, the stronger its gravity; the stronger its gravity, the greater its collapse. As the gravity of the star continues to increase, even the neutrons collapse. Neutrons are not solid particles but consist of a central nucleus surrounded by particles called pions. Neutrons are made up of a lot of empty space, as shown in Figure 15.15.

According to this theory, a star larger than our sun can shrink to a size smaller than a grain of sand but still retain all of its original mass of about 1×10^{12} tons. Such a particle is called a black hole, an object with such a large gravitational pull that light cannot escape its surface.

FIGURE 15.15 Neutron

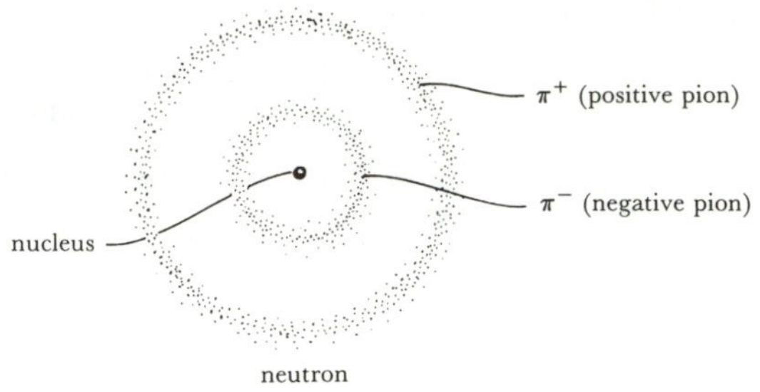

Within the confines of a black hole, the coordinates of space and time become warped, and time is believed to stop.

A black hole can wander through the universe consuming other stars that cross its path. Some scientists believe that 50% to 90% of the mass of the universe today may be tied up inside black holes. Perhaps, as some believe, the entire universe will collapse into one black hole, another singularity (see Figure 15.16).

FIGURE 15.16 Lifespan of our universe

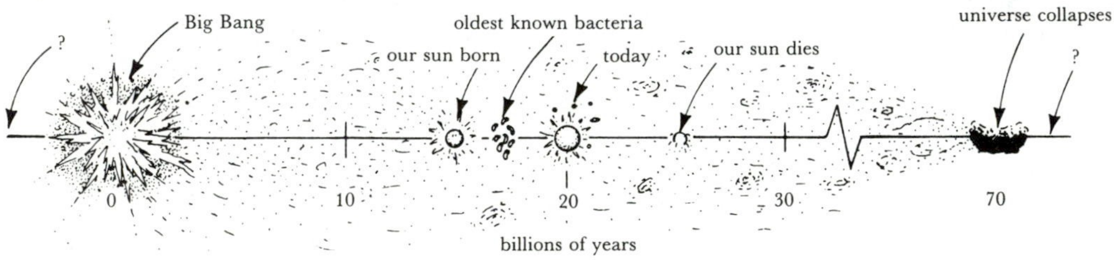

PROBLEM SET 15

The problems in Problem Set 15 parallel the examples in Chapter 15. For example, if you should have trouble working Problem 5, go back to Example 5 in this chapter to get help. The correct answers to these problems are given at the end of this book.

1. How many protons are present in the nuclei of the atoms $_2He$, $_3Li$, and $_4Be$?
2. What is the atomic number for helium? Find Z for lithium and beryllium.
3. Which element has an atomic number of 6? How many electrons are in the atomic orbitals of an atom of this element?
4. If 2 protons were added to the nucleus of a carbon atom, what new element would result?
5. Find the mass numbers for the atoms ^1H, ^2H, and ^3H.
6. Find the values of A, Z, and N for $^{19}_{8}O$.
7. Find the nuclidic mass, the mass number, the atomic number, the number of neutrons, the number of protons, and the number of electrons in one atom of the isotope $^{60}_{28}Ni$.
8. Find the value of A, Z, and N for $^{27}_{13}Al$. Also, how many electrons are in one atom of this isotope, and what is its nuclidic mass?
9. The members of the human race come in two varieties: male and female. Assume that the human race is exactly

60% male and 40% female. Also, assume that all men weigh exactly 300 lb, and that all women weigh exactly 200 lb. What is the average weight of a human?

10. Assume that the human race is exactly 85% female and 15% male. All males weigh exactly 230 lb, and all females weigh exactly 95 lb. What is the average weight of a person?
11. Assume that each person in a random sample of exactly 1000 humans was weighed. Assume that this sample contains a normal distribution of males and females as described in Problem 10. What is the total weight of this sample of 1000 people?
12. The element chlorine exists in two isotopic forms, ^{35}Cl and ^{37}Cl. About 75.5% of all chlorine atoms are chlorine-35. What is the average weight of a chlorine atom?
13. About 99.76% of all vanadium exists as the ^{51}V isotope. The rest exists as ^{50}V. What is the atomic weight of vanadium?
14. The element chromium exists in four naturally occurring isotopic forms: Cr-50 (4.31%), Cr-52 (83.76%), Cr-53 (9.55%), and Cr-54 (2.38%). What atomic weight is reported on a periodic table for chromium?
15. The fermium isotope $^{248}_{100}Fm$ is unstable and decays by emitting an alpha particle from its nucleus. What element does fermium-248 turn into after losing an alpha particle?
16. Einsteinium-249 decays by giving off an alpha particle. What element forms when Es-249 decays?
17. The nitrogen isotope ^{16}N is unstable and decays by emitting a beta particle. What element forms as a result of this decay?
18. Oxygen-20 is unstable and decays by giving off a beta particle. What element forms as a result of this decay?
19. Titanium-44 is unstable and undergoes K-capture. What element forms as a result of the K-capture?
20. Copper-64 is unstable and undergoes K-capture. What new element forms as a result of this K-capture?
21. Magnesium-22 decays by emitting a gamma ray. Write an equation showing this decay.
22. Aluminum-30 decays by releasing gamma radiation. Write an equation showing this decay process.
23. When the hypothetical radioactive element X undergoes radioactive decay, its atomic number and the number of neutrons in its nucleus decrease. What type of decay would account for these changes?
24. When the hypothetical radioactive element Y undergoes radioactive decay, the mass number remains constant and the atomic number increases. What type of decay would account for these changes?
25. When the hypothetical radioactive element Z undergoes radioactive decay, the mass number and the atomic number decrease. What type of decay occurs?
26. It has been found that a sample of radium-226 gives off 5.562×10^{14} helium ions per minute, and that the volume of helium gas collected at STP over a period of 100 days is 2.9808 mL. Calculate the value of Avogadro's number.
27. A beryllium-7 atom has a mass of 7.016 931 amu. Compare its mass with the total mass of the protons, neutrons, and electrons from which Be-7 is made.
28. The mass of boron-8 is 8.024 612 amu. Find the binding energy of boron-8.
29. The mass of carbon-10 is 10.016 83 amu. Find the binding energy for carbon-10.
30. Determine which nucleus is more stable, C-10 or B-8.
31. Calculate the binding energy per nucleon of N-12. The mass of an N-12 atom is 12.018 71 amu.
32. Calculate the binding energy per nucleon for O-14. The mass of one O-14 atom is 14.008 597 amu.
33. Which is more stable, O-14 or N-12?
34. A gold sample contains 4.00×10^{30} atoms of Au-191 and decays at a rate of 8.00×10^{29} atoms/hr. Find the rate constant for gold-191.
35. A 4.0-g sample of tin-120 decays at a rate of 8.6×10^{20} tin-120 atoms/min. Find the rate constant for Sn-120.
36. Nickel-59 has a rate constant of 2.466×10^{-8}/day. Find the rate of decay of a 10.0-g sample of Ni-59.
37. The rate constant for aluminum-25 is 5.70/min. Find the mass of a sample of aluminum-25 if it has a decay rate of 4.8×10^5 atoms/min.
38. A sample of nitrogen-13 contains 500 atoms of N-13. How long must you wait for the sample to drop down to 350 atoms of N-13 if the rate constant for N-13 is 0.0693/min?
39. A sample of nitrogen-13 weighs 58.0 g. How long must you wait for the sample to weigh 22.0 g? The rate constant for N-13 is 0.0693/min.
40. A sample of oxygen-19 weighs 3.0 g. How much will this sample weigh after 10.0 sec? The rate constant for oxygen-19 is 0.0236/sec.
41. A sample of oxygen-19 has an activity of 6000 cps. What will the activity of this sample be after 12.0 sec? The rate constant for O-19 is 0.0236/sec.
42. Gold-191 has a half-life of 204 min. Find the rate constant for gold-191.
43. Tin-120 has a half-life of 9.6×10^2 sec. Find the rate constant for tin-120.
44. The half-life of nickel-59 is 8×10^4 yr. Find the rate constant for nickel-59.
45. A sample of oxygen-19 weighs 3.0 g. How much will this sample weigh after 12.0 sec? The half-life of oxygen-19 is 29.4 sec.
46. A sample of oxygen-19 contains 7000 atoms of O-19. How many atoms of O-19 will be present after 12.0 sec? The half-life of O-19 is 29.4 sec.
47. A hospital has 400 g of Co-60. How much will they have 1.00 yr from today? The half-life of Co-60 is 5.27 yr.
48. A hospital has a 400-g supply of Co-60. Additional Co-60 must be purchased whenever the supply of Co-60 is reduced to 300 g. How many years will pass before the hospital's supply of Co-60 is reduced to 300 g?
49. A sample of wood is found to have an activity of 6.00 cpm/gC. If fresh wood has an activity of 15.3 cpm/gC,

and the half-life of carbon-14 is 5730 yr, how old is the wood sample?

50. Campfire ashes were found to have an activity of 1.10 cpm/gC. How long ago was the campfire lit?
51. How many years must pass for a piece of cloth to lose 90% of its original activity?
52. Corncobs are found that only have three-fourths of their original activity. How old are the corncobs?
53. If Noah's ark were found today, it wild be about 4400 years old. What would its carbon-14 activity be?
54. If the Shroud of Turin is a hoax and it is only 500 years old, what would its present carbon-14 activity be?
55. The organic material found in a broken strata of the next to the oldest layer of sedimentary layers had a carbon-14 activity of 13.05 cpm/gC. Calculate the age of this layer.
56. A hair sample shows a radioactivity of 4000 disintegrations per second of arsenic-80. How many grams of arsenic-80 are present in the hair sample if the half-life of arsenic-80 is 15.0 sec?

······ Appendix

How to Take Examinations

REDUCING EXAM ANXIETY

The following true story describes a frequently recurring situation: An examination made up of five questions was presented to a class. The first question was the most difficult. Very few students were expected to answer it correctly. Bob, one of the brightest students in the class, began to work on it immediately. Some of the other students also looked at the first problem and continued to stare at it for several minutes without any progress. They didn't know where to begin and could never have answered the question even with all the time in the world. Frank, an above-average student, realized how difficult the first question was and moved on to questions two through five, finishing them with time to spare. He then used this extra time to work on the first question. In the meantime, Bob had managed to solve the first question, but had only a few minutes left with which to complete the remaining four questions. He panicked and did poorly on these last four questions as he rushed to complete them. (They were easy questions that he had solved a dozen times before.)

Finally, time was up and the exams were collected, corrected, and returned. Bob, for all his intelligence, knowledge, and ability, received a grade of 70%. He had received full credit (20 points) for the first question and partial credit (another 50 points) on the remaining four questions. Frank's grade was an 80%. He missed the first question completely but did the last four questions correctly. Frank's grade of 80% was the highest grade in the class. One other student managed to get a 70%. The remaining grades were scattered between 25% and 65% (D's and F's).

What do you suppose the first hint is?

Look for the easy problems first. There are practical and psychological advantages to this approach. The practical advantage should be obvious to you after reading the preceding story. The psychological advantage to this method of attacking an exam is that it calms you down. Many students do poorly on examinations simply because they are nervous. Finding questions with which you are familiar on an examination is a calming experience.

It's probably obvious to you that showing up on time for an exam is better than running into an exam ten minutes late. In the latter case, you would be out of breath and nervous, and you would find yourself rushing through the exam trying to make up for lost time. The overall result would be identical to Bob's experience in the previous story.

However, showing up on time is not enough. Whenever possible, show up early. Then, use your free time to get involved in some light conversation with other students or, if you arrive at the examination room so early that no other students are there yet, just sit and relax. Do not spend these few precious moments before an examination frantically trying to cram (learn?) more material. If you feel the need to cram, then you are probably not well prepared for the examination.

As a result of arriving at the examination room early, you will feel calm. You will be able to think better in this relaxed (but alert) mental and physical state.

ADVANCE PREPARATION

There is a big difference between studying and reviewing. You study material to familiarize yourself with the material, to learn it, to understand it. When you review material, you go over information that you already understand, refreshing your memory, solidifying the ideas that you already know. With this distinction in mind, never study the night before an examination. You can review material then, but don't expect to do well when you try to learn new material, for the first time, the night before an examination.

Studying for an examination should be a daily activity consisting (ideally) of an hour or two to go over each day's lecture notes and homework assignments. Avoid going to a lecture without thoroughly understanding your notes from the previous lecture.

You will find it helpful to avoid letting the lecture be your first exposure to new ideas. Just skimming over the upcoming lecture material and vocabulary in your textbook can be beneficial.

When an examination is announced, begin reviewing your lecture notes (which you have already studied) at least a week in advance. You will find that your first review takes the greatest amount of time. You should try to review your notes several times. Each review will take less time. Eventually you will be turning the pages of your notebook very quickly as your eyes spot and your mind registers the most important ideas on each page.

A frequent comment made by students to their professors is "It looks easy when you solve problems, but when I try to do it on exams, I can't." There is a difference between being able to follow the solution to a problem and solving that problem yourself. You may think you understand material because you understand how your professor is doing the calculations or explaining the theories. However, unless you can do the calculations yourself or correctly *write* (with a pencil, not mentally) explanations and definitions, then you don't really understand the material. This is why, when you read this book, it is very important for you to become actively involved by actually doing the calculations or writing brief paragraphs to answer questions. You might find it beneficial to ask your professor to check a few of your essay answers to show you how to express yourself more clearly and correctly.

Answers to Problem Sets

PROBLEM SET 1

1. The exponential part is 10^2; the nonexponential part is 7.4; the exponent is 2.
2. 5.2×10^3 **3.** 8.34×10^6
4. 1×10^9 **5.** 7×10^{-1} **6.** 1.8×10^{-3}
7. 5000 **8.** 0.000 04 **9.** 3.27×10^5
10. 7.6×10^1
11. 5.4×10^1; 0.54×10^2; 0.054×10^3; 0.0054×10^4; 54×10^0; 540×10^{-1}; 5400×10^{-2}; $54\,000 \times 10^{-3}$
12. 2.6×10^4 **13.** 3.006×10^7
14. 6.994×10^5
15. $30 = 3 \times 10^1$; $300 = 3 \times 10^2$; product $= 9 \times 10^3$
16. 1.2×10^6 **17.** +7 or just 7
18. −11 **19.** +4 or just 4 **20.** −2
21. $3 + 4 = 7$; $-8 + -3 = -11$; $-3 + 7 = 4$; $-9 + 7 = -2$
22. 10 **23.** 10 **24.** −2
25. 6×10^4; 2×10^2; 3×10^2 **26.** 4×10^7
27. 2×10^4 **28.** 10^3 or 1×10^3
29. 10^3 or 1×10^3 **30.** 10^8 or 1×10^8
31. 5×10^4 **32.** 3×10^3 **33.** 4×10^1
34. 4×10^2 **35.** 2×10^{-20}
36. 4.1×10^{-5} **37.** 8.1×10^{-15}
38. 8.1×10^{33} **39.** 4 **40.** −6 **41.** 0
42. 0.7 **43.** 0.1 **44.** 0.65 **45.** 0.35
46. 0.28 **47.** 1.60 **48.** 4.6
49. −4.35 **50.** 3.98
51. 10^5 or 100 000
52. 10^8 or 100 000 000 **53.** 10^{14}
54. 10^{-6} or 0.000 001 **55.** 1 **56.** 5.0
57. 6.4 **58.** 9.0 **59.** 2.5×10^5
60. 2×10^3 **61.** 9×10^7
62. 6.4×10^{-9} **63.** 3.2×10^{-5}
64. 1.15×10^{-9} **65.** 2.4×10^5 or 240 000
66. 630 **67.** 0.4 **68.** 200 **69.** 30
70. integer **71.** measured **72.** 3
73. 3 leading, 1 middle, 2 trailing **74.** 3
75. 7 **76.** 18 **77.** 234.4 in.
78. 1530 g **79.** 1530 g **80.** 1541 g
81. 0.3 **82.** 0.286 **83.** 39.7 in.^{2}
84. 0.294 83
85. 2 482 848.55; 2 482 858.5; 2 842 859; $2.482\,86 \times 10^6$; 2.4829×10^6; 2.483×10^6; 2.48×10^6; 2.5×10^6; 2×10^6.

PROBLEM SET 2

1. The number part is 4 and the unit or dimension part is feet.
2. Your measurement does not have a unit.
3. $3 \times 3 = 3^1 \times 3^1 = 3^2 = 9$
4. inches^2 **5.** $(\not{4} \times 5)/\not{4} = 5$ **6.** feet
7. $(3^2 \times 4)/3 = (\not{3} \times 3 \times 4)/\not{3} = 3 \times 4 = 12$
8. feet × seconds **9.** inches per yard
10. 16 ft^2 **11.** in.^{3} **12.** 5 oz
13. 2 ft **14.** 2.1×10^4 ft
15. 1.8×10^2 min **16.** 0.33 day
17. 1.9×10^2 hr^2/day is incorrect; 0.33 day is correct
18. 2.160×10^7 sec^2/hr is incorrect; 1.667 hr is correct
19. Step 1: 2.0-week vacation; Step 2: days/week; Step 3: 7 days/week; Step 4: (2.0-week vacation) (7 days/week) = 14 days vacation.
20. Step 1: 2.0-lb object; Step 2: oz/lb; Step 3: 16 oz/lb; Step 4: (2.0-lb object)(16 oz/lb) = 32-oz object

21. Step 1: 150 lb; Step 2: tons/lb; Step 3: 1 ton/2000 lb; Step 4: (150 lb)(1 ton/2000 lb) = 0.0750 ton
22. Step 1: 5 ft; Step 2: mi/ft; Step 3: 1.894×10^{-4} mi/ft; Step 4: (5 ft)(1.894×10^{-4} mi/ft) = 9×10^{-4} mi
23. Step 1: 3 people; Step 2: lb of snail sauce enhancer/people; Step 3: 2.0 lb snail sauce enhancer/5 people; Step 4: (3 people) (2.0 lb snail sauce enhancer/5 people) = 1.2 lb snail sauce enhancer
24. 5.0×10^2 mi 25. 2.15 gal gasoline
26. 6.0×10^2 oz paper 27. 1.6×10^2 dolls
28. $1.20 29. 1×10^3 days
30. 3×10^4 hr 31. 2×10^6 min
32. \$$3 \times 10^7$ 33. 6.46×10^{-6} mi^2
34. 3.7×10^{-6} mi^2 35. 1.5×10^8 km
36. 37 mi 37. 2.4×10^5 mi
38. 0.031 km 39. 0.9 km/sec
40. 3×10^5 g Uncle Harvey
41. 1.4×10^8 mg Aunt Tillie
42. 1.4×10^{-6} Mg cockroach
43. 5×10^{21} Mg black hole
44. 9.1×10^{-31} kg electron
45. 3.2×10^{-29} oz positron
46. 7.47 L basketball
47. 2.24×10^4 mL hydrogen
48. 20.0 qt oxygen

PROBLEM SET 3

1. (50 g soln)(5.0 g salt/100 g soln) = 2.5 g salt
2. (20 kg pig)(90 kg water/100 kg pig) = 18 kg water
3. (300 buttons)(20 brass buttons/100 buttons) = 60 brass buttons
4. (2/5)(100/centum) = 40/centum = 40%
5. (3/10)(100/centum) = 30/centum = 30%
6. (2 brown chickens/8 chickens)(100 chickens/centum chickens) = 25 brown chickens/centum chickens = 25% brown chickens
7. 2.70 g/cm^3 8. 2.70 g/cm^3
9. 168 lb/ft^3 10. 1204 lb/ft^3
11. 4.024×10^{-2} ft^3 Au 12. 72.9 g Al
13. solid 100% Al 14. 1 g/cm^3
15. 8.45 16. 0.75 17. 20 g/cm^3
18. 62.4 lb water/ft^3 water 19. 3.21
20. 20.0 cal; 83.7 J 21. 80 cal; 3.3×10^2 J
22. 4.2×10^6 cal lost; 1.8×10^7 J lost
23. The cooler beaker gives off more heat; the cooler beaker loses 2500 cal; the warmer beaker loses 150 cal
24. 2000 cal lost; 8368 J lost
25. 7.4×10^4 cal 26. 2×10^4 cal
27. 7.5 hr 28. −218.8°C; 54.4 K
29. 2970°C; 3243 K 30. 1945°F
31. −297.4°F
32. 5.6°C temperature increase; 5.6 K temperature increase
33. 4500°F temperature difference; 2500 K temperature difference
34. 1.49 lb candy 35. 246.2°F
36. 977°F 37. 4.8×10^2 cal; 0°C
38. 13 kJ 39. 2×10^1 cal
40. 3×10^4 cal 41. 4.0×10^4 cal
42. 5.9×10^4 cal

PROBLEM SET 4

1. 2×10^{24} molecules SO_3
2. 5×10^{23} molecules O_2 3. 50 mol H_2
4. 2×10^{15} yr 5. 1×10^4 yr
6. 7×10^1 L He 7. 0.45 mol N_2
8. 1.86×10^{-21} L Ar 9. 48 amu
10. 36 amu 11. 2 amu 12. 34 amu
13. 82 amu 14. 58.5 amu
15. 294 amu 16. 2 g deuterium
17. 36 g C 18. 20 g Br 19. 64 g O
20. 294 g $Al_2(SO_3)_3$
21. 1.02×10^{21} formula units of $Al_2(SO_3)_3$
22. 0.08 mol sugar
23. 5×10^{22} molecules sugar
24. 3.12 mol O_2 25. 3 g O_2
26. 0.1 g N_2 27. 66 L NH_3
28. 9.5×10^{23} NH_3 molecules
29. 8.4×10^{-23} g NH_3
30. 1.1×10^{-22} L NH_3 31. 24 g CO_2
32. 9.4×10^{23} SO_2 molecules
33. Chemically combine 1 mol of sulfur with 3 mol of atomic oxygen (O, not O_2).
34. 2 mol of carbon and 4 mol of atomic hydrogen (H, not H_2)
35. 6 mol of O (atomic oxygen, not O_2, which is molecular oxygen)
36. 12 mol of Cl (atomic chlorine, not Cl_2, which is molecular chlorine)
37. 1.2 mol Na
38. 2 atoms of carbon; 6 atoms of hydrogen
39. CH_3
40. 10 carbon atoms; 22 hydrogen atoms
41. C_5H_{11} 42. C_3H_8 43. C_3H_8
44. CH_3 45. C_3H_8 46. CO
47. The empirical formula is CH_3; the molecular formula is C_2H_6.
48. SO_3 49. Al_2O_3

PROBLEM SET 5

1. 44 psi; 3.0×10^2 kPa
2. 1.05 atm; 106 kPa 3. 3×10^4 mm Hg
4. 6 L gas 5. 60 mL gas 6. 27 L N_2
7. 27 L NH_3 8. 68 L He
9. 3 L lung volume
10. Your lungs would burst while trying to expand to 30 L.
11. 4.14 L gas 12. 7.0 kL CO_2
13. 610 mL N_2
14. volume decreases to 330 mL
15. volume increases to 844 mL
16. 3.3 atm 17. 1060 mm Hg
18. 220 kPa 19. 30.4 atm 20. 9 atm
21. P_p(He) = 1.2 atm; P_p(Ar) = 2.8 atm
22. 705 mm Hg = 1.2 atm
23. 83.7 kPa 24. $P_p(O_2)$ = 1.2 atm 25. 75 L gas
26. 9.50 L gas 27. 696 mL 28. 2.00 L H_2
29. 30.625 amu 30. 80 amu
31. 14 amu 32. 52.2 amu
33. 157 amu 34. 31 L gas 35. 1.5 atm
36. 24 K 37. 0.975 mol gas
38. 44.8 L gas 39. 0.0821 L atm/1.00 mol K

PROBLEM SET 6

1. solid 2. gas
3. The coefficient of K is 2; the coefficient of Cl_2 is 1; the coefficient of KCl is 2; 2 atoms of K reacts with 1 molecule of Cl_2 to form 2 formula units of KCl.
4. The coefficient of H is 2; the coefficient of O_2 is 1; the coefficient of H_2O is 2; 2 molecules of H_2 reacts with 1 molecule of O_2 to form 2 molecules of H_2O.
5. yes 6. yes 7. no 8. $H_2 + F_2 \rightarrow 2HF$
9. $2C + 2H_2 \rightarrow C_2H_4$
10. $3H_2CO_3 + 2Al(OH)_3 \rightarrow Al_2(CO_3)_3 + 6H_2O$
11. $2H_3PO_3 + 3Ba(OH)_2 \rightarrow Ba_3(PO_3)_2 + 6H_2O$
12. $2AlBr_3 + 3Na_2CO_3 \rightarrow Al_2(CO_3)_3 + 6NaBr$
13. 2 mol of S reacts with 3 mol of O_2 to form 2 mol of SO_3
14. 64 g S 15. 96 g O_2 16. 160 g SO_3
17. 1.2×10^{24} S atoms
18. 1.8×10^{24} O_2 molecules
19. 67.2 L O_2 20. 44.8 L SO_3
21.

	Column 1		Column 2		Column 3
Line 1	2S(s)	+	$3O_2$(g)	→	$2SO_3$(g)
Line 2	2 mol S	+	3 mol O_2	→	2 mol SO_3
Line 3	64 g S	+	96 g O_2	→	160 g SO_3
Line 4	~0 L S	+	67.2 L O_2	→	44.8 L SO_3
Line 5	1.2×10^{24} S atoms	+	1.8×10^{24} O_2 molecules	→	1.2×10^{24} SO_3 molecules

22. 44.8 L SO_3 23. 1.2×10^{24} S atoms
24. "Two mol of S reacts with 96 g of O_2 to produce 44.8 L of SO_3" is one of several possible interpretations.
25. "Sixty-four grams of S reacts with 67.2 L of O_2 to produce 1.2×10^{24} SO_3 molecules" is one of several possible interpretations.
26. "Two mol of S reacts with 96 g of O_2 to produce 1.2×10^{24} SO_3 molecules" is one of several possible interpretations.
27. 6 mol O_2 28. 80 g SO_3
29. 192 g O_2 30. 9×10^{23} O_2 molecules
31. 89.6 L SO_3 32. 3.5 mol O_2 reacts
33. 4 mol H_2O forms
34. 15 mol O_2 is consumed
35. 8 mol of ZnS is required
36. 1.8 mol H_2S forms 37. 4 L O_2 reacts
38. 12 L of NO forms
39. 4.0 L of CH_4 reacts
40. 60.0 L of O_2 reacts
41. 55 g of CO_2 forms
42. 16 g O_2 is used up
43. 398 g Na_2CO_3 is required
44. 21.4 g $Ca(OH)_2$ reacts
45. 9.71 g H_2 is needed
46. 30.5 tons O_2 is used
47. 64 lb of O_2 is needed 48. 39 oz NO_2 forms
49. 5.1 g C burned 50. 47 kcal produced
51. 4.50 g H_2 burned
52. 5.7×10^3 kJ released 53. 1.8×10^3 kJ
54. 1.1 mol H_2O
55. 1.3×10^2 L O_2 at STP
56. 2.12×10^{24} H_2O molecules
57. 2.1×10^{22} O_2 molecules
58. 1.06×10^{-20} g O_2
59. 1.0×10^2 L O_2 consumed
60. 60 L of H_2 consumed 61. 37.5 g SiO_2 forms
62. 3.4×10^2 g H_2O forms 63. 77 L NO forms

PROBLEM SET 7

1. Honey is the solute; water is the solvent; the combination of honey dissolved in water is the solution.

2. Either A or B can be the solute; if A is chosen as the solute, then B is the solvent; if B is chosen as the solute, then A is the solvent.
3. 23% by weight HCl
4. 44% by weight D 5. 320 g sulfuric acid
6. 68.4 g A 7. 0.25 8. 0.75
9. 0.27 10. 0.0051 11. 3 M NaOH
12. This solution is 1 M NaOH; the molarity of this NaOH solution is 1; this is a 1 molar NaOH solution.
13. 12 M NaOH 14. 0.5 M NaOH
15. 2.0 M NaOH 16. 0.964 M NaOH
17. 4 m NaOH 18. 2.50 m NaOH
19. 1.0 m NaOH 20. 0.63 m NaOH
21. 0.5266 M H_2SO_4 22. 0.537 m H_2SO_4
23. 0.00957 24. 0.527 M H_2SO_4
25. 0.538 m H_2SO_4 26. 0.00958

PROBLEM SET 8

1. freezing point is −1.86°C; boiling point is 100.51°C
2. 1.00 m sugar
3. 1.00 m sugar; freezing point is −1.86°C; boiling point is 100.51°C
4. freezing point is −0.47°C; boiling point is 100.13°C
5. freezing point is −3.72°C; boiling point is 101.02°C
6. freezing point is −7.44°C; boiling point is 102.04°C
7. freezing point is −3.72°C; boiling point is 101.02°C
8. freezing point is −9.49°C; boiling point is 102.60°C
9. 100.84°C 10. 20 m sugar
11. 6.8×10^3 g sugar
12. 1.8×10^4 g sugar
13. 2.9×10^4 g sugar 14. −33.3°C
15. 25 amu 16. 60 amu 17. 18.5 amu
18. 62.8 amu 19. 12.3 atm
20. 4.09 amu 21. 4.90 atm
22. 3.00×10^4 amu 23. 4.53×10^4 amu

PROBLEM SET 9

1. $H_2O + CO \rightleftharpoons H_2 + CO_2$; $K = \dfrac{[H_2][CO_2]}{[H_2O][CO]}$
2. $2NO_2 \rightleftharpoons 2NO + O_2$; $K = \dfrac{[NO]^2[O_2]}{[NO_2]^2}$
3. $K = \dfrac{[C]^4[D]}{[A]^2[B]^3}$
4. $K = 0.63$ 5. $K = 0.63$ 6. 0.15 M
7. $K = 999$. The forward reaction occurs to a much greater extent than the reverse reaction.
8. $K = 0.00100$. Very little G reacts to form H.
9. $K = 1.0$. About one-half of X reacts to form Y.
10. it will try to decrease the PCl_3 concentration
11. shifts to the left
12. to produce more Cl_2 13. shifts to the right
14. to the left 15. to the right
16. to the right 17. to the right
18. it will try to warm up
19. shifts to the right
20. it tries to cool down
21. shifts to the left 22. to the right
23. it increases 24. gives off heat
25. it decreases
26. it tries to increase the pressure
27. shifts to the left
28. it tries to decrease the pressure
29. shifts to the right 30. to the right
31. to the left 32. doesn't shift either way
33. doesn't shift either way
34. doesn't shift either way 35. to the right
36. shifts to the right; pressure increases
37. shifts to the left; heat is absorbed by the reaction, cooling the surroundings
38. shifts to the left; pressure increases; reaction absorbs heat

PROBLEM SET 10

1. −1 2. −4 3. −1.5 4. 5
5. 3.5 6. 10 000 7. 0.000 1
8. 0.001 9. 2.5×10^{-6}
10. 2.0×10^{-4} 11. 4; acidic
12. 7; neutral 13. 3.7; acidic
14. 7.2; basic 15. 10^{-9} 16. 5×10^{-9}
17. 1.3×10^{-7} 18. 4.5×10^{-4} 19. 8.7
20. 4.3 21. 1×10^{-6} 22. 2×10^{-6}
23. 2 24. 3 25. 0.7 26. 1×10^{-9}
27. 9 28. 0.01%
29. yes; K is very small 30. 10^{-3} 31. 3
32. 10% 33. 8×10^{-9} 34. 8.1
35. 0.02% 36. 10^{-2} 37. 3 38. 1%
39. 1 40. 10% 41. 1×10^{-8} 42. 5
43. 3

PROBLEM SET 12

1. −250 kcal; exothermic
2. 130 kcal; endothermic
3. 15 kJ; endothermic 4. 350 kcal/mol E
5. 1600 kcal/mol A 6. 563 kJ
7. 2.0 kcal 8. −152 kcal/mol CaO(s)

9. -24 kcal/mol PbS(s) 10. 286 kJ
11. -12.6 kcal 12. -6.3 kcal/mol HI(g)
13. -174 kJ 14. -19 cal/K
15. -10 J/K; products are less disordered than reactants
16. -50 J/K; disorder of system decreases
17. 82.5 cal/mol E K; E contains less disorder than the total disorder of A plus B
18. -4.0 cal/K
19. 11 J/K; SO_2 is more disordered than the total disorder of the reactants
20. 51 cal/K; disorder increases; you could predict disorder increase by observing the phases of the products and reactants
21. 1.6 cal/g O_2 K 22. state 2 23. state 2

PROBLEM SET 13

1. -200 kJ; spontaneous
2. 800 kcal; not spontaneous
3. -800 kcal; spontaneous
4. 825 kJ/mol E
5. 94 kcal; not spontaneous
6. -300 kJ/mol $SO_2(g)$
7. 94 kcal; not spontaneous
8. 72 kcal; not spontaneous
9. 6 kcal; not spontaneous
10. 6.9 kcal; not spontaneous
11. 55 kcal; not spontaneous
12. 2.9×10^{-8} 13. 1.6×10^{-53}
14. 1.2×10^{-69} 15. 384 cal/day
16. -45 kcal 17. 0.045 kcal/K
18. -58.5 kcal 19. 7.9×10^{42}
20. The reaction is exothermic.
21. The reaction is spontaneous.
22. The disorder of the system increases; the disorder of the universe increases; the disorder of the surroundings may have increased or decreased.

PROBLEM SET 14

1. -4.8×10^4 C 2. 4.68×10^{22} e^-
3. 1×10^{19} e^- 4. 2×10^{20} e^-
5. 3×10^{23} e^- 6. 2.657×10^{-20} faraday
7. 0.42 faraday 8. 3.9×10^4 sec
9. 1×10^4 A 10. 4 sec 11. 8 g Cu
12. 0.39 g Zn 13. 1.1×10^2 days
14. 1.5×10^2 mL H_2
15. 1.7×10^2 mL Cl_2 16. $Cu + Cl_2 \rightarrow Cu^{2+}(aq) + 2Cl^-(aq)$ 17. $Zn + 2H^+(aq) \rightarrow Zn^{2+}(aq) + H_2$
18. -0.76 V 19. 0.76 V 20. 1.66 V
21. -2.37 V
22. $Mg + Zn^{2+}(aq) \rightarrow Mg^{2+}(aq) + Zn$; $E° = 1.61$ V
23. $Fe + Co^{2+}(aq) \rightarrow Fe^{2+}(aq) + Co$; $E° = 0.16$ V
24. $Fe + Cl_2 \rightarrow Fe^{2+}(aq) + 2Cl^-(aq)$; $E° = 1.80$ V
25. $Cu + \frac{1}{2}O_2 + 2H^+(aq) \rightarrow Cu^{2+}(aq) + H_2O$; $E° = 0.89$ V
26. 2.03 V 27. 0.88 V 28. 0.38 V
29. -0.74 V 30. 0.3 g Zn dissolves

PROBLEM SET 15

1. He has 2 protons, Li has 3 protons, and Be has 4 protons.
2. Helium has an atomic number of 2; $Z = 3$ for lithium, and $Z = 4$ for beryllium.
3. carbon; 6 electrons 4. oxygen
5. $A = 1, 2$, and 3, respectively, for ^1H, ^2H, and ^3H
6. $A = 19$; $Z = 8$; $N = 11$
7. nuclidic mass = 60 amu; $A = 60$; $Z = 28$; 32 neutrons; 28 protons; 28 electrons
8. $A = 27$; $Z = 13$; $N = 14$; 13 electrons; nuclidic mass = 27 amu
9. 260 lb 10. 115.25 lb 11. 115 250 lb
12. 35.5 amu 13. 51.0 amu 14. 52.1 amu
15. Cf(Fm $\rightarrow {}^{244}_{98}Cf + {}^{4}_{2}He$)
16. Bk (${}^{249}_{99}Es \rightarrow {}^{245}_{97}Bk + {}^{4}_{2}He$)
17. O(${}^{16}_{7}N \rightarrow {}^{16}_{8}O + {}^{0}_{-1}\beta$)
18. F (${}^{20}_{8}O \rightarrow {}^{20}_{9}F + {}^{0}_{-1}\beta$)
19. Sc (${}^{44}_{22}Ti + {}^{0}_{-1}\beta \xrightarrow{\text{K-capture}} {}^{44}_{21}Sc$)
20. Ni (${}^{64}_{29}Cu + {}^{0}_{-1}\beta \xrightarrow{\text{K-capture}} {}^{64}_{28}Ni$)
21. ${}^{22}_{12}Mg \rightarrow {}^{22}_{12}Mg + \gamma$
22. ${}^{30}_{13}Al \rightarrow {}^{30}_{13}Al + \gamma$
23. alpha 24. beta 25. alpha
26. 6.03×10^{23} He atoms/mol He
27. mass of 4e + 4p + 3n = 7.057 326 amu; more massive than the mass of a Be-7 atom
28. 37.8 MeV 29. 60.4 MeV 30. C-10
31. 6.17 MeV/nucleon
32. 7.06 MeV/nucleon 33. O-14
34. 0.200/hr 35. 0.043/min
36. 2.52×10^{15} Ni-59 atoms/day
37. 3.5×10^{-18} g Al-25 38. 5.14 min
39. 14.0 min 40. 2.4 g O-19
41. 4.52×10^3 cps 42. 3.40×10^{-3}/min
43. 7.2×10^{-4}/sec 44. 9×10^{-6}/yr
45. 2.3 g O-19 46. 5.27×10^3 O-19 atoms
47. 351 g Co-60 48. 2.19 yr
49. 7.74×10^3 yr 50. 2.18×10^4 yr
51. 1.90×10^4 yr 52. 2.37×10^3 yr
53. 8.99 cpm/gC 54. 14.40 cpm/gC
55. 1315 yr 56. 1.15×10^{-17} g ^{80}As

Index